Advances on Theoretical and Methodological Aspects of Probability and Statistics

Advances on Theoretical and Methodological Aspects of Probability and Statistics

Edited by
N. Balakrishnan

McMaster University
Hamilton, Canada

CRC Press
Taylor & Francis Group
Boca Raton London New York

CRC Press is an imprint of the
Taylor & Francis Group, an **informa** business
A TAYLOR & FRANCIS BOOK

ADVANCES ON THEORETICAL AND METHODOLOGICAL ASPECTS OF PROBABILITY AND STATISTICS

CRC Press
Taylor & Francis Group
6000 Broken Sound Parkway NW, Suite 300
Boca Raton, FL 33487-2742

First issued in paperback 2020

ISBN 13: 978-0-367-57852-7 (pbk)
ISBN 13: 978-1-56032-981-7 (hbk)

Cover design by Ellen Seguin.

A CIP catalog record for this book is available from the British Library.

Library of Congress Cataloging-in-Publication Data is available from the publisher.

**Visit the Taylor & Francis Web site at
http://www.taylorandfrancis.com**

**and the CRC Press Web site at
http://www.crcpress.com**

CONTENTS

Part VI   Regression Methods

PREFACE

This is one of two volumes consisting of 32 invited papers presented at the *International Indian Statistical Association Conference* held during October 10–11, 1998, at McMaster University, Hamilton, Ontario, Canada. This Second International Conference of IISA was attended by about 240 participants and included around 170 talks on many different areas of Probability and Statistics. All the papers submitted for publication in this volume were refereed rigorously. The help offered in this regard by the members of the Editorial Board listed earlier and numerous referees is kindly acknowledged. This volume, which includes 32 of the invited papers presented at the conference, focuses on *Advances on Theoretical and Methodological Aspects of Probability and Statistics*.

For the benefit of the readers, this volume has been divided into seven parts as follows:

Part I	Stochastic Processes and Inference
Part II	Distributions and Characterizations
Part III	Inference
Part IV	Bayesian Inference
Part V	Selection Methods
Part VI	Regression Methods
Part VII	Methods in Health Research

I sincerely hope that the readers of this volume will find the papers to be useful and of interest. I thank all the authors for submitting their papers for publication in this volume.

Special thanks go to Ms. Arnella Moore and Ms. Concetta Seminara-Kennedy (both of Gordon and Breach) and Ms. Stephanie Weidel (of Taylor & Francis) for supporting this project and also for helping with the production of this volume. My final thanks go to Mrs. Debbie Iscoe for her fine typesetting of the entire volume.

I hope the readers of this volume enjoy it as much as I did putting it together!

N. BALAKRISHNAN   MCMASTER UNIVERSITY
                                  HAMILTON, ONTARIO, CANADA

LIST OF CONTRIBUTORS

Alsaleh, Jamal A., Department of Statistics, Kuwait University,
P.O. Box 5969, Kuwait 13060

Aoshima, Makoto, Institute of Mathematics, University of Tsukuba,
Ibaraki 305-8571, Japan
aoshima@math.tsukuba.ac.jp

Arnold, B. C., Department of Statistics, University of California,
Riverside, CA 92521, U.S.A.
barry.arnold@ucr.edu

Balakrishnan, N., Department of Mathematics and Statistics,
McMaster University, Hamilton, Ontario, Canada L8S 4K1
bala@mcmail.cis.mcmaster.ca

Basak, Prasanta, Department of Mathematics, Penn State University,
Altoona, PA 16001-3760, U.S.A.
fkv@psu.edu

Basawa, I. V., Department of Statistics, The University of Georgia,
Athens, GA 30602-1952, U.S.A.
ishwar@stat.uga.edu

Boukai, Benzion, Department of Mathematical Sciences, Indiana
University–Purdue University, Indianapolis, IN 46202-3216, U.S.A.
boukai@math.uipui.edu

Chan, Schultz, Department of Statistics, University of Florida,
Gainesville, FL 32611, U.S.A.

Chen, Pinyuen, Department of Mathematics, Syracuse University,
Syracuse, NY 13244-1150, U.S.A.

Chen, William, W. H., Intenal Revenue Service, P.O. Box 2608, Washington, DC 20013-2608, U.S.A.
william-chen@soi.irs.gov

Cheong, Young-Ho, Department of Statistics, The University of Western Ontario, London, Ontario, Canada N6A 5B7

Datta, Somnath, Department of Statistics, The University of Georgia, Athens, GA 30602-1952, U.S.A.
datta@stat.uga.edu

Datta, Sujay, Department of Mathematics and Computer Science, Northern Michigan University, Marquette, MI 49855, U.S.A.
sdatta@nmu.edu

Datta, Susmita, Department of Mathematics and Computer Science, Georgia State University, Atlanta, GA 30303-3083, U.S.A.
sdatta@cs.gsu.edu

Dey, J., Department of Statistics and Applied Probability, Michigan State University, East Lansing, MI 48824, U.S.A.

Draghici, L., Department of Statistics and Applied Probability, Michigan State University, East Lansing, MI 48824, U.S.A.

Ghosh, Kaushik, Department of Statistics, George Washington University, Washington, DC 20052, U.S.A.
ghosh@gwu.edu

Ghosh, Malay, Department of Statistics, University of Florida, Gainesville, FL 32611, U.S.A.
ghoshm@stat.ufl.edu

Gupta, Shanti S., Department of Statistics, Purdue University, West Lafayette, IN 47907, U.S.A.
sgupta@stat.purdue.edu

Hamon, Agnés, Laboratory SABRES, Université de Bretagne Sud, 56000 Vannes, France

Hussein, Khaled, Department of Mathematics, Southern Illinois University, Carbondale, IL 62901-4408, U.S.A.

Jandhyala, Venkata K., Department of Pure and Applied Mathematics and Program in Statistics, Washington State University, Pullman,

WA 99164-3113, U.S.A.
jandhyala@wsu.edu

Jørgensen, Bent, Department of Statistics, University of British Columbia, Vancouver, British Columbia, Canada V6T 1Z2
bent@stat.ubc.ca

Kallianpur, G., Department of Statistics, Center for Stochastic Processes, University of North Carolina, Chapel Hill, NC 27599-3260, U.S.A.
gk@stat.unc.edu

Khan, Rasul A., Department of Mathematics, Cleveland State University, Cleveland, OH 44114-4680, U.S.A.
khan@math.scuohio.edu

Kundu, Subrata, Department of Statistics, George Washington University, Washington, DC 20052, U.S.A.
kundu@gwu.edu

Lee, Yi-Tzu, Department of Mathematics and Statistics, University of Maryland Baltimore County, Baltimore, MD 21250, U.S.A.
math.umbc.edu

Liang, TaChen, Department of Mathematics, Wayne State University, Detroit, MI 48202, U.S.A.

Lin, Xun, Department of Statistics, Purdue University, West Lafayette, IN 47907, U.S.A.

Lin, Zhengyan, Department of Mathematics, Hangzhou University, Hangzhou, China 310028

Ma, Yimin, Department of Mathematics and Statistics, University of Regina, Regina, Saskatchewan, Canada

Mandal, Pranab Kumar, EURANDOM / LG 1.21, P. O. Box 513, 5600 MB Eindhoven, The Netherlands
mandal@eurandom.tue.nl

Mathai, A. M., Department of Mathematics and Statistics, McGill University, Montreal, Quebec H3A 2K6, Canada
mathai@math.mcgill.ca

Mathew, Thomas, Department of Mathematics and Statistics, University of Maryland at Baltimore County, Baltimore, MD 21250,

U.S.A.
math.umbc.edu

Mehlman, Marc H., Department of Mathematics, University of
Pittsburgh, Johnstown, PA 15904, U.S.A.
mehlman+@pitt.edu

Mesbah, Mounir, Laboratory SABRES, Université de Bretagne Sud,
56000 Vannes, France
mounir.mesbah@univ-ubs.fr

Mudholkar, Govind S., Department of Statistics, University of Rochester,
Rochester, NY 14627, U.S.A.
govind@metro.bst.rochester.edu

Mukhopadhyay, Nitis, Department of Statistics, University of
Connecticut, Storrs, CT 06269-3102, U.S.A.
mukhop@uconnvm.uconn.edu

Nayak, Tapan K., Department of Statistics, George Washington
University, Washington, DC 20052, U.S.A.
tapan@gwu.edu

Panchapakesan, S., Department of Mathematics, Southern Illinois
University, Carbondale, IL 62901-4408, U.S.A.
kesan@math.siu.edu; spkesan@prodigy.net

Provost, Serge B., Department of Statistics, The University of Western
Ontario, London, Ontario, Canada N6A 5B7
provost@stats.uwo.ca

Ramamoorthi, R. V., Department of Statistics and Applied Probability, Michigan State University, East Lansing, MI 48824, U.S.A.
ramamoor@stt.msu.edu

Rao, Poduri S. R. S., Department of Statistics, University of Rochester,
Rochester, NY 14627, U.S.A.
raos@troi.cc.rochester.edu

Roberston, C. A., Department of Statistics, University of California,
Riverside, CA 92521, U.S.A.

Satten, Glen A., Division of HIV/AIDS Prevention: Surveillance and
Epidemiology, National Center for HIV, STD and TB Prevention,
Centers for Disease Control and Prevention, Atlanta, GA, U.S.A.

Shore, Haim, Department of Industrial Engineering, Ben-Gurion University of the Negev, Beer-Sheva 84105, Israel
shore@bgumail.bgu.ac.il

Srivastava, Deo Kumar, St. Jude Children's Research Hospital, Memphis, TN 38105-2794, U.S.A.
kumar.srivastava@stjude.org

Swift, Randall J., Department of Mathematics, Western Kentucky University, Bowling Green, KY 42101-3576, U.S.A.
randall.swift@wku.edu

Takada, Yoshikazu, Department of Mathematics, Kumamoto University, Kumamoto 860-8555, Japan

Thavaneswaran, A., Department of Statistics, University of Manitoba, Winnipeg, Manitoba, Canada R3T 2N2
thavane@cc.umanitoba.ca

Thompson, M. E., Department of Statistics and Actuarial Science, University of Waterloo, Waterloo, Ontario, Canada N2L 3G1
methomps@uwaterloo.ca

Vinogradov, Vladimir, Department of Mathematics, Ohio University, Athens, OH 45701, U.S.A.
vinograd@oaks.cats.ohiou.edu

LIST OF TABLES

LIST OF FIGURES

Part I

Stochastic Processes and Inference

CHAPTER 1

NONLINEAR FILTERING WITH STOCHASTIC DELAY EQUATIONS

G. KALLIANPUR   P. K. MANDAL

University of North Carolina, Chapel Hill, NC

Abstract: We consider a model where the coefficient function 'h' appearing in the observation model depends not only on the instantaneous value of the signal X_t, but also on the past signal values. The signal process is modeled by a stochastic delay differential equation (SDDE). The signal process is characterized as the unique solution to an appropriate martingale problem. A Zakai-type stochastic differential equation (SDE) is obtained for the optimal filter corresponding to the nonlinear filtering problem and the filter is characterized as the unique solution to the Zakai equation.

Keywords and phrases: Nonlinear filtering, Zakai equation, stochastic delay equations, martingale problem

1.1 INTRODUCTION

The general filtering problem can be described as follows. The *signal* or *system process* X_t, $(0 \leq t \leq T)$, is unobservable. Information about (X_t) is obtained by observing another process Y which is a function of X corrupted by noise. The usual model for Y is

$$Y_t = \int_0^t h(s, X_s) \, ds + W_t, \qquad (1.1.1)$$

where h is a measurable function and (W_t) is a standard Wiener process. The observation σ-field $\mathcal{F}_t^Y = \sigma\{Y_s, \ 0 \leq s \leq t\}$ contains all the available

3

information about X_t. The primary aim of filtering theory is to get an estimate of X_t based on the information $\mathcal{F}_t^Y$. This is given by the conditional distribution ν_t of X_t given $\mathcal{F}_t^Y$, or equivalently, the conditional expectation $E(f(X_t)|\mathcal{F}_t^Y)$ for a rich enough class of functions f. Since this estimate minimizes the squared error loss, ν is called the *optimal filter*.

It is known that the *non-linear filter* ν_t satisfies a stochastic differential equation (SDE) widely known as the *Kushner* or the *Fujisaki-Kallianpur-Kunita* (FKK) equation. See Kushner (1967), Fujisaki, Kunita and Kallianpur (1972) and Kallianpur (1980) . When the signal process is a Markov process satisfying the SDE

$$dX_t = A(t, X_t)\, dt + B(t, X_t)\, d\tilde{W}_t,$$

where $\tilde{W}$ is another Brownian motion independent of W, Zakai (1969) obtained an equivalent stochastic differential equation for a measure valued process σ_t, called the unnormalized conditional distribution of X_t given $\mathcal{F}_t^Y$, such that $\nu_t = \frac{\sigma_t}{<\sigma_t, 1>}$.

In this article we consider the case where the coefficient 'h' in the observation model (1.1.1) depends not only on the current state of the signal but also on the values from the past of length $r > 0$. In particular, we consider

$$dY_t = h(t, \pi_t X)\, dt + dW_t, \tag{1.1.2}$$

where $\pi_t X$, $(0 \le t \le T)$, is a $C([-r, 0], \mathbb{R}) \equiv$ C-valued process defined by

$$\pi_t X(u) = X(t + u), \quad -r \le u \le 0.$$

Also, unlike the usual theory, we consider the signal process to be non-Markov. In a recent paper, Bhatt and Karandikar (1996) studied the non-linear filtering problem corresponding to a non-Markov signal process where they allowed the coefficients to depend on the past values of the observation but dependence on the signal is through instantaneous values only. We take the signal process to be governed by a so called Stochastic Delay Differential Equation (SDDE):

$$X(t) = \begin{cases} \eta(0) + \int_0^t a(\pi_u X)\, du + \int_0^t b(\pi_u X)\, d\tilde{W}(u), & 0 \le t \le T, \\ \eta(t), & -r \le t \le 0, \end{cases} \tag{1.1.3}$$

where $r > 0$, η is a C-valued square integrable random variable, $(\tilde{W}(t), 0 \le t \le T)$ is a standard Brownian motion, independent of W and a and b are two continuous functionals on C satisfying the Lipschitz condition :

$$|a(\theta) - a(\tilde{\theta})| + |b(\theta) - b(\tilde{\theta})| \le K\|\theta - \tilde{\theta}\|_C, \quad \theta, \tilde{\theta} \in C \tag{1.1.4}$$

for some constant $K > 0$.

Stochastic delay differential equations were first studied by Ito and Nisio (1964) for the case of infinite delay ($r = \infty$). Recently, Mohammed (1984) has done an extensive investigation of stochastic functional differential equations with finite delay. Although the solution of a SDDE is not Markov, properly picked *slices* of the solution paths (namely the C-valued process $\pi_t X$) constitute a Markov process. See, for example, Mohammed (1984. Theorem III.1.1).

The main objective of this paper is to obtain a Zakai-type equation for the above filtering problem and to show that the optimal filter is characterized as the unique solution of that equation. We do this by applying the ideas and, in some cases, extending the results of Mohammed (1984).

We organize this article as follows. In Section 1.2, we start with some known results on martingale problems and their connections to Markov processes. Also, we introduce the notation and definitions we will follow throughout this article.

The main results on SDDE needed for our analysis are discussed in Section 1.3. A few of the results in this section are new and some are generalizations of the results of Mohammed (1984). We show that for any solution $(X_t, -r \leq t \leq T)$ of the SDDE (1.1.3), the process $(\pi_t X, 0 \leq t \leq T)$ can be characterized as the unique solution to a martingale problem corresponding to a suitable operator A^0. Then the martingale problem techniques are used to prove the Markov property of $\pi_t X$ as given in Theorem 1.3.4. Also, the latter result is more general than Theorem IV.4.3 of Mohammed (1984) in that we do not require the boundedness assumption on the coefficients a and b to obtain the explicit form of the generator.

Section 1.4 deals with the filtering problem with delay equations. Here we deduce a stochastic differential equation for the so called unnormalized conditional expectation of $\pi_t X$ given $\mathcal{F}_t^Y$. The corresponding Zakai type equation for the unnormalized conditional distribution of $\pi_t X$ given $\mathcal{F}_t^Y$ is obtained in Section 1.5 and the uniqueness of the solution to the Zakai equation is also proved there.

1.2 PRELIMINARIES

Suppose S is a complete, separable metric space and B is an operator on $C(S)$, the space of continuous functionals on S, with domain $\mathcal{D}(B) \subset C_b(S)$, the space of bounded continuous functionals on S.

For a sequence of functions $\phi_m \in C_b(S)$, $m = 1, 2, \ldots$ and $\phi \in C_b(S)$, we say that ϕ is the *bounded pointwise* limit of ϕ_m if $\sup_m \|\phi_m\| < \infty$ and $\phi_m(x) \to \phi(x) \forall x \in S$. We write $\phi = \text{bp-}\lim_{m \to \infty} \phi_m$.

We impose the following conditions on the operator B.

C1. There exists $\Phi \in C(S)$, satisfying

$$|Bf(x)| \leq K_f \Phi(x), \quad \forall f \in \mathcal{D}(B), x \in S,$$

where K_f is a constant depending on f.

C2. There exists a countable subset $\{f_n\} \subset \mathcal{D}(B)$ such that

$$\text{bp-closure}\left(\{(f_n, \Phi^{-1} B f_n) : n \geq 1\}\right) \supset \{(f, \Phi^{-1} B f) : f \in \mathcal{D}(B)\},$$

where "bp-closure" means the bounded pointwise closure.

C3. $\mathcal{D}(B)$ is an algebra that separates points in S and contains the constant functions.

Definition Suppose μ is a probability measure on S. A process Z_t, $0 \leq t \leq T$, defined on some probability space $(\Omega, \mathcal{F}, P)$ and taking values in S is said to be a solution to the martingale problem for (B, μ) if:

(i) $P \circ Z_0^{-1} = \mu$;

(ii) $\int_0^t E\Phi(Z_s) \, ds < \infty$, for every $t \leq T$;

(iii) for all $f \in \mathcal{D}(B)$,

$$M_t^f := f(Z_t) - f(Z_0) - \int_0^t (Bf)(Z_u) \, du$$

is a martingale.

Definition The martingale problem for (B, μ) is said to be *well posed* in a class of processes $\mathcal{C}$ if there exists a solution $Z^1 \in \mathcal{C}$ to the martingale problem for (B, μ) and if $Z^2 \in \mathcal{C}$ is also a solution to the martingale problem for (B, μ), then Z^1 and Z^2 have the same probability distributions.

We will assume the following additional conditions.

C4. The martingale problem for (B, δ_z) is well posed in the class of r.c.l.l. solutions for every $z \in S$.

C5. For all $\mu \in \mathcal{P}(S)$, the space of probability measures on S, any progressively measurable solution to the martingale problem for (B, μ) admits a r.c.l.l. modification.

The following result says that the uniqueness of the solution of a martingale problem always implies the Markov property [see Theorems IV.4.2 and IV.4.6 of Ethier and Kurtz (1986) and Remark 2.1 of Horowitz and Karandikar (1990)].

Lemma 1.2.1 *Suppose B satisfies the conditions C1, C2 and C4. Then the solution Z to the martingale problem for (B, μ) is a Markov process. Further, if A is the generator of Z, then $\mathcal{D}(B) \subset \mathcal{D}(A)$ and A and B coincide on $\mathcal{D}(B)$.*

We will denote by $\mathcal{C}_b$ the Banach space of all bounded continuous functions $\phi : C \to \mathbf{R}$ with the supremum norm

$$\|\phi\|_{\mathcal{C}_b} = \sup\{|\phi(\theta)| : \theta \in C\}.$$

Define a weak topology on $\mathcal{C}_b$ as follows : Let $\mathcal{M}(C)$ be the Banach space of all finite regular measures on $\mathcal{B}(C)$, the Borel sets of C, given the total variation norm. Consider the continuous bilinear form $< \cdot, \cdot > : \mathcal{C}_b \times \mathcal{M}(C) \to \mathbb{R}$ given by

$$< \phi, \mu >= \int_C \phi(\theta) \, d\mu(\theta), \quad \phi \in \mathcal{C}_b, \quad \mu \in \mathcal{M}(C).$$

A family $\{\phi_t : t > 0\}$ in $\mathcal{C}_b$ is said to *converge weakly* to $\phi \in \mathcal{C}_b$ as $t \to 0+$ if $\lim_{t \to 0+} < \phi_t, \mu >=< \phi, \mu >$ for all $\mu \in \mathcal{M}(C)$. We write $\phi = \text{w-}\lim_{t \to 0+} \phi_t$.

The following result states the relationship between the weak convergence and the bounded pointwise convergence [see Proposition IV.3.1 of Mohammed (1984)].

Proposition 1.2.1 *Suppose for each $t > 0$, $\phi_t \in \mathcal{C}_b$ and also $\phi \in \mathcal{C}_b$. Then $\phi = \text{w-}\lim_{t \to 0+} \phi_t$ if and only if $\{\|\phi_t\|_{\mathcal{C}_b} : t > 0\}$ is bounded and $\phi_t(\theta) \to \phi(\theta)$ as $t \to 0+$ for each $\theta \in C$, that is, ϕ_t converges to ϕ bounded pointwise.*

1.3 STOCHASTIC DELAY DIFFERENTIAL EQUATIONS

Let $(\Omega, \mathcal{F}, P)$ be a complete probability space and $W = (W(t))_{0 \le t \le T}$ be a real valued Wiener process defined on it.

Suppose $(\mathcal{F}_t)_{0 \le t \le T}$ is a family of increasing P-complete sub-σ-fields of $\mathcal{F}$ such that for each $t \in [0, T]$,

(E0) $\mathcal{F}_t^W \subseteq \mathcal{F}_t$ and $\mathcal{F}_t \perp\!\!\!\perp \sigma\{W(v) - W(u), t \le u \le v \le T\}$.

Suppose $\mathsf{C} \equiv \mathsf{C}([-r,0],\mathbf{R})$ is the class of all continuous functions from $[-r,0]$ to $\mathbf{R}$. For $0 \leq s \leq t \leq T$, and a C-valued random variable η on $(\Omega,\mathcal{F},P)$, let

$$\mathcal{F}_t^{\eta,W} = \sigma\{\eta,W(u),0 \leq u \leq t\} \vee [P\text{-null sets in } \mathcal{F}]$$

and

$$\mathcal{F}_t^{\eta,W,s} = \sigma\{\eta,W(u)-W(s),s \leq u \leq t\} \vee [P\text{-null sets in } \mathcal{F}].$$

For any Banach space $\mathbf{B}$ with norm $\|\cdot\|_{\mathbf{B}}$ we denote by $L^2(\Omega,\mathbf{B})$ the space of all random variables ξ taking values in $\mathbf{B}$ such that $E\|\xi\|_{\mathbf{B}}^2 < \infty$. For each sample path of a real valued process $\xi(t)$, $-r \leq t \leq T$, define

$$\pi_t\xi : [-r,0] \to \mathbf{R}, \ 0 \leq t \leq T \qquad \text{by}$$

$$\pi_t\xi(u) = \xi(t+u), \quad \text{for } -r \leq u \leq 0.$$

The following theorem on the existence and the uniqueness of the solution of an SDDE has been proved by Mohammed (1984). See, for example, Theorem II.2.1, Lemma III.1.2 and Remark V.2.2(ii) on page 143.

Theorem 1.3.1 *Suppose that $(\Omega,\mathcal{F},P),W,\{\mathcal{F}_t\}$ are given as above. Suppose $a,b : [0,T] \times \mathsf{C} \to \mathbf{R}$ are two Borel measurable functions satisfying the following Lipschitz and growth conditions. For all $t \in [0,T]$ and $\theta,\tilde{\theta} \in \mathsf{C}$,*

(E1) $|a(t,\theta) - a(t,\tilde{\theta})|^2 + |b(t,\theta) - b(t,\tilde{\theta})|^2 \leq K\|\theta - \tilde{\theta}\|_{\mathsf{C}}^2,$

(E2) $|a(t,\theta)|^2 + |b(t,\theta)|^2 \leq K(1 + \|\theta\|_{\mathsf{C}}^2),$

for some positive constant K independent of $t \in [0,T], \theta, \tilde{\theta}$. Suppose $0 \leq s \leq T$ and η is a $\mathcal{F}_s$-measurable C-valued random variable. Then the stochastic delay differential equation (SDDE) with the initial process η, given by

$$\xi_s(t)$$
$$= \begin{cases} \eta(0) + \int_s^t a(u,\pi_u\xi_s)\,du + \int_s^t b(u,\pi_u\xi_s)\,dW(u), & s \leq t \leq T, \\ \eta(t-s), & s-r \leq t \leq s, \end{cases}$$

$$(1.3.5)$$

possesses a unique continuous strong solution $(\xi_{s,\eta}(t), s-r \leq t \leq T)$ such that for each $t \geq s$, $\xi_{s,\eta}(t)$ is $\mathcal{F}_t^{\eta,W,s}$-measurable.

Remark Suppose $a,b : \mathsf{C} \to \mathbf{R}$ are two continuous functionals satisfying the Lipschitz condition (1.1.4). Consider $a(t,\theta) \equiv a(\theta)$ and $b(t,\theta) \equiv b(\theta)$, $t \in$

$[0, T], \theta \in C$. Then

$$
\begin{aligned}
|a(t, &\theta)|^2 + |b(t, \theta)|^2 \\
&\leq \; 2\left\{|a(\theta) - a(0)|^2 + |a(0)|^2 + |b(\theta) - b(0)|^2 + |b(0)|^2\right\} \\
&\leq \; 2\left\{|a(0)|^2 + |b(0)|^2\right\} + 2K\|\theta\|_C^2, \quad \text{by (1.1.4)} \\
&\leq \; K_1(1 + \|\theta\|_C^2),
\end{aligned}
$$

for some constant $K_1 > 0$ independent of t and θ. Hence, under (1.1.4), $a(t, \theta)$ and $b(t, \theta)$ satisfy the conditions (E1) and (E2) of Theorem 1.3.1. Therefore there exists a unique strong solution to the SDDE (1.1.3).

In the filtering problem of Section 1.4 we will need the assumption that the initial process η is square integrable. It then follows that

$$
\int_0^T E\left[|a(\pi_t X)|^2 + |b(\pi_t X)|^2\right] dt < \infty. \tag{1.3.6}
$$

Now we will proceed to obtain an operator A^0 with its domain $\mathcal{D}(A^0) \subset C_b$ such that if $(X(t), -r \leq t \leq T)$ is the solution to the SDDE (1.1.3) and $\phi \in \mathcal{D}(A^0)$, then $(\pi_t X, 0 \leq t \leq T)$ is a solution to the martingale problem corresponding to A^0. This will be one of the main tools in dealing with the nonlinear filtering problem with delay equations in the next section. First we prove the following

Lemma 1.3.1 *Suppose $(X(t), -r \leq t \leq T)$ is the solution to the SDDE (1.1.3) and $F \in C_b(\mathbf{R}), g \in C^1[-r, 0]$. Then*

$$
\begin{aligned}
\int_{-r}^0 &F(\pi_t X(s))g(s)\, ds - \int_{-r}^0 F(\pi_0 X(s))g(s)\, ds \\
&= \int_0^t \left[F(\pi_u X(0))g(0) - F(\pi_u X(-r))g(-r) - \int_{-r}^0 F(\pi_u X(s))g'(s)\, ds\right] du
\end{aligned} \tag{1.3.7}
$$

PROOF Note that

$$
\begin{aligned}
\int_0^t \int_{-r}^0 F(\pi_u X(s))g'(s)\, ds\, du &= \int_0^t \int_{-r}^0 F(X(u+s))g'(s)\, ds\, du \\
&= \int_0^t \int_{u-r}^u F(X(\alpha))g'(\alpha - u)\, d\alpha\, du \\
&= \int_{-r}^t F(X(\alpha)) \left[\int_{\alpha \vee 0}^{(\alpha+r) \wedge t} g'(\alpha - u)\, du\right] d\alpha
\end{aligned}
$$

$$= \int_{-r}^{t} F(X(\alpha)) \left[g(0 \wedge \alpha) - g(-r \vee (\alpha - t)) \right] d\alpha$$

$$= \int_{-r}^{0} F(X(\alpha))g(\alpha) \, d\alpha + \int_{0}^{t} F(X(\alpha))g(0) \, d\alpha$$

$$- \int_{-r}^{t-r} F(X(\alpha))g(-r) \, d\alpha - \int_{t-r}^{t} F(X(\alpha))g(\alpha - t) \, d\alpha$$

$$= \int_{-r}^{0} F(X(s))g(s) \, ds + \int_{0}^{t} F(X(u))g(0) \, du$$

$$- \int_{0}^{t} F(X(u-r))g(-r) \, du - \int_{-r}^{0} F(X(t+s))g(s) \, ds$$

$$= \int_{0}^{t} F(\pi_u X(0))g(0) \, du - \int_{0}^{t} F(\pi_u X(-r))g(-r) \, du$$

$$+ \int_{-r}^{0} F(\pi_0 X(s))g(s) \, ds - \int_{-r}^{0} F(\pi_t X(s))g(s) \, ds,$$

which gives rise to the equation (1.3.7).   □

Definition *Quasi-tame Function* [Mohammed (1984, Definition IV.4.2, pp. 105)] A function $\phi : C \to R$ is said to be a quasi-tame function if there exist $k > 0$, C^∞-bounded maps $f : R^k \to R$; $F_j : R \to R$ and piecewise C^1 functions $g_j : [-r, 0] \to R$ with g'_j absolutely integrable for $j = 1, \ldots, k-1$, such that

$$\phi(\theta) = f \left(\int_{-r}^{0} F_1(\theta(s))g_1(s) \, ds, \cdots, \int_{-r}^{0} F_{k-1}(\theta(s))g_{k-1}(s) \, ds, \theta(0) \right)$$
$$\tag{1.3.8}$$

for $\theta \in C$ with the understanding that when $k = 1$, $\phi(\theta) = f(\theta(0))$. Let the space of quasi-tame functions be denoted by τ_q.

Now suppose $\phi \in \tau_q$ is given by (1.3.8). Then SDDE (1.1.3), identity (1.3.7) and an application of the Ito formula yield that

$$\phi(\pi_t X) - \phi(\pi_0 X)$$

$$= \sum_{j=1}^{k-1} \int_{0}^{t} \left[\frac{\partial f}{\partial x_j} \left(\int_{-r}^{0} F_1(\pi_u X(s))g_1(s) \, ds, \right. \right.$$

$$\cdots, \int_{-r}^{0} F_{k-1}(\pi_u X(s))g_{k-1}(s) \, ds, \pi_u X(0)$$

$$\times \{ F_j(\pi_u X(0))g_j(0) - F_j(\pi_u X(-r))g_j(-r)$$

$$-\int_{-r}^{0} F_j(\theta(s))g_j'(s)\, ds \bigg\}\bigg]\, du$$

$$+\int_{0}^{t} \frac{\partial f}{\partial x_k} a(\pi_u X)\, du + \int_{0}^{t} \frac{\partial f}{\partial x_k} b(\pi_u X)\, d\tilde{W}(u) + \frac{1}{2}\int_{0}^{t} \frac{\partial^2 f}{\partial x_k^2} b^2(\pi_u X)\, du.$$

$$(1.3.9)$$

Define an operator A^0 on $\mathcal{C}_b$ with $\mathcal{D}(A^0) = \tau_q$ as follows. Let $\phi \in \mathcal{D}(A^0)$ be of the form (1.3.8). Then

$$(A^0\phi)(\theta)$$

$$:= \sum_{j=1}^{k-1} \left[\frac{\partial f}{\partial x_j} \left(\int_{-r}^{0} F_1(\theta(s))g_1(s)\, ds, \cdots, \int_{-r}^{0} F_{k-1}(\theta(s))g_{k-1}(s)\, ds, \theta(0) \right) \right.$$

$$\left. \times \left\{ F_j(\theta(0))g_j(0) - F_j(\theta(-r))g_j(-r) - \int_{-r}^{0} F_j(\theta(s))g_j'(s)\, ds \right\} \right]$$

$$+ \frac{\partial f}{\partial x_k} a(\theta) + \frac{1}{2}\frac{\partial^2 f}{\partial x_k^2} b^2(\theta), \quad \theta \in \mathsf{C}.$$

$$(1.3.10)$$

Then it is easy to see that the following theorem holds. See, for example, Mohammed (1998, p. 26).

Theorem 1.3.2 *Suppose* $(X(t), -r \le t \le T)$ *is given by the SDDE (1.1.3) with the coefficients* a, b *satisfying the Lipschitz condition (1.1.4). Suppose* $\phi \in \tau_q$. *Then*

$$M_t^\phi = \phi(\pi_t X) - \phi(\eta) - \int_{0}^{t} (A^0\phi)(\pi_u X)\, du$$

is a $(\mathcal{F}_t)$-*martingale.*

Let us note the following properties of the operator A^0.

Proposition 1.3.1 *Suppose* A^0 *is defined as above. Then* A^0 *satisfies the conditions C1-C3 of Section 1.2.*

PROOF Suppose $\phi \in \mathcal{D}(A^0)$ is given by

$$\phi(\theta) = f\left(\int_{-r}^{0} F_1(\theta(s))g_1(s)\, ds, \cdots, \int_{-r}^{0} F_{k-1}(\theta(s))g_{k-1}(s)\, ds, \theta(0) \right),$$

where $f \in C_b^\infty(\mathbf{R}^k), F_j \in C_b^\infty(\mathbf{R}), g_j \in C^1([-r,0]), j = 1, \ldots, k - 1$. Then for $\theta \in C$,

$$
\begin{aligned}
|(A^0\phi)(\theta)| \\
\leq \sum_{j=1}^{k-1} &\left[\left| \frac{\partial f}{\partial x_j} \left(\int_{-r}^0 F_1(\theta(s))g_1(s)\,ds, \right. \right. \right. \\
&\qquad \left. \ldots, \int_{-r}^0 F_{k-1}(\theta(s))g_{k-1}(s)\,ds, \theta(0) \right) \Bigg| \\
&\times \left| F_j(\theta(0))g_j(0) - F_j(\theta(-r))g_j(-r) - \int_{-r}^0 F_j(\theta(s))g_j'(s)\,ds \right| \Bigg] \\
&+ \left| \frac{\partial f}{\partial x_k} \right| |a(\theta)| + \frac{1}{2} \left| \frac{\partial^2 f}{\partial x_k^2} \right| |b^2(\theta)| \\
\leq\ & C_\phi \Phi(\theta), \hspace{5cm} (1.3.11)
\end{aligned}
$$

where C_ϕ is a constant depending on $f, F_j, g_j, j = 1, \ldots, k - 1$ and

$$
\Phi(\theta) = 1 + |a(\theta)| + |b^2(\theta)|. \hspace{2cm} (1.3.12)
$$

Therefore, C1 is satisfied by A^0.

To see that C2 holds, note that

$C_b^\infty(\mathbf{R}^k)$ is separable in $\|f\| + \|f'\| + \|f''\|$ norm,

$C_b^\infty(\mathbf{R})$ is separable in $\|F\|$ norm,

$C^1[-r, 0]$ is separable in $\|g\| + \|g'\|$ norm.

This will imply the existence of a countable set $\{\phi_n\} \subset \tau_q \equiv \mathcal{D}(A^0)$ such that

$$
\text{bp-closure}\left(\left\{\left(\phi_n, \Phi^{-1}(A^0\phi_n)\right) : n \geq 1\right\}\right) \supset \left\{\left(\phi, \Phi^{-1}(A^0\phi)\right) : \phi \in \mathcal{D}(A^0)\right\}
$$

and hence C2 follows.

That $\mathcal{D}(A^0)$ is an algebra follows from Mohammed (1984, p. 107). It is also easy to check that $\mathcal{D}(A^0)$ separates points in $C[-r, 0]$ and contains the constant functions which implies that A^0 satisfies C3.   □

From Theorem 1.3.2 we then have that $\pi_t X$ is a solution to the martingale problem corresponding to (A^0, η). We now show that it is the unique solution.

Theorem 1.3.3 Suppose η is a square integrable C-valued random variable and A^0 is as given by (1.3.10). Then the martingale problem for (A^0, η) is well posed.

PROOF Let Z_t, defined on a probability space $(\hat{\Omega}, \hat{\mathcal{F}}_t, \hat{P})$, be a progressively measurable solution to (A^0, η)-martingale problem, i.e. for $\phi \in \tau_q$, $\phi(Z_t)$ is a semi-martingale, given by

$$\phi(Z_t) = \phi(Z_0) + \int_0^t (A^0 \phi)(Z_u)\, du + M_t^\phi \qquad (1.3.13)$$

and $Z_0 = \eta$. We shall show that $Z_t = \pi_t \hat{X}$ for some continuous process $\{\hat{X}(t), -r \leq t \leq T\}$ satisfying a SDDE of the form (1.1.3). Then by the uniqueness of the solution to the SDDE (1.1.3) we will have that the distribution of Z_t is the same as that of $\pi_t X$, proving the well-posedness of the martingale problem for (A^0, η).

From (1.3.13) it follows that

$$< M^\phi >_t = \int_0^t \Gamma(\phi, \phi)(Z_u)\, du, \qquad (1.3.14)$$

where

$$\Gamma(\phi, \phi) = A^0(\phi^2) - 2\phi(A^0 \phi). \qquad (1.3.15)$$

Also, applying the Ito formula to (1.3.13) we have for $\beta \in C^1[0,T]$ and $\phi \in \tau_q$,

$$\begin{aligned}
\beta(t)\phi(Z_t) &= \beta(0)\phi(Z_0) + \int_0^t \beta'(u)\phi(Z_u)\, du \\
&\quad + \int_0^t \beta(u)(A^0\phi)(Z_u)\, du + \int_0^t \beta(u)\, dM_u^\phi. \quad (1.3.16)
\end{aligned}$$

Now suppose $F \in C_b^\infty(\mathbf{R})$, $g \in C^1[-r, 0]$. Let $\Delta = \Delta(F, g)$ be a bound for the integral $\int_{-r}^0 F(\theta(s))g(s)\, ds$, $\theta \in C$. Suppose $\iota_\Delta(x)$ is a C_b^∞-bump function [Hirsch (1976, pp. 41–42)] such that

$$\begin{aligned}
\iota_\Delta(x) &= 1, && \text{if} && |x| \leq \Delta, \\
0 < \iota_\Delta(x) &< 1, && \text{if} && \Delta < |x| < \Delta + 1, \\
\iota_\Delta(x) &= 0, && \text{if} && |x| \geq \Delta + 1.
\end{aligned}$$

Suppose $f(x) = x\iota_\Delta(x)$, so that $f \in C_b^\infty(\mathbf{R})$ and $f(x) = x$, for $|x| \leq \Delta$. Also, let $f^* \in C_b^\infty(\mathbf{R}^2)$ be given by $f^*(x_1, x_2) = f(x_1)$. Consider a quasi-tame function of the form (1.3.8) with $k = 2$, given by,

$$\begin{aligned}
\phi(\theta) &:= f^*\left(\int_{-r}^0 F(\theta(s))g(s)\, ds, \theta(0)\right) = f\left(\int_{-r}^0 F(\theta(s))g(s)\, ds\right) \\
&= \int_{-r}^0 F(\theta(s))g(s)\, ds. \qquad (1.3.17)
\end{aligned}$$

Then from (1.3.10), we have

$$A^0\phi(\theta)$$

$$= \frac{\partial f^*}{\partial x_1}\left(\int_{-r}^0 F(\theta(s))g(s)\,ds, \theta(0)\right)$$

$$\times \left\{F(\theta(0))g(0) - F(\theta(-r))g(-r) - \int_{-r}^0 F(\theta(s))g'(s)\,ds\right\}$$

$$+ \frac{\partial f^*}{\partial x_2}a(\theta) + \frac{1}{2}\frac{\partial^2 f^*}{\partial x_2^2}b^2(\theta)$$

$$= f'\left(\int_{-r}^0 F(\theta(s))g(s)\,ds\right)$$

$$\times \left\{F(\theta(0))g(0) - F(\theta(-r))g(-r) - \int_{-r}^0 F(\theta(s))g'(s)\,ds\right\}$$

$$= F(\theta(0))g(0) - F(\theta(-r))g(-r) - \int_{-r}^0 F(\theta(s))g'(s)\,ds \qquad (1.3.18)$$

and similarly,

$$(A^0\phi^2)(\theta)$$

$$= 2f\left(\int_{-r}^0 F(\theta(s))g(s)\,ds\right) \times f'\left(\int_{-r}^0 F(\theta(s))g(s)\,ds\right)$$

$$\times \left\{F(\theta(0))g(0) - F(\theta(-r))g(-r) - \int_{-r}^0 F(\theta(s))g'(s)\,ds\right\}$$

$$= 2\phi(\theta)(A^0\phi(\theta)).$$

Then from (1.3.15), $\Gamma(\phi,\phi)(\theta) = (A^0\phi^2)(\theta) - 2\phi(\theta)(A^0\phi(\theta)) = 0$ and hence, from (1.3.14), $< M^\phi >_t = 0$. Therefore $M_t^\phi = 0$ a.s. $\forall t$. From (1.3.16), we then have for $t' \geq t \geq 0$,

$$\beta(t')\phi(Z_{t'}) = \beta(t)\phi(Z_t) + \int_t^{t'} \beta'(u)\phi(Z_u)\,du + \int_t^{t'} \beta(u)(A^0\phi)(Z_u)\,du.$$

Using the special forms of ϕ and $A^0\phi$, given by (1.3.17), and (1.3.18), respectively, for $t' \geq t \geq 0$, we have

$$\int_{-r}^0 F(Z_{t'}(s))\beta(t')g(s)\,ds - \int_{-r}^0 F(Z_t(s))\beta(t)g(s)\,ds$$

$$= \int_t^{t'}\int_{-r}^0 F(Z_u(s))\beta'(u)g(s)\,ds\,du$$

$$+ \int_t^{t'} \beta(u) \left[F(Z_u(0))g(0) - F(Z_u(-r))g(-r) \right.$$

$$\left. - \int_{-r}^0 F(Z_u(s))g'(s)\, ds \right] du$$

$$= \int_t^{t'} F(Z_u(0))\beta(u)g(0)\, du - \int_t^{t'} F(Z_u(-r))\beta(u)g(-r)\, du$$

$$+ \int_t^{t'} \int_{-r}^0 F(Z_u(s)) \left[\beta'(u)g(s) - \beta(u)g'(s) \right] ds\, du.$$

Letting $G(t,s) = \beta(t)g(s)$ for $t \in [0,T], s \in [-r,0]$, we may rewrite the above equation in the following form

$$\int_{-r}^0 F(Z_{t'}(s))G(t',s)\, ds - \int_{-r}^0 F(Z_t(s))G(t,s)\, ds$$

$$= \int_t^{t'} F(Z_u(0))G(u,0)\, du - \int_t^{t'} F(Z_u(-r))G(u,-r)\, du$$

$$+ \int_t^{t'} \int_{-r}^0 F(Z_u(s)) \left(\frac{\partial G}{\partial u} - \frac{\partial G}{\partial s} \right) ds\, du. \qquad (1.3.19)$$

By linearity we will then have equation (1.3.19) for all functions G of the form $G(t,s) = \sum_{i=1}^m \beta_i(t)g_i(s)$, where $\beta_i \in C^1[0,T], g_i \in C^1[-r,0]$, $i = 1,\ldots,m$. Then by standard limiting arguments it can be shown that (1.3.19) holds for all $G \in C^{1,1}([0,T] \times [-r,0])$.

Define

$$\hat{X}(t) = \begin{cases} Z_t(0), & \text{if } 0 \le t \le T, \\ Z_0(t) \equiv \eta(t), & \text{if } -r \le t \le 0. \end{cases} \qquad (1.3.20)$$

To show that $Z_t = \pi_t \hat{X}$ it suffices to show that for $t_1, t_2 \in [0,T], s_1, s_2 \in [-r,0]$,

$$Z_{t_1}(s_1) = Z_{t_2}(s_2), \quad \text{if } t_1 + s_1 = t_2 + s_2. \qquad (1.3.21)$$

For, if $t \ge 0, -r \le s \le 0$,

$$\pi_t \hat{X}(s) = \hat{X}(t+s) = \begin{cases} Z_{t+s}(0), & \text{if } 0 \le t+s \le T, \\ Z_0(t+s), & \text{if } -r \le t+s \le 0. \end{cases} \Big\} = Z_t(s).$$

First let us consider the case when $-r < s_2 < s_1 < 0$. It suffices to show that for some $0 < \delta < s_2 + r$,

$$Z_t(s) = Z_{t'}(s+t-t') \quad \forall s_2 \le s \le s_1 \quad \text{whenever } t < t' < t + \delta. \qquad (1.3.22)$$

Because, if (1.3.22) holds, then letting k to be the largest integer smaller than $[2(t_2 - t_1)/\delta]$, we have

$$
\begin{aligned}
Z_{t_1}(s_1) &= Z_{t_1+\delta/2}(s_1 - \delta/2) = Z_{t_1+\delta}(s_1 - \delta) \\
&= \cdots = Z_{t_1+k\delta/2}(s_1 - k\delta/2) = Z_{t_2}(s_1 + t_1 - t_2) = Z_{t_2}(s_2).
\end{aligned}
$$

Note that $s_2 \leq s_1 - j\delta/2 \leq s_1$, for all $j = 1, \ldots, k$.

To prove (1.3.22) suppose $\epsilon > 0$ is such that $-r < s_2 - \epsilon < s_1 + \epsilon < 0$. Let $\delta = \min(r + (s_2 - \epsilon), -(s_1 + \epsilon))$ and $g^* \in C^1[-r, 0]$ be supported on $[s_2 - \epsilon, s_1 + \epsilon]$. Fix a $t \in [0, T]$ and let $t < t' < t + \delta$. Taking $G(u, s) = g^*(u + s - t)$ in (1.3.19), we see that for $t \leq u \leq t'$,

$$
\begin{aligned}
G(u, 0) &= g^*(u - t) = 0, \text{ since } u - t \geq 0 > s_1 + \epsilon. \\
G(u, -r) &= g^*(u - r - t) = 0, \\
&\qquad \text{since } u - r - t \leq t' - t - r < \delta - r \leq s_2 - \epsilon. \\
\frac{\partial G}{\partial u} - \frac{\partial G}{\partial s} &= 0.
\end{aligned}
$$

Hence from (1.3.19)

$$
\begin{aligned}
\int_{-r}^{0} F(Z_{t'}(s))G(t', s)\, ds &= \int_{-r}^{0} F(Z_t(s))G(t, s)\, ds, \\
\Rightarrow \int_{-r}^{0} F(Z_{t'}(s))g^*(t' + s - t)\, ds &= \int_{-r}^{0} F(Z_t(s))g^*(s)\, ds, \\
\Rightarrow \int_{-r}^{0} F(Z_{t'}(\alpha + t - t'))g^*(\alpha)\, d\alpha &= \int_{-r}^{0} F(Z_t(\alpha))g^*(\alpha)\, d\alpha,
\end{aligned}
$$

$$(1.3.23)$$

putting $t' + s - t = \alpha$. Note that during the change of variable of integration in (1.3.23), the boundary points for α lie outside the support of g^* because $t' - r - t < \delta - r \leq s_2 - \epsilon$ and $t' - t \geq 0 > s_1 + \epsilon$.

Since (1.3.23) holds for all $g^* \in C^1[s_2 - \epsilon, s_1 + \epsilon]$, we have for any $\alpha \in (s_2 - \epsilon, s_1 + \epsilon)$,

$$
F(Z_{t'}(\alpha + t - t')) = F(Z_t(\alpha)).
$$

But this being true for all $F \in C_b^\infty(\mathbb{R})$, we have $Z_{t'}(\alpha + t - t') = Z_t(\alpha)$, $\forall \alpha \in [s_2, s_1]$, which is (1.3.22). Hence we have proved (1.3.21) when $-r < s_2 < s_1 < 0$.

If $-r \leq s_2 < s_1 < 0$, then take a sequence $s_2^m > -r$ which decreases to s_2. Then for large m, so that $s_1 - s_2 + s_2^m < 0$, we have

$$
Z_{t_2}(s_2^m) = Z_{t_1}(s_1 - s_2 + s_2^m) \text{ since } t_1 + s_1 - s_2 + s_2^m = t_2 + s_2 - s_2 + s_2^m = t_2 + s_2^m.
$$

Taking the limit as $m \to \infty$, by continuity of $Z_t(\cdot)$, we then have $Z_{t_2}(s_2) = Z_{t_1}(s_1)$.

If $-r < s_2 < s_1 \leq 0$, then taking a sequence $s_1^m < 0$ increasing to s_1 we have for large m (so that $s_2 - s_1 + s_1^m > -r$),

$$Z_{t_2}(s_2 - s_1 + s_1^m) = Z_{t_1}(s_1^m), \text{ since } t_2 + s_2 - s_1 + s_1^m = t_1 + s_1 - s_1 + s_1^m = t_1 + s_1^m.$$

Again by continuity of $Z_t(\cdot)$, taking the limit as $m \to \infty$, we get $Z_{t_2}(s_2) = Z_{t_1}(s_1)$.

Finally if $s_2 = -r$ and $s_1 = 0$, then $Z_{t_2}(-r) = Z_{t_2-r/2}(-r/2) = Z_{t_2-r}(0) = Z_{t_1}(0)$.

Thus, we have proved that $Z_t = \pi_t \hat{X}$. It is easy to check that $\{\hat{X}(t), -r \leq t \leq T\}$ is a continuous process and hence, so is $(Z_t, 0 \leq t \leq T)$. Now it remains to show that the process $\hat{X}$ satisfies a SDDE of the form (1.1.3).

For $f \in C_b^\infty(\mathbb{R})$, taking $\phi(\theta) = f(\theta(0))$ in (1.3.13), we have

$$f(Z_t(0)) - f(Z_0(0)) - \int_0^t f'(Z_u(0))a(Z_u)\,du - \frac{1}{2}\int_0^t f''(Z_u(0))b^2(Z_u)\,du$$

is a martingale. That is,

$$f(\hat{X}(t)) - f(\hat{X}(0)) - \int_0^t f'(\hat{X}(u))a(\pi_u\hat{X})\,du - \frac{1}{2}\int_0^t f''(\hat{X}(u))b^2(\pi_u\hat{X})\,du$$

is a martingale for all $f \in C_b^\infty(\mathbb{R})$. Then using standard arguments [see, for example, Theorems 13.55 and 14.80 of Jacod (1979) and Theorem 4.5.2 of Stroock and Varadhan (1979)] we conclude that $\hat{X}$ satisfies a SDDE of the form (1.1.3) for some Brownian motion $\tilde{W}$. This implies that the law of $\hat{X}$ and hence of $Z_t = \pi_t\hat{X}$ is uniquely determined. Thus the martingale problem for (A^0, η) is well posed.   □

Remark In the course of proving Theorem 1.3.3 we have proved that for any probability measure μ on Borel sets of $C([-r, 0], \mathbb{R})$, the martingale problem for (A^0, μ) is well posed in the class of progressibly measurable solutions and any progressively measurable solution to (A^0, μ) has a continuous modification. In particular, we have that the conditions C4 and C5 of Section 1.2 hold for A^0.

The next result is on the Markov property of $\pi_t X$ and has been proved by Mohammed (1984) using a different method. The form of its weak generator is also obtained there but under the additional assumption that 'a' and 'b' are bounded. See Theorems III.1.1 and IV.4.3.

Theorem 1.3.4 *Suppose* $(X(t), -r \leq t \leq T)$ *satisfies the SDDE (1.1.3) with the coefficients* a, b *satisfying the Lipschitz condition (1.1.4). Then*
(i) $\pi_t X$ *is a Markov process.*
(ii) τ_q *is in the domain,* $\mathcal{D}(A)$, *of the weak generator* A *of* $(\pi_t X)$ *and the restriction of* A *on* τ_q *is the same as* A^0.
(iii) τ_q *is weakly dense in* $\mathcal{C}_b$.

PROOF   By Theorem 1.3.2, $\pi_t X$ is a solution to the martingale problem for (A^0, η). From Proposition 1.3.1 and the remark following the proof of Theorem 1.3.3, A^0 satisfies conditions C1, C2, C4. Then an application of Lemma 1.2.1 with $B = A^0$ proves (i) and (ii). To prove part (iii) it suffices to prove that the tame functions can be approximated by quasi-tame functions. For, by Theorem IV.4.1 of Mohammed (1984), the tame functions are weakly dense in $\mathcal{C}_b$. So let ϕ be a tame function and has the representation

$$\phi(\theta) = f(\theta(s_1), \ldots, \theta(s_k))$$

where $f : \mathbb{R}^k \to \mathbb{R}$ is C^∞-bounded and $s_1 < \cdots < s_k \in [-r, 0]$. We have to find a sequence $\phi_m, m = 1, 2, \ldots$ of quasi tame functions such that $\phi = \text{w-}\lim_{m \to \infty} \phi_m$.
Choose a sequence $\{\iota_m\}, m = 1, 2, \ldots$ of C_b^∞-bump functions [Hirsch (1976, pp. 41–42)] with the properties

$$\iota_m(x) = 1, \quad \text{if} \quad |x| \leq m,$$
$$0 \leq \iota_m(x) \leq 1, \quad \text{if} \quad m < |x| < m + 1,$$
$$\iota_m(x) = 0, \quad \text{if} \quad |x| \geq m + 1.$$

Define $F^m(x) = x\iota_m(x)$. Then each F^m is a C^∞ bounded function. Also define $g_j^m(t) = mg(m(t - s_j))$ for $t \in [-r, 0]$, where

$$g(t) = e^{\frac{1}{(t-1)^2 - 1}}, 0 < t < 2, \text{ and } = 0 \text{ otherwise.}$$

Note that $g(t)$ is a C^∞ function satisfying $\int_0^2 g(t)\, dt = 1$. (In fact, any continuously differentiable g on $(-\infty, \infty)$, supported on a bounded interval, such that $\int_{-\infty}^{\infty} g(t)\, dt = 1$ will do the work.)
   Now let $\epsilon > 0$ be given. Suppose $-r \leq s_j < 0$ and $\theta \in C$. By uniform continuity of θ find $\delta_j > 0$ such that $|\theta(s_j) - \theta(s_j + t)| < \epsilon$ for $|t| < \delta_j$. Also suppose K is a bound for $\theta(t), t \in [-r, 0]$. Then

$$\left| \int_{-r}^{0} F^m(\theta(s)) g_j^m(s)\, ds - \theta(s_j) \right|$$

$$= \left| \int_{-r}^{0} \theta(s) \iota_m(\theta(s)) mg(m(s - s_j))\, ds - \theta(s_j) \right|$$

$$= \left| \int_{-r}^{0} \theta(s) mg(m(s - s_j))\, ds - \theta(s_j) \right|,$$

for $m > K$, so that $\iota_m(\theta(s)) = 1$

$$= \left| \int_{m(-r-s_j)}^{-ms_j} \theta(s_j + t/m) g(t)\, dt - \theta(s_j) \right|, \text{ putting } t = m(s - s_j)$$

$$= \left| \int_{0}^{2} \theta(s_j + t/m) g(t)\, dt - \theta(s_j) \right|, \text{ for large } m, \text{ such that } -ms_j > 2$$

$$= \left| \int_{0}^{2} \{\theta(s_j + t/m) - \theta(s_j)\} g(t)\, dt \right|, \text{ since } \int_{0}^{2} g(t)\, dt = 1$$

$$\leq \int_{0}^{2} |\theta(s_j + t/m) - \theta(s_j)| g(t)\, dt$$

$$\leq \epsilon, \text{ for large } m, \text{ such that } t/m < \delta_j.$$

Hence for any $\theta \in C$,

$$\int_{-r}^{0} F^m(\theta(s)) g_j^m(s)\, ds \to \theta(s_j) \text{ as } m \to \infty. \qquad (1.3.24)$$

We consider two cases.

Case 1 $s_k = 0$. For $\theta \in C$ define

$$\phi_m(\theta) = f\left(\int_{-r}^{0} F^m(\theta(s)) g_1^m(s)\, ds, \ldots, \int_{-r}^{0} F^m(\theta(s)) g_{k-1}^m(s)\, ds, \theta(0) \right).$$

Then $\phi_m, m = 1, 2, \ldots,$ are quasi-tame functions and since f is continuous, from (1.3.24), we have

$$\phi_m(\theta) \to f(\theta(s_1), \ldots, \theta(s_{k-1}), \theta(0)) = f(\theta(s_1), \ldots, \theta(s_k)) = \phi(\theta),$$

$$\text{as } m \to \infty.$$

Case 2 $s_k < 0$. For $\theta \in C$ define

$$\phi_m(\theta) = f^*\left(\int_{-r}^{0} F^m(\theta(s)) g_1^m(s)\, ds, \ldots, \int_{-r}^{0} F^m(\theta(s)) g_k^m(s)\, ds, \theta(0) \right),$$

where $f^*(x_1, \ldots, x_k, x_{k+1}) = f(x_1, \ldots, x_k)$. Clearly, $\phi_m, m = 1, 2, \ldots$, are quasi-tame functions. Since f and hence f^* is continuous, from (1.3.24), we have

$$\phi_m(\theta) \to f^*(\theta(s_1), \ldots, \theta(s_k), \theta(0)) \;=\; f(\theta(s_1), \ldots, \theta(s_k)) = \phi(\theta),$$

$$\text{as } m \to \infty.$$

Also in both the cases,

$$|\phi_m(\theta)| \le \sup\{|f(z)| : z \in \mathbf{R}^k\}, \text{ for all } m, \theta.$$

Hence, by Proposition 1.2.1, $\phi = \text{w-} \lim_{m \to \infty} \phi_m$.   □

Remark As noted earlier, Mohammed (1984) has given alternative proofs of the parts *(i)* and *(ii)* of the theorem. But the form of A^0 is obtained with the further assumption that the coefficients a, b are globally bounded (see, e.g., Theorem III.1.1 and Theorem IV.4.3). The proof of part *(iii)* given by Mohammed (1984, Theorem IV.4.2(iii)) is not entirely correct. After the completion of the article it has come to the authors' attention that a corrected proof of *(iii)* is given in the errata to be found at http://salah.math.siu.edu/sfde.html.

1.4   THE FILTERING PROBLEM

Suppose the signal process $(X_t, -r \le t \le T)$ is governed by the SDDE (1.1.3) with η a $\mathcal{F}_0^{\tilde{W}}$-measurable square integrable C-valued random variable and $a, b : \mathsf{C} \to \mathbf{R}$ satisfying the following Lipschitz condition :

$$|a(\theta) - a(\tilde{\theta})| + |b(\theta) - b(\tilde{\theta})| \le K \|\theta - \tilde{\theta}\|_\mathsf{C}, \; \theta, \tilde{\theta} \in \mathsf{C}$$

for some constant $K > 0$.

Suppose the observation process $(Y_t, 0 \le t \le T)$ is given by (1.1.2). We assume that $h : [0, T] \times \mathsf{C} \to \mathbf{R}$ is a Borel measurable function satisfying

$$\int_0^T E|h(s, \pi_s X)|^2 \, ds < \infty. \tag{1.4.25}$$

Suppose all the processes are defined on the probability space $(\Omega, \mathcal{F}(\mathcal{F}_t), P)$. First, we are going to obtain an analog of the Bayes formula due to Kallianpur and Striebel (1968), in our setup. Our treatment is similar to the one in the appendix of Kallianpur and Karandikar (1988).

Define a measure Q on $(\Omega, \mathcal{F})$ by $dQ = \rho_T^{-1} dP$ where

$$\rho_t = \exp\left\{ \int_0^t h(s, \pi_s X)\, dY_s - \frac{1}{2} \int_0^t |h(s, \pi_s X)|^2\, ds \right\}, 0 \le t \le T. \quad (1.4.26)$$

It then follows that the measure Q is a probability measure under which (Y_t) is a Wiener process independent of (X_t) and the distribution of (X), under Q, is the same as that under P. It also follows that for any $\mathcal{F}_T^X$ measurable integrable function g,

$$E_P\left(g \,\middle|\, \mathcal{F}_t^Y\right) = \frac{E_Q\left(g\rho_t \,\middle|\, \mathcal{F}_t^Y\right)}{E_Q\left(\rho_t \,\middle|\, \mathcal{F}_t^Y\right)}. \quad (1.4.27)$$

See Theorem 11.3.1 of Kallianpur (1980) and the appendix of Kallianpur and Karandikar (1988).

For the convenience of later analysis, we will consider the probability space $(\Omega \times \Omega, \mathcal{F} \otimes \mathcal{F}, \mu)$, where $\mu = Q \otimes P$. We denote a typical point in $\Omega \times \Omega$ by (ω, ω'). For any $\mathcal{F}$-measurable function ζ on Ω, $\zeta^\dagger$ will denote the function on $\Omega \times \Omega$ defined by $\zeta^\dagger(\omega, \omega') = \zeta(\omega)$ which clearly is an $\mathcal{F} \otimes \mathcal{F}$-measurable function. Also, ζ' will denote the function on $\Omega \times \Omega$ given by $\zeta'(\omega, \omega') = \zeta(\omega')$. Thus, $X_t^\dagger(\omega, \omega') = X_t(\omega)$ and $X_t'(\omega, \omega') = X_t(\omega')$. So, under μ, X_t' is an independent copy of $X_t^\dagger$. Also, with this notation $E_Q(\zeta|\mathcal{F}_t^Y)(\omega) = E_\mu(\zeta^\dagger|\mathcal{F}_t^{Y^\dagger})(\omega, \omega')$ for $(\omega, \omega') \in \Omega \times \Omega$.

Hence, from (1.4.27), we have

$$E_P\left(g \,\middle|\, \mathcal{F}_t^Y\right)(\omega) = \frac{E_\mu\left(g^\dagger \rho_t^\dagger \,\middle|\, \mathcal{F}_t^{Y^\dagger}\right)(\omega, \omega')}{E_\mu\left(\rho_t^\dagger \,\middle|\, \mathcal{F}_t^{Y^\dagger}\right)(\omega, \omega')} \quad (1.4.28)$$

for $(\omega, \omega') \in \Omega \times \Omega$. Since (X_t) is independent of (Y_t) under Q, $(X_t^\dagger)$ is independent of $(Y_t^\dagger)$ under μ. Also, since, under μ, (X_t') has the same distribution as $(X_t^\dagger)$ the joint distribution of (X_t') and $(Y_t^\dagger)$, under μ, is the same as that of $(X_t^\dagger)$ and $(Y_t^\dagger)$. Hence

$$E_\mu\left(g^\dagger \rho_t^\dagger \,\middle|\, \mathcal{F}_t^{Y^\dagger}\right) = E_\mu\left(g' R_t \,\middle|\, \mathcal{F}_t^{Y^\dagger}\right), \quad (1.4.29)$$

where R_t is obtained by changing X and Y to X' and $Y^\dagger$, respectively, in the formula (1.4.26) of ρ_t, so that

$$R_t = \exp\left\{ \int_0^t h(s, \pi_s X')\, dY_s^\dagger - \frac{1}{2} \int_0^t |h(s, \pi_s X')|^2\, ds \right\}. \quad (1.4.30)$$

Also, using the independence of (X_t') and $(Y_t^\dagger)$, it is easy to check that

$$E_\mu \left(g' R_t \big| \mathcal{F}_t^{Y^\dagger} \right) (\omega, \omega') = \int_\Omega g(\omega'') R_t(\omega, \omega'') P(d\omega''). \qquad (1.4.31)$$

Hence from (1.4.28), (1.4.29) and (1.4.31), we have the following

Theorem 1.4.1 *Let g be an $\mathcal{F}_T^X$-measurable integrable random variable. Then for $t \in [0, T]$,*

$$E_P \left(g \big| \mathcal{F}_t^Y \right) (\omega) = \frac{\int_\Omega g(\omega') R_t(\omega, \omega') P(d\omega')}{\int_\Omega R_t(\omega, \omega') P(d\omega')}, \qquad (1.4.32)$$

where R_t is given by (1.4.30).

Then for $\phi \in \mathcal{C}_b$, we have

$$E_P \left(\phi(\pi_t X) \big| \mathcal{F}_t^Y \right) = \frac{\int_\Omega \phi(\pi_t X(\omega')) R_t(\omega, \omega') P(d\omega')}{\int_\Omega R_t(\omega, \omega') P(d\omega')} \qquad (1.4.33)$$

$$= \frac{\sigma_t(\phi, Y)}{\sigma_t(1, Y)}, \qquad (1.4.34)$$

where

$$\sigma_t(\phi, Y)(\omega) = \int_\Omega \phi(\pi_t X(\omega')) R_t(\omega, \omega') P(d\omega'), \quad 0 \le t \le T. \qquad (1.4.35)$$

Because of the relation (1.4.34), $\sigma_t(\phi, Y)$ is called the unnormalized conditional expectation of $\phi(\pi_t X)$ given $\mathcal{F}_t^Y$.

Now we obtain a stochastic differential equation for $\sigma_t(\phi, Y)$ which can then be used to get the Zakai-type equation for the unnormalized conditional distribution of $\pi_t X$ given $\mathcal{F}_t^Y$.

Theorem 1.4.2 *Suppose $\phi \in \mathcal{C}_b$ is a quasi-tame function of the form (1.3.8). Suppose the following condition holds.*

$$E_P \left[\left(\int_0^T \{ |a(\pi_s X)| + b^2(\pi_s X) \} \, ds \right)^2 \int_0^T |h(s, \pi_s X)|^2 \, ds \right] < \infty. \qquad (1.4.36)$$

Then $\sigma_t(\phi, Y)$, the unnormalized conditional expectation of $\phi(\pi_t X)$ given $\mathcal{F}_t^Y$, satisfies the following stochastic differential equation

$$\sigma_t(\phi, Y) = E_P(\phi(\eta)) + \int_0^t \sigma_s(A^0\phi, Y) \, ds + \int_0^t \sigma_s(\phi \, h_s, Y) \, dY_s, \qquad (1.4.37)$$

where A^0 is as defined in (1.3.10).

PROOF Define $g_t = \phi(\pi_T X) - \int_t^T (A^0\phi)(\pi_s X)\, ds$. Since, by Theorem 1.3.2, $M_t^\phi := \phi(\pi_t X) - \phi(\eta) - \int_0^t (A^0\phi)(\pi_u X)\, du$ is a martingale, we have

$$
\begin{aligned}
E_P\left(g_t \,|\, \mathcal{F}_t^X\right) &= E_P\left(\phi(\pi_T X) - \int_t^T (A^0\phi)(\pi_s X)\, ds \,\Big|\, \mathcal{F}_t^X\right) \\
&= E_P\left(M_T^\phi - M_t^\phi + \phi(\pi_t X) \,|\, \mathcal{F}_t^X\right) \\
&= \phi(\pi_t X).
\end{aligned}
$$

Also, note that for fixed ω, as a function of ω',

$$
R_t(\omega, \omega') = \exp\left\{\int_0^t h(s, \pi_s X(\omega'))\, dY_s(\omega) - \frac{1}{2}\int_0^t |h(s, \pi_s X(\omega'))|^2\, ds\right\}
$$

is $\mathcal{F}_t^X$-measurable. Then, from (1.4.35), we have for $0 \le t \le T$,

$$
\begin{aligned}
\sigma_t(\phi, Y)(\omega) &= \int_\Omega E_P\left(g_t \,|\, \mathcal{F}_t^X\right)(\omega') R_t(\omega, \omega')\, P(d\omega') \\
&= \int_\Omega E_P\left(g_t(\cdot) R_t(\omega, \cdot) \,|\, \mathcal{F}_t^X\right)(\omega')\, P(d\omega') \\
&= \int_\Omega g_t(\omega') R_t(\omega, \omega')\, P(d\omega'). \qquad (1.4.38)
\end{aligned}
$$

Now applying the Ito formula to g_t' and R_t, we get

$$
\begin{aligned}
dg_t' &= (A^0\phi)(\pi_t X')\, dt, \\
dR_t &= R_t h(t, \pi_t X')\, dY_t^\dagger, \\
\text{and hence} \quad d[g_t' R_t] &= R_t (A^0\phi)(\pi_t X')\, dt + g_t' R_t h(t, \pi_t X')\, dY_t^\dagger.
\end{aligned}
$$

So for $0 \le t \le T$,

$$
g_t' R_t = g_0' R_0 + \int_0^t (A^0\phi)(\pi_s X') R_s\, ds + \int_0^t g_s' h(s, \pi_s X') R_s\, dY_s^\dagger.
$$

Hence

$$
\begin{aligned}
g_t(\omega') R_t(\omega, \omega') &= g_0(\omega') + \int_0^t (A^0\phi)(\pi_s X(\omega')) R_s(\omega, \omega')\, ds \\
&\quad + \int_0^t g_s(\omega') h(s, \pi_s X(\omega')) R_s(\omega, \omega')\, dY_s(\omega).
\end{aligned}
$$

$$(1.4.39)$$

Using (1.4.39) in (1.4.38), we get

$$
\begin{aligned}
\sigma_t(\phi, Y)(\omega) &= \int_\Omega g_0(\omega') \, P(d\omega') \\
&\quad + \int_\Omega \int_0^t (A^0\phi)(\pi_s X(\omega')) R_s(\omega, \omega') \, ds \, P(d\omega') \\
&\quad + \int_\Omega \int_0^t g_s(\omega') h(s, \pi_s X(\omega')) R_s(\omega, \omega') \, dY_s(\omega) \, P(d\omega') \\
&= I_1 + I_2 + I_3, \qquad \text{say.} \tag{1.4.40}
\end{aligned}
$$

Then, by Theorem 1.3.2, we have

$$
I_1 = \int_\Omega g_0(\omega') \, P(d\omega') = E_P g_0 = E_P \left(M_T^\phi + \phi(\eta) \right) = E_P(\phi(\eta)). \tag{1.4.41}
$$

Also, an interchange of the integral signs and (1.4.35) yields

$$
\begin{aligned}
I_2 &= \int_\Omega \int_0^t (A^0\phi)(\pi_s X(\omega')) R_s(\omega, \omega') \, ds \, P(d\omega') \\
&= \int_0^t \int_\Omega (A^0\phi)(\pi_s X(\omega')) R_s(\omega, \omega') \, P(d\omega') \, ds \tag{1.4.42} \\
&= \int_0^t \sigma_s(A^0\phi, Y) \, ds. \tag{1.4.43}
\end{aligned}
$$

In (1.4.42) above to interchange the integral signs we have used Fubini's theorem which is justified because of the following.

$$
\int_\Omega \int_0^t \int_\Omega |(A^0\phi)(\pi_s X(\omega'))| \, R_s(\omega, \omega') \, P(d\omega') \, ds \, Q(d\omega)
$$

$$
\leq C_\phi \int_\Omega \int_0^t \int_\Omega \Phi(\pi_s X(\omega')) R_s(\omega, \omega') \, P(d\omega') \, ds \, Q(d\omega), \quad \text{by (1.3.11)}
$$

$$
= C_\phi \int_0^t \int_\Omega \int_\Omega \Phi(\pi_s X(\omega')) R_s(\omega, \omega') \, Q(d\omega) \, P(d\omega') \, ds
$$

$$
= C_\phi \int_0^t \int_\Omega \Phi(\pi_s X(\omega'))
$$
$$
\times \left\{ \int_\Omega e^{\int_0^t h(s, \pi_s X(\omega')) \, dY_s(\omega) - \frac{1}{2} \int_0^t |h(s, \pi_s X(\omega'))|^2 \, ds} \, Q(d\omega) \right\} P(d\omega') \, ds
$$

$$
= C_\phi \int_0^t \int_\Omega \Phi(\pi_s X(\omega')) \, P(d\omega') \, ds
$$

$$= C_\phi \int_0^t E_P \left(1 + |a(\pi_s X)| + |b(\pi_s X)|^2\right) ds, \quad \text{by the form (1.3.12) of } \Phi$$

$$< \infty, \quad \text{by (1.3.6).}$$

Therefore,

$$\int_0^t \int_\Omega |(A^0 \phi)(\pi_s X(\omega'))| \, R_s(\omega, \omega') \, P(d\omega') \, ds < \infty,$$

Q-a.s. and hence P-a.s., since $P \equiv Q$. So the interchange of the integral signs in (1.4.42) is justified.

To complete the proof of the theorem it remains to show that

$$I_3 = \int_0^t \sigma_s(\phi \, h_s, Y) \, dY_s,$$

where

$$I_3 = \int_\Omega \int_0^t g_s(\omega') h(s, \pi_s X(\omega')) R_s(\omega, \omega') \, dY_s(\omega) \, P(d\omega').$$

Note that I_3 and $\sigma_t(\phi, Y)$ can be expressed as conditional expectations w.r.t. μ on the product space $(\Omega \times \Omega, \mathcal{F} \otimes \mathcal{F})$, namely

$$I_3(\omega) = E_\mu \left(\int_0^t g_s' h(s, \pi_s X') R_s \, dY_s^\dagger \, \Big| \, \mathcal{F}_t^{Y^\dagger} \right) (\omega, \omega') \qquad (1.4.44)$$

and

$$\sigma_t(\phi, Y)(\omega) = E_\mu \left(\phi(\pi_t X') R_t \, \Big| \, \mathcal{F}_t^{Y^\dagger} \right) (\omega, \omega'). \qquad (1.4.45)$$

To simplify I_3 we need to use a Fubini theorem for stochastic integrals, which we state below [see, for example, Liptser and Shiryayev (1977, Theorem 5.14)].

Lemma 1.4.1 Let $W_t, 0 \le t \le T$, be a Wiener martingale with respect to $(\mathcal{F}_t), 0 \le t \le T$. Let A_s be $\mathcal{F}_s$-progressively measurable such that $M_t := \int_0^t A_s \, dW_s$ is an $(\mathcal{F}_t)$-martingale. Assume that (A_s) satisfy the following further conditions.

(i) $\quad \int_0^T |A_s|^2 \, ds < \infty \qquad$ a.s.

(ii) $\quad \int_0^T \left\{ E \left(|A_s| \, | \, \mathcal{F}_s^W \right) \right\}^2 \, ds < \infty \qquad$ a.s.

Then

$$E \left(\int_0^t A_s \, dW_s \, \Big| \, \mathcal{F}_t^W \right) = \int_0^t E \left(A_s \, | \, \mathcal{F}_s^W \right) \, dW_s \qquad \text{a.s.}$$

We apply the lemma on $Y_t^\dagger, 0 \le t \le T$, defined on the product space $(\Omega \times \Omega, \mathcal{F} \otimes \mathcal{F}, \mu)$, and $\mathcal{G}_t = \mathcal{F}_t^Y \otimes \mathcal{F}$. We take $A_s = g_s' h(s, \pi_s X') R_s$. To verify the assumptions of the lemma recall that (Y_t) is a Wiener process on $(\Omega, \mathcal{F}, Q)$ and $\mu = Q \otimes P$. Hence $(Y_t^\dagger, \mathcal{G}_t)$ is a Wiener martingale on $(\Omega \times \Omega, (\mathcal{G}_t), \mu)$. Clearly, A_s is $\mathcal{G}_s$-progressibly measurable. To see that $M_t := \int_0^t A_s \, dY_s^\dagger$ is a martingale first note that

$$|g_t| = \left| \phi(\pi_t X) - \int_t^T (A^0 \phi)(\pi_s X) \, ds \right| \le C_\phi \left(1 + \int_0^T \Phi(\pi_s X) \, ds \right)$$

(1.4.46)

for some constant C_ϕ. Also, since

$$R_t = e^{\int_0^t h(s, \pi_s X') \, dY_s^\dagger - \frac{1}{2} \int_0^t |h(s, \pi_s X')|^2 \, ds} = 1 + \int_0^t h(s, \pi_s X') R_s \, dY_s^\dagger,$$

(1.4.47)

it follows that R_t is a continuous $(\mathcal{G}_t)$-local martingale. Hence, being positive, R_t is a supermartingale. On the other hand,

$$\begin{aligned} E_\mu R_T &= \int_\Omega \int_\Omega \exp \left\{ \int_0^T h(s, \pi_s X(\omega')) \, dY_s(\omega) \right. \\ &\qquad \left. - \frac{1}{2} \int_0^T |h(s, \pi_s X(\omega'))|^2 \, ds \right\} Q(d\omega) \, P(d\omega') \\ &= 1. \end{aligned}$$

Hence $R_t, 0 \le t \le T$, is a $(\mathcal{G}_t)$-martingale. Now

$$\begin{aligned} M_t &= \int_0^t A_s \, dY_s^\dagger = \int_0^t g_s' h(s, \pi_s X') R_s \, dY_s^\dagger \\ &= \int_0^t g_s' \, dR_s, \qquad \text{by (1.4.47)} \\ &= g_t' R_t - g_0' R_0 - \int_0^t R_s \, dg_s' = g_t' R_t - g_0' - \int_0^t R_s (A^0 \phi)(\pi_s X') \, ds. \end{aligned}$$

Then from (1.4.46) we have

$$|M_t| \le C_\phi \left(1 + \int_0^T \Phi(\pi_s X') \, ds \right) (1 + R_t) + C_\phi \int_0^t R_s \Phi(\pi_s X') \, ds.$$

(1.4.48)

Note that

$$E_\mu \int_0^T R_s \Phi(\pi_s X') \, ds$$

$$= \int_\Omega \int_\Omega \int_0^T R_s(\omega, \omega') \Phi(\pi_s X(\omega')) \, ds \, Q(d\omega) \, P(d\omega')$$

$$= \int_0^T \int_\Omega \left\{ \int_\Omega R_s(\omega, \omega') \, Q(d\omega) \right\} \Phi(\pi_s X(\omega')) \, P(d\omega') \, ds$$

$$= \int_0^T \int_\Omega \Phi(\pi_s X(\omega')) \, P(d\omega') \, ds \; = \; \int_0^T E_P \Phi(\pi_s X) \, ds$$

$$< \; \infty,$$

by (1.3.12) and (1.3.6). Similarly,

$$E_\mu \left[R_t \int_0^T \Phi(\pi_s X') \, ds \right] \; = \; E_P \int_0^T \Phi(\pi_s X) \, ds \; < \; \infty,$$

$$E_\mu \int_0^T \Phi(\pi_s X') \, ds \; = \; E_P \int_0^T \Phi(\pi_s X) \, ds < \infty.$$

Also $E_\mu R_t = 1$. Hence we have, from (1.4.48), $E|M_t| < \infty$. Now

$$E_\mu (M_t | \mathcal{G}_s) \; = \; E_\mu \left(\int_0^t g'_u h(u, \pi_u X') R_u \, dY_u^\dagger \middle| \mathcal{G}_s \right)$$

$$= \; M_s + E_\mu \left(\int_s^t g'_u h(u, \pi_u X') R_u \, dY_u^\dagger \middle| \mathcal{G}_s \right).$$

Therefore, to prove that M_t is a $\mathcal{G}_t$-martingale it suffices to show that

$$E_\mu \left(\int_s^t g'_u h(u, \pi_u X') R_u \, dY_u^\dagger \middle| \mathcal{G}_s \right) = 0. \qquad (1.4.49)$$

So let $B = B_s^y \times B^x \in \mathcal{G}_s = \mathcal{F}_s^Y \otimes \mathcal{F}$, where $B_s^y \in \mathcal{F}_s^Y$ and $B^x \in \mathcal{F}$.

$$\int_B \int_s^t g'_u h(u, \pi_u X') R_u \, dY_u^\dagger \, d\mu$$

$$= \int_{B^x} \int_\Omega \left\{ \int_s^t g_u(\omega') h(u, \pi_u X(\omega')) R_u(\omega, \omega') \, dY_u(\omega) \right\}$$

$$\times 1_{B_s^y}(\omega) \, Q(d\omega) \, P(d\omega')$$

$$= \int_{B^x} \int_\Omega \int_s^t g_u(\omega') h(u, \pi_u X(\omega'))$$

$$\times \left\{ e^{\int_0^u h(v,\pi_v X(\omega'))\, dY_v(\omega) - \frac{1}{2}\int_0^u |h(v,\pi_v X(\omega'))|^2\, dv}\, dY_u(\omega) \right\}$$

$$\times 1_{B_s^y}(\omega)\, Q(d\omega)\, P(d\omega')$$

$$= \int_{B^x}\int_\Omega \left\{ \left[\int_s^t g_u(\omega')h(u,\pi_u X(\omega')) \right.\right.$$

$$\left.\times\ e^{\int_s^u h(v,\pi_v X(\omega'))\, dY_v(\omega) - \frac{1}{2}\int_0^u |h(v,\pi_v X(\omega'))|^2\, dv}\, dY_u(\omega) \right\}$$

$$\times \left\{ e^{\int_0^s h(v,\pi_v X(\omega'))\, dY_v(\omega)} 1_{B_s^y}(\omega) \right\} Q(d\omega)\, P(d\omega')$$

$$= \int_{B^x} E_Q \left\{ \int_s^t g_u(\omega')h(u,\pi_u X(\omega')) \right.$$

$$\left.\times\ e^{\int_s^u h(v,\pi_v X(\omega'))\, dY_v - \frac{1}{2}\int_0^u |h(v,\pi_v X(\omega'))|^2\, dv}\, dY_u \right\}$$

$$\times E_Q \left\{ e^{\int_0^s h(v,\pi_v X(\omega'))\, dY_v} 1_{B_s^y} \right\} P(d\omega') \qquad (1.4.50)$$

since for fixed ω', the quantities inside the two pairs of curly braces are $\mathcal{F}_t^{Y,s}$ and $\mathcal{F}_s^Y$ measurable, respectively, where recall that $\mathcal{F}_t^{Y,s} = \sigma\{Y_u - Y_s, s \le u \le t\}$ and under Q, $\mathcal{F}_t^{Y,s} \perp\!\!\!\perp \mathcal{F}_s^Y$. We will show that the first expectation $E_Q\{\cdots\}$ in (1.4.50) is zero. Note that for P-almost all ω',

$$\int_s^t E_Q \left\{ g_u(\omega')h(u,\pi_u X(\omega'))e^{\int_s^u h(v,\pi_v X(\omega'))\, dY_v - \frac{1}{2}\int_0^u |h(v,\pi_v X(\omega'))|^2\, dv} \right\}^2 du$$

$$= \int_s^t g_u^2(\omega')h^2(u,\pi_u X(\omega'))e^{2\int_s^u h^2(v,\pi_v X(\omega'))\, dv - \int_0^u |h(v,\pi_v X(\omega'))|^2\, dv}\, du$$

$$\le \exp\left\{ 3\int_0^T h^2(v,\pi_v X(\omega'))\, dv \right\} \int_s^t g_u^2(\omega')h^2(u,\pi_u X(\omega'))\, du\ <\ \infty$$

$$(1.4.51)$$

since, by (1.4.25), $\int_0^T h^2(v,\pi_v X(\omega'))\, dv$ is finite for P-almost all ω' and so is $\int_s^t g_u^2(\omega')h^2(u,\pi_u X(\omega'))\, du$ because

$$\int_\Omega\int_0^T g_u^2(\omega')h^2(u,\pi_u X(\omega'))\, du\, P(d\omega') = E_P \int_0^T g_u^2 h^2(u,\pi_u X)\, du$$

$$\le\ C_\phi^2 E_P \left[\left(1 + \int_0^T \Phi(\pi_u X)\, du \right)^2 \int_0^T h^2(s,\pi_s X)\, ds \right]$$

$$\le\ K_1 E_P \int_0^T h^2(s,\pi_s X)\, ds$$

$$+ K_2 \, E_P \left[\left(\int_0^T \{ |a(\pi_u X)| + b^2(\pi_u X) \} \, du \right)^2 \int_0^T |h(s, \pi_s X)|^2 \, ds \right],$$

$$\text{by (1.3.12)}$$

$$< \infty, \qquad\qquad (1.4.52)$$

by (1.4.25) and (1.4.36). Then from (1.4.51), we have

$$E_Q \left\{ \int_s^t g_u(\omega') h(u, \pi_u X(\omega')) \right.$$

$$\left. \times \, e^{\int_s^u h(v, \pi_v X(\omega')) \, dY_v - \frac{1}{2} \int_0^u |h(v, \pi_v X(\omega'))|^2 \, dv} \, dY_u \right\} = 0.$$

Hence, from (1.4.50), we get for $B \in \mathcal{G}_s$,

$$\int_B \int_s^t g_u' h(u, \pi_u X') R_u \, dY_u^\dagger \, d\mu = 0,$$

that is, equation (1.4.49) holds. Hence M_t is a $(\mathcal{G}_t)$-martingale. Now

$$\int_0^T |A_s|^2 \, ds \leq \sup_{0 \leq u \leq T} R_u^2 \int_0^T |g_s' h(s, \pi_s X')|^2 \, ds < \infty \quad \text{a.s. } [\mu],$$

because R_t is continuous and by (1.4.52)

$$E_\mu \int_0^T |g_s' h(s, \pi_s X')|^2 \, ds = E_P \int_0^T g_u^2 h^2(u, \pi_u X) \, du < \infty.$$

So condition (i) holds. Finally for condition (ii) note that, from the Bayes formula, given by (1.4.28) and (1.4.29), we have

$$E_\mu \left(|A_s| \, \Big| \mathcal{F}_s^{Y^\dagger} \right) (\omega, \omega')$$

$$= \; E_\mu \left(|g_s' h(s, \pi_s X')| \, R_s \, \Big| \mathcal{F}_s^{Y^\dagger} \right) (\omega, \omega')$$

$$= \; E_\mu \left(R_s \, \Big| \mathcal{F}_s^{Y^\dagger} \right) (\omega, \omega') \times E_P \left(|g_s h(s, \pi_s X)| \, \Big| \mathcal{F}_s^Y \right) (\omega)$$

so that, defining $\zeta_s(\omega) := E_P \left(|g_s h(s, \pi_s X)| \, \Big| \mathcal{F}_s^Y \right) (\omega)$ we have

$$\int_0^T \left\{ E_\mu \left(|A_s| \, \Big| \mathcal{F}_s^{Y^\dagger} \right) \right\}^2 \, ds \leq \sup_{0 \leq s \leq T} \left\{ E_\mu \left(R_s \Big| \mathcal{F}_s^{Y^\dagger} \right) \right\}^2 \times \int_0^T (\zeta_s^\dagger)^2 \, ds.$$

$$(1.4.53)$$

Now since $(R_t, \mathcal{G}_t)$ is a μ-martingale and $\mathcal{F}_t^{Y^\dagger} \subseteq \mathcal{G}_t$, $E_\mu(R_s|\mathcal{F}_s^{Y^\dagger})$ is a $\mathcal{F}_s^{Y^\dagger}$-martingale and therefore, has a continuous path (or a continuous version). This is so because $Y_t^\dagger$ is a Wiener process and hence every $\mathcal{F}_t^{Y^\dagger}$-adapted martingale can be represented as a stochastic integral with respect to $(Y_t^\dagger)$. Hence the first term on the RHS of (1.4.53) is μ-a.s. finite. For the second term, note that

$$
\begin{aligned}
E_P \int_0^T (\zeta_s)^2 \, ds &\leq E_P \int_0^T \left\{ E_P \left(|g_s h(s, \pi_s X)| \, \big| \mathcal{F}_s^Y \right) \right\}^2 ds \\
&\leq E_P \int_0^T E_P \left(|g_s h(s, \pi_s X)|^2 \, \big| \mathcal{F}_s^Y \right) ds \\
&= \int_0^T E_P |g_s h(s, \pi_s X)|^2 \, ds \; < \; \infty,
\end{aligned}
$$

by (1.4.52). So $\int_0^T (\zeta_s)^2 \, ds$ is P-a.s. and hence Q-a.s. finite. Therefore, $\int_0^T (\zeta_s^\dagger)^2 \, ds$ is μ-a.s. finite. Then condition (ii) follows from (1.4.53).

Now with the help of the Lemma 1.4.1 above, taking the conditional expectation in (1.4.44) inside the integral sign, we have

$$
\begin{aligned}
I_3(\omega) &= E_\mu \left(\int_0^t g_s' h(s, \pi_s X') R_s \, dY_s^\dagger \, \bigg| \mathcal{F}_t^{Y^\dagger} \right) (\omega, \omega') \\
&= \int_0^t E_\mu \left(g_s' h(s, \pi_s X') R_s \, \big| \mathcal{F}_t^{Y^\dagger} \right) dY_s^\dagger \\
&= \int_0^t \int_\Omega g_s(\omega') h(s, \pi_s X(\omega')) R_s(\omega, \omega') \, P(d\omega') \, dY_s(\omega) \\
&= \int_0^t \int_\Omega E_P \left(g_s \, \big| \mathcal{F}_s^X \right)(\omega') h(s, \pi_s X(\omega')) R_s(\omega, \omega') \, P(d\omega') \, dY_s(\omega) \\
&= \int_0^t \int_\Omega \phi(\pi_s X(\omega')) h(s, \pi_s X(\omega')) R_s(\omega, \omega') \, P(d\omega') \, dY_s(\omega) \\
&= \int_0^t \sigma_s(\phi \, h_s, Y) \, dY_s. \qquad (1.4.54)
\end{aligned}
$$

The theorem follows from (1.4.40), after substituting the forms (1.4.41), (1.4.43) and (1.4.54) of I_1, I_2 and I_3, respectively. □

Remark Note that if the coefficients 'a' and 'b' in the signal model (1.1.3) are bounded, then the extra integrability condition (1.4.36) that is imposed on the coefficient 'h' in Theorem 1.4.2 is satisfied because of (1.4.25).

1.5 ZAKAI EQUATION AND UNIQUENESS

In this section we will deduce an analog of the Zakai equation for the filtering problem with stochastic delay equations. We will obtain a measure valued stochastic differential equation for what is called the unnormalized conditional distribution of $\pi_t X$ given $\mathcal{F}_t^Y$ and prove the uniqueness of the solution. To prove the uniqueness we will use the uniqueness result of Bhatt, Kallianpur and Karandikar (1995).

Recall, from (1.4.27), that

$$E_P\left(g\,\big|\mathcal{F}_t^Y\right) = \frac{E_Q\left(g\rho_t\,\big|\mathcal{F}_t^Y\right)}{E_Q\left(\rho_t\,\big|\mathcal{F}_t^Y\right)}. \tag{1.5.55}$$

Suppose Y^t denotes the path of Y up to time t, i.e., Y^t is a $C[0,t]$-valued random variable where $Y^t(s) = Y_s$ for $0 \le s \le t$. Let $\tilde{v}_t : C[-r,0] \times C[0,t] \to \mathbf{R}$ be the measurable function such that

$$\tilde{v}_t(\pi_t X, Y^t) = E_Q\left(\rho_t\,\big|\mathcal{F}_t^Y \vee \mathcal{F}(\pi_t X)\right). \tag{1.5.56}$$

Define for $\omega \in \Omega$, a measure $v_t(\cdot, Y^t(\omega))$ on $(C, \mathcal{B}(C))$ by

$$v_t(d\theta, Y^t) = \tilde{v}_t(\theta, Y^t)\, Q \circ (\pi_t X)^{-1}(d\theta). \tag{1.5.57}$$

Now for any $\phi \in \mathcal{C}_b$,

$$\begin{aligned}
E_Q\left(\phi(\pi_t X)\rho_t\,\big|\mathcal{F}_t^Y\right)(\omega) &= E_Q\left[\phi(\pi_t X)E_Q\left(\rho_t\,\big|\mathcal{F}_t^Y \vee \mathcal{F}(\pi_t X)\right)\big|\mathcal{F}_t^Y\right](\omega) \\
&= E_Q\left[\phi(\pi_t X)\,\tilde{v}_t(\pi_t X, Y^t)\big|\mathcal{F}_t^Y\right](\omega).
\end{aligned}$$

Since (X) and (Y) are independent under Q, the conditional expectation on the RHS of the equation above can be evaluated as

$$\begin{aligned}
&E_Q\left\{\phi(\pi_t X)\,\tilde{v}_t(\pi_t X, Y^t)\big|\,\mathcal{F}_t^Y\right\}(\omega) \\
&= \int_\Omega \phi(\pi_t X(\omega'))\,\tilde{v}_t\left(\pi_t X(\omega'), Y^t(\omega)\right) Q(d\omega') \\
&= \int_C \phi(\theta)\,\tilde{v}_t(\theta, Y^t(\omega))\, Q \circ (\pi_t X)^{-1}(d\theta) \\
&= \int_C \phi(\theta)\, v_t(d\theta, Y^t(\omega)).
\end{aligned}$$

Hence we have

$$E_Q\left(\phi(\pi_t X)\rho_t\,\big|\mathcal{F}_t^Y\right) = \int_C \phi(\theta)\, v_t(d\theta, Y^t). \tag{1.5.58}$$

From (1.5.58), it follows immediately that $v_t(\cdot, Y^t)$ is a.e. a finite measure because $v_t(\mathsf{C}, Y^t) = E_Q(\rho_t|\mathcal{F}_t^Y)$ and $E_Q(\rho_t) = 1$. $v_t(\cdot, Y^t)$ is called the *unnormalized conditional distribution* of $\pi_t X$ given $\mathcal{F}_t^Y$, because for any $\phi \in \mathcal{C}_b$, from (1.5.55) and (1.5.58),

$$E_P\left(\phi(\pi_t X)\,|\mathcal{F}_t^Y\right) = \frac{\int_\mathsf{C} \phi(\theta)\, v_t(d\theta, Y^t)}{\int_\mathsf{C} v_t(d\theta, Y^t)}. \tag{1.5.59}$$

Indeed, if $P(\cdot, Y^t)$ is the conditional distribution of $\pi_t X$ given $\mathcal{F}_t^Y$, then from (1.5.59), we have

$$P(\cdot, Y^t) = \frac{v_t(\cdot, Y^t)}{\int_\mathsf{C} v_t(d\theta, Y^t)}. \tag{1.5.60}$$

Recall that

$$E_P\left(\phi(\pi_t X)\,|\mathcal{F}_t^Y\right) = \frac{\sigma_t(\phi, Y)}{\sigma_t(1, Y)}. \tag{1.5.61}$$

Hence comparing (1.5.59) and (1.5.61), we have for $\phi \in \mathcal{C}_b$,

$$\langle v_t(\cdot, Y^t), \phi \rangle := \int_{mathsf C} \phi(\theta)\, v_t(d\theta, Y^t) = \sigma_t(\phi, Y). \tag{1.5.62}$$

Then from the SDE (1.4.37) satisfied by $\sigma_t(\phi, Y)$, we get the following measure-valued Zakai equation for $v_t(\cdot, Y^t)$. For a quasi-tame function $\phi \in \mathcal{C}_b$,

$$\langle v_t(\cdot, Y^t), \phi \rangle$$
$$= E_P(\phi(\eta)) + \int_0^t \langle v_s(\cdot, Y^s), A^0\phi \rangle\, ds + \int_0^t \langle v_s(\cdot, Y^s), \phi h_s \rangle\, dY_s \tag{1.5.63}$$

The following is the desired uniqueness result for the solution to the Zakai equation (1.5.63).

Theorem 1.5.1 *Suppose the signal process $\{X(t), -r \le t \le T\}$ is given by the SDDE (1.1.3) where the coefficients a, b satisfies the Lipschitz condition (1.1.4). Suppose the observation process Y is given by (1.1.2). Assume that the condition (1.4.25) holds. Suppose A^0 is as in (1.3.10) and Φ, given by (1.3.12), is as in C1 of Section 1.2 corresponding to A^0. Further assume that the condition (1.4.36) holds.*

If ρ_t is an $\mathcal{F}_t^Y$-adapted (positive, finite) measure valued process satisfying

$$< \rho_t, \phi > = E_P(\phi(\eta)) + \int_0^t < \rho_u, A^0\phi >\, du + \int_0^t < \rho_u, \phi h_u >\, dY_u, \quad \phi \in \tau_q \tag{1.5.64}$$

and

$$E_Q \int_0^T < \rho_t, \Phi > dt < \infty, \qquad (1.5.65)$$

then $\rho_t = v_t$, for all $0 \le t \le T$ a.s., where v_t, given by (1.5.62) and (1.4.35), is the unnormalized conditional distribution of $\pi_t X$ given $\mathcal{F}_t^Y$.

PROOF First note, from (1.5.63), that v_t satisfies the Zakai equation (1.5.64). Recall, from (1.4.35) and (1.4.30), that

$$\sigma_t(\phi, Y)(\omega) = \int_\Omega \phi(\pi_t X(\omega')) R_t(\omega, \omega') \, P(d\omega'), \quad 0 \le t \le T,$$

where

$$R_t(\omega, \omega') = \exp\left\{ \int_0^t h(s, \pi_s X(\omega')) \, dY_s(\omega) - \frac{1}{2} \int_0^t |h(s, \pi_s X(\omega'))|^2 \, ds \right\}.$$

Then

$$E_Q \int_0^T < v_t, \Phi > dt \; = \; E_Q \int_0^T \sigma_t(\Phi, Y) \, dt$$

$$= \; E_Q \int_0^T \int_\Omega \Phi(\pi_t X(\omega')) R_t(\cdot, \omega') \, P(d\omega') \, dt$$

$$= \; \int_\Omega \int_0^T \int_\Omega \Phi(\pi_t X(\omega')) R_t(\omega, \omega') \, P(d\omega') \, dt \, Q(d\omega)$$

$$= \; \int_\Omega \int_0^T \Phi(\pi_t X(\omega')) \int_\Omega$$
$$\times \, e^{\int_0^t h(s, \pi_s X(\omega')) \, dY_s(\omega) - \frac{1}{2} \int_0^t |h(s, \pi_s X(\omega'))|^2 \, ds} \, Q(d\omega) \, dt \, P(d\omega')$$

$$= \; \int_\Omega \int_0^T \Phi(\pi_t X(\omega')) \, dt \, P(d\omega'), \quad \text{since } Y \text{ is a standard BM under } Q$$

$$= \; E_P \int_0^T \left[1 + |a(\pi_t X)| + |b(\pi_t X)^2| \right] dt, \qquad \text{by (1.3.12)}$$

$$< \; \infty, \qquad \text{by (1.3.6).}$$

Hence v_t satisfies the condition (1.5.65) also.

Since X and the observation noise W are independent, so are $\pi_t X$ and W. From Theorem 1.3.2 and Theorem 1.3.3, we have that $\pi_t X$ is the unique solution to the martingale problem for (A^0, η) Also, from Proposition 1.3.1 and the remark following the proof of Theorem 1.3.3, A^0 satisfies the conditions C1-C5. Then an application of Theorem 4.1 of Bhatt, Kallianpur and

Karadikar (1995) on the modified signal process $\pi_t X$ completes the proof.
$\square$

Using the unique solution ρ_t of the measure valued Zakai equation (1.5.64) one can, in principle, calculate the unnormalized conditional expectation of $f(X_t) = f(\pi_t X(0))$ given $\mathcal{F}_t^Y$ and hence can obtain $E(f(X_t)|\mathcal{F}_t^Y)$. In fact, one can, in principle, calculate the conditional expectations of the form $E(g(t, \pi_t X)|\mathcal{F}_t^Y)$ as

$$E(g(t, \pi_t X)|\mathcal{F}_t^Y) = \frac{< \rho_t, g(t, \cdot) >}{< \rho_t, 1 >}. \tag{1.5.66}$$

We then have the following formula, given in Theorem 1.5.2 below, for $E(f(X_t)|\mathcal{F}_t^Y)$ where the various conditional expectations on the right hand side of (1.5.67) is obtained as in (1.5.66).

Theorem 1.5.2 *Suppose $f : \mathbf{R} \to \mathbf{R}$ is C^∞-bounded. Let E^t denote the conditional expectation (given $\mathcal{F}_t^Y$). Then*

$$E^t[f(X_t)]$$
$$= E[f(X_0)] + \int_0^t \left\{ E^s\left[f'(X_s)a(\pi_s X)\right] + \frac{1}{2}E^s\left[f''(X_s)b^2(\pi_s X)\right] \right\} ds$$
$$+ \int_0^t \left\{ E^s\left[f(X_s)h(s, \pi_s X)\right] - E^s[f(X_s)]E^s[h(s, \pi_s X)] \right\}$$
$$\times \left\{ dY_s - E^s[h(s, \pi_s X)] ds \right\}. \tag{1.5.67}$$

PROOF Let $f : \mathbf{R} \to \mathbf{R}$ be a C^∞-bounded function. Consider the quasi-tame function $\bar{f}$ given by $\bar{f}(\theta) = f(\theta(0))$ for $\theta \in C$. Then

$$E\left(f(X_t)|\mathcal{F}_t^Y\right) = E\left(\bar{f}(\pi_t X)|\mathcal{F}_t^Y\right) = \frac{< v_t, \bar{f} >}{< v_t, 1 >}.$$

Also observe that A^0 operated on $\bar{f}$ takes the following form :

$$A^0 \bar{f}(\theta) = f'(\theta(0))a(\theta) + \frac{1}{2}f''(\theta(0))b^2(\theta) = \overline{f'}(\theta)a(\theta) + \frac{1}{2}\overline{f''}(\theta)b^2(\theta), \quad \theta \in C.$$

Hence it follows from (1.5.63) that

$$d < v_t, \bar{f} > = \left\langle v_t, \left(\overline{f'}a + \frac{1}{2}\overline{f''}b^2\right) \right\rangle dt + < v_t, \bar{f} \, h_t > dY_t,$$
$$d < v_t, 1 > = < v_t, \, h_t > dY_t.$$

Then a simple application of the Ito formula to $\frac{<v_t,\bar{f}>}{<v_t,1>}$ yields the formula (1.5.67) for $E(f(X_t)|\mathcal{F}_t^Y)$. □

Acknowledgements Research of G. Kallianpur was partially supported by Army Research Office Grant DAAL-0392-G-0008.

REFERENCES

Bhatt, A. G., Kallianpur, G. and Karandikar, R. L. (1995). Uniqueness and robustness of solution of measure valued equations of nonlinear filtering, *Annals of Probability*, **23**, 1895–1938.

Bhatt, A. G. and Karandikar, R. L. (1996). Characterization of the optimal filter: The non Markov case, Preprint.

Dynkin, E. B. (1965). *Markov Processes*, Vols.I, II, Springer-Verlag, Berlin.

Ethier, S. N. and Kurtz, T. G. (1986). *Markov Processes : Characterization and Convergence*, John Wiley & Sons, New York.

Fujisaki, M., Kallianpur, G. and Kunita, H. (1972). Stochastic differential equations for the nonlinear filtering problem, *Osaka Journal of Mathematics*, **9**, 19–40.

Hirsch, M. W. (1976). *Differential Topology*, Springer-Verlag, New York.

Horowitz, J. and Karandikar, R. L. (1990). Martingale problems associated with the Boltzman equation, In *Seminar on Stochastic Processes, 1989* (Ed., E. Cinlar *et al.*), pp. 75–122, Birkhaüser, Boston.

Ito, K. and Nisio, M. (1964). On stationary solutions of a stochastic differential equation, *J. Math. Kyoto University*, **4**, 1–75.

Jacod, J. (1979). *Calcul Stochastique et Problemes de Martingales*, Lecture Notes in Mathematics **714**, Springer-Verlag, Berlin.

Kallianpur, G. (1980). *Stochastic Filtering Theory*, Springer-Verlag, New York.

Kallianpur, G. and Karandikar, R. L. (1988). *White Noise Theory of Prediction, Filtering and Smoothing*, Gordon and Breach, New York.

Kallianpur, G. and Striebel, C. (1968). Estimations of stochastic processes : Arbitrary system process with additive white noise observation errors, *Annals of Mathematical Statistics*, **39**, 785–801.

Kushner, H. (1967). Dynamical equations for optimal nonlinear filtering, *Journal of Differential Equations*, **3**, 179–190.

Liptser, R. S. and Shiryayev, A. N. (1977). *Statistics of Random Processes*, Vol. **1**, Springer-Verlag, New York.

Mohammed, S-E. A. (1984). *Stochastic Functional Differential Equations*, Pitman, Boston.

Mohammed, S-E. A. (1998). Stochastic differential systems with memory: theory, examples and applications, In *Stochastic Ananlysis and Related Topics VI, The Geilo Workshop, 1996* (Eds., L. Decreusfond, J. Gjerde, B. Oksendal and A. Ustunel), pp. 1–77, Birkhaüser, Boston.

Stroock, D. W. and Varadhan, S. R. S. (1979). *Multidimensional Diffusion Processes*, Springer-Verlag, Berlin.

Zakai, M. (1969). On the optimal filtering of diffusion processes, *Zeitschrift Wahrscheinlichkeitstheorie und Verwandte Gebiete*, **11**, 230–243

CHAPTER 2

SIGMA OSCILLATORY PROCESSES

RANDALL J. SWIFT

Western Kentucky University, Bowling Green, KY

Abstract: The class of harmonizable processes is a natural extension of the class of stationary processes. In 1965, Priestley introduced the class of oscillatory processes and the concept of their evolutionary spectrum as a tool for the frequency analysis of these processes. Recently, Swift (1997), introduced the class of oscillatory harmonizable processes as an extension of Priestley's oscillatory class. In this paper, it is shown that these classes of processes are not closed with respect to independent elements. A broader class of nonstationary processes, termed sigma oscillatory harmonizable, is introduced and shown to be closed with respect to the sum of independent elements.

Keywords and phrases: Harmonizable processes, oscillatory processes, evolutionary spectra, sigma oscillatory processes

2.1 SOME CLASSES OF NONSTATIONARY PROCESSES

In the following work, there is always an underlying probability space, (Ω, Σ, P).

To set the stage, we consider second order stochastic processes. More specifically, mappings $X : \mathbb{R} \to L_0^2(P)$ where for $p \geq 1$, $L_0^p(P)$ is the set of all complex valued $f \in L^p(\Omega, \Sigma, P)$ such that $E(f) = 0$, with $E(f) = \int_\Omega f(\omega) \, dP(\omega)$ the expectation.

Recall that a stochastic process $X : \mathbb{R} \to L_0^2(P)$ is *stationary* (stationary in the wide or Khintchine sense) if its covariance $r(s,t) = E(X(s)\overline{X(t)})$ is

37

continuous and is a function of the difference of its arguments, so that

$$r(s,t) = \tilde{r}(s - t).$$

An equivalent and useful definition of a stationary process is one whose covariance function can be represented as

$$\tilde{r}(\tau) = \int_{\mathbb{R}} e^{i\lambda\tau} dF(\lambda), \qquad (2.1.1)$$

for a unique non-negative bounded Borel measure $F(\cdot)$. This alternate definition is a consequence of a classical theorem of Bochner's c.f., Gihman and Skorohod (1974), and motivates the following definition.

Definition 2.1.1 A stochastic process $X : \mathbb{R} \to L_0^2(P)$ is *weakly harmonizable* if its covariance $r(\cdot,\cdot)$ is expressible as

$$r(s,t) = \int_{\mathbb{R}} \int_{\mathbb{R}} e^{i\lambda s - i\lambda' t} dF(\lambda, \lambda') \qquad (2.1.2)$$

where $F : \mathbb{R} \times \mathbb{R} \to \mathbb{C}$ is a positive semi-definite bimeasure, hence of finite Fréchet variation.

The integrals in (2.1.2) are strict Morse-Transue, Chang and Rao (1986). A stochastic process, $X(\cdot)$, is *strongly harmonizable* if the bimeasure $F(\cdot,\cdot)$ in (2.1.2) extends to a complex measure and hence is of bounded Vitali variation. In either case, $F(\cdot,\cdot)$ is termed the *spectral bi-measure* (or *spectral measure*) of the harmonizable process.

Comparison of equation (2.1.2) with equation (2.1.1) shows that when $F(\cdot,\cdot)$ concentrates on the diagonal $\lambda = \lambda'$, both the weak and strong harmonizability concepts reduce to the stationary concept. The power and usefulness of harmonizable processes is now clear. They retain the powerful Fourier analytic methods inherent with stationary processes, as seen in Bochner's theorem, (2.1.1); but they relax the requirement of stationarity.

The structure and properties of harmonizable processes has been investigated and developed extensively by M.M. Rao and others. A recent detailed account of harmonizable processes and some of their applications may be found in Swift (1997a).

A broader class of nonstationary processes which contains the weakly harmonizable class was introduced by Swift (1997b). This class is defined as follows.

Definition 2.1.2 *A stochastic process* $X : \mathbb{R} \to L_0^2(P)$ *is oscillatory weakly harmonizable, if its covariance has representation*

$$r(s,t) = \int_{\mathbb{R}} \int_{\mathbb{R}} A(s,\lambda)\overline{A(t,\lambda')} \, e^{i\lambda s - i\lambda' t} \, dF(\lambda,\lambda') \qquad (2.1.3)$$

where $F(\cdot,\cdot)$ *is a function of bounded Fréchet variation, and*

$$A(t,\lambda) = \int_{\mathbb{R}} e^{itx} H(\lambda,dx) \qquad (2.1.4)$$

with $H(\cdot,B)$ *a Borel function on* $\mathbb{R}, H(\lambda,\cdot)$ *a signed measure and* $A(t,\lambda)$ *having an absolute maximum at* $\lambda = 0$ *independent of* t.

Note that if $A(t,\lambda) = 1$, this class coincides with the weakly harmonizable class of processes.

In 1965, M.B. Priestley, introduced and studied a generalization of the class of stationary processes. In particular, a stochastic process $X : \mathbb{R} \to L_0^2(P)$ is *oscillatory* if it has representation

$$X(t) = \int_{\mathbb{R}} A(t,\lambda) \, e^{i\lambda t} dZ(\lambda) \qquad (2.1.5)$$

where $Z(\cdot)$ is a stochastic measure with orthogonal increments and $A(t,\lambda)$ satisfies (2.1.4). Using this representation, the covariance of an oscillatory process is

$$r(s,t) = \int_{\mathbb{R}} A(s,\lambda)\overline{A(t,\lambda)} \, e^{i\lambda(s-t)} d\Phi(\lambda);$$

so that in the same fashion as Priestley's oscillatory processes extend the class of stationary processes, the oscillatory weakly harmonizable processes extend the weakly harmonizable class.

One further observes that if $F(\cdot,\cdot)$, the spectral bimeasure of a oscillatory weakly harmonizable process concentrates on the diagonal $\lambda = \lambda'$, the oscillatory processes are obtained. Thus the oscillatory harmonizable processes also provide an extension to the class of oscillatory processes, which we will now term *oscillatory stationary*.

Using the definition of an oscillatory harmonizable process and a version of Karhunen's theorem, Swift (1997b) obtained the spectral representation of an oscillatory weakly harmonizable process as

$$X(t) = \int_{\mathbb{R}} A(t,\lambda) \, e^{i\lambda t} dZ(\lambda) \qquad (2.1.6)$$

where $Z(\cdot)$ is a stochastic measure satisfying

$$E(Z(B_1)\overline{Z(B_2)}) = F(B_1, B_2)$$

with $F(\cdot, \cdot)$ a function of bounded Fréchet variation.

A general class of nonstationary processes which extends the classes of processes introduced above was first considered by Cramér in 1952. A refined definition of these processes is due to Rao (1984), who gave the following definition.

Definition 2.1.3 *A second-order process* $X : T \to L^2(P)$ *is of Cramér class (or class (C)) if its covariance function* $r(\cdot, \cdot)$ *is representable as*

$$r(t_1, t_2) = \int_S \int_S g(t_1, \lambda)\overline{g(t_2, \lambda')} \, dF(\lambda, \lambda') \tag{2.1.7}$$

relative to a family $\{g(t, \cdot), t \in T\}$ *of Borel functions and a positive definite function* $F(\cdot, \cdot)$ *of locally bounded variation on* $S \times S$, *[S will be in the classical case* $\hat{T}$ *the dual of an LCA group* T, *and generally* $(S, \mathbf{B})$ *is a measurable space] with each* g *satisfying the (Lebesgue) integrability condition:*

$$0 \le \int_S \int_S g(t_1, \lambda)\overline{g(t_2, \lambda')} \, dF(\lambda, \lambda') < \infty, \ t \in T.$$

If $F(\cdot, \cdot)$ has a locally finite Fréchet variation, then the integrals in equation (2.1.7) are in the sense of (strict) Morse-Transue and the corresponding concept is termed *weak class (C)*.

Chang and Rao (1986), obtained the integral representation of weak class (C) processes as given by the following Theorem.

Theorem 2.1.1 *If* $X : T \to L_0^2(P)$ *is of weak Cramér class relative to a family* $\{g(t, \cdot), t \in T\}$ *of Borel functions and a positive definite bimeasure* $F(\cdot, \cdot)$ *of locally bounded Fréchet variation on* $S \times S$, *then there exists a stochastic measure* $Z : \mathbf{B} \to L_0^2(P)$, $\mathbf{B}$ *a* σ-*algebra of* S, *such that*

$$X(t) = \int_S g(t, \lambda)dZ(\lambda) \tag{2.1.8}$$

where
$$E(Z(A)\overline{Z(B)}) = F(A, B) \text{ for } (A, B) \in \mathbf{B} \times \mathbf{B}.$$

Conversely, if $X(\cdot)$ *is a second-order process defined by (2.1.8) then it is a process of weak class (C).*

If there exists a representation of a weak class (C) process of the form (2.1.8) then there is a multitude of different representations of the process, each representation based upon a different family of functions. Clearly, one valid choice is the complex exponential family given by

$$g(t, \lambda) = e^{i\lambda t}$$

with which, the representation (2.1.8) reduces to the spectral representation of a weakly harmonizable process.

The complex exponential family provides the well-known spectral decomposition of the process and forms the basis of the physical interpretation of spectral analysis as an "energy distribution over frequency".

The oscillatory weakly harmonizable processes are intimately connected with weak class (C) processes. Swift (1997b) showed the following proposition which follows from setting $g(t, \lambda) = e^{it\lambda} A(t, \lambda)$ in the spectral representation of weak class (C) processes.

Proposition 2.1.1 *The class of oscillatory weakly harmonizable processes coincides with the class of weak class (C) processes indexed on* $\mathbb{R}$.

2.2 SIGMA OSCILLATORY PROCESSES

In the previous section, it was mentioned that for an oscillatory weakly harmonizable process $X(\cdot)$ there will, in general, be a large number of different families of functions in terms of each of which the process $X(\cdot)$ has a representation of the form (2.1.6), with each family inducing a different measure $Z(\cdot)$.

Swift recently gave the following definition.

Definition 2.2.1 Let $\mathcal{F}$ denote a particular family of functions of the form $g(t, \lambda) = A(t, \lambda)e^{it\lambda}$, and let $X(\cdot)$ be an oscillatory weakly harmonizable process with integral representation of the form (2.1.6) in terms of the family $\mathcal{F}$. The evolutionary spectrum of $X(t)$ with respect to the family $\mathcal{F}$ is given by

$$dG(s, t, \lambda, \lambda') = A(s, \lambda)\overline{A(t, \lambda')} \, dF(\lambda, \lambda'). \tag{2.2.9}$$

One notes that if the spectral measure $F(\cdot, \cdot)$ concentrates on the diagonal $\lambda = \lambda'$, then the evolutionary spectrum (2.2.9) becomes

$$d\tilde{G}(t, \lambda) = |A(t, \lambda)|^2 \, d\tilde{F}(\lambda),$$

which is the evolutionary spectra for an oscillatory process as given by Priestley (1965).

Using this definition, it is possible to extend much of Priestley's results, as was recently done by Swift. However, a difficulty with the above definition of the evolutionary spectrum, (as well as with Priestley's definition in the oscillatory stationary case), arises when one considers the sum of two independent oscillatory weakly harmonizable processes. This difficulty can be seen by considering the evolutionary spectrum of the process

$$Y(t) = X_1(t) + X_2(t),$$

where $X_1(t)$ and $X_2(t)$ are two independent oscillatory weakly harmonizable processes, with respective evolutionary spectrums

$$dG_1(s,t,\lambda,\lambda') = A_1(s,\lambda)\overline{A_1(t,\lambda')}\,dF_1(\lambda,\lambda')$$

and

$$dG_2(s,t,\lambda,\lambda') = A_2(s,\lambda)\overline{A_2(t,\lambda')}\,dF_2(\lambda,\lambda').$$

Does there exist a family of functions $\mathcal{F}_Y = \{A_Y(t,\lambda)\,e^{i\lambda t}\}$ with respect to which an evolutionary spectrum for $Y(t)$, say $dG(s,t,\lambda,\lambda')$, may be defined? Further, does this spectrum satisfy

$$dG(s,t,\lambda,\lambda') = dG_1(s,t,\lambda,\lambda') + dG_1(s,t,\lambda,\lambda')?$$

The answer is in the negative, as may be seen by assuming such a family existed. It follows from this assumption that

$$A_Y(s,\lambda)\overline{A_Y(t,\lambda')}\,dF_Y(\lambda,\lambda') = A_1(s,\lambda)\overline{A_1(t,\lambda')}\,dF_1(\lambda,\lambda')$$

$$+ A_2(s,\lambda)\overline{A_2(t,\lambda')}\,dF_2(\lambda,\lambda'),$$

so that if

$$A_Y(s,\lambda)\overline{A_Y(t,\lambda')}\,dF_Y(\lambda,\lambda') \neq 0$$

then

$$dF_Y(\lambda,\lambda') = \frac{A_1(s,\lambda)\overline{A_1(t,\lambda')}}{A_Y(s,\lambda)\overline{A_Y(t,\lambda')}}\,dF_1(\lambda,\lambda') + \frac{A_2(s,\lambda)\overline{A_2(t,\lambda')}}{A_Y(s,\lambda)\overline{A_Y(t,\lambda')}}\,dF_2(\lambda,\lambda').$$

Thus,

$$\frac{A_1(s,\lambda)\overline{A_1(t,\lambda')}}{A_Y(s,\lambda)\overline{A_Y(t,\lambda')}} \quad \text{and} \quad \frac{A_2(s,\lambda)\overline{A_2(t,\lambda')}}{A_Y(s,\lambda)\overline{A_Y(t,\lambda')}}$$

do not depend upon s or t. Hence, the ratio

$$\frac{A_1(s,\lambda)\overline{A_1(t,\lambda')}}{A_2(s,\lambda)\overline{A_2(t,\lambda')}}$$

also does not depend upon s or t, so that $X_1(t)$ and $X_2(t)$ have time proportional evolutionary spectra.

To circumvent this difficulty, the following definition is made.

Definition 2.2.2 A stochastic process $Y : \mathbb{R} \to L_0^2(P)$ is sigma oscillatory weakly harmonizable if it can be represented as

$$Y(t) = \sum_{i=1}^{n} X_i(t)$$

where the $X_i(\cdot), i = 1, \ldots, n$ are pairwise independent oscillatory weakly harmonizable processes.

An immediate consequence of this definition is the following proposition whose proof is obvious.

Proposition 2.2.1 *A sigma oscillatory weakly harmonizable process* $Y : \mathbb{R} \to L_0^2(P)$ *has spectral representation*

$$Y(t) = \sum_{i=1}^{n} \int_{\mathbb{R}} A_i(t, \lambda) \, e^{i\lambda t} dZ_i(\lambda) \qquad (2.2.10)$$

where $Z_i(\cdot), i = 1, \ldots, n$ *are stochastic measures satisfying*

$$E(Z_i(B_1)\overline{Z_i(B_2)}) = F_i(B_1, B_2)$$

with $F_i(\cdot, \cdot), i = 1, \ldots, n$ *functions of bounded Fréchet variation.*

The evolutionary spectrum of each of the oscillatory weakly harmonizable processes $X_i(t), i = 1, \ldots, n$ is defined by

$$d\,G_i(s, t, \lambda, \lambda') = A_i(s, \lambda)\overline{A_i(t, \lambda')} \, d\,F_i(\lambda, \lambda'),$$

with respect to the family of functions

$$\mathcal{F}_i = \{A_i(t, \lambda)e^{it\lambda}\}.$$

Using the pairwise independence of the $X_i(t)'s$, the covariance of $Y(t)$ can be computed as

$$r(s, t) = \sum_{i=1}^{n} \int_{\mathbb{R}} \int_{\mathbb{R}} A_i(s, \lambda)\overline{A_i(t, \lambda')} \, e^{i\lambda s - i\lambda' t} \, dF_i(\lambda, \lambda'), \qquad (2.2.11)$$

so that the evolutionary spectrum of $Y(\cdot)$ can be defined with respect to the vector family

$$\mathcal{F} = \{\mathcal{F}_1, \mathcal{F}_2, \ldots, \mathcal{F}_n\}$$

by

$$dG(s,t,\lambda,\lambda') = \sum_{i=1}^{n} A_i(s,\lambda)\overline{A_i(t,\lambda')}\, dF_i(\lambda,\lambda'). \qquad (2.2.12)$$

One notes here that question posed above now has a positive answer. That is, the sum of two sigma oscillatory weakly harmonizable processes is a sigma oscillatory weakly harmonizable process whose spectrum is the sum of the spectra of each process.

2.3 DETERMINATION OF THE EVOLUTIONARY SPECTRA

Conditions on the vector family of functions

$$\mathcal{F} = \{\mathcal{F}_1, \mathcal{F}_2, \ldots, \mathcal{F}_n\}$$

provide useful information about the evolutionary spectra of the sigma oscillatory weakly harmonizable process $Y(t)$. For each family $\mathcal{F}_i$, define the function $B_{\mathcal{F}_i}(\lambda)$ by

$$B_{\mathcal{F}_i}(\lambda) = \int_{\mathbb{R}} \mid x \mid \mid H_i \mid (\lambda, dx)$$

where each $H_i(\cdot,\cdot)$ is the signed measure in the representation of the oscillatory weakly harmonizable process $X_i(t)$, cf., (2.1.4). Each $B_{\mathcal{F}_i}(\lambda)$ is a measure of the width of $|H_i(\lambda, dx)|$.

Using this notation, we have the following definition.

Definition 2.3.1 A vector family of functions

$$\mathcal{F} = \{\mathcal{F}_1, \mathcal{F}_2, \ldots, \mathcal{F}_n\}$$

has bounded-width if for each $i = 1, \ldots, n$

$$B_{\mathcal{F}_i}(\lambda) < \infty \text{ for all } \lambda \in \mathbb{R}.$$

The characteristic width of the vector family $\mathcal{F}$ is given by

$$B_{\mathcal{F}} = \min_{1 \le i \le n} \left\{ \frac{1}{\sup_{\lambda} B_{\mathcal{F}_i}(\lambda)} \right\}.$$

Using this definition, a subclass of oscillatory weakly harmonizable processes is defined as follows.

Definition 2.3.2 A process $Y : \mathbb{R} \to L_0^2(P)$ is a bounded-width sigma oscillatory weakly harmonizable process if there exists a bounded-width vector family $\mathcal{F}$ in terms of which $Y(\cdot)$ has a spectral representation given by (2.2.10).

If $\mathcal{C}$ is the class of vector families $\mathcal{F}$ with respect to which a process $Y(\cdot)$ admits the spectral representation (2.2.10), then the *characteristic width* of the process $Y(\cdot)$ is given by

$$B_Y = \min_{1 \leq i \leq n} \left\{ \sup_{\mathcal{F} \in \mathcal{C}} B_{\mathcal{F}} \right\}.$$

Let $\alpha : \mathbb{R} \to \mathbb{R}$ be a linear filter with transfer function

$$\Gamma(\lambda) = \int_{\mathbb{R}} \alpha(u)\, e^{-iu\lambda}\, du$$

normalized so that

$$2\pi \int_{\mathbb{R}} \alpha^2(u)\, du = \int_{\mathbb{R}} |\Gamma(\lambda)|^2\, d\lambda = 1. \qquad (2.3.13)$$

The *width* of $\alpha(\cdot)$ is given by

$$B_\alpha = \int_{\mathbb{R}} |u|\, |\alpha(u)|\, du. \qquad (2.3.14)$$

Consider for a fixed frequency ω_0, the following linear transformation of the bounded-width sigma oscillatory weakly harmonizable process $Y(\cdot)$,

$$L(t) = \int_{\mathbb{R}} \alpha(u)\, Y(t-u)\, e^{-i(t-u)\omega_0}\, du$$

$$= \sum_{i=1}^{n} \int_{\mathbb{R}} \alpha(u)\, X_i(t-u)\, e^{-i(t-u)\omega_0}\, du \qquad (2.3.15)$$

where each of the $X_i(\cdot), i = 1, \ldots, n$ is a bounded-width oscillatory weakly harmonizable process. Thus, applying the following result of Swift,

Theorem 2.3.1 *Let $\alpha : \mathbb{R} \to \mathbb{R}$ be a linear filter, then the covariance of a linear transformation*

$$L(t) = \int_{\mathbb{R}} \alpha(u)\, X(t-u)\, e^{-i(t-u)\omega_0}\, du$$

of a bounded-width oscillatory weakly harmonizable process $X(\cdot)$ is

$$\rho(s,t) = \int_{\mathbb{R}} \int_{\mathbb{R}} |\Gamma(\theta)|^2 \, dG(s,t,\lambda+\omega_0,\lambda'+\omega_0) + O(\varepsilon),$$

where $O(\varepsilon) \to 0$ as B_α, the width of the linear filter $\alpha(\cdot)$, goes to zero.

to each of the $X_i(\cdot)'s$, together with the independence conditions, yields the following Proposition.

Proposition 2.3.1 *Let $\alpha : \mathbb{R} \to \mathbb{R}$ be a linear filter, then the covariance of a linear transformation*

$$L(t) = \int_{\mathbb{R}} \alpha(u) \, Y(t-u) \, e^{-i(t-u)\omega_0} \, du$$

of a bounded-width sigma oscillatory weakly harmonizable process $Y(\cdot)$ is

$$\rho(s,t) = \int_{\mathbb{R}} \int_{\mathbb{R}} |\Gamma(\theta)|^2 \, dG_Y(s,t,\lambda+\omega_0,\lambda'+\omega_0) + O(\varepsilon),$$

where $O(\varepsilon) \to 0$ as B_α, the width of the linear filter $\alpha(\cdot)$, goes to zero, and $G_Y(\cdot,\cdot,\cdot,\cdot)$ is the evolutionary spectrum of the bounded-width sigma oscillatory weakly harmonizable process $Y(\cdot)$.

Acknowledgments The author expresses his thanks to Professor M. M. Rao for his continuing advice, encouragement and guidance during the work of this project. The author also expresses his gratitude to Western Kentucky University for a sabbatical leave during the Fall 1998 semester, during which this work was completed.

REFERENCES

Chang, D. K. and Rao, M. M. (1986). Bimeasures and nonstationary processes, In *Real and Stochastic Analysis*, pp. 7–118, John Wiley & Sons, New York.

Cramér, H. (1952). A contribution to the theory of stochastic processes, *Proceedings of the Second Berkeley Symposium Mathematics, Statistics and Probability*, **2**, pp. 55–77, University of California Press, Berkeley.

Gihman, I. I. and Skorohod, A. V. (1974). *The Theory of Stochastic Processes I*, Springer-Verlag, New York.

Priestley, M. B. (1965). Evolutionary spectra and nonstationary processes, *Journal of the Royal Statistical Society, Series B*, **27**, 204–237.

Rao, M. M. (1984). Harmonizable processes: Structure theory, *L'Enseign Math*, **28**, 295–351.

Swift, R. J. (1997a). Some aspects of harmonizable processes and fields, In *Real and Stochastic Analysis: Recent Advances* (Ed., M. M. Rao), pp. 303–365, CRC Press, Boca Raton.

Swift, R. J. (1997b). An operator characterization of oscillatory harmonizable processes, In *Stochastic Processes and Functional Analysis* (Eds., J. Goldstein, N. Gretsky and J. J. Uhl), pp. 235–243, Marcel Dekker, New York.

Swift, R. J. (2000). The evolutionary spectra of a harmonizable process, *Journal of Applied Statistical Science* (to appear).

CHAPTER 3

SOME PROPERTIES OF HARMONIZABLE PROCESSES

MARC H. MEHLMAN

University of Pittsburgh, Johnstown, PA

Abstract: Harmonizable processes are considered as Fourier transforms of vector measures. The incremental processes derived from harmonizable processes are examined here. In particular, they are seen to be harmonizable too, and are used to establish results concerning the derivatives and definite integrals of harmonizable processes (both of which turn out to be harmonizable too).

The nth moment of a vector measure is introduced and it is suggested that there may be a theory that connects harmonizable processes to their corresponding vector measures much as characteristic functions are connected to their corresponding probability measures.

Finally, a result concerning the moving averages of harmonizable processes with continuous parameter is updated to include moving average representations of the derivative of these harmonizable processes too.

Keywords and phrases: Harmonizable processes, incremental processes, moving averages:

3.1 INTRODUCTION

Convention 3.1.1 Let (Ω, Σ, P) be a probability space. Let $\mathbf{D}$ represent either $\mathbf{R}$, the reals, or $\mathbf{Z}$, the integers. The unit circle, $\mathbf{T}$, will be thought of as $[-\pi, \pi)$. Note that $\hat{\mathbf{D}}$, the topological dual group of $\mathbf{D}$, is either $\mathbf{R}$ or $\mathbf{T}$, depending on whether $\mathbf{D}$ is $\mathbf{R}$ or $\mathbf{T}$ respectively.

Definition 3.1.2 Denote $\mathbf{L}_0^2(P)$ to be the Hilbert space of all complex valued functions $f \in \mathbf{L}^2(P)$ such that $\mathbf{E}(f) = 0$, where $\mathbf{E}(f) \stackrel{\text{def}}{=} \int_\Omega f(x) \, P(dx)$, the expectation of f.

Definition 3.1.3 An n–dimensional vector measure, $Z : \Sigma \to \left[\mathbf{L}_0^2(P)\right]^n$ has *orthogonal increments* iff $\mathbf{E}(Z(\Delta)Z(\Delta')^*) = \mathbf{0}_n{}^1$ for all $\Delta, \Delta' \in \Sigma$ such that $\Delta \cap \Delta' = \emptyset$.

Definition 3.1.4 An n–dimensional process, $X : \mathbf{D} \to \left[\mathbf{L}_0^2(P)\right]^n$, is *weakly harmonizable* iff X_t can be written as the following Dunford–Schwartz integral:

$$X_t = \int_{\hat{\mathbf{D}}} e^{i\lambda t} \, Z_X(d\lambda).$$

If $Z_X(d\lambda)$ has orthogonal increments, X_t is *stationary* (stationary in the wide or Khinchine sense).

It is somewhat traditional to define weakly harmonizable and stationary processes in terms of their covariance functions, rather than as Fourier transforms of a vector measure. However, the following Theorem states that such definitions are equivalent to the one above.

Theorem 3.1.5 *A process is harmonizable iff its covariance function is expressible as the Morse–Transue integral[2],*

$$r_X(s, t) = \int_{\hat{\mathbf{D}} \times \hat{\mathbf{D}}} e^{is\lambda - it\lambda'} \, F_X(d\lambda, d\lambda'). \tag{3.1.1}$$

A process is stationary iff its covariance function is expressible as

$$r_X(s, t) = \tilde{r}(t - s) = \int_{\hat{\mathbf{D}}} e^{i(s-t)\lambda} \, F_X(d\lambda).$$

Kolmogorov (1941) proved the above Theorem for the stationary case. The weakly harmonizable case is obtained by Rao (1982) and, in a more general case, the above Theorem is derived in Chang and Rao (1986). The "only if" part of the above Theorem is somewhat easy to see. The "if" part in the weakly harmonizable case is a easy consequence of the Dilation Theorem [see Chang and Rao (1986)] and the fact the above Theorem holds in the stationary case.

---

[1]Here * denotes the adjoint operator, i.e., the conjugate transpose operator.

[2]For definition of a Morse–Transue integral, see Chang and Rao (1986).

Definition 3.1.6 The *spectral bimeasure* of a weakly harmonizable process, X_t, is the positive semi–definite bimeasure, $F_X(\cdot, \cdot)$ in (3.1.1). A weakly harmonizable random process is *strongly harmonizable* iff its spectral bimeasure extends to a measure on the Borel σ–algebra of $\Sigma \times \Sigma$. All strongly harmonizable process are weakly harmonizable with the integral in (3.1.1) just an ordinary Lebesgue integral.

3.2   INCREMENTAL PROCESSES

Definition 3.2.1 Given a process, X_t, and $\tau \in \mathbf{D}$, its *increment process*, $(I_\tau X)_t$, is defined as

$$(I_\tau X)_t \overset{\text{def}}{=} X_{t+\tau} - X_t.$$

Theorem 3.2.2 *If X_t is harmonizable (stationary) and $\tau \in \mathbf{D}$ then $(I_\tau X)_t$ is harmonizable (stationary). Furthermore, if $\mathbf{D} = \mathbf{R}$ then*

1. *if $X_t' \overset{\text{def}}{=} \lim_{\tau \to 0} \frac{1}{t}(I_\tau X)_t$ exists (here convergence is in the mean square sense), then $X_t' = \int_{\mathbf{R}} e^{it\lambda} i\lambda \, Z_X(d\lambda)$ and hence is harmonizable (stationary).*

2. *$\int_0^s X_t \, dt = \int_{\mathbf{R}} \left(\frac{e^{i\lambda s} - 1}{i\lambda} \right) Z_X(d\lambda)$ is harmonizable also (here the integrand equals s when $\lambda = 0$).*

PROOF   If X_t is harmonizable, then

$$(I_\tau X)_t = \int_{\hat{\mathbf{D}}} e^{i(t+\tau)\lambda} Z_X(d\lambda) - \int_{\hat{\mathbf{D}}} e^{it\lambda} Z_X(d\lambda) = \int_{\hat{\mathbf{D}}} e^{it\lambda} (e^{i\tau\lambda} - 1) Z_X(d\lambda).$$

Letting $Z_{I_\tau X}(d\lambda) \overset{\text{def}}{=} (e^{i\tau\lambda} - 1) Z_X(d\lambda)$, one sees that $(I_\tau X)_t$ is harmonizable too. If X_t is stationary, by observing the covariance function of $(I\tau X)_t$ one sees $(I_\tau X)_t$ is stationary too.

If X_t is harmonizable (stationary) with continuous parameter and X_t' exists, then

$$
\begin{aligned}
X_t' &= \lim_{\tau \to 0} \frac{1}{\tau}(I_\tau X)_t \\
&= \lim_{\tau \to 0} \int_{\mathbf{R}} e^{it\lambda} \left(\frac{e^{i\tau\lambda} - 1}{\tau} \right) Z_X(d\lambda) \\
&= \int_{\mathbf{R}} e^{it\lambda} \lim_{\tau \to 0} \left(\frac{e^{i\tau\lambda} - 1}{\tau} \right) Z_X(d\lambda) \\
&= \int_{\mathbf{R}} e^{it\lambda} i\lambda \, Z_X(d\lambda)
\end{aligned}
$$

which is again harmonizable (stationary). One can pass the limit inside the integral sign since the integrand converges uniformly on compact sets.

Similarly, if X_t is harmonizable (stationary) with continuous parameter, then

$$
\begin{aligned}
\int_0^s X_t \, dt &= \int_0^s \left(\int_{\mathbf{R}} e^{i\lambda t} \, Z_X(d\lambda) \right) dt \\
&= \int_{\mathbf{R}} \left(\int_0^s e^{i\lambda t} \, dt \right) Z_X(d\lambda) \\
&= \int_{\mathbf{R}} \left(\frac{e^{i\lambda s} - 1}{i\lambda} \right) Z_X(d\lambda) \quad\quad (3.2.2)
\end{aligned}
$$

where $\frac{e^{i\lambda s}-1}{i\lambda}$ equals s when $\lambda = 0$.[3] Again, switching the order of integration is justified by noticing the integrand converges uniformly on compact sets.   □

Swift (1996, Theorem 2.1) states that (3.2.2) is the spectral representation of a second order mean square differentiable process with strongly harmonizable increments. This result may be obtained from the above Theorem after noticing that "harmonizable" is not destroyed by integration or differentiation and that the increments of a harmonizable process are harmonizable. In particular, since $X_s = \int_0^s X_t' \, dt + X_0$ one can rewrite (3.2.2) to obtain

$$
X_s = \int_{\mathbf{R}} \left(\frac{e^{i\lambda s} - 1}{i\lambda} \right) Z_{X'}(d\lambda) + X_0.
$$

3.3   MOMENTS OF HARMONIZABLE PROCESSES

Definition 3.3.1 The nth *moment* of a vector measure, Z, (if it exists) is

$$
M_n(Z) \stackrel{\text{def}}{=} \int_{\hat{D}} \lambda^n \, Z(d\lambda).
$$

The nth *moment* of a harmonizable process, $X_t = \int_{\hat{D}} e^{it\lambda} Z_X(d\lambda)$, (if it exists) is $M_n(Z_X)$.

---

[3]One could use $\tilde{X}_t \stackrel{\text{def}}{=} X_t - Z_X(\{0\})$ instead of X_t and notice that $Z_{\tilde{X}}(\{0\}) = 0$. Then

$$
\int_0^s X_t \, dt = \int_{\mathbf{R}-\{0\}} \left(\frac{e^{i\lambda s} - 1}{i\lambda} \right) Z_X(d\lambda).
$$

In the discrete parameter case, moments of all orders exist. Furthermore,

$$
\begin{aligned}
X_t &= \int_{\mathbf{T}} e^{i\lambda t} Z_X(d\lambda) \\
&= \int_{\mathbf{T}} \left(\sum_{j=0}^{\infty} \frac{(i\lambda t)^j}{j!} \right) Z_X(d\lambda) \\
&= \sum_{j=0}^{\infty} \frac{(it)^j}{j!} \int_{\mathbf{T}} \lambda^j Z_X(d\lambda) \\
&= \sum_{j=0}^{\infty} \left(\frac{i^j}{j!} M_j(Z_X) \right) t^j
\end{aligned}
$$

For the continuous parameter case, one needs a condition to obtain the analogous result. The following Theorem is not hard to prove (derive the $n = 0$ case first).

Theorem 3.3.2 *Let X_t be a continuous parameter harmonizable process with moments of all orders and assume*

$$
\lim_{n \uparrow \infty} \int_{\mathbf{R}} \left(\sum_{j=n}^{\infty} \frac{(i\lambda t)^j}{j!} \right) Z(d\lambda) = 0.
$$

Then

$$
X_t^{(n)} = \sum_{j=n}^{\infty} \left(\frac{i^j}{(j-n)!} M_j(Z_X) \right) t^{j-n}
$$

is a continuous parameter harmonizable process with moments of all orders too.

It maybe possible to employ techniques more similar to Paul Lévy's work with characteristic functions than ordinary Fourier Analysis, to answer questions of inversion and convergence for the continuous parameter case. Harmonizable processes would play the role of characteristic functions and spectral measures would play the role of distributions. For instance, the following Theorem and its proof mimics Rao (1984, Proposition 4.2.6).

Theorem 3.3.3 *If the harmonizable process, X_t, with continuous parameter, has $p \in \mathbf{Z}^+$ derivatives at $t = 0$, then X_t has $2[p/2]$ moments, where $[x]$ is the largest integer not exceeding x. On the other hand, if X_t has $p \in \mathbf{Z}^+$ moments, then X_t is p times continuously differentiable.*

3.4 VIRILE REPRESENTATIONS

Definition 3.4.1 An n-dimensional harmonizable process, X_t, has *factorizable spectral measure (f.s.m.)* iff its covariance function can be represented as

$$r_X(s,t) = \iint_{\hat{\mathbf{D}}\times\hat{\mathbf{D}}} e^{is\lambda - it\lambda'} \underbrace{c(\lambda)c^*(\lambda')\,\mu(d\lambda, d\lambda')}_{F_X(d\lambda, d\lambda')}, \qquad (3.4.3)$$

where $c(\cdot)$ is an $n \times m$ matrix valued function with components in $\mathbf{L}^2(d\lambda)$ and $\mu(d\lambda, d\lambda')$ is a one dimensional bimeasure. A process with f.s.m. is an *f.s.m. process*. The f.s.m. covariance representation (3.4.3) has *full rank m* iff $c(\lambda)c^*(\lambda')$ has matrix rank m a.e. with respect to $\mu(\cdot, \cdot)$. If $n = m$, the f.s.m. covariance representation (3.4.3) of full rank n has *maximal rank*. An n dimensional f.s.m. process, X_t has a *virile covariance representation*, iff

1. $c(\cdot)$ is equal everywhere to the inverse Fourier transform of its Fourier transform.

2. For $N \in \mathbf{Z}^+$, letting

$$c_N(\lambda) \stackrel{\text{def}}{=} \begin{cases} \displaystyle\sum_{|j|>N} \hat{c}(j)e^{ij\lambda} & \text{discrete case} \\[2mm] \displaystyle\int_{s\geq N} \hat{c}(s)e^{is\lambda}\,ds & \text{cont. case} \end{cases}$$

then

$$\lim_{N\uparrow\infty} \iint_{\hat{\mathbf{D}}\times\hat{\mathbf{D}}} c_N(\lambda)c_N^*(\lambda')\mu(d\lambda, d\lambda') = \mathbf{0}_n.$$

Every stationary process is a f.s.m. process since its spectral measure is positive definite. There exists examples of strongly harmonizable processes that are not f.s.m. processes [Mehlman (1991, Example 5.2)].

If A is the set where $c(\cdot)$ and the inverse Fourier transform of its Fourier transform differ and if $|\mu|(A, A) = 0$, then one can use the inverse Fourier transform of the Fourier transform instead of the original $c(\cdot)$. In particular, if a strongly harmonizable f.s.m. process has its spectral measure equal to Lebesgue measure (on $\hat{\mathbf{D}} \times \hat{\mathbf{D}}$ or on the diagonal of $\hat{\mathbf{D}} \times \hat{\mathbf{D}}$) then it has a virile covariance representation.

Rank is not defined for all f.s.m. processes. However, rank (if it exists) is independent of f.s.m. representation. Even if the $c(\lambda)c^*(\lambda')$ has constant rank p, there need not be an f.s.m. covariance representation with full rank p.

Definition 3.4.2 A *moving average representation* of a n-dimensional random dom process, X_t, is a representation

$$X_t = \begin{cases} \sum_{j \in \mathbf{Z}} \hat{c}(j - t)\xi_j & \text{discrete case} \\ \int_{\mathbf{R}} \hat{c}(\lambda - t)\xi_\lambda \, d\lambda & \text{cont. case} \end{cases} \tag{3.4.4}$$

where

1. $\hat{c}(\cdot)$ is the Fourier transform of an $\mathbf{L}^2(d\lambda)$ function $c : \hat{\mathbf{D}} \to \mathcal{M}_{n,m}$ where $\mathcal{M}_{n,m}$ are all $n \times m$ matrices and

2. $r_\xi(s,t) = \rho(s,t)\mathbf{I}_m$ where $\rho(\cdot,\cdot)$ is the covariance function of a one dimensional process.

A moving average representation (3.4.4) has *full rank m* iff $c(\lambda)$ has rank m for all $\lambda \in \mathbf{T}$. A harmonizable moving average is a *virile moving average* iff ξ_t is harmonizable with spectral bimeasure $\mu_\xi(d\lambda, d\lambda')\mathbf{I}_m$ and for $N \in \mathbf{Z}^+$, letting

$$c_N(\lambda) \stackrel{\text{def}}{=} \begin{cases} \sum_{|j| > N} \hat{c}(j)e^{ij\lambda} & \text{discrete case} \\ \int_{s \geq N} \hat{c}(s)e^{is\lambda} \, ds & \text{cont. case} \end{cases}$$

then

$$\lim_{N \uparrow \infty} \int\int_{\hat{\mathbf{D}} \times \hat{\mathbf{D}}} c_N(\lambda)c_N^*(\lambda') \, \mu_\xi(d\lambda, d\lambda') = \mathbf{0}_n.$$

If A is the set where $c(\cdot)$ and the inverse Fourier transform of its Fourier transform differ and if $|\mu|(A, A) = 0$, then one can use the inverse Fourier transform of the Fourier transform instead of the original $c(\cdot)$. In particular, if a strongly harmonizable f.s.m. process has its spectral measure equal to Lebesgue measure (on $\hat{\mathbf{D}} \times \hat{\mathbf{D}}$ or on the diagonal of $\hat{\mathbf{D}} \times \hat{\mathbf{D}}$) then it has a virile covariance representation.

Theorem 3.4.3 Let X_t be an n-dimensional, continuous parametered, strongly harmonizable process with a strongly harmonizable virile moving average representation with full rank m,

$$X_t = \int\int_{\mathbf{R} \times \mathbf{R}} e^{is\lambda - it\lambda'} c(\lambda)c^*(\lambda') \, \mu(d\lambda, d\lambda')$$

Assume

$$\int\int_{\mathbf{R} \times \mathbf{R}} (\lambda\lambda')^j e^{i\lambda s - i\lambda' t} F_X(\lambda, \lambda')$$

exists for all $s, t \in \mathbf{R}$ *and some* $j \in \{0, 1, 2, \cdots\}$. *Then for all* $k \leq j$,

$$X_t^{(k)} = \int_{\hat{\mathbf{R}}} (i\lambda)^k e^{i\lambda t} Z_X(d\lambda) = \int_{\hat{\mathbf{R}}} (i\lambda)^k e^{i\lambda t} c(\lambda) Z_\xi(d\lambda)$$

exists and is a strongly harmonizable process with a strongly harmonizable virile moving average representation with full rank m,

$$X_t^{(k)} = \int_{\mathbf{R}} \hat{c}(\lambda - t) \xi_\lambda^k \, d\lambda$$

where $r_{\xi^k}(s, t) = \iint_{\mathbf{R} \times \mathbf{R}} e^{is\lambda - it\lambda'} (\lambda\lambda')^k \mu(d\lambda, d\lambda') \mathbf{I}_m$.

The above Theorem is proved for the case $k = 0$ in Mehlman (1991, Theorem 6.5). The above Theorem now follows from part one of Theorem 3.2.2 and the repetitive use of the $k = 0$ case.

Acknowledgements Some of the ideas for this talk originated during talks with Dr. Randall Swift. I would also like to thank Dr. Alan Krinik for encouraging me to attend the October IISA conference.

REFERENCES

Chang, D. K. and Rao, M.M. (1986). Bimeasures and Nonstationary Processes, In *Real and Stochastic Analysis* (Ed., M. M.Rao), pp. 7–118, John Wiley & Sons, New York.

Kolmogorov, A. (1941). Stationary sequences in Hilbert spaces, *Bull. Math. Univ. Moscow*, **2** (in Russian).

Mehlman, M. H. (1991). Structure and moving average representation for multidimensional strongly harmonizable processes, *Stochastic Analysis and Applications*, **9**, 323–361.

Rao, M. M. (1982). Harmonizable processes: Structure theory, *L'Enseignement Math.*, **28**, 295–351.

Rao, M. M. (1984). *Probability Theory with Applications*, Academic Press, Orlando.

Swift, R. (1996). Stochastic processes with harmonizable increments, *Journal of Combinatorics, Information & System Sciences*, **21**, 47–60.

CHAPTER 4

INFERENCE FOR BRANCHING PROCESSES

I. V. BASAWA

University of Georgia, Athens, Georgia

Abstract: Problems of inference for Galton-Watson (G-W) branching processes are discussed. It is shown that the G-W branching process belongs to a local asymptotic mixed normal family. Consequently, the limit distribution of the maximum likelihood estimator of the offspring mean parameter is non-normal. The usual test statistics such as the score, Wald and likelihood ratio statistics do not have the same limit distribution and are not asymptotically uniformly most powerful. Some of these difficulties regarding efficiency can be overcome via a conditional approach. Quasilikelihood, Bayes and empirical Bayes estimators are discussed briefly.

Keywords and phrases: Inference for stochastic processes, branching processes, asymptotic inference, maximum likelihood estimation, quasilikelihood estimation, Bayes estimator, empirical Bayes estimator, asymptotic tests, conditional inference

4.1 INTRODUCTION

Inference for branching processes is an important area of research with applications in cell kinetics, genetics, population growth and biostatistics among others. See Guttorp (1991) for an excellent review and references. In this paper, we give an over-view of asymptotic inference problems for a Galton-Watson branching process. Local asymptotic mixed normal

57

(LAMN) family is used as a unified frame-work to discuss diverse statistical issues such as estimation efficiency, test efficiency, confidence bounds, conditional inference, etc. Quasilikelihood, Bayes and empirical Bayes estimators are also discussed briefly. Basawa and Scott (1983) have reviewed the basic theory for the LAMN family in general.

Preliminary background for G-W branching processes is given in Section 4.2. The LAMN formulation and some statistical consequences are presented in Section 4.3. Section 4.4 shows that the G-W branching process is an example of a LAMN model. Asymptotic efficiency of estimators, tests and confidence bounds are discussed in Sections 4.5–4.7. Section 4.8 is concerned with a conditional approach to resolve difficulties in efficiency comparisons of tests and confidence bounds. Prediction and a test of fit based on prediction errors are presented in Section 4.9. Section 4.10 gives an outline of quasilikelihood approach to estimation. Bayes and empirical Bayes estimators are discussed briefly in Section 4.11 and Section 4.12 contains some concluding remarks.

4.2   GALTON-WATSON BRANCHING PROCESS: BACKGROUND

Let $\{\xi_{ni}\}$, $i = 1, 2, \ldots$, for each $n = 1, 2, \ldots$, denote a sequence of independent and identically distributed random variables taking values in $S = (0, 1, 2, \ldots)$ with probabilities (offspring distribution) $p_k = P(\xi_{ni} = k)$, $k \in S$, $0 \le p_k < 1$, $\sum_k p_k = 1$ and $p_0 + p_1 < 1$. Define $Z_{n+1} = \sum_{i=1}^{Z_n} \xi_{ni}$. The sequence $\{Z_n\}$, $n = 0, 1, 2, \ldots, Z_0 = 1$, is then a Markov chain defined on the state space S, with transition probabilities

$$
p_{jk} = P(Z_{n+1} = k | Z_n = j) = P(\sum_{i=1}^{j} \xi_{ni} = k)
$$

$$
= \begin{cases} p_k^{*(j)}, & \text{if } j > 0 \\ \delta_{jk}, & \text{if } j = 0, \end{cases}
$$

where $\{p_k^{*(j)}\}$ denotes the j-fold convolution of $\{p_k\}$ and δ_{jk} is the Kronecker delta (i.e. δ_{jk} is equal to 1 for $j = k$ and zero otherwise).

Let ξ_{nj} denote the number of offspring produced by the jth member of the nth generation and Z_n, the generation size of the nth generation. We shall refer to $\{Z_n\}$ as the Galton-Watson (G-W) branching process. The G-W branching process has a long history and it has a wide spectrum of applications. The main goal of this paper is to review statistical inference issues concerning the G-W branching process.

Denote $m = \sum_{k=0}^{\infty} k p_k$, and $\sigma^2 = \sum_{k=0}^{\infty} (k - \mu)^2 p_k$, the mean and the variance of the offspring distribution. It is easily verified that

$$E(Z_{n+1}|Z_n) = m Z_n \quad \text{and} \quad Var(Z_{n+1}|Z_n) = \sigma^2 Z_n.$$

The asymptotic behavior of $\{Z_n\}$ depends crucially on whether $m < 1$ (subcritical), $m = 1$ (critical) and $m > 1$ (supercritical). In particular, for $m > 1$, and $\sigma^2 < \infty$, there exists a random variable $W \geq 0$ such that

$$Z_n m^{-n} \xrightarrow{\text{a.s.}} W, \quad \text{as } n \to \infty.$$

We shall focus on inference problems concerning the offspring mean m.

4.3 LOCALLY ASYMPTOTIC MIXED NORMAL (LAMN) FAMILY

Let $Y(n) = (Y_1, Y_2, \ldots, Y_n)$ denote a vector of observations with joint density $p_n(y(n); \theta)$, $\theta \in \Omega \subset R^p$, and denote the likelihood function $L_n(\theta) \propto p_n(Y(n); \theta)$. The log-likelihood ratio is defined as

$$\Lambda_n(\theta_n, \theta) = \log[L_n(\theta_n)/L_n(\theta)],$$

where $\theta_n = \theta + h \delta_n^{-1}(\theta)$, h is a $(p \times 1)$ vector of real numbers, and $\delta_n(\theta)$ is a $(p \times p)$ symmetric, non-singular and non-random matrix such that $\|\delta_n(\theta)\| \to \infty$ as $n \to \infty$. The likelihood score vector and the sample information matrix are defined respectively by

$$S_n(\theta) = \frac{d \log L_n(\theta)}{d\theta}, \quad F_n(\theta) = -\frac{d^2 \log L_n(\theta)}{d\theta d\theta^T}.$$

The LAMN model is specified by the following conditions:

LAMN Conditions
  As $n \to \infty$,

(a) $\Lambda_n(\theta_n, \theta) = h^T \Delta_n(\theta) - \frac{1}{2} h^T W_n(\theta) h + o_p(1)$, where $\Delta_n(\theta) = \delta_n^{-1}(\theta) S_n(\theta)$ and $W_n(\theta) = \delta_n^{-1}(\theta) F_n(\theta) \delta_n^{-1}(\theta)$.

(b) $W_n(\theta) \xrightarrow{p} W(\theta)$, where $W(\theta)$ is an almost sure non-negative definite random matrix.

(c) $\Delta_n(\theta) \xrightarrow{d} N_p^*(0, W(\theta))$, where N_p^* denotes a variance mixture of normals.

Remarks: Note that if $W(\theta)$ is degenerate (i.e. non-random) the above conditions reduce to the usual local asymptotic normality (LAN) conditions and N_p^* reduces to a normal distribution. See Basawa and Scott (1983) for details on LAMN models for non-degenerate $W(\theta)$.

Some Consequences of LAMN Assumptions

From now on, we shall assume that $W(\theta)$ is a non-degenerate random variable. Under LAMN conditions, the maximum likelihood estimator $\hat{\theta}$ of θ has a non-normal limit distribution. The usual test statistics such as score, Wald and likelihood ratio statistics have non-standard limit distributions. Moreover, asymptotically uniformly most powerful tests for θ do not exist. One can, however, use conditional inference (conditional on $W(\theta)$ or on its estimate) to resolve most of these inferential problems. We refer to Basawa and Scott (1983) for a detailed discussion of these issues.

In this paper, we illustrate the above mentioned statistical questions and other related matters via the example of a G-W branching process.

4.4   G-W BRANCHING PROCESS AS A PROTO-TYPE EXAMPLE OF A LAMN MODEL

Suppose $(Z_0 = 1, Z_1, \ldots, Z_n)$ denote the generation sizes of a G-W branching process. We assume that the offspring distribution belongs to a power series family. More specifically,

$$P(\xi_{nj} = k) = P(Z_1 = k) = a_k \theta^k / A(\theta),$$

where $A(\theta) = \sum_{k=0}^{\infty} a_k \theta^k$. We then have $m(\theta) = E(Z_1) = \theta \frac{A'(\theta)}{A(\theta)}$, and $\sigma^2(\theta) = Var(Z_1) = \theta m'(\theta)$. The log-likelihood function is

$$\log L_n(\theta) = \sum_{i=1}^{n} \log a_{Z_i} + (\sum_{i=1}^{n} Z_i) \log \theta - (\sum_{i=1}^{n} Z_{i-1}) \log A(\theta),$$

and the likelihood score function is given by

$$
\begin{aligned}
S_n(m) &= \frac{d \log L_n}{dm} = \left(\frac{\partial \log L_n}{\partial \theta}\right) \left(\frac{d\theta}{dm}\right) \\
&= \sigma_{(\theta)}^{-2} (\sum_{1}^{n} Z_i - m \sum_{1}^{n} Z_{i-1}).
\end{aligned}
$$

The maximum likelihood (ML) estimator $\hat{m}$ of m is obtained by solving $S_n(m) = 0$ for m and

$$\hat{m} = (\sum_1^n Z_i)/(\sum_1^n Z_{i-1}).$$

The sample and the expected Fisher informations are given respectively by

$$F_n(m) = -S_n'(m) = \sigma_{(\theta)}^{-2} \sum_1^n Z_{i-1} - (\frac{d}{dm}\sigma_{(\theta)}^{-2})(\sum_1^n Z_i - m \sum_1^n Z_{i-1}),$$

$$I_n(m) = E(F_n(m)) = \sigma_{(\theta)}^{-2}(\frac{m^n - 1}{m - 1}).$$

Define $m_n = m + hI_n^{-1/2}(m)$. If we suppose that $P(Z_1) = 0$, and $m > 1$, it is easily verified that the G-W branching process belongs to the LAMN family with

$$\Lambda_n(m_n, m) = hI_n^{-1/2}(m)S_n(m) - \frac{1}{2}h^2 I_n^{-1}(m)F_n(m) + o_p(1),$$

where $I_n^{-1}(m)F_n(m) \xrightarrow{p} W > 0$, and W is a non-degenerate random variable. If, in particular, the offspring distribution is geometric, W has an exponential distribution with mean unity. In general,

$$m^{-n}Z_n \xrightarrow{a.s.} W.$$

A consistent estimator of σ^2 is given by

$$\hat{\sigma}^2 = \frac{1}{n}\sum_{i=1}^n (Z_i - \hat{m}Z_{i-1})^2 Z_{i-1}^{-1}.$$

4.5   ESTIMATION EFFICIENCY

Let $J_n(m) = \sigma_{(\theta)}^{-2}\sum_1^n Z_{i-1}$. Then, $I_n(m) = E(J_n(m))$. We then have $J_n(m)/I_n(m) \xrightarrow{p} W$. It can then be shown that
 (a) $J_n^{1/2}(m)(\hat{m} - m) \xrightarrow{d} N(0, 1)$, and
 (b) $I_n^{1/2}(m)(\hat{m} - m) \xrightarrow{d} N^*(0, W^{-1})$.
   See Heyde (1975, 1977, 1978) and Basawa and Scott (1976). Note that the limit distribution in (b) is non-normal. In particular, if the offspring distribution is geometric, the limit distribution $N^*(0, W^{-1})$ reduces to the

Student's t distribution with 2 degrees of freedom whose variance is infinite. In other words, for the geometric offspring distribution the asymptotic variance of the ML estimator of m is infinite! The question then is how does one establish the asymptotic efficiency of the ML estimator.

This question was resolved by Heyde (1977) using the random norm as follows:

If T_N is any estimator of m such that

$$J_n^{1/2}(m)(T_n - m) \xrightarrow{d} N(0, \nu(m)),$$

then $\nu(m) \geq 1$, for all m (> 1), and the equality $\nu(m) = 1$ is attained for the ML estimator $\hat{m}$.

The above result establishes the asymptotic efficiency of the ML estimator in some sense. However, a more satisfactory resolution of the efficiency question is in terms of the limit distribution using a non-random norm. One such result is as follows:

If T_n is any "regular" estimator of m such that $I_n^{1/2}(m)(T_n - m)$ has a limit distribution, then, under regularity conditions,

$$\lim_{n \to \infty} P(I_n^{1/2}(m)|T_n - m| \leq c) \leq P(|Z^*| \leq c), \quad \text{for any } c > 0,$$

where Z^* is a random variable distributed as $N^*(0, W^{-1})$. The equality in the above inequality is attained for the ML estimator.

See Heyde (1978) and Basawa and Scott (1983) for details. A more general result is given in Basawa and Scott (1983). Note that the above result based on the limiting probability of concentration, viz.,

$$\lim P(I_n^{1/2}(m)|T_n - m| \leq c),$$

does not require the finiteness of the asymptotic variance.

4.6　TEST EFFICIENCY

Consider the problem of testing

$$H : m = m_0, (m_0 > 1), \quad \text{against} \quad K : m > m_0.$$

Recall that the likelihood score function is given by

$$S_n(m) = J_n(m)(\hat{m} - m).$$

Consider the following three test statistics:

$$T_1 = I_n^{-1/2}(m_0)S_n(m_0),$$

$$T_2 = I_n^{1/2}(m_0)(\hat{m} - m_0),$$

and

$$T_3 = J_n^{1/2}(m_0)(\hat{m} - m_0).$$

It can be verified (see Basawa and Scott (1976)) that, under H, we have

$$T_1 \xrightarrow{d} ZW^{1/2}, T_2 \xrightarrow{d} ZW^{-1/2}, \quad \text{and} \quad T_3 \xrightarrow{d} Z,$$

where Z is a standard normal random variable independent of W. In particular, if the offspring distribution is geometric, then $ZW^{1/2}$ and $ZW^{-1/2}$ are distributed, respectively, as a double exponential and a $t(2)$ distribution.

Define the test functions:

$$\phi_n^{(i)} = \begin{cases} 1, & T_i \geq k^{(i)}(\alpha) \\ 0, & \text{otherwise,} \end{cases}$$

$i = 1, 2, 3$, where the constants $k^{(i)}(\alpha)$ are chosen such that $\lim \beta_{\phi_n^{(i)}}(m_0) = \alpha$, where $\beta_{\phi_n^{(i)}}(m) = E_m \phi_n^{(i)}$ denotes the power function of the test $\phi_n^{(i)}$.

A **locally most powerful** (LMP) test maximizes the local power $\beta'_{\phi_n}(m_0)$. From standard theory, T_1 is a LMP test. Basawa and Scott (1976) have compared the tests $\phi_n^{(i)}$, $i = 1, 2, 3$ on the basis of the local power and concluded that

$$\beta'_{\phi_n^{(1)}}(m_0) > \beta'_{\phi_n^{(3)}}(m_0) > \beta'_{\phi_n^{(2)}}(m_0).$$

However, it is more desirable to use limiting power at contiguous alternatives rather than the local power for comparisons. Sweeting (1978) compared the three tests using such a criterion. An asymptotically most powerful test maximizes the limiting power $\lim \beta_{\phi_n}(m_n)$, where $m_n = m_0 + h I_n^{-1/2}(m_0)$ for all $h > 0$. Sweeting's (1978) comparisons based on $\lim \beta_{\phi_n}(m_n)$ are inconclusive. This is not surprising because, for any given h, the Neyman-Pearson most powerful test statistic $\Lambda_n(m_n, m_0)$ for testing $m = m_0$ against $m = m_n$, depends on h, even asymptotically. This is seen by the following:

$$\begin{aligned} \Lambda_n(m_n, m_0) &= h I_n^{-1/2}(m_0) S_n(m_0) - \frac{1}{2} h^2 I_n^{-1}(m_0) F_n(m_0) + o_p(1) \\ &= h T_1 - \frac{1}{2} h^2 W + o_p(1) \end{aligned}$$

which depends on h even asymptotically because the second term in the expansion is a non-degenerate random variable. See Basawa and Scott (1983) for details. It, therefore, follows that for the G-W branching process, there does not exist an asymptotically uniformly most powerful test based on the criterion $\lim \beta_{\phi_n}(m_n)$. In particular, none of the three tests $\phi_n^{(i)}$, $i = 1, 2, 3$, dominates the others for all h in the sense of maximizing $\lim \beta_{\phi_n}(m_n)$.

4.7 CONFIDENCE BOUNDS

We now consider the problem of constructing confidence bounds for the mean m of the offspring distribution. As before, it will be assumed that $m > 1$ and $P(Z_1 = 0) = 0$ to ensure that the process grows exponentially and does not get extinct.

Let V_n denote an asymptotic lower confidence bound for m satisfying

$$\lim_{n \to \infty} P_m(V_n \leq m) = 1 - \alpha, \quad \text{for all} \ \ m > 1.$$

Using the statistic T_1 of Section 4.6, viz.,

$$T_1 = I_n^{-1/2}(m_0)S_n(m_0),$$

we have

$$\lim P_{m_0}(T_1 \geq c_1(\alpha)) = \alpha,$$

where $c_1(\alpha)$ is given by

$$P(ZW^{1/2} \geq c_1(\alpha)) = \alpha.$$

We then have

$$\lim P_{m_0}(I_n^{-1/2}(m_0) J_n(m_0)(\hat{m} - m_0) \geq c_1(\alpha)) = \alpha,$$

and hence,

$$\lim P_m(m \geq \hat{m} - c_1(\alpha)I_n^{1/2}(\hat{)}J_n^{-1}(\hat{m})) = 1 - \alpha, \quad \text{for all} \ \ m > 1.$$

Consequently

$$V_n^{(1)} = \hat{m} - c_n(\alpha)I_n^{1/2}(\hat{m})J_n^{-1}(\hat{m})$$

is an asymptotic $(1 - \alpha)$ lower confidence bound for m. Similarly, using the statistics T_2 and T_3 of Section 4.6, we can construct alternative lower confidence bounds as:

$$V_n^{(2)} = \hat{m} - c_2(\alpha)I_n^{-1/2}(\hat{m}), \quad \text{where } c_2(\alpha) \text{ is given by}$$

$$P(ZW^{-1/2} \geq c_2(\alpha)) = \alpha,$$

and

$$V_n^{(3)} = \hat{m} - c_3(\alpha) J_n^{-1/2}(\hat{m}), \quad \text{where } c_3(\alpha) \text{ is given by}$$

$$P(Z \geq c_3(\alpha)) = \alpha.$$

Similarly, upper confidence bounds and confidence intervals can be constructed.

It may be noted that the sample and expected Fisher information, $J_n(m)$ and $I_n(m)$ depend on σ^2 which can be replaced by its estimate $\hat{\sigma}^2$ given in Section 4.4. This does not affect the asymptotic results discussed above.

Efficiency of Confidence Bounds

V_n is said to be asymptotically uniformly most accurate lower confidence bound for m if it minimizes

$$\lim P_m(m_n \geq V_n), \quad \text{for all } m > 1,$$

where $m_n = m - hI_n^{-1/2}$, $h > 0$.

The existence of uniformly must accurate confidence bounds is related to that of asymptotically uniformly most powerful tests. As seen in Section 4.6, there does not exist asymptotically uniformly most powerful test for m, and consequently we do not have asymptotically most accurate confidence bounds. In particular, none of the bounds $V_n^{(i)}$, $i = 1, 2, 3$, dominates the others in the sense of the optimality criterion given above.

The problem of non-existence of asymptotically most powerful tests and confidence bounds can be resolved if we use a conditional approach discussed in Section 4.8 below.

4.8 CONDITIONAL INFERENCE

The problems related to the efficiency of tests and confidence bounds for m are mainly due to the fact that the almost sure limit W of $m^{-n}Z_n$ is a nondegenerate random variable. Conditional on $W = w$, the G-W branching process can be viewed as a member of the (conditional) local asymptotic normal (LAN) family. It is well known that asymptotic uniformly most powerful tests and confidence bounds exist for the LAN family. This motivates the conditional approach to resolve the efficiency questions. We will illustrate below this approach for the special case of a G-W branching process with geometric offspring distribution.

Suppose

$$P(Z = z) = m^{-1}(1 - m^{-1})^{z-1}, \quad z = 1, 2, \ldots,$$

$(m > 1)$. We then have $m^{-n}Z_n \xrightarrow{\text{a.s.}} W$, where W is an exponential random variable with $E(W) = 1$. The joint density of $Z(n) = (Z_1, \ldots, Z_n)$, conditional on $W = w$, assuming $Z_0 = 1$, is given by

$$p_n^c(z(n); \theta, \eta) \propto \exp[\theta U_n + \eta V_n - k_n(\theta, \eta)],$$

where $\theta = \log n$, $\eta = \log[m^{-1}(m-1)w]$, $U_n = \sum_{j=1}^n j(z_j - z_{j-1})$, and $V_n = \sum_{j=1}^n (z_j - z_{j-1})$. See Keiding (1974) for the details. Note that (m, w) are reparammeterized as (θ, η). Testing $m = m_0$ against $m > m_0$ is equivalent to testing $H : \theta = \theta_0$ against $K : \theta > \theta_0$, treating η as an unknown nuisance parameter. It can then be verified that the uniformly most powerful unbiased level$-\alpha$ test for testing H against K is given by

$$\phi(U_n, V_n) = \begin{cases} 1, & \text{if } U_n > k(V_n) \\ a(V_n), & \text{if } U_n = k(V_n) \\ 0, & \text{if } U_n < k(V_n), \end{cases}$$

where $k(V_n)$ and $a(V_n)$ are determined such that $0 \leq a(V_n) \leq 1$, and $E_{m_0}(\phi(U_n, V_n)|V_n = v_n) = \alpha$ for almost all v_n. See Basawa (1981a) for details.

It can be shown that the three test statistics $T_n^{(i)}$, $i = 1, 2, 3$, discussed in Section 4.6 are asymptotically equivalent and asymptotically uniformly most powerful conditionally on $W = w$. See Basawa and Scott (1983) for details. The asymptotic equivalence and efficiency of the three confidence bounds discussed in Section 4.7 follows from the results discussed by Basawa (1981b).

An alternative conditional approach proposed by Feigin and Reiser (1979) and Sweeting (1978, 1986, 1992) is based on conditioning on W_n where

$$W_n = I_n^{-1}(\hat{m}) J_n(\hat{m}).$$

Asymptotically, this approach also leads to the same conclusions as that based on conditioning on the limiting random variable W.

4.9　PREDICTION AND TEST OF FIT

Denote $Z(n) = (Z_1, \ldots, Z_n)$, and $Y(p) = (Z_{n+1}, \ldots, Z_{n+p})$, with $Z_0 = 1$. Consider the problem of predicting $Y(p)$ given the sample $Z(n)$. The minimum mean square error predictor of $Y(p)$ is given by

$$\begin{aligned} \hat{Y}_m(p) &= E(Y(p)|Z(n)) \\ &= (mZ_n, m^2 Z_n, \ldots, m^p Z_n) \end{aligned}$$

which depends on the unknown parameter m. Replacing m by its estimator $\hat{m}_n = (\sum_1^n Z_{t-1})^{-1}(\sum_1^n Z_t)$, we get an estimated predictor $\hat{Y}_{\hat{m}_n}(p)$ of $Y(p)$. Consider

$$
\begin{aligned}
D_n &= \eta_n^{-1/2}(\hat{Y}_{\hat{m}_n}(p) - Y(p)) \\
&= \eta_n^{-1/2}(\hat{Y}_m(p) - Y(p)) + \eta_n^{-1/2}(\hat{Y}_{\hat{m}_n}(p) - \hat{Y}_m(p)) \\
&= U_n + V_n, \quad \text{say,}
\end{aligned}
$$

where $\eta_n = \sum_1^n Z_{t-1}$. Basawa (1987) has shown that U_n and V_n are asymptotically independent normal with mean zero and covariance matrices Σ_1 and Σ_2. See Basawa (1987) for the expressions for Σ_1 and Σ_2. It then follows that the limit distribution of the (estimated) prediction error is given by

$$
D_n = \eta_n^{-1/2}(\hat{Y}_{\hat{m}_n}(p) - Y(p)) \xrightarrow{d} N_p(0, \Sigma_1 + \Sigma_2).
$$

Note that Σ_1 is a contribution to the prediction mean square error from prediction and Σ_2 from estimation.

A test statistic for goodness of fit can be constructed as

$$
Q_n = D_n^T(\Sigma_1 + \Sigma_2)^{-1}D_n,
$$

which has an asymptotic $\chi^2(p)$ distribution under the null hypothesis that $\{Z_t\}$ is a G-W branching process with $m > 1$.

4.10 QUASILIKELIHOOD ESTIMATION

Recall that $E(Z_t|Z_{t-1},\ldots,Z_1) = mZ_{t-1}$, and $Var(Z_t|Z_{t-1},\ldots,Z_1) = \sigma^2 Z_{t-1}$. If we choose

$$
g_t = (Z_t - mZ_{t-1})
$$

as the elementary estimating function, we may consider the class of estimating functions

$$
G_n = \sum_{t=1}^n a_t g_t,
$$

where the weights a_t are to be determined appropriately. The optimum choice of $\{a_t\}$ according to the Godambe criterion (see, for instance, Godambe (1985) and Heyde (1997)) is given by

$$
a_t = g_t'(Var(g_t|Z_{t-1},\ldots,Z_1))^{-1}
$$

$$
= -\sigma^{-2}.
$$

The optimum estimating function (also known as a quasi-score function) is therefore given by

$$G_n^0 = -\sigma^{-2} \sum_{1}^{n} (Z_t - mZ_{t-1}).$$

Note that if the offspring distribution is assumed to belong to a power series family (see Section 4.4), the quasi-score function G_n^0 coincides with the likelihood score function as seen in Section 4.4. The quasilikelihood estimator of m obtained as the solution of the equation $G_n^0 = 0$, is the same as the maximum likelihood estimator $\hat{m}_n$ obtained in Section 4.4. Note, however, that we do not assume a power series offspring distribution in the derivation of the quasilikelihood estimator. Basawa and Prakasa Rao (1980) give a "least-squares" derivation of $\hat{m}$.

We now consider k independent G-W branching processes $\{Y_t(s)\}$, $s = 1, \ldots, k$, where $Y_t(s)$ denotes the generation size of the sth branching process. Let $m(s)$ and $\sigma^2(s)$ denote the offspring mean and variance respectively of the sth process. For a given link function $u(\cdot)$ suppose $u(m(s)) = x^T(s)\beta$, where β is a $(p \times 1)$ vector of unknown parameters and $x(s)$, a $(p \times 1)$ vector of fixed covariates. Denote

$$\begin{aligned}
Y_t &= (Y_t(1), \ldots, Y_t(k))^T, \quad \mu_t = E(Y_t | Y_{t-1}, \ldots, Y_1), \quad \text{and} \\
V_t &= Cov(Y_t | Y_{t-1}, \ldots, Y_1).
\end{aligned}$$

We then have

$$\mu_t = (m(1)Y_{t-1}(1), M(2)Y_{t-1}(2), \ldots, m(k)Y_{t-1}(k))^T,$$

and

$$V_t = Diag\{\sigma^2(1)Y_{t-1}(1), \sigma^2(2)Y_{t-1}(2), \ldots, \sigma^2(k)Y_{t-1}(k)\}.$$

The quasi-score function for estimating β is given by

$$G^0(\beta) = \sum_{t=1}^{n} \left(\frac{d\mu_t}{dp}\right)^T V_t^{-1} (Y_t - \mu_t(\beta)).$$

If $\hat{\beta}$ is a solution of the equation $G^0(\beta) = 0$, one can show, under regularity conditions, that

$$F_n^{1/2}(\beta)(\hat{\beta} - \beta) \xrightarrow{d} N_p(0, I),$$

where

$$F_n(\beta) = \sum_{t=1}^{n} \left(\frac{d\mu_t}{d\beta}\right) V_t^{-1} \left(\frac{d\mu_t}{d\beta}\right).$$

4.11 BAYES AND EMPIRICAL BAYES ESTIMATION

Consider the G-W branching process $\{Z_t\}$ with the power series offspring distribution

$$P(Z_1 = z) = a_z \theta^z / A(\theta).$$

The likelihood function based on the sample $(Z_1, \ldots, Z_n)$ is given by

$$L_n(\theta) = \theta^{\sum_1^n Z_i} (A(\theta))^{-\sum_1^n Z_{i-1}}.$$

Assume a conjugate prior for θ with density

$$\pi(\theta) \propto \theta^\alpha (A(\theta))^{-\beta} / K(\alpha, \beta), \quad \theta > 0.$$

The posterior density of θ given $Z(n) = (Z_1, \ldots, Z_n)$ is:

$$p(\theta | Z(n)) \propto \theta^{\alpha + \sum_1^n Z_i} (A(\theta))^{-(\beta + \sum_1^n Z_{i-1})}.$$

The maximum posterior probability (MPP) estimator of m is obtained by solving $\frac{d \log p(\theta | Z(n))}{dm} = 0$. Note that

$$
\begin{aligned}
\frac{d \log p(\theta | Z(n))}{dm} &= \left(\frac{d \log p(\theta | Z(n))}{d\theta} \right) \left(\frac{d\theta}{dm} \right) \\
&= \sigma^{-2} [(\alpha + \sum_1^n Z_i) - (\beta + \sum_1^n Z_{i-1})m],
\end{aligned}
$$

since $m = \theta A'(\theta)/A(\theta)$ and $\sigma^2 = \theta m'(\theta)$. Finally, the MPP estimator of m is given by

$$\hat{m}_B = \frac{\alpha + \sum_1^n Z_i}{\beta + \sum_1^n Z_{i-1}}.$$

The estimator $\hat{m}_B$ can be viewed as the Bayes estimator with respect to the zero-one loss function. The marginal likelihood based on $Z(n)$ is given by

$$\bar{L}_n(\alpha, \beta) \propto \int L_n(\theta) \pi(\theta) d\theta = K^{-1}(\alpha, \beta) K(\alpha + \sum_1^n Z_i, \beta + \sum_1^n Z_{i-1}).$$

Let $\hat{\alpha}$ and $\hat{\beta}$ denote the maximum likelihood estimators of α and β based on $\bar{L}_n(\alpha, \beta)$. An empirical Bayes estimator of m is then given by

$$\hat{m}_{EB} = \frac{\hat{\alpha} + \sum_1^n Z_i}{\hat{\beta} + \sum_1^n Z_{i-1}}.$$

It can be shown that, for $m > 1$,

$$(\sum_1^n Z_{i-1})^{1/2}(\hat{m}_B - m) \text{ and } (\sum_1^n Z_{i-1})^{1/2}(\hat{m}_{EB} - m)$$

have the same limit distribution, and the limit distribution is $N(0, \sigma^2)$. It may be recalled that this limit distribution is also the same as that for the maximum likelihood estimator $\hat{m} = (\sum_1^2 Z_{i-1})^{-1}(\sum_1^n Z_i)$ discussed in Section 4.4.

4.12   CONCLUDING REMARKS

The G-W branching process is a proto-type example of the local asymptotic mixed normal family. The asymptotic properties of the estimator, confidence bounds, and tests for the offspring mean $m(m > 1)$ are non-standard since the limit distributions are non-normal. Asymptotic efficiency of the ML estimator is established via the limiting probability of concentration in the unconditional set-up. The large sample efficiency of tests and confidence bounds are studied in a conditional frame-work. Quasilikelihood, Bayes and empirical Bayes estimators are also discussed briefly. It would be of interest to investigate asymptotic inference problems for multi-type branching processes using a similar unified approach.

REFERENCES

Basawa, I. V. (1981a). Efficiency of conditional tests for mixture experiments with applications to the birth and branching processes, *Biometrika*, **68**, 153–165.

Basawa, I. V. (1981b). Efficiency of conditional maximum likelihood estimators and confidence limits for mixtures of exponential families, *Biometrika*, **68**, 515–523.

Basawa, I. V. (1987). Asymptotic distributions of prediction errors and related tests of fit for non-stationary processes, *Annals of Statistics*, **15**, 46–58.

Basawa, I. V. and Prakasa Rao, B. L. S. (1980). *Statistical Inference for Stochastic Processes*, Academic Press, London.

Basawa, I. V. and Scott, D. J. (1976). Efficient tests for branching processes, *Biometrika*, **63**, 531–536.

Basawa, I. V. and Scott, D. J. (1983). Asymptotic optimal inference for non-ergodic models, In *Lecture Notes in Statistics*, Vol. 17, Springer-Verlag, New York.

Feigin, P. D. and Reiser, B. (1979). On asymptotic ancillarity and inference for Yule and regular non-ergodic processes, *Biometrika*, **66**, 279–284.

Godambe, V. P. (1985). The foundations of finite sample estimation in stochastic processes, *Biometrika*, **72**, 419–428.

Guttorp, P. (1991). *Statistical Inference for Branching Processes*, John Wiley & Sons, New York.

Heyde, C. C. (1975). Remarks on efficiency in estimation for branching processes, *Biometrika*, **62**, 49–55.

Heyde, C. C. (1977). An optimal property of maximum likelihood with application to branching process estimation, *Bulletin Institute of Statistics*, **47**, 407–417.

Heyde, C. C. (1978). On an optimal property of maximum likelihood estimator of a parameter from a stochastic process, *Stochastics Processes and its Applications*, **8**, 1–9.

Heyde, C. C. (1997). *Quasilikelihood and Its Applications*, Springer-Verlag, New York.

Keiding, N. (1974). Estimation in the birth process, *Biometrika*, **61**, 71–80.

Sweeting, T. J. (1978). On efficient tests for branching processes, *Biometrika*, **65**, 123–128.

Sweeting, T. J. (1986). Asymptotic conditional inference for the offspring mean of a supercritical Galton-Watson process, *Annals of Statistics*, **14**, 925–933.

Sweeting, T. J. (1992). Asymptotic ancillarity and conditional inference for stochastic processes, *Annals of Statistics*, **20**, 580–589.

Part II
Distributions and Characterizations

CHAPTER 5

THE CONDITIONAL DISTRIBUTION OF X GIVEN $X = Y$ CAN BE ALMOST ANYTHING!

B. C. ARNOLD   C. A. ROBERSTON

University of California, Riverside, CA

Abstract: Suppose that X and Y are independent absolutely continuous random variables. We consider the problem of deciding what is an appropriate choice for the conditional density of X given $X = Y$. It has been earlier noted [see, for example, Rao (1993)] that ambiguous responses to this question are possible. In the present note we argue that the situation is actually worse than this might suggest. In fact the conditional density of X given $X = Y$ is essentially arbitrary. It can be whatever you wish!

Keywords and phrases: Monotone transformations, null events, arbitrary conditional densities

5.1 INTRODUCTION

It has long been known that conditioning on events of probability zero can be tricky business. The following example, taken from Rao (1993) is not atypical. Consider X and Y to be two independent standard exponential random variables. Define Z and W as follows

$$Z = (Y - 2)/X \tag{5.1.1}$$

75

$$W = Y - X . \tag{5.1.2}$$

Observe next that the events $\{Z = 1\}$ and $\{W = 2\}$ are equivalent, and could be denoted by a common label B. What then should be the conditional distribution of X given the event B? We could approach the question by considering the joint distribution of X and Z, obtaining from this the conditional densities of X given $Z = z$ for any z, and then setting $z = 1$. This leads to the putative answer

$$\begin{aligned} f_{X|B}(x) &= f_{X|Z}(x|1) \\ &= \frac{x}{4}e^{-2x}I(x > 0) , \end{aligned} \tag{5.1.3}$$

a gamma density with shape parameter 2 and scale parameter $1/2$. Alternatively, we could consider the joint density of X and W. We could obtain from it, the conditional density of X given $W = w$ for any w and then set $w = 2$. The result of this exercise is as follows.

$$\begin{aligned} f_{X|B}(x) &= f_{X|W}(x|2) \\ &= 2e^{-2x}I(x > 0) , \end{aligned} \tag{5.1.4}$$

a gamma density with shape parameter 1 and scale parameter $1/2$. The conditional density of X given B is apparently not well defined. Kolmogorov (1956) and more recently Rao (1993) argue that reporting a single conditional distribution such as (5.1.3) or (5.1.4), rather than a family of conditional densities such as $f_{X|Z}(x|z)$, $z \, \epsilon \, \mathbf{R}$ or $f_{X|W}(x|w)$, $w \, \epsilon \, \mathbf{R}$, is inadmissible. We are not proposing to question this decision. We do wish however to point out that conditional probabilities such as $f_{X|B}(x)$ above are not just not uniquely determined. They, in fact, can pretty much be anything. In this light it becomes even more important to emphasize the dangers of dealing with conditional densities given events of measure zero.

5.2   THE DISTRIBUTION OF X GIVEN $X = Y$ CAN BE ALMOST ANYTHING

Consider X and Y two independent random variables with absolutely continuous distribution functions. Suppose that the possible values of X form an interval $\mathcal{X}$ (possibly of infinite length) while the possible values of Y form an interval $\mathcal{Y}$ (again, possibly of infinite length). We assume that $\mathcal{X} \cap \mathcal{Y}$ is a non-empty interval. It is thus logically possible that $X = Y$ even though this event has zero probability.

We focus on the conditional distribution of X given that $X = Y$. We claim that this conditional density can arguably be almost <u>any</u> density supported on $\mathcal{X} \cap \mathcal{Y}$. To see this, consider an arbitrary strictly increasing function $h : \mathbf{R} \to \mathbf{R}$. The event $\{X = Y\}$ is of course then equivalent to the event $\{h(X) = h(Y)\}$. Now consider two random variables

$$U = X \tag{5.2.5}$$

$$V = h(X) - h(Y) . \tag{5.2.6}$$

The conditional density of U given $V = 0$ then provides an expression that could be interpreted as the conditional density of X given $\{X = Y\}$. Now assume that h^{-1} is differentiable so that a straightforward Jacobian argument yields

$$f_{U,V}(u, v) = f_X(u) f_Y(h^{-1}(h(u) - v)) h^{-1'}(h(u) - v) . \tag{5.2.7}$$

From this expression we find

$$f_{U|V}(u|0) \propto f_X(u) f_Y(u) h^{-1'}(h(u)) , \tag{5.2.8}$$

an expression which is non-zero on $\mathcal{X} \cap \mathcal{Y}$. The missing normalizing constant in (5.2.8) is such that the given conditional density integrates to 1 as it should.

If for example we take $h(x) = x$ in (5.2.6) then we would get

$$f_{U|V}(u|0) \propto f_X(u) f_Y(u) \tag{5.2.9}$$

since in this case $h^{-1'}(t) \equiv 1$.

If we take $h(x) = \log x$ assuming that $X > 0, Y > 0$ we would find $h^{-1'}(t) = e^t$, so $h^{-1'}(h(u)) = u$ and

$$f_{U|V}(u|0) \propto u f_X(u) f_Y(u) I(u > 0) . \tag{5.2.10}$$

If we take $h(x) = x^3$, we find $h^{-1'}(h(u)) = 1/3u^2$ and

$$f_{U|V}(u|0) \propto u^{-2} f_X(u) f_Y(u) . \tag{5.2.11}$$

Other monotone functions lead to other conditional densities.

Suppose that $g(u)$ is an integrable positive function on $\mathcal{X} \cap \mathcal{Y}$. We claim that there exists a suitable choice of monotone function $h(x)$ in (5.2.6) to yield

$$f_{U|V}(u|0) \propto g(u) I(u \in \mathcal{X} \cap \mathcal{Y}) \tag{5.2.12}$$

so that, as claimed, this conditional density can be quite arbitrary. In order for (5.2.12) to hold, referring to (5.2.8), it must be true that

$$h^{-1'}(h(u)) = \frac{g(u)}{f_X(u) f_Y(u)}, \quad u \in \mathcal{X} \cap \mathcal{Y} . \tag{5.2.13}$$

Let us denote the known function on the right of (5.2.13) by $w(u)$. We then seek a strictly monotone differentiable function h on $\mathcal{X} \cap \mathcal{Y}$ such that $h^{-1'}(h(u)) = w(u)$, on $\mathcal{X} \cap \mathcal{Y}$. Let us denote

$$k(v) = h^{-1}(v) \qquad (5.2.14)$$

then we seek $k(\cdot)$ such that

$$k'(k^{-1}(u)) = w(u) . \qquad (5.2.15)$$

If we write

$$u = k(k^{-1}(u))$$

and differentiate, we confirm that

$$k'(k^{-1}(u)) = 1/k^{-1'}(u) . \qquad (5.2.16)$$

Consequently we seek a function k such that

$$\frac{1}{w(u)} = k^{-1'}(u) = h'(u) . \qquad (5.2.17)$$

It follows that a suitable choice for $h(\cdot)$ is

$$h(u) = \int_0^u \frac{1}{w(v)} dv . \qquad (5.2.18)$$

With this choice of monotone function $h(\cdot)$, Equation (5.2.13) holds and consequently

$$f_{U|V}(u|0) \propto g(u)I(u \in \mathcal{X} \cap \mathcal{Y}) . \qquad (5.2.19)$$

Whatever g is desired can be produced by suitable choice of monotone function h. We reasonably then say that the conditional density of X given $X = Y$ can be anything.

5.3   DEPENDENT VARIABLES

It is reasonable to ask whether the assumption of independence of X and Y is crucial in the above development. In fact it is not. Instead of assuming that X and Y are independent we may assume that (X, Y) is absolutely continuous with an arbitrary joint density $f_{X,Y}(x,y)$. Again assume that the possible values of X are denoted by $\mathcal{X}$ and the possible values of Y by $\mathcal{Y}$. It is still true that the conditional density of X given $X = Y$ can be almost any density supported on $\mathcal{X} \cap \mathcal{Y}$. As before let h be an increasing function and define U and V as in (5.2.5)–(5.2.6). Since we no longer assume X and Y are independent, (5.2.7) and (5.2.8) must be replaced by

$$f_{U,V}(u,v) = f_{X,Y}(u, h^{-1}(h(u) - v))h^{-1'}(h(u) - v) \qquad (5.3.20)$$

and

$$f_{U|V}(u|0) \propto f_{X,Y}(u,u)h^{-1'}(h(u)) . \qquad (5.3.21)$$

If we wish to have

$$f_{U|V}(u|0) \propto g(u)I(u\varepsilon\mathcal{X} \cap \mathcal{Y}) \qquad (5.3.22)$$

we must select h such that

$$h^{-1'}(h(u)) = \frac{g(u)}{f_{X,Y}(u,u)}, \quad u\varepsilon\mathcal{X} \cap \mathcal{Y} . \qquad (5.3.23)$$

But this can obviously be achieved, as described in Section 5.2, by choosing h as in (5.2.18) with $w(u)$ now defined to be the right hand expression in (5.3.23) [instead of (5.2.13)]. So in the dependent case also, the conditional density of X given $X = Y$ can be anything.

5.4   RELATED EXAMPLES

A large number of related paradoxes can be recounted. We will mention two.

(a) Consider two i.i.d. normal $(0, 1)$ random variables. We wish to evaluate the conditional density of $X^2 + Y^2 = Z$ given $X = Y$. Method 1 consists of writing $Z = U^2 + V^2$ where $U = (X + Y)/\sqrt{2}$ and $V = (X - Y)/\sqrt{2}$. Of course U, V are i.i.d. normal $(0, 1)$ random variables and the event $X = Y$ is equivalent to $V = 0$. Consequently given $V = 0$ (i.e. given $X = Y$), the distribution of Z will be like that of U^2, i.e. χ_1^2.

Method 2 involves writing $Z = R^2$ where $R = \sqrt{X^2 + Y^2}$ and $\Theta = tan^{-1}(Y/X)$. It is well known that R and Θ are independent random variables $R^2 \sim \chi_2^2$ and $\Theta \sim$ uniform$(-\pi, \pi)$. The event $X = Y$ is equivalent to $\Theta = \pi/4$ or $-3\pi/4$. But since R^2 and Θ are independent, the conditional distribution of $Z = R^2$ given $\Theta = \pi/4$ or $-3\pi/4$ (i.e. given $X = Y$) will be χ_2^2.

So the distribution of Z given $X = Y$ is, take your pick, χ_1^2 or χ_2^2. Or something else, if you wish.

(b) An example described by Kolmogorov (1956) and attributed to Borel. Consider points uniformly distributed on the surface of the unit 3 dimensional sphere. Such points can be identified by the latitude (angular distance in radians above or below the equator) denoted by X and their longitude (angular distance in radians from a fixed prime meridian, a half circle through the poles, to a half great circle through the point and the poles) denoted by Y. There are however other ways of describing points on the sphere. One could associate with a point, its latitude, X, as above and its prime meridian latitude, Z, the angular distance in radians above or below the prime meridian. Kolmogorov was interested in the distribution of the point given that it was on the prime meridian great circle. Since a uniform distribution on the sphere will be uniform on any subset of the sphere, it might be argued that the distribution of the location of the point (i.e. its latitude) given that it is on the prime meridian great circle should be uniform. Such an argument is appealing, but since the prime meridian great circle is a subset of the sphere surface of zero measure, it is not rigorous or compelling. Indeed in the spirit of the present paper, the conditional distribution of the latitude given that the point is on the prime meridian great circle can be quite arbitrary.

If we consider (as did Kolmogorov) the joint distribution of (X, Y) (i.e. latitude and longitude), we find easily that Y is uniform $(-\pi, \pi)$ while the distribution of X is not uniform. In fact

$$f_X(x) = \frac{1}{2}\cos x, \quad -\pi/2 < x < \pi/2 . \tag{5.4.24}$$

In addition X and Y are independent random variables. The condition that the point lie on the prime meridian great circle is equivalent to $Y = 0$ or $Y = \pi$. The conditional distribution of latitude given that we are on the prime meridian, i.e. the distribution of X given $Y = 0$ or $Y = \pi$ is by independence the same as the unconditional distribution of X given in (5.4.24). It is clearly not uniform.

Alternatively if we consider the joint distribution of (X, Z) (i.e. latitude and prime meridian latitude), it is evident that X and Z are identically distributed with common distribution given by (5.4.24) but they are dependent. The point will be on the prime meridian great circle iff $Z = 0$. So the conditional distribution of the latitude, given we are on the prime meridian great circle will, in this formulation, be the conditional distribution of X given $Z = 0$. In this case the conditional density is uniform, i.e.

$$f_{X|Z}(x|0) = \frac{1}{\pi}, \quad -\pi/2 < x < \pi/2 . \qquad (5.4.25)$$

(Note that the conditional density of X given $Z = z$ (for $z \neq 0$) is not uniform).

Other ways of specifying the location of a point on the surface of the sphere can be expected to yield other conditional densities for the latitude given that the point is on the prime meridian great circle. There seems to be nothing inherently "correct" about either (5.4.24) or (5.4.25) as a conditional density and, essentially, anything would be just as good.

REFERENCES

Kolmogorov, A. N. (1956). *Foundations of the Theory of Probability*, Chelsea, New York.

Rao, M. M. (1993). *Conditional Measures and Applications*, Marcel Dekker, New York.

CHAPTER 6

AN APPLICATION OF RECORD RANGE AND SOME CHARACTERIZATION RESULTS

PRASANTA BASAK

Penn State University, Altoona, PA

Abstract: Let $Z_1, Z_2, \cdots$ be a sequence of independent and identically distributed random variables with absolutely continuous distribution function F. For $n \geq 1$, we denote by $Z_{1,n} \leq Z_{2,n} \leq \cdots \leq Z_{n,n}$ the order statistics of $Z_1, Z_2, \cdots, Z_n$. Consider the sequence of sample ranges $W_i = Z_{i,i} - Z_{1,i}, i = 2, 3, 4, \cdots$. Then define $W(n), n = 1, 2, \cdots$ to be the nth record range in the sequence of sample ranges W_is. Observe that a new record range is attained as soon as a new upper or lower record is observed. This article introduces a general sequential method for model choice and outlier detection involving the record range. The procedure require only limited information about the sample, namely the suitably chosen stopping times. Also, we give a characterization of the exponential distributions.

Keywords and phrases: Exponential distributions, record values, stopping times

6.1 INTRODUCTION

Let $Z_1, Z_2, \cdots$ be a sequence of independent and identically distributed (iid) random variables with absolutely continuous distribution function (df) F, and probability density function (pdf) f. For $n \geq 1$, we denote the order statistics of $Z_1, Z_2, \cdots, Z_n$ by

$$Z_{1,n} \leq Z_{2,n} \leq \cdots \leq Z_{n,n}. \tag{6.1.1}$$

83

Since F is continuous, removing a null set will give us $Z_i \neq Z_j; i \neq j$. That is we get the strict inequality in (6.1.1).

Let $X(0) = Y(0) = Z_1$, and a new record (either an upper or lower) occurs at time j if either $Z_j > \max \{Z_1, Z_2, \cdots, Z_j \}$, or $Z_j < \min \{Z_1, Z_2, \cdots, Z_j \}$. Let $X(n)$ be the lower record value and $Y(n)$ be the upper record value at the nth record in this sequence. Then for $n \geq 1$, define $W(n) = Y(n) - X(n)$ to be the nth record range. Also, it can be defined by looking at the sample ranges $W_i = Z_{i,i} - Z_{1,i}, i = 2, 3, 4, \cdots$. By this definition, we get $W(1) = W_2$ to be the first record range. Observe that a record range is encountered whenever either a new lower or upper record value in Z_is is observed.

Properties of record values and order statistics of iid random variables have been extensively studied in the literature. See, for example, Ahsanullah (1995), Arnold, Balakrishnan, and Nagaraja (1998), Deheuvels (1984), and bibliographies in Galambos and Kotz (1978), Nagaraja (1988), Nevzorov (1987). DasGupta, Rinott and Vidakovic (1998) intoduced some methods for model choice and outlier detection using order statistics.

The goal of this article is to introduce a technique for model choice that uses a sequential approach. Of course, the well known SPRT due to Wald already does that, and it is also well known that Wald's SPRT is strongly admissible. The hope in introducing new method is that one can save, in many cases, very substantially on the expected sample size by sacrificing minimally on error rates. Also, in this article, we give some characterizations of the exponential distributions.

The method proposed here apply particularly well to two specific types of problems: choosing between tails and investigating the effect of outliers. Therefore, these will be emphasized consistently. We define the following stopping time.

- For iid samples from an absolutely continuous distribution with df F (we assume unbounded interval of support) and density f, we will let N denote the stopping time

$$N = \inf \{n > 0 : W(n) > c\}, c > 0 \text{ is fixed but arbitrary.}$$

N thus simply gives the waiting time until the record range of an iid sample exceeds a given value c. Observe that here the waiting time N is defined in terms of number of records (either upper or lower) and not in terms of the number of observations. Note that N is quite evidently constructed for applications to problems of choosing tails. The idea is that for populations with thicker tails, N would tend to be smaller, giving rise to the possibility that one can choose the right tail without sampling for too long. With this introductory description, we will now study the stopping times on a theoretical basis in section 6.2. In section 6.3, we give some characterizations of the exponential distributions based on record range time and the lower record value at that time.

6.2   THE STOPPING TIME N

6.2.1   The Mean and the Variance of N

It is known [see Houchens (1984)] that, for $n \geq 1$, the pdf of $W(n)$ is given by

$$f_{W(n)}(w) = \frac{2^n}{\Gamma(n)} \int_{-\infty}^{\infty} [-\log(1 - F(x+w) + F(x))]^{n-1} f(x+w)f(x) \, dx.$$

To see that N is a stopping time, we simply use the identity

$$P(N > n) = P(W(n) < c) = \int_0^c f_{W(n)}(w) dw$$

$$= 2^n \int_0^c \int_{-\infty}^{\infty} \frac{[-\log(1 - F(x+w) + F(x))]^{n-1}}{\Gamma(n)} f(x+w)f(x) \, dx \, dw$$

$$= 2^n \int_{-\infty}^{\infty} \left\{ \int_0^c \frac{[-\log(1 - F(x+w) + F(x))]^{n-1}}{\Gamma(n-1)} f(x+w) \, dw \right\}$$
$$\cdot f(x) \, dx$$

$$= 2^n \int_{-\infty}^{\infty} \left\{ 1 - \sum_{j=0}^{n-1} \frac{[-\log(1 - F(x+c) + F(x))]^j}{j!} \right.$$
$$\left. \cdot (1 - F(x+c) + F(x)) \right\} f(x) \, dx. \qquad (6.2.2)$$

in applying the formula (6.2.2), little care has to be taken if the support of F is finite. Now use the fact that $X(n) \to \infty$ a.s., while $Y(n) \to -\infty$ a.s, implying $W(n)$ converges in probability to ∞, giving $P(N > n) \to 0$, as $n \to \infty$, meaning N is proper.

Theorem 6.2.1 *For every $0 < c < \infty$, expected value of N is finite and is given by*

$$E(N) = \int_{-\infty}^{\infty} \frac{2f(x)}{1 - F(x+c) + F(x)} \, dx - 1.$$

PROOF For non-negative integer valued random variable N, it is known that

$$E(N) = \sum_{n=0}^{\infty} P(N > n).$$

So,

$$
\begin{aligned}
E(N) &= \sum_{n=0}^{\infty} P(N > n) = 1 + \sum_{n=1}^{\infty} P(N > n) \\
&= 1 + \sum_{n=1}^{\infty} \int_0^c \int_{-\infty}^{\infty} \frac{2^n \left[-\log(1 - F(x+w) + F(x)) \right]^{n-1}}{\Gamma(n)} \cdot \\
&\qquad f(x+w) \, f(x) \, dx \, dw \\
&= 1 + 2 \int_0^c \int_{-\infty}^{\infty} \sum_{n=1}^{\infty} \frac{\left[-2\log(1 - F(x+w) + F(x)) \right]^{n-1}}{\Gamma(n)} \cdot \\
&\qquad f(x+w) \, f(x) \, dx \, dw \\
&= 1 + 2 \int_0^c \int_{-\infty}^{\infty} \frac{f(x+w)f(x)}{[1 - F(x+w) + F(x)]^2} \, dx \, dw \\
&= 1 + 2 \int_{-\infty}^{\infty} \left\{ \int_0^c \frac{f(x+w)}{[1 - F(x+w) + F(x)]^2} \, dw \right\} f(x) \, dx \\
&= 1 + 2 \int_{-\infty}^{\infty} \left\{ \frac{1}{1 - F(x+c) + F(x)} - 1 \right\} f(x) \, dx \\
&= \int_{-\infty}^{\infty} \frac{2f(x)}{1 - F(x+c) + F(x)} \, dx - 1. \qquad (6.2.3)
\end{aligned}
$$

To see that the expectation is finite, we can invoke Tonelli's theorem. The right side of equation (6.2.3) can be written as

$$2 \sum_{j=0}^{\infty} \left[F(x+c) - F(x) \right]^j f(x) \, dx.$$

This is easily shown to be finite since

$$\sup_x \left[F(x+c) - F(x) \right] < 1.$$

The case of bounded support can be dealt in similar fashion. □

Theorem 6.2.2 *For every* $0 < c < \infty$, *variance of N is finite and is given by*

$$Var(N) = 2 - E(N) - E^2(N) - 8 \int_{-\infty}^{\infty} \frac{\log(1 - F(x+c) + F(x))}{1 - F(x+c) + F(x)} f(x) \, dx$$

PROOF   First we find $E(N(N-1))$, which is known to be

$$\sum_{n=1}^{\infty} 2n \, P(N > n)$$

for non-negative integer valued random variable.

$$E(N(N-1)) = 2 \sum_{n=1}^{\infty} nP(N > n)$$

$$= 2 \sum_{n=1}^{\infty} \int_0^c \int_{-\infty}^{\infty} \frac{n2^n \left[-\log(1 - F(x+w) + F(x))\right]^{n-1}}{\Gamma(n)}$$
$$\cdot f(x+w)f(x) \, dx \, dw$$

$$= 4 \int_0^c \int_{-\infty}^{\infty} \sum_{n=1}^{\infty} \frac{n \left[-2\log(1 - F(x+w) + F(x))\right]^{n-1}}{\Gamma(n)}$$
$$\cdot f(x+w)f(x) \, dx \, dw$$

$$= 4 \int_0^c \int_{-\infty}^{\infty} \frac{1 - 2\log(1 - F(x+w) + F(x))}{[1 - F(x+w) + F(x)]^2} f(x+w)f(x) \, dx \, dw$$

$$= 4 \int_0^c \int_{-\infty}^{\infty} \frac{1}{[1 - F(x+w) + F(x)]^2} f(x+w)f(x) \, dx \, dw$$
$$-8 \int_0^c \int_{-\infty}^{\infty} \frac{\log(1 - F(x+w) + F(x))}{[1 - F(x+w) + F(x)]^2} f(x+w)f(x) \, dx \, dw$$

$$= 4 \left(\frac{E(N) - 1}{2} \right)$$
$$-8 \int_{-\infty}^{\infty} \left[\int_0^c \frac{\log(1 - F(x+w) + F(x))}{[1 - F(x+w) + F(x)]^2} f(x+w) \, dw \right] f(x) \, dx$$

$$= 2E(N) - 2 - 8 \int_{-\infty}^{\infty} \left[\frac{1 + \log(1 - F(x+c) + F(x))}{1 - F(x+c) + F(x)} - 1 \right] f(x) \, dx$$

$$= 2E(N) - 2 - 8 \int_{-\infty}^{\infty} \frac{f(x)}{1 - F(x+c) + F(x)} \, dx$$

$$-8\int_{-\infty}^{\infty} \frac{\log(1 - F(x+c) + F(x))}{1 - F(x+c) + F(x)} f(x)\ dx + 8$$

$$= 2E(N) + 6 - 8\left(\frac{E(N)+1}{2}\right)$$

$$-8\int_{-\infty}^{\infty} \frac{\log(1 - F(x+c) + F(x))}{1 - F(x+c) + F(x)} f(x)\ dx$$

$$= 2 - 2E(N) - 8\int_{-\infty}^{\infty} \frac{\log(1 - F(x+c) + F(x))}{1 - F(x+c) + F(x)} f(x)\ dx$$

So,

$$\mathrm{Var}(N) = E(N(N-1)) + E(N) - E^2(N)$$
$$= 2 - E(N) - E^2(N)$$
$$-8\int_{-\infty}^{\infty} \frac{\log(1 - F(x+c) + F(x))}{1 - F(x+c) + F(x)} f(x)\ dx.$$

□

Example 1 The standard normal (with cdf $\Phi(0,1)$, and pdf $\phi(\cdot)$) and standard Laplace (also known as standard double exponential) (with pdf $\frac{1}{2}e^{-|x|}$) cases. Application of the above formulas, on numerical integration, gives the values for the mean and the standard deviation of N as in Tables 6.1 and 6.2 respectively. Here, we standardize the standard Laplace distribution, so that comparisons with standard normal will be more meaningful.

TABLE 6.1 Mean and standard deviation of N under standard normal

c	1	2	3	4	5	6	7	8	9	10
$E(N)$	1.76	2.95	4.60	6.70	9.26	12.29	15.79	19.76	24.22	29.16
sd(N)	0.90	1.45	1.97	2.47	2.95	3.44	3.92	4.40	4.88	5.37

TABLE 6.2 Mean and standard deviation of N under standard Laplace

c	1	2	3	4	5	6	7	8	9	10
$E(N)$	1.98	3.26	4.64	6.05	7.46	8.87	10.29	11.70	13.11	14.53
sd(N)	1.07	1.65	2.06	2.39	2.68	2.93	3.16	3.38	3.58	3.77

Note that standard Laplace does not have quartiles equal to those of $\Phi(0,1)$, but the effect of a thicker tail is transparent from the numbers in Table 6.2. Some care is needed in interpreting the the two tables, although the effect of a thicker tail is clear from the respective expected values. It will take on average 15.79 records from the standard normal for the record

range to hit 7, but only 10.29 on the averge for the same range in the case
of Laplace. Care is needed though, because 10.29 is within two sigma of
the mean value of N for the normal case. However, if one was sampling
from the Laplace, almost certainly one would hit the value 7 in 17 records,
and the chance of doing so for normal population is low. Thus a sequential
sample is likely to give information regarding the model one is sampling
from.

One can do formal tests by setting up $H_0 : F$ is $\Phi(\theta, 1)$, for some θ,
versus $H_1 : F$ is $Laplace(\theta, 1)$, for some θ. A reasonable procedure seems
to be: Reject H_0 if $N > n$. The choice of c and this later threshold value n is
an optimality problem, and many formulations is possible. For instance, one
can ask for the optimal pair (c, n) that minimizes some weighted average
of the Type 1 and Type 2 errors. Incidentally, these two errors can be
calculated immediately because they both involve only the event $\{N > n\}$
and we have seen how to evaluate $P(N > n)$ for any absolutely CDF.

6.2.2   Behavior for Large c: Almost Sure Limits

It will now be necessary to explicitly emphasize the role of the c in the
definition of N. Consequently, we will for now write N as $N(c)$. It is
evident that $N(c) \to \infty$ almost surely as $c \to \infty$. One can give just a
sample point argument. One would expect, however, that the rates of
convergence would depend on the tail.

Theorem 6.2.3 $\frac{R^{-1}(N(c))}{c} \overset{a.s.}{\to} 1$ as $c \to \infty$, where $R(x) = -\log \bar{F}(x) = -\log(1 - F(x))$. In other words, $N(c)$ is of order $R(c)$.

PROOF   To arrive at the stated assertion, we use the inequalities

$$W(N(c) - 1) \leq c \leq W(N(c)),$$

where for any n, $W(n)$ denotes $Y(n) - X(n)$. This is immediate from the
definition of $N(c)$. Consequently,

$$\frac{W(N(c) - 1)}{R^{-1}(N(c))} \leq \frac{c}{R^{-1}(N(c))} \leq \frac{W(N(c))}{R^{-1}(N(c))}.$$

Now use the following three facts.

1. $\frac{Y(n)}{R^{-1}(n)}$ and $\frac{-X(n)}{R^{-1}(n)}$ converge almost surely to $\frac{1}{2}$.

    The above almost sure convergence is true for the distribution func-
    tions satisfying the conditions (6.2.4) below [Resnick (1973, Theorem
    4)]

Theorem 6.2.4 *A sufficient condition for* $\lim_{n \to \infty} \frac{V(n)}{R^{-1}(n)} \overset{a.s.}{\to} 1$ *is*

$$\lim_{s \to \infty} \frac{R^{-1}(s + t\sqrt{s \log \log s})}{R^{-1}(s)} = 1 \quad \text{for all real } t, \qquad (6.2.4)$$

where, $V(n)$ is the usual nth upper record value.

Resnick (1973) also showed that (6.2.4) is equivalent to

$$\lim_{t \to \infty} \frac{R(tx) - R(t)}{\sqrt{2R(t) \log \log R(t)}} = \infty \,\, \forall x > 1. \qquad (6.2.5)$$

If R^{-1} is regularly varying with exponent α, $0 \leq \alpha \leq \infty$ [equivalent to R regularly varying with exponent α^{-1}] then (6.2.4) or (6.2.5) is satisfied.

2. $N(c) \to \infty$ almost surely as $c \to \infty$, which was discussed before.

3. If $W(n)$ is a sequence of random variables and $a(n)$ is a sequence of real numbers such that $W(n)/a(n) \to 1$ almost surely as $n \to \infty$, then so does $W(N(c))/a(N(c))$ as $n \to \infty$. $\qquad \square$

It now follows from our inequalities above that $\frac{R^{-1}(N(c))}{c} \to 1$ almost surely, and the stated result follows.

Example 1 (Normal case) Conventional wisdom says that the $\Phi(0,1)$ distribution has thin tails; furthermore the values for $E(N)$ in normal case of Example 1 suggest $N(c)$ has to be very large for large c. For standard normal distribution, $x\bar{\Phi}(x) \simeq \phi(x)$ for large values of x using Mill's ratio. That is, $R(x) \simeq \frac{x^2}{2}$ implying $R^{-1}(x) = \sqrt{2x}$. So,

$$\lim_{n \to \infty} \frac{Y(n)}{R^{-1}(n)} = \lim_{n \to \infty} \frac{Y(n)}{\sqrt{2n}} = \frac{1}{2} \text{ a.s.}$$

That is, $N(c)$ is of order $R(c) = c^2/2$ for large c.

Arguments entirely parallel to the normal case imply that in the case of Laplace $N(c)$ to be of order of c. Notice the slower rate at which N grows in comparison to the normal case. It will also be intersting to see at what rate does $E[N(c)]$ grow. We will work out an example in greater detail.

Example 2 For nonnegative random variables interesting variation of stopping time N can be given although N itself still makes sense. Consider the case of the exponential density $\exp(-x)$, $x \geq 0$. Define a new stopping time N_1 as

$$N_1 = \inf\{n > 0 : Y(n)/X(n) > c\}.$$

On transforming to the distribution of a log-exponential random variable (the *negative extreme value distribution*) this corresponds to the stopping time N.

We will, however, study the expectation of this new stopping time without transforming, as it seems better to do so. One can check by using now familiar arguments that $E(N_1)$ is finite and equals

$$\int_0^\infty \frac{2f(x)}{1 - F(cx) + F(x)} dx,$$

which in this case reduces to

$$\int_0^1 \frac{2}{1 - x + x^c} dx,$$

on the change of variable in the integral. The question is at which rate this expectation goes to infinity. We use an old technique known as method of Laplace to do this.

The function $1 - x + x^c$ has unique minimum at $x = (1/c)^{\frac{1}{c-1}}$; this goes to 1 as $c \to \infty$. But the dependence on c creates hurdles in the subsequent analysis. Instead we make the change of variable $y = cx^{c-1}$. The integral now becomes

$$\frac{2c}{c-1} \int_0^c \frac{y^{\frac{2-c}{c-1}}}{c^{\frac{c}{c-1}} - cy^{\frac{1}{c-1}} + y^{\frac{c}{c-1}}} dy.$$

The contribution from $[0, 1]$ remains bounded; the contribution from $(1, c)$ determines the rate of convergence. On expanding the term in the denominator of the integrand around $y = 1$, and using the fact that $c^{\frac{1}{c-1}} - 1$ is of the order $\frac{\log c}{c}$, one gets $E(N_1)$ equivalent to $\frac{c}{(1+\log c)} \log c$, i.e., c.

6.3   CHARACTERIZATION RESULTS

In this section, we give a characterization of the exponential distributions in terms of record range time and lower record value at that time. The record range times, $I(n)$, are defined as follows. Let $W_m = Z_{m:m} - Z_{1:m}, m \geq 1$ and define

$$I(1) = 2, \text{ and } I(n+1) = \min\{j : W_j > W_{I(n)}\}, n \geq 1.$$

$I(n)$ is the nth record range time, i.e., the time when the nth record value (either lower or upper) is observed. Observe that $I(n)$ is the index of the observation creating the nth record range. Let $X(n)$ be the lower record value at time $I(n)$.

Houchens (1984) shows that the pmf of $I(n)$ is given by

$$P\left(I(n) = m\right) = \frac{2^n}{m!}|S_{m-2}^{n-1}|; \quad m \geq n+1, n \geq 1,$$

where S_m^n are the Stirling numbers of the first kind, defined by

$$x(x-1)\cdots(x-m+1) = \sum_{n \geq 0} S_m^n x^n.$$

In order to prove our main characterizing results, we will need the following lemma.

Lemma 6.3.1 $P\left(I(n)X(n) > x\right) = \displaystyle\sum_{m=n+1}^{\infty} \frac{2^n|S_{m-2}^{n-1}|}{m!}\bar{F}^m\left(\frac{x}{m}\right).$

PROOF

$$
\begin{aligned}
P\left(I(n)X(n) > x\right) &= \sum_{m=n+1}^{\infty} P\left(X(n) > \frac{x}{m}, I(n) = m\right) \\
&= \sum_{m=n+1}^{\infty} P(I(n) = m) \cdot P\left(X(n) > \frac{x}{m}\,\Big|\, I(n) = m\right) \\
&= \sum_{m=n+1}^{\infty} P(I(n) = m) \cdot P\left(Z_{1:m} > \frac{x}{m}\right) \\
&= \sum_{m=n+1}^{\infty} \frac{2^n|S_{m-2}^{n-1}|}{m!}\bar{F}^m\left(\frac{x}{m}\right).
\end{aligned}
$$

Then we the have the following theorem.

Theorem 6.3.2 *Suppose that*

$$\lim_{x \to 0^+} \frac{F(x)}{x} = \theta, 0 < \theta < \infty.$$

Then $I(n)X(n) \overset{d}{=} Z_1$ if and only if $\bar{F}(x) = exp(-\theta x), x > 0$, for some $\theta >$ 0.

PROOF (*If Part*) Easy to verify.

(*Only if Part*) Let $I(n)X(n) \stackrel{d}{=} Z_1$ and define $u(x) = \frac{R(x)}{x}$, $x > 0$. Since $lim_{x\to0+} \frac{F(x)}{x} = \theta$, we have $u(0) = u(0^+)$. Then

$$\sum_{m=n+1}^{\infty} \frac{2^n |S_{m-2}^{n-1}|}{m!} e^{-mxu(x/m)} = e^{-xu(x)}, \quad x > 0. \qquad (6.3.6)$$

The theorem is proved if it can be shown that $u(x)$ in equation (6.3.6) is a constant. Following Ahsanullah and Kirmani (1991) we show that $u(x)$ is indeed a constant. Define for any given $t > 0$,

$$a_0 \;=\; \min_{x\in[0,t]} u(x), \quad x_0 = \inf\{x \in [0,t] | u(x) = a_0\},$$
$$a_1 \;=\; \max_{x\in[0,t]} u(x), \quad x_1 = \inf\{x \in [0,t] | u(x) = a_1\}.$$

By continuity of $u(x)$, we have $x \in [0,t]$ and $u(x_0) = a_0$. Therefore,

$$u(x_0) \le u\left(\frac{x_0}{m}\right) \text{ for all } m \ge 1. \qquad (6.3.7)$$

If in (6.3.7) equality holds for all $m > n$ then $u(x_0) = u(0)$, which gives $x_0 = 0$ by the definition of x_0. Now suppose that $x_0 > 0$. Then in (6.3.7) we must have strict inequality for at least one value of $m > n$ and hence

$$e^{-mx_0u(x_0)} - \sum_{m=n+1}^{\infty} \frac{2^n |S_{m-2}^{n-1}|}{m!} e^{-mx_0u(x_0/m)} =$$
$$\sum_{m=n+1}^{\infty} \frac{2^n |S_{m-2}^{n-1}|}{m!} \left\{ e^{-mx_0u(x_0)} - e^{-mx_0u(x_0/(n+1))} \right\} > 0,$$

which contradicts (6.3.6). Therefore, $x_0 = 0$. Similarly, it can be shown that $x_1 = 0$. This proves that $\min_{x\in[0,t]} u(x) = \max_{x\in[0,t]} u(x)$ and therefore $u(x)$ is constant.

Theorem 6.3.3 *Suppose that*

$$lim_{x\to0+} \frac{F(x)}{x} = \theta, 0 < \theta < \infty.$$

Then $\bar{F}(x) = exp(-\theta x)$ if and only if, for some $m \ge 2$, the conditional distribution of $I(n)X(n)$ given $I(n) = m$ is identical with the unconditional distribution of $I(n)X(n)$.

PROOF The conditional survival function of $I(n)X(n)$ given $I(n) = m$ is given by

$$P\left(I(n)X(n) > x|I(n) = m\right) = \bar{F}^m(x/m).$$

So $K(x) = P(I(n)X(n) \leq x)$ is independent of m if and only if, for all $x > 0$, $\bar{K}(x) = \bar{F}^m(x/m)$. Using the method of Galambos and Kotz (1978, pp. 39–40), we get

$$\bar{K}(x) = \bar{F}^m(x/m) = \lim_{r \to \infty} \bar{F}^{m^r}(x/m^r) = exp(-\theta x).$$

It follows that $\bar{F}(x) = exp(-\theta x), x > 0$. $\square$

Our next characterization deals with distributions having harmonic new better (worse) than used in expectation property. A distribution function $F(\cdot)$ with $F(0) = 0$ and $\mu = E(Z_1) < \infty$ is said to be harmonic new better (worse) than used in expectation abbreviated HNBUE (HN-WUE), if $\int_t^\infty \bar{F}(x)dx \leq (\geq)\mu e^{-t/\mu}, t \geq 0$ [see Basu and Ebrahimi (1986)]. These distributions contain all new better (worse) than used in expectation (NBUE, NWUE) distributions and also IFR (DFR), IFRA (DFRA), and NBU (NWU) distributions.

Theorem 6.3.4 *Suppose the underlying distribution $F(\cdot)$ is HNBUE (HN-WUE). Then $E\left[I(n)X(n)\right] = E\left[Z_1\right]$ if and only if $F(\cdot)$ is exponential.*

PROOF Let $\mu = E\left[Z_1\right]$ and suppose $F(\cdot)$ is HNBUE (HNWUE). Then we have

$$
\begin{aligned}
E\left[I(n)X(n)\right] &= \sum_{m=n+1}^{\infty} \frac{2^n|S_{m-2}^{n-1}|}{m!} \int_0^\infty \bar{F}^m(x/m)dx \\
&= \sum_{m=n+1}^{\infty} \frac{2^n|S_{m-2}^{n-1}|}{(m-1)!} \int_0^\infty \bar{F}^m(y)dy \\
&\geq (\leq) \sum_{m=n+1}^{\infty} \frac{2^n|S_{m-1}^{n-1}|}{m!}\mu,
\end{aligned}
$$

with equality if and only if

$$\int_0^\infty \bar{F}^m(y)dy = \frac{\mu}{m} \quad \text{for all } m \geq n+1. \tag{6.3.8}$$

But (6.3.8) is a necessary and sufficient condition for $F(\cdot)$ to be exponential when $F(\cdot)$ is HNBUE (HNWUE) [see Basu and Kirmani (1986)]. $\square$

Acknowledgement The author is indebted to the referee for constructive criticisms, which led to improved presentation of the article.

REFERENCES

Ahsanullah, M. and Kirmani, S. N. U. A. (1991). Characterizations of the exponential distribution through a lower record, *Communications in Statistics—Theory and Methods*, **20**, 1293–1299.

Ahsanullah, M. (1995). *Record Statistics*, Nova Science Publishers, Inc., Commack, New York.

Arnold, B. C., Balakrishnan, N. and Nagaraja, H. N. (1998). *Records*, John Wiley & Sons, New York.

Basu A. P. and Ebrahimi, N. (1986). HNBUE and HNWUE distributions - a survey, In *Reliability and Quality Control* (Ed., A. P. Basu), pp. 33–46, Elsevier Science Publishers, New York.

Basu, A. P. and Kirmani, S. N. U. A. (1986). Some results involving HNBUE distributions, *Journal of Applied Probability*, **23**, 1038-1044.

DasGupta, A., Rinott, Y. and Vidakovic, B. (1998). Stopping times related to diagnostics and outliers, Preprint.

Deheuvels, P. (1984). The characterization of distributions by order statistics and record values – a unified approach, *Journal of Applied Probability,* **21**, 326–334.

Galambos, J. and Kotz, S. (1978). *Characterizations of Probability Distributions*, Springer-Verlag, New York.

Galambos, J. (1978). *The Asyptotic Theory of Extreme Order Statistics*, John Wiley & Sons, New York.

Houchens, R. L. (1984). Record value theory and inference, *Ph.D. Dissertation*, University of California, Riverside, CA.

Nagaraja, H. N. (1988). Record values and related statistics – a review, *Communications in Statistics—Theory and Methods*, **17**, 2223–2238.

Nevzorov, V. B. (1987). Records, *Probability Theory and Applications*, **13**, 201–228.

Resnick, S. I. (1973). Reord values and maxima, *The Annals of Probability*, **1**, 650–662.

CHAPTER 7

CONTENTS OF RANDOM SIMPLICES AND RANDOM PARALLELOTOPES

A. M. MATHAI

McGill University, Montreal, Quebec, Canada
The University of Texas at El Paso, TX

Abstract: General moments and distributions of the volume contents of random simplices and random parallelotopes in R^n, the n-dimensional Euclidean space, when some general probability measures are associated with the vertices, are considered in this article. Usual methods available in the literature for dealing with such problems depend heavily on results from integral and differential geometry. Algebraic procedures based on properties of Jacobians of matrix transformations and functions of matrix argument, and no results from integral and differential geometry, will be used in the present article. Various types of results will be obtained when the vertices are preselected points or when the points arrive according to some stochastic processes.

Keywords and phrases: Random simplex, random parallelotope, random volume, exact distributions, moments

7.1 INTRODUCTION

A few known results from basic linear algebra and a few results on Jacobians of matrix transformations are needed in the discussions later on. Hence these will be introduced here as lemmas.

7.1.1 Some Basic Results from Linear Algebra

Consider a parallelogram in a plane. The 3-dimensional analogue is the parallelepiped. In each of (x_1, x_2)-plane, (x_1, x_3)-plane, and (x_2, x_3)-plane it is a parallelogram. The n-dimensional analogue of a parallelogram is the parallelotope. Consider a rectangular coordinate system with the origin denoted by O. Let $P_1 = (x_{11}, x_{12})$ and $P_2 = (x_{21}, x_{22})$ be two points. Then a parallelogram can be generated by these points and O. The area of the parallelogram is twice the area of the triangle OP_1P_2 or $2\left(\frac{1}{2}ha_1\right)$ where a_1 is the length of OP_1 and h is the perpendicular distance from P_2 to the line OP_1. Then the area is $|a_1a_2 \sin\theta|$ where θ is the angle between OP_1 and OP_2. But

$$
\begin{aligned}
|a_1 a_2 \sin\theta| &= a_1 a_2 \sqrt{1 - \cos^2\theta} \\
&= \sqrt{a_1^2 a_2^2 - (a_1, a_2)^2}
\end{aligned}
$$

where (a_1, a_2) is the dot product of the vectors $\vec{OP_1}$ and $\vec{OP_2}$ or in terms of the coordinates

$$
(a_1, a_2) = x_{11}x_{21} + x_{12}x_{22}, \ a_1^2 = x_{11}^2 + x_{12}^2, \ a_2^2 = x_{21}^2 + x_{22}^2.
$$

Substituting the coordinates, the expression for the area reads as

$$
\begin{aligned}
|a_1 a_2 \sin\theta| &= \sqrt{(x_{11}x_{22} - x_{12}x_{21})^2} = |x_{11}x_{22} - x_{21}x_{12}| \\
&= \begin{vmatrix} x_{11} & x_{12} \\ x_{21} & x_{22} \end{vmatrix}_+
\end{aligned}
$$

where the $+$ indicates that the absolute value of the determinant is taken. With three points $P_1(x_{11}, x_{12}, x_{13})$, $P_2(x_{21}, x_{22}, x_{23})$, $P_3(x_{31}, x_{32}, x_{33})$ in a 3-dimensional space we can create a parallelepiped. The volume is given by the cross product of the vectors $\vec{OP_1}$, $\vec{OP_2}$ and $\vec{OP_3}$ which is the determinant

$$
\text{Vol} = \begin{vmatrix} x_{11} & x_{12} & x_{13} \\ x_{21} & x_{22} & x_{23} \\ x_{31} & x_{32} & x_{33} \end{vmatrix}_+ .
$$

(a) The content of a parallelotope

Generalizing this concept the volume of an n-dimensional parallelotope is given by

$$\text{Vol (parallelotope)} = \begin{vmatrix} x_{11} & \cdots & x_{1n} \\ \vdots & \vdots & \vdots \\ x_{n1} & \cdots & x_{nn} \end{vmatrix}_+ \qquad (7.1.1)$$

where the n points which created the parallelotope are

$$P_j(x_{j1}, ..., x_{jn}), \ j = 1, ..., n.$$

Lemma 7.1.1 *Let $P_j = P_j(x_{j1}, ..., x_{jn})$, $j = 1, ..., n$ be n linearly indepen-dent points in a rectangular coordinate system with the origin O. Let θ_{ij} be the angle between the vectors $\vec{OP_i}$ and $\vec{OP_j}$, $i \neq j$ and a_j the length of $\vec{OP_j}$, $j = 1, ..., n$. Then the volume content of the parallelotope created by the vectors $\vec{OP_1}, ..., \vec{OP_n}$ are given in (1).*

(b) The content of a simplex

Consider 3 noncollinear points P_1, P_2, P_3 in a plane. Let the coordinates of these points be $P_1(x_{11}, x_{12})$, $P_2(x_{21}, x_{22})$, $P_3(x_{31}, x_{32})$. The area of this triangle can be computed by shifting the origin to one of the points, say P_3. When the origin is shifted to P_3 the other points, with reference to the new origin, are $P_1(x_{11} - x_{31}, x_{12} - x_{32})$ and $P_2(x_{21} - x_{31}, x_{22} - x_{32})$. Then the area of the triangle, A_2, is the area of the corresponding parallelogram divided by 2. That is,

$$A_2 = \frac{1}{2} \begin{vmatrix} x_{11} - x_{31} & x_{12} - x_{32} \\ x_{21} - x_{31} & x_{22} - x_{32} \end{vmatrix}_+ = \frac{1}{2} \begin{vmatrix} x_{11} & x_{12} & 1 \\ x_{21} & x_{22} & 1 \\ x_{31} & x_{32} & 1 \end{vmatrix}_+.$$

The last expression is more convenient since it gives the coordinates of the individual points on separate rows. Now consider 4 points in a 3-dimensional space such that no point lies on the points, lines and the plane generated by the other 3 points. Consider the solid body having these four points as vertices. All the sides of this body are triangles. Six such identical bodies can be packed into a parallelepiped. Then the volume content, by proceeding as above and denoting it by A_3, is given by

$$A_3 = \frac{1}{3!} \begin{vmatrix} x_{11} & x_{12} & x_{13} & 1 \\ x_{21} & x_{22} & x_{23} & 1 \\ x_{31} & x_{32} & x_{33} & 1 \\ x_{41} & x_{42} & x_{43} & 1 \end{vmatrix}_+.$$

Lemma 7.1.2 *Consider* $n+1$ *points in an n-dimensional Euclidean space* R^n, $P_j = P_j(x_{j1}, ..., x_{jn})$, $j = 1, ..., n+1$, *so that no point lies on the* 0-*dimensional*, 1-*dimensional*,...,$(n-1)$-*dimensional bodies created by the other n points. The convex hull created by these* $n+1$ *points is a simplex and the volume content, denoting by* A_n, *is given by*

$$A_n = \frac{1}{n!} \begin{vmatrix} x_{11} & \cdots & x_{1n} & 1 \\ x_{21} & \cdots & x_{2n} & 1 \\ \vdots & \vdots & \vdots & \vdots \\ x_{n1} & \cdots & x_{nn} & 1 \\ x_{n+1\,1} & \cdots & x_{n+1\,n} & 1 \end{vmatrix}_+ . \qquad (7.1.2)$$

From the expression for A_2 note that if all the sides of the triangle are equal to some number a then the area is $A_2 = \frac{\sqrt{3}}{4}a^2$.

As a corollary we have the following result: If all the edges of the regular simplex in Lemma 1.2 are of equal lengths a then

$$A_n = \frac{a^n}{n!} \left(\frac{n+1}{2^n} \right)^{\frac{1}{2}}.$$

Thus we have for

$$n = 1, \quad A_1 = a = \text{length of a line segment};$$

$$n = 2, \quad A_2 = \frac{\sqrt{3}}{4}a^2 = \text{area of an equilateral triangle};$$

$$n = 3, \quad A_3 = \frac{a^3}{6\sqrt{2}} = \text{the volume of a regular tetrahedron};$$

$$n = 4, \quad A_4 = \frac{\sqrt{5}}{96}a^4 = \text{the volume of a regular simplex in } R^4.$$

From (7.1.2) observe that by subtracting the last row from each row the following results are obtained:

$$A_n = \frac{1}{n!} \begin{vmatrix} x_{11} - x_{n+1\,1} & \cdots & x_{1n} - x_{n+1\,n} & 0 \\ \vdots & \vdots & \vdots & \vdots \\ x_{n1} - x_{n+1\,1} & \cdots & x_{nn} - x_{n+1\,n} & 0 \\ x_{n+1\,1} & \cdots & x_{n+1\,n} & 1 \end{vmatrix}$$

$$= \frac{1}{n!} \begin{vmatrix} x_{11} - x_{n+1\,1} & \cdots & x_{1n} - x_{n+1\,n} \\ \vdots & \vdots & \vdots \\ x_{n1} - x_{n+1\,1} & \cdots & x_{nn} - x_{n+1\,n} \end{vmatrix}$$

$$= \frac{1}{n!} |YY'|^{\frac{1}{2}} \qquad\qquad (7.1.3)$$

where $|YY'|$ denotes the determinant of YY' and Y is the matrix $Y = (y_{ij})$, $y_{ij} = x_{ij} - x_{n+1\,j}$. When the matrix Y is an identity matrix which is the same as saying that

$$(x_{i1} - x_{n+1\,1}, ..., x_{in} - x_{n+1\,n}) = (0, ..., 1, ...0),$$

the i-th unit vector, $i = 1, ..., n$ then

$$A_n = \frac{1}{n!}.$$

But the right side of (3), excluding $\frac{1}{n!}$, is the content or volume of an n-parallelotope formed in the rectangular coordinate system with the origin at the point P_{n+1}. Hence we have the following result:

Lemma 7.1.3 *Let X_j be an $n \times 1$ real vector, $j = 0, 1, ..., r$. Thus $X_0, X_1, ..., X_r$ are $r + 1$ points in the Euclidean n-space R^n, $r \le n$. Let Δ_n denote the volume content or r-content of the r-simplex created by the ordered and linearly independent points $X_0, ..., X_r$. Let ∇_n be the volume or the r-content of the r-parallelotope created by the vectors $X_j \overset{\rightarrow}{-} X_0$, for $j = 1, ..., r$ or the r points taking X_0 as the origin of the coordinate system. Then*

$$\nabla_n = r!\, \Delta_n, \quad \Delta_n = \frac{1}{r!} \nabla_n \qquad\qquad (7.1.4)$$

where

$$\nabla_n = |Y'Y|^{\frac{1}{2}}, \; Y = (Y_1, ..., Y_r), \; Y_j = X_j - X_0, \; j = 1, ..., r. \qquad (7.1.5)$$

Note that ∇_n^2 can be interpreted as Wilks' sample generalized variance when $E(Y) = O$, where E denotes the expected value and O the null matrix. Properties and applications of this concept of generalized variance may be seen from books on multivariate statistical analysis, see for example Anderson (1984). A critical examination of this concept and showing that it cannot be interpreted as a generalized variance may be seen from Mathai (1968).

7.1.2 Some Basic Results on Jacobians of Matrix Transformations

When dealing with the distributions and moments of Δ_n and ∇_n of (7.1.4) and (7.1.5) we will need to consider one-to-one transformations involving matrices and the associated Jacobians. Jacobians of frequently used matrix transformations are available from Mathai (1997). Some Jacobians that we need later on are listed here as lemmas. In the following discussion $\|(\cdot)\|$ denotes the norm or distance or length in $(\cdot)$, and the symmetric positive definiteness of a matrix X will be denoted by $X = X' > 0$. All matrices appearing are $p \times p$ real symmetric positive definite unless stated otherwise. $0 < A < X < B$ means $A = A' > 0$, $B = B' > 0$, $X = X' > 0$, $B - X > 0$, $X - A > 0$ and $\int_A^B f(X)\mathrm{d}X$ will mean the integral of the real scalar function f of the matrix X integrated over $0 < A < X = X' < B$ and $\mathrm{d}X$ denotes the volume element. That is, for $X = (x_{ij})$,

$$\mathrm{d}X = \wedge_{i,j}\mathrm{d}x_{ij}$$

if all x_{ij}'s are functionally independent and it is $\wedge_{i\geq j}\mathrm{d}x_{ij}$ if $X = X'$ and all the elements are functionally independent subject to the condition $X = X'$. A real matrix-variate gamma, $\Gamma_p(\alpha)$, is defined as

$$\Gamma_p(\alpha) \;=\; \pi^{p(p-1)/4}\, \Gamma(\alpha)\Gamma\left(\alpha - \frac{1}{2}\right)\Gamma(\alpha - 1)...\Gamma\left(\alpha - \frac{p-1}{2}\right),$$

$$\Re(\alpha) > \frac{p-1}{2} \qquad (7.1.6)$$

where $\Re(\cdot)$ denotes the real part of $(\cdot)$. It is easy to show , see for example Mathai (1993), that for a constant matrix $B = B' > 0$

$$|B|^{-\alpha}\Gamma_p(\alpha) \;=\; \int_{X=X'>0} |X|^{\alpha - \frac{p+1}{2}} e^{-\mathrm{tr}(BX)}\mathrm{d}X, \;\; \Re(\alpha) > \frac{p-1}{2}$$

$$(7.1.7)$$

where tr denotes the trace. When $B = I$, an identity matrix, (7.1.7) gives an integral representation for $\Gamma_p(\alpha)$ which can also be taken as a definition for $\Gamma_p(\alpha)$. The integral is evaluated with the help of the following two results. These and other results that are needed in our discussion will be stated as lemmas. A detailed treatment of the Jacobians of matrix transformations may be found in Mathai (1997).

Lemma 7.1.4 *Let X be a $p \times p$ symmetric positive definite matrix of functionally independent real variables. Let $T = (t_{ij})$ be a lower triangular* *matrix, $t_{ij} = 0$, $i < j$, with positive diagonal elements, $t_{jj} > 0$, $j = 1, ..., p$. Then,*

$$X = TT' \Rightarrow dX = 2^p \left\{ \prod_{j=1}^{p} t_{jj}^{p+1-j} \right\} dT. \qquad (7.1.8)$$

It is easy to see that the transformation is one to one when $t_{jj} > 0$, $j = 1, ..., p$.

Lemma 7.1.5 *Let X be a $p \times p$ symmetric matrix of functionally independent real variables and A a nonsingular constant matrix. Then, ignoring the sign,*

$$Y = AXA' \Rightarrow dY = |A|^{p+1} dX \qquad (7.1.9)$$

By using (7.1.7), (7.1.8) and (7.1.9) one can establish the following results, which are also known as the type-1 and type-2 real matrix-variate beta integrals.

$$B_p(\alpha, \beta) = \frac{\Gamma_p(\alpha)\Gamma_p(\beta)}{\Gamma_p(\alpha + \beta)} = \int_{0 < X = X' < I} |X|^{\alpha - \frac{p+1}{2}} |I - X|^{\beta - \frac{p+1}{2}} dX \qquad (7.1.10)$$

$$= \int_{X = X' > 0} |X|^{\alpha - \frac{p+1}{2}} |I + X|^{-(\alpha + \beta)} dX \qquad (7.1.11)$$

for $\Re(\alpha) > \frac{p-1}{2}$, $\Re(\beta) > \frac{p-1}{2}$.

When $X_j \in R^n$, $j = 1, ..., p$ are mutually independently distributed as standard normal or Gaussian, $X_j \sim N_n(0, I_n)$, then the $n \times p$ matrix $X = (X_1, ..., X_p)$, $n \geq p$, of full rank p, has a standard matrix-variate Gaussian distribution with the density

$$f(X) = c \exp\{-\frac{1}{2} \text{tr}(X'X)\}. \qquad (7.1.12)$$

The normalizing constant $c = (2\pi)^{-np/2}$ can be obtained either by directly integrating x_{ij}'s over the real line $-\infty < x_{ij} < \infty$, observing that $\text{tr}(X'X) = \sum_{i,j} x_{ij}^2$, or with the help of the following result:

Lemma 7.1.6 *Let X be an $n \times p$, $n \geq p$, matrix of functionally independent np real variables. Let $T = (t_{ij})$ be a lower triangular matrix with positive* *diagonal elements. Let U be an $n \times p$, $n \geq p$, semiorthonormal matrix, $U'U = I_p$ or U in the Stiefel manifold $V_{p,n}$, $U \in V_{p,n}$. Then*

$$X' = TU' \Rightarrow dX = \left\{ \prod_{j=1}^{p} t_{jj}^{n-j} \right\} dT \, dU^* \qquad (7.1.13)$$

where

$$dU^* = \wedge_{j=1}^{p} \wedge_{k+j+1}^{n} U_k'(dU_j), \qquad (7.1.14)$$

(dU_j) *indicates the $n \times 1$ vector of differentials of the elements of U_j, the j-th column of U.*

If the density in (7.1.12) is now evaluated by transforming X to T and U and integrating out $-\infty < t_{ij} < \infty$, $i > j$, $0 < t_{jj} < \infty$, $j = 1, ..., p$ then one has the following result:

$$\int_{V_{p,n}} dU^* = \frac{2^p \pi^{np/2}}{\Gamma_p\left(\frac{n}{2}\right)} \qquad (7.1.15)$$

If $n = p$ then $V_{p,n}$ is the full orthogonal group $O(p)$ and

$$\int_{O(p)} dU^* = \frac{2^p \pi^{p^2/2}}{\Gamma_p\left(\frac{n}{2}\right)}. \qquad (7.1.16)$$

From Lemma 7.1.4, and equations (7.1.13) and (7.1.14) we have the following result:

Lemma 7.1.7 *Let X, T and U be as defined in Lemma 7.1.6 and let $A = X'X$. Then*

$$dX = 2^{-p}|A|^{\frac{n}{2} - \frac{p+1}{2}} dA \, dU^* \qquad (7.1.17)$$

where dU^* is defined in (7.1.14), with the integral in (7.1.15).

7.1.3 Some Practical Situations

The two-dimensional simplex is a triangle and some of the practical situations where random triangles appear will be mentioned here.

(a) Density of plants in a forest

Any study of the density of plants in a forest requires measuring the distance from a plant to the nearest plant. Trees are assumed to draw the nutrients from circular areas centered at their bases. Thus in the long-run a natural forest will have trees at the lattice points of equilateral triangles of sides l where l is the distance between the centres of tightly packed circles of radius $l/2$ each. Study of such problems in forestry may be seen from the references in Holgate (1965).

(b) Growth of crystals

Consider the problem of crystal growth on a wire of length l. Suppose that at random locations on the wire and at random times crystals are born along the wire. These can be taken initially as random points on a line segment of length l, each crystal growing with constant speed v on both sides in such a way that when a crystal touches its neighbor the growth stops on that side where the contact is made. New crystals can be born only on uncovered parts of the wire. We can show that the problem is equivalent to studying random points in an isosceles triangle if we start with one crystal at each end of the wire. This and other related problems are examined in Krengel (1967) for the cases when the crystals are born at random points on the wire and when the points arrive according to a Poisson process.

(c) Travel distance in a triangular city core

Another situation that one can consider is the problem of travel distances to a city core which is triangular in nature. A vehicle arrives at the entry point to the city core. This entry point could be one of the vertices of the triangle or a point on one of the sides. Taking into consideration all possible destination points inside the city core a typical travel can be taken as the random path from a vertex or from a random point on a side to a random point inside the triangle. In this case the triangle is fixed and one is interested in random points inside this triangle. Items of interest will then be the expected travel distance and the density and moments of this random distance. Such problems of travel distances in circular cities are considered by many authors, see some of the references from Mathai (1998b,d) and Mathai and Moschopoulos (1998). Random distances in Gaussian random fields may be seen from Provost and Barnwal (1993) and Provost and Cheong (1997).

(d) Sylvester problem

The famous problem known in the geometrical probability literature as Sylvester's four-point problem is the problem of evaluating the probability that four points selected at random in a given finite plane convex region, such as a circle of finite radius in a plane, form a convex quadrilateral. This can be evaluated by computing the probability that the fourth point selected at random is inside the triangle formed by the first three randomly selected points. In this case the quadrilateral is not convex. A generalization of the problem will be to compute the probability that a point selected at random within a well-defined convex region in R^n will fall inside the random simplex formed by $n + 1$ random points in R^n.

The above ones are a few of the practical problems where one has to deal with a random simplex or the volume content, in the above examples the areas, of a random simplex. Specific cases when the vertices of the simplex have some specified distributions are considered by many authors. We will consider some very general families of distributions for the vertices and then look at the distribution of the volume contents of such random simplices and parallelotopes. Miles (1971), Ruben (1979) and Ruben and Miles (1980) considered Gaussian distributed independent random points. When X_j, $j = 1, ..., p$ are iid (independently and identically distributed) real random $n \times 1$ vectors with $E(X_j) = 0$ then the determinant $|X'X|$ is the square of the random p-content of the p-parallelotope in R^n. This remains invariant under orthogonal transformations or under rotations of the orthogonal coordinate axes. Independence and isotropy of the random points in R^n, not necessarily identically distributed, are usually the basic assumptions in the literature when dealing with the distributions of random p-contents. Various techniques available in the literature use almost exclusively results from differential and integral geometry, see for example, Coleman (1969), Kendall and Moran (1963), Kingman (1969), Ruben (1979), Santaló (1976) and Solomon (1978). Some aspects of spherically symmetric and elliptically contoured distributions are available in Fang and Anderson (1990), Fang, Kotz and Ng (1990) and Fang and Zhang (1990).

In the present paper we consider the case of the joint distribution of the elements of the $n \times p$ matrix X belonging to some general classes of distributions, where the X_j's need not be independently or identically distributed. Particular cases such as independently distributed isotropic random points are shown to give rise to the results obtained by Ruben (1979) and Miles (1971). Our methods will be based on algebraic procedures making use of the lemmas listed above and no results from integral geometry and decomposition of measures will be used. One major drawback for procedures based on integral geometry seems to be that only integer moments, that is for $h = 0, 1, ...$ when the h-th moment of the p-content is evaluated, are coming out from such techniques. Our procedure will provide the h-th moment for an arbitrary h, real or complex, whenever the moment exists.

7.2   DISTRIBUTION OF THE VOLUME OR CONTENT OF A RANDOM PARALLELOTOPE IN R^n

Let X_j, $j = 1, ..., p$ be an ordered set of random points in R^n for a prefixed p and $X = (X_1, ..., X_p)$ the $n \times p$, $n \geq p$, matrix of rank p. Let $U = (U_1, ..., U_p)$ be an $n \times p$ semiorthonormal matrix, $U'U = I_p$, $U \in V_{p,n}$. Let $T = (t_{ij})$ be a lower triangular matrix with positive diagonal elements. Then from elementary considerations the following results are easily established.

$$X' = TU' \Rightarrow X_1' = t_{11}U_1' \Rightarrow \|X_1\| = t_{11} \Rightarrow U_1 = \frac{X_1}{\|X_1\|} \quad (7.2.18)$$

where $\|(\cdot)\|$ denotes the length of $(\cdot)$. Again, from the transformation we have

$$X_2' = t_{21}U_1' + t_{22}U_2' \Rightarrow t_{21} = X_2'U_1 = \frac{X_2'X_1}{\|X_1\|}$$

which then yields

$$t_{22}U_2' = X_2' - \left[\frac{X_2'X_1}{\|X_1\|^2}\right]X_1'.$$

Let

$$Z_1 = X_1 = t_{11}U_1, \ Z_2 = t_{22}U_2, \ ...$$

Then the transformation in (7.2.18) implies the Gram-Schmidt orthogonalization process, producing a set of p mutually orthogonal $n \times 1$ vectors $Z_1, ..., Z_p$ where $\|Z_j\| = t_{jj}$, $j = 1, ..., p$. Then the p-content $\nabla_{p,n}$ of the p-parallelotope in R^n determined by the vectors $\vec{OX}_1, ..., \vec{OX}_p$, which is also $p!$ times the p-content $\Delta_{p,n}$ of the p-simplex with the vertices $O, X_1, .., X_p$, is given by

$$\nabla_{p,n} = p! \, \Delta_{p,n} = |X'X|^{\frac{1}{2}} = |T'U'UT|^{\frac{1}{2}} \quad (7.2.19)$$

$$= |TT'|^{\frac{1}{2}} = t_{11}t_{22}...t_{pp} = \|Z_1\|...\|Z_p\|. \quad (7.2.20)$$

Hence an arbitrary h-th moment of $\nabla_{p,n}$ is available from the h-th moment of the product $t_{11}...t_{pp}$. The exact distribution of $\nabla_{p,n}$ is available from that of $t_{11}...t_{pp}$.

One of the simplest cases that we can deal with is when the joint distribution of the elements in the real $n \times p$, $n \geq p$, matrix X has an absolutely continuous distribution with the density function $f(X)$ which can be expressed as a function of $X'X$. In Sections 7.2 and 7.3 we consider the case of a fixed or preselected p. Let

$$f(X) = g(X'X)$$

where $g(X'X) > 0$ with probability 1 on the support of $X'X$. Let the columns of X be linearly independent so that X is of full rank p. Let $W = X'X$. With the help of the lemmas we have

$$
\begin{aligned}
1 &= \int_X f(X)\mathrm{d}X = \int_X g(W)\mathrm{d}X \\
&= \frac{\pi^{np/2}}{\Gamma_p\left(\frac{n}{2}\right)} \int_{W=W'>0} |W|^{\frac{n}{2}-\frac{p+1}{2}} g(W)\mathrm{d}W
\end{aligned}
$$

and

$$E_n|W|^h = \frac{\pi^{np/2}}{\Gamma_p\left(\frac{n}{2}\right)} \int_{W>0} |W|^{\frac{n}{2}+h-\frac{p+1}{2}} g(W)\mathrm{d}W \qquad (7.2.21)$$

provided the integral on the right converges, where $E_n|W|^h$ denotes the h-th moment of $|W|$ in the density of X. As an example, if

$$f(X) = c\,|X'X|^\alpha\, e^{-\mathrm{tr}(BX'X)}, \quad B = B' > 0, \quad \Re(\alpha) > -1 \qquad (7.2.22)$$

where c is the normalizing constant and B is a constant matrix, then from (7.2.21) we have the following result:

Theorem 7.2.1 *When $W = X'X$ with X having the density in (7.2.22) then the h-th moment of $|W|$ in the density (7.2.22), $E_n|W|^h$, is given by*

$$E_n|W|^h = |B|^{-h}\frac{\Gamma_p\left(\alpha+\frac{n}{2}+h\right)}{\Gamma_p\left(\alpha+\frac{n}{2}\right)}, \quad \Re(h) > -\frac{n}{2}+\frac{p-1}{2}-\Re(\alpha).$$

When the columns of X are independently distributed standard normal then $\alpha = 0$, $B = \frac{1}{2}I_n$ and in this case Theorem 7.2.1 gives

$$E_n|W|^h = 2^{ph}\frac{\Gamma_p\left(\frac{n}{2}+h\right)}{\Gamma_p\left(\frac{n}{2}\right)}, \quad \Re(h) > -\frac{n}{2}+\frac{p-1}{2}$$

which is the h-th moment of what is known in the statistical literature as Wilks' generalized variance. From Theorem 7.2.1 observe that

$$E_n|BW|^h = \frac{\Gamma_p\left(\alpha + \frac{n}{2} + h\right)}{\Gamma_p\left(\alpha + \frac{n}{2}\right)}$$

$$= \prod_{j=1}^{p} \frac{\Gamma\left(\alpha + \frac{n}{2} - \frac{j-1}{2} + h\right)}{\Gamma\left(\alpha + \frac{n}{2} - \frac{j-1}{2}\right)} = \prod_{j=1}^{p} E(w_j^h)$$

where w_j is a real scalar gamma random variable with the shape parameter $\alpha + \frac{n}{2} - \frac{j-1}{2}$ and the scale parameter 1. Hence $|BW|$ is structurally a product of p independent real gamma variables and the exact density of such a product is available in terms of a Meijer's G-function or in explicit series form, see Mathai (1993). For the sake of completeness we will list this result as a theorem.

Theorem 7.2.2 *When $W = X'X$ with X having the density in (7.2.22) then $y = |BW|$ has the density, denoted by $f_y(y)$, given by*

$$f_y(y) = \left\{ \prod_{j=1}^{p} \Gamma\left(\alpha + \frac{n}{2} - \frac{j-1}{2}\right) \right\}^{-1} y^{-1} G_{0,p}^{p,0}[y|_{\alpha + \frac{n}{2} - \frac{j-1}{2}, \ j=1,...,p}],$$

$0 < y < \infty$ *and zero elsewhere.*

Theory and applications of G and H-functions are available from Mathai and Saxena (1973, 1978) and Mathai (1993). Since the definition of G or H-function will take up too much space it will not be given here.

7.2.1   Matrix-Variate Type–1 Beta Distribution

Many types of matrix-variate distributions are discussed in Mathai (1997). One of the simplest of such distributions is a type–1 beta distribution. Let the $n \times p$, $n \geq p$, real random matrix X of full rank p have the density

$$f(X) = c\,|X'X|^\alpha|I - X'X|^{\beta - \frac{p+1}{2}} \tag{7.2.23}$$

for $0 < X'X < I$ with probability 1, $\Re(\alpha) > -1$, $\Re(\beta) > \frac{p-1}{2}$ where c is the normalizing constant. Then from the lemmas we have the following result:

Theorem 7.2.3 *Let $W = X'X$ with X having the density in (7.2.23) and let $E_n|W|^h$ denote the h-th moment of $|W|$ in this density. Then*

$$E_n|W|^h = \frac{\Gamma_p\left(\alpha + \frac{n}{2} + h\right)}{\Gamma_p\left(\alpha + \frac{n}{2}\right)} \frac{\Gamma_p\left(\alpha + \beta + \frac{n}{2}\right)}{\Gamma_p\left(\alpha + \beta + \frac{n}{2} + h\right)}$$

$$\Re(h) > -\frac{n}{2} + \frac{p-1}{2} - \Re(\alpha).$$

The right side can be written as the h-th moment of a product of p independent real scalar type-1 beta random variables. That is,

$$E_n|W|^h = \prod_{j=1}^{p} E(w_j^h)$$

where w_j is type-1 beta with the parameters $\left(\alpha + \frac{n}{2} - \frac{j-1}{2}, \beta\right)$. Thus, structurally, $|W|$ is a product of the independent variables $w_1, ..., w_p$. The exact density of such a structure can be written as a G-function.

Theorem 7.2.4 *For the W in Theorem 7.2.3 let $z = |W|$ with the density of z denoted by $f_z(z)$. Then*

$$f_z(z) = \frac{\Gamma_p\left(\alpha + \beta + \frac{n}{2}\right)}{\Gamma_p\left(\alpha + \frac{n}{2}\right)} z^{-1} G_{p,p}^{p,0}\left[z \Big|_{\alpha + \frac{n}{2} - \frac{i-1}{2}, \; j=1,...,p}^{\alpha + \beta + \frac{n}{2} - \frac{i-1}{2}, \; j=1,...,p}\right], \quad 0 \le z \le 1$$

and zero elsewhere.

7.2.2 Matrix-Variate Type–2 Beta Density

Another simple case one can consider is when the density of X belongs to a general type–2 beta family. Let the $n \times p$, $n \ge p$, real random matrix of rank p have the density

$$f(X) = c\, |X'X|^\alpha |I + X'X|^{-(\alpha+\beta)} \tag{7.2.24}$$

with $X'X > 0$ almost surely, $\Re(\alpha) > -1$, $\Re(\beta) > p$ and c is the normalizing constant. Then from the lemmas we have

Theorem 7.2.5 *Let X have the density in (7.2.24), $W = X'X$ and $E_n|W|^h$ the h-th moment of $|W|$ in (7.2.24). Then*

$$E_n|W|^h = \frac{\Gamma_p\left(\alpha + \frac{n}{2} + h\right)}{\Gamma_p\left(\alpha + \frac{n}{2}\right)} \frac{\Gamma_p\left(\beta - \frac{n}{2} - h\right)}{\Gamma_p\left(\beta - \frac{n}{2}\right)},$$

$$-\frac{n}{2} + \frac{p-1}{2} - \Re(\alpha) < \Re(h) < \Re(\beta) - \frac{n}{2} - \frac{p-1}{2}.$$

Thus, only a few moments satisfying the above inequality will exist. From the structure of the gamma product in $E_n|W|^h$ of Theorem 7.2.5 it is easy to note that $|A|$ is structurally a product of p independent real scalar type-2 beta random variables and then the exact density of $|W|$ can be written as follows:

Theorem 7.2.6 *When the density of X is as in (24) with the moment expression given in Theorem 7.2.5 the exact density of $x = |W|$ is given by*

$$f_x(x) = \left[\Gamma_p\left(\alpha + \frac{n}{2}\right)\Gamma_p\left(\beta - \frac{n}{2}\right)\right]^{-1} x^{-1} G_{p,p}^{p,p}\left[x\,\Big|\,{\substack{-\beta + \frac{n}{2} + \frac{i+1}{2},\ j=1,\dots,p \\ \alpha + \frac{n}{2} - \frac{j-1}{2},\ j=1,\dots,p}}\right],$$

$0 < x < \infty$ *and zero elsewhere.*

Various types of computable series representations for G-functions of the types $G_{0,p}^{p,0}$, $G_{p,p}^{p,0}$, $G_{p,p}^{p,p}$, and other forms, are available from Mathai (1984) as solutions of certain integral equations.

7.3 SPHERICALLY SYMMETRIC DISTRIBUTIONS

Let the $n \times 1$ real vector random variable X_j have a spherically symmetric distribution then, by definition, the density of X_j can be written as a positive function of $\|X_j\|^2 = X_j'X_j$. Then going through a triangular decomposition for $X = (X_1, \dots, X_p)$ as in (18) we have

$$X_j'X_j = t_{j1}^2 + \dots + t_{jj}^2.$$

As an example let the p columns X_j, $j = 1, \dots, p$ of the $n \times p$, $n \geq p$, matrix X be independently distributed with X_j having the density of the form

$$f_j(X_j) = c_j(1 - X_j'X_j)^{\beta_j - 1}, \ \beta_j > 0, \ 0 < X_j'X_j < 1, \quad (7.3.25)$$

and zero elsewhere, where c_j is the normalizing constant. This is a multivariate type-1 beta form. Note that

$$f_j(X_j) = c_j[1 - (t_{j1}^2 + \dots + t_{jj}^2)]^{\beta_j - 1}. \quad (7.3.26)$$

The procedure goes through also when $f_j(X_j)$ contains a factor of the type $\prod_{k=1}^{j}(t_{jk}^2)^{\gamma_{jk}}$ with $\gamma_{jk} \geq 0$, $k = 1, \dots, j-1$. Hence the condition that X_j, $j = 1, \dots, p$ are independent and isotropic is not necessary in this example as long as the joint density of $X_1, \dots, X_p$ can be written finally in the form

$$f(X) = c \prod_{j=1}^{p}\left\{\prod_{k=1}^{j}(t_{jk}^2)^{\gamma_{jk}}\right\}[d_j - (t_{j1}^2 + \dots + t_{jj}^2)]^{\beta_j - 1} \quad (7.3.27)$$

where c is the normalizing constant, $\gamma_{jk} \geq 0$, $d_j > 0$, $\beta_j > 0$, $(d_j - t_{j1}^2 - \ldots - t_{jj}^2) > 0$ for all j and k.

In (7.3.26), $-1 < t_{jk} < 1$, $k = 1, \ldots, j - 1$, $0 < t_{jj} < 1$ as per the assumptions in the matrix T. Integrating out t_{j1} from (7.3.26) gives the following: For convenience write $t_{j1}^2 + \ldots + t_{jj}^2 = t_{12\ldots j}^2$. Then

$$\int_{-1}^{1} [1 - t_{12\ldots j}^2]^{\beta_j - 1} dt_{j1}$$

$$= 2 \int_{0}^{1} [1 - t_{12\ldots j}^2]^{\beta_j - 1} dt_{j1}$$

$$= 2[1 - t_{2\ldots j}^2]^{\beta_j - 1} \int_{0}^{\sqrt{1 - t_{2\ldots j}^2}} \left[1 - \frac{t_{j1}^2}{1 - t_{2\ldots j}^2} \right]^{\beta_j - 1} dt_{j1}$$

$$= \frac{\Gamma\left(\frac{1}{2}\right) \Gamma(\beta_j)}{\Gamma\left(\beta_j + \frac{1}{2}\right)} [1 - t_{2\ldots j}^2]^{\beta_j + \frac{1}{2} - 1}.$$

Successive integrations of $t_{j1}, \ldots, t_{jj-1}$ yield the final form

$$(\sqrt{\pi})^{j-1} \frac{\Gamma(\beta_j)}{\Gamma\left(\beta_j + \frac{j-1}{2}\right)} [1 - t_{jj}^2]^{\beta_j + \frac{j-1}{2} - 1}. \qquad (7.3.28)$$

Multiplying (7.3.28) by t_{jj}^{n-j+2h} and integrating gives

$$\gamma_j \frac{\Gamma\left(\frac{n}{2} - \frac{j-1}{2} + h\right)}{\Gamma\left(\frac{n}{2} + \beta_j + h\right)}, \quad h > -\frac{n}{2} + \frac{j-1}{2} \qquad (7.3.29)$$

where γ_j depends only on j and it is free of h. Then we have

Theorem 7.3.1 *Let the columns X_j, $j = 1, \ldots, p$ of the $n \times p$, $n \geq p$, matrix X be independently distributed and having the density in (7.3.25). Let $W = X'X$. Then the h-th moment of $|W|$ is given by*

$$E_n |W|^h = \prod_{j=1}^{p} \left\{ \frac{\Gamma\left(\frac{n}{2} - \frac{j-1}{2} + h\right)}{\Gamma\left(\frac{n}{2} - \frac{j-1}{2}\right)} \frac{\Gamma\left(\frac{n}{2} + \beta_j\right)}{\Gamma\left(\frac{n}{2} + \beta_j + h\right)} \right\},$$

$$\Re(h) > -\frac{n}{2} + \frac{p-1}{2}. \qquad (7.3.30)$$

From the right side in (7.3.30) it is clear that $|W|$ is structurally a product of p independent real type-1 beta random variables with the parameters $\left(\frac{n}{2} - \frac{j-1}{2}, \beta_j + \frac{j-1}{2}\right)$, $j = 1, \ldots, p$. (7.3.30) is Equation (33) of Ruben (1979) who obtained it with the help of several results from integral geometry. As indicated earlier, the density of $|W|$ of (7.3.30) can be written as a G-function of the type $G_{p,p}^{p,0}$. Several results of the type in Theorem 7.3.1, covering the cases when X_j's have some general classes of distributions, are worked out in Mathai (1998c) by using only Jacobians of matrix transformations and some properties of functions of matrix argument. Hence further cases of this type will not be given here.

7.4   ARRIVAL OF POINTS BY A POISSON PROCESS

When points arrive into a well-defined region according to some random process then the moments and densities obtained in Sections 7.2 and 7.3 can be taken as conditional moments and densities with p given. Consider the case of Poisson arrivals of points. Let p have the following probability function:

$$f_p(p) \;=\; \frac{\lambda^p}{p!}\, e^{-\lambda}, \; \lambda > 0, \; p = 0, 1, ... \qquad (7.4.31)$$

and zero elsewhere. With a pre-selected rectangular coordinate system we need at least one point to create a non-zero length or volume. If no such pre-selected system is there then we need at least two points to create a non-zero length or volume for the p-parallelotope and p-simplex. If points arrive according to (7.4.31) then the h-th moment in (7.3.30) is the conditional h-th moment of $|W|$, given p. The unconditional h-th moment of $|W|$ is given by the following, observing that when $p = 0$ the volume is zero:

$$
\begin{aligned}
E_n|W|^h \;&=\; E_n[E_n(|W|^h|p)] \\
&=\; \sum_{p=1}^{\infty} \frac{\lambda^p}{p!}\, e^{-\lambda} \left\{ \prod_{j=1}^{p} \frac{\Gamma\left(\frac{n}{2} - \frac{j-1}{2} + h\right) \Gamma\left(\frac{n}{2} + \beta_j\right)}{\Gamma\left(\frac{n}{2} - \frac{j-1}{2}\right) \Gamma\left(\frac{n}{2} + \beta_j + h\right)} \right\}.
\end{aligned}
$$
$$\qquad (7.4.32)$$

Since the summation in (7.4.32) is over p and since the product of gammas is also over p we cannot simplify (7.4.32) in terms of some standard special functions. New categories of functions are to be defined if (7.4.32) is to be written in a more simplified form. All the results in Sections 7.2 and 7.3 can be rewritten in the light of the above discussion if p is a random variable having its own distribution. Further discussion of this aspect will not be done here.

REFERENCES

Anderson, T. W. (1984). *An Introduction to Multivariate Statistical Analysis*, John Wiley & Sons, New York.

Coleman, R. (1969). Random path through convex bodies, *Journal of Applied Probability*, **6**, 430–441.

Fang, K.-I. and Anderson. T. W. (1990). *Statistical Inference in Elliptically Contoured and Related Distributions*, Allerton Press, New York.

Fang, K.-T., Kotz, S., and Ng, K. W. (1990). *Symmetric Multivariate and Related Distributions. Monograph on Statistics and Applied Probability*, Chapman and Hall, New York.

Fang, K.-T. and Zhang, Y.-T. (1990). *Generalized Multivariate Analysis*, Springer-Verlag, Berlin.

Holgate, P. (1965). The distance from a random point to the nearest point of a closely packed lattice, *Biometrika*, **52**, 261–263.

Kendall, M. G. and Moran. P. A. P. (1963). *Geometrical Probability*, Griffin, London.

Kingman, J. F. C. (1969). Random secants of a convex body, *Journal of Applied Probability*, **6**, 660–672.

Krengel, U. (1967). A problem of random points in a triangle, *The American Mathematical Monthly*, **74**, 8–14.

Mathai, A. M. (1968). Some limit theorems in terms of dispersion, *Metron*, **27**, 125–135.

Mathai, A. M. (1982). On a conjecture in geometric probability regarding asymptotic normality of a random simplex, *Annals of Probability*, **10**, 247–251.

Mathai, A. M. (1984). Extensions of Wilks' integral equations and distributions of test statistics, *Annals of the Institute of Statistical Mathematics*, **36**, 271–288.

Mathai, A. M. (1993). *A Handbook of Generalized Special Functions for Statistical and Physical Sciences*, Oxford University Press, Oxford.

Mathai, A. M. (1997). *Jacobians of Matrix Transformations and Functions of Matrix Argument*, World Scientific Publishing, New York.

Mathai, A. M. (1998a). *An Introduction to Geometrical Probability: Distributional Aspects with Applications*, Gordon and Breach, Newark.

Mathai, A. M. (1998b). Random distances associated with triangles, *International Journal of Mathematical and Statistical Sciences*, 7, 77–96.

Mathai, A. M. (1998c). Random p-content of a p-parallelotope in Euclidean n-space, *Advances in Applied Probability*, 31 (to appear).

Mathai, A. M. (1998d). Pollution by vehicular travels from the suburbs to the city core, *Environmetrics*, 9, 617–628.

Mathai, A. M. and Moschopoulos, P. G. (1998). Pollution by vehicular travels from satellite townships to the city, *Environmetrics* (to appear).

Mathai, A. M. and Saxena, R. K. (1973). *Generalized Hypergeometric Functions with Applications in Statistics and Physical Sciences*, Lecture Notes No. 348. Springer-Verlag, Heidelberg.

Mathai, A. M. and Saxena, R. K. (1978). *The H-function with Applications in Statistics and Other Disciplines*, Wiley Halsted, New York.

Miles, R. E. (1971). Isotropic random simplices, *Advances in Applied Probability*, 3, 353–382.

Provost, S. B. and Barnwal, R. K. (1993). A probabilistic model for the determination of acid rain levels, *Water Pollution Research Journal of Canada*, 28, 337–353.

Provost, S. B. and Cheong, Y.-H. (1997). The probability content of cones in isotropic random fields, *Journal of Multivariate Analysis*, 66, 237–254.

Ruben, H. (1979). The volume of an isotropic random parallelotope, *Journal of Applied Probability*, 16, 84–94.

Ruben, H. and Miles, R. E. (1980). A canonical decomposition of the probability measure of sets of isotropic random points in R^n, *Journal of Multivariate Analysis*, 10, 1–18.

Santaló, L. A. (1976). *Integral Geometry and Geometric Probability*, Addison-Wesley, Reading.

Solomon, H. (1978). *Geometric Probability*, SIAM, Philadelphia.

CHAPTER 8

THE DISTRIBUTION OF FUNCTIONS OF ELLIPTICALLY CONTOURED VECTORS IN TERMS OF THEIR GAUSSIAN COUNTERPARTS

YOUNG-HO CHEONG   SERGE B. PROVOST

The University of Western Ontario, London, Ontario, Canada

Abstract: No general representations of the density or the distribution function of quadratic forms in elliptically contoured vectors are currently available in the literature. As a result, the random vectors associated with quadratic forms have been assumed—almost invariably and at times inappropriately—to follow a multivariate normal distribution in statistical applications. It is shown in this paper that a certain scale mixture representation of the density function of a central elliptically contoured vector, as well as its extension to non-central random vectors yield computable expressions for the moments, the density and the distribution function of quadratic forms in elliptically contoured vectors. The same approach can readily be used to determine the distribution of other functions of elliptically contoured vectors in terms of their Gaussian counterparts.

Key words and phrases: Elliptically contoured distributions, spherically symmetric distributions, quadratic forms, exact distribution, moments

8.1 INTRODUCTION AND NOTATION

The class of elliptically contoured distributions which includes the multivariate normal distribution possesses several of its properties while allowing for increased flexibility in modeling random processes. Several fields of application involve distributions belonging to this class, including for example, filtering and stochastic control: Chu (1973); random input signal: McGraw and Wagner (1968); financial analysis: Zellner (1976) and the references therein; the analysis of stock market data: Mandelbrot (1963) and Fama (1965); and Bayesian Kalman filtering: Girón and Rojano (1994). Studies on the robustness of statistical procedures when the probability model departs from the multivariate normal distribution to the broader class of elliptically contoured distributions were carried out by King (1980) and Osiewalski and Steel (1993). Results related to regression analysis can be found for example in Fraser and Ng (1980). Several multivariate applications are also discussed in Devlin, Gnanadesikan and Kettenring (1976). Heavy-tailed time series models were recently discussed in Resnick (1997).

A p-dimensional vector $\mathbf{X}$ has an *elliptically contoured (or elliptical) distribution* with mean vector $\boldsymbol{\mu}$ and scale parameter matrix Σ if its characteristic function $\phi(\mathbf{t})$ can be written as

$$\phi(\mathbf{t}) = e^{i\mathbf{t}'\boldsymbol{\mu}}\xi(\mathbf{t}'\Sigma\,\mathbf{t})$$

where $\boldsymbol{\mu}$ is a p-dimensional real vector, Σ is a $p \times p$ nonnegative definite matrix and $\xi(\cdot)$ is a nonnegative function, see Cambanis, Huang and Simons (1981); this will be denoted

$$\mathbf{X} \sim \mathcal{C}_p(\boldsymbol{\mu}, \Sigma, \xi).$$

Moreover, the densities associated with elliptically contoured vectors $\mathbf{X}$ are of the form $h((\mathbf{x} - \boldsymbol{\mu})'\Sigma^{-1}(\mathbf{x} - \boldsymbol{\mu}))$ where $h(\cdot)$ is a density function defined on $(0, \infty)$ whose $(p/2 - 1)$-th moment exists, see Fang, Kotz and Ng (1990). In particular, when $\boldsymbol{\mu}$ is the null vector and Σ is the identity matrix of order p, $\mathbf{X}$ is said to have a *spherically symmetric (or spherical) distribution*; this will be denoted

$$\mathbf{X} \sim \mathcal{S}_p(\xi).$$

In fact, whenever $\mathbf{Y} \sim \mathcal{C}_p(\boldsymbol{\mu}, \Sigma, \xi)$ and Σ is a positive definite matrix, $\Sigma^{-\frac{1}{2}}(\mathbf{Y} - \boldsymbol{\mu}) \sim \mathcal{S}_p(\xi)$, where $\Sigma^{-1/2}$ denotes the inverse of the symmetric square root of Σ. Furthermore, spherical distributions are invariant under orthogonal transformations: For any orthogonal matrix P, $\mathbf{X}$ and $P\mathbf{X}$ are identically distributed—this will be denoted $\mathbf{X} \simeq P\mathbf{X}$. Other characterizations and properties of elliptical distributions are available from Chmielewski (1981), Fang, Kotz and Ng (1990) and Mathai, Provost and Hayakawa (1995), among others.

A representation of the density function of a central elliptically contoured vector in terms of a scale mixture of normal variates involving a certain weighting function is given in Section 8.2 and used in Section 8.3 to obtain integral representations for both the density and the distribution function of quadratic forms in elliptically contoured vectors. Computable expressions for the moments of quadratic forms are obtained from an extension of a theorem due to Chu (1973) in Section 8.4. The proposed technique can be readily used to obtain similar distributional results for other functions of elliptically distributed vectors in terms of their Gaussian counterparts. A numerical example is provided in Section 8.5.

The distributional results obtained in this paper for quadratic forms in elliptically contoured random vectors not only extend but make use of their Gaussian counterparts. As pointed out at the beginning of this section, elliptically contoured distributions are used as models in a host of applications. Quadratic forms being ubiquitous in statistics, the results derived in this paper should prove useful in a variety of contexts and lead to the development of improved statistical inference techniques.

8.2 A REPRESENTATION OF THE DENSITY FUNCTION OF ELLIPTICAL VECTORS

Chu (1973) showed that the density of a central elliptically contoured random vector has the mixture representation given in the following theorem.

Theorem 8.2.1 *If* $\mathbf{X}$ *is a p-dimensional central elliptically contoured random vector with scale parameter matrix* Σ *and density function* $g(\mathbf{x})$, *then, under certain regularity conditions, there exists a scalar function* $w(t)$ *defined on* $(0, \infty)$ *such that*

$$g(\mathbf{x}) = \int_0^\infty w(t)\eta_{\mathbf{x}}(t^{-1}\Sigma)dt$$

where $\eta_{\mathbf{X}}(t^{-1}\Sigma)$ *denotes the density function of* $\mathbf{X} \sim N_p(\mathbf{0}, t^{-1}\Sigma)$, *a p-dimensional Gaussian random vector with mean* $\mathbf{0}$ *and covariance matrix* $t^{-1}\Sigma$, *and*

$$w(t) = (2\pi)^{p/2}|\Sigma|^{1/2}t^{-p/2}\mathcal{L}^{-1}(f(s)),$$

$\mathcal{L}^{-1}(f(s))$ *denoting the inverse Laplace transform of* $f(s)$ *with* $f(s) = g(\mathbf{x})$ *when* $s = \mathbf{x}'\Sigma^{-1}\mathbf{x}/2$.

In fact, $\mathcal{L}^{-1}(f(s))$ exists whenever $f(s)$ is an analytic function and $f(s)$ is $O(s^{-k})$ as $s \to \infty$ for $k > 1$; for additional properties of the Laplace transform and its inverse, one may refer to Gradshteyn and Ryzhik (1980, Chapter 17).

On integrating $g(\mathbf{x})$ as defined in Theorem 8.2.1 over $\mathcal{R}^p$ and changing the order of integration, one can easily show that $w(t)$ integrates to 1. Hence, $w(t)$ can be regarded as a weighting function. The weighting functions associated with certain p-dimensional elliptically contoured distributions are given explicitly in Table 8.1. When $w(t)$ is a non-negative function (as in most cases of interest), it is a density function. It follows from Theorem 8.2.1 that a central elliptical distribution is completely specified by its scale parameter matrix Σ and its weighting function $w(t)$ whenever the latter exists.

TABLE 8.1 Some elliptical distributions and their weighting functions

Distribution	Density function	Weighting function
Gaussian	$\frac{1}{(2\pi)^{p/2}\|\Sigma\|^{1/2}}e^{-s}$ with $s = \mathbf{x}'\Sigma^{-1}\mathbf{x}/2$ throughout	$\delta(t-1)$ The Dirac delta function
Contaminated Normal	$\frac{1}{(2\pi)^{p/2}\|\Sigma\|^{1/2}}$ $\{\phi\lambda^{p/2}e^{-\lambda s}$ $+(1-\phi)e^{-s}\}$	$\phi\,\delta(t-1)+$ $(1-\phi)\,\delta(1-\sigma^2)$
Type VII Pearson	$\frac{\Gamma(m)\|\Sigma\|^{-1/2}}{q^{p/2}\pi^{p/2}\Gamma(m-p/2)}$ $\times(1+2s/q)^{-m}$	$\frac{t^{m-p/2-1}e^{-q\,t/2}}{(q/2)^{p/2-m}\,\Gamma(m-p/2)}$
t-distribution with ν d.f.	$\frac{\nu^{\nu/2}\Gamma((\nu+p)/2)}{\pi^{p/2}\Gamma(\nu/2)}\|\Sigma\|^{-1/2}$ $\times(\nu+2s)^{-(\nu+p)/2}$	$\frac{\nu(\nu t/2)^{(\nu/2)-1}e^{-\nu t/2}}{2\Gamma(\nu/2)}$
Multivariate Analog of the Bilateral Exponential Density	$\frac{\Gamma(p/2)}{2^{p+1}\pi^{p/2}\Gamma(p)\|\Sigma\|^{1/2}}$ $e^{-\sqrt{2s}}$	$\frac{\Gamma(p/2)}{\Gamma(p)2^{(k+5)/2}\pi}$ $\times e^{-\frac{1}{8t}}/t^{(p+3)/2}$
The Generalized Slash Distribution	$\frac{\nu}{(2\pi)^{p/2}}s^{-p/2-v}$ $\|\Sigma\|^{-1/2}\{\Gamma(p/2+v)$ $-\Gamma(p/2+v,s)\}$	$\nu\,t^{\nu-1}$
'Sinusoidal' weight $(p=1$ or $3)$	$-\frac{2^{p/2}}{p\,\Gamma(p/2)\cos(p\,\pi/4)}$ $\times\{1+(2s)^2\}^{-1}$	$-\frac{2^{p/2}}{p\,\Gamma(p/2)\cos(p\,\pi/4)}$ $\times t^{-p/2}\sin(t/2)$

8.3 THE EXACT DISTRIBUTION OF QUADRATIC FORMS

We now turn our attention to quadratic forms in elliptically contoured vectors. Quadratic forms play a major role in statistical inference; several applications are described for instance in Mathai and Provost (1992), Chapter 7. However, few results are available in the literature for the case of elliptically contoured vectors. Some particular results, including the case where the matrix of the quadratic form is idempotent, may be found for example in Fang and Zhang (1990), Anderson and Fang (1987), Li (1987), Fan (1986), Hsu (1990), and Fang, Fan and Xu (1987). Many distributional results on quadratic forms in *normal* vectors are readily available in the literature, see for example Mathai and Provost (1992), Provost and Rudiuk (1992, 1994, 1995), and the references therein.

Let $\mathbf{X} \sim C_p(\boldsymbol{\mu}, \Sigma, \xi)$, Σ be a positive definite matrix, $\boldsymbol{\alpha}$ be a p-dimensional real vector, and A be a real symmetric matrix. The exact distribution function of $Q = (\mathbf{X} - \boldsymbol{\alpha})'A(\mathbf{X} - \boldsymbol{\alpha})$, denoted by $F_Q(q)$, can be obtained as follows:

$$
\begin{aligned}
F_Q(q) &= \Pr((\mathbf{X} - \boldsymbol{\alpha})'A(\mathbf{X} - \boldsymbol{\alpha}) \leq q) \\
&= \Pr((\mathbf{X} - \boldsymbol{\mu} + \boldsymbol{\mu} - \boldsymbol{\alpha})'\Sigma^{-1/2}PA^*P'\Sigma^{-1/2}(\mathbf{X} - \boldsymbol{\mu} + \boldsymbol{\mu} - \boldsymbol{\alpha}) \leq q)
\end{aligned}
$$

where $A^* = P'\Sigma^{1/2}A\Sigma^{1/2}P$, $\Sigma^{1/2}$ denotes the symmetric square root of Σ and $P = (\mathbf{v}_1, \dots, \mathbf{v}_p)$ is an orthogonal matrix such that

$$
P'\Sigma^{1/2}A\Sigma^{1/2}P = \mathrm{diag}(\lambda_1, \dots, \lambda_p),
$$

$\lambda_1, \dots, \lambda_p$ being the characteristic roots of $\Sigma^{1/2}A\Sigma^{1/2}$ (or equivalently those of $A\Sigma$) and $\mathbf{v}_1, \dots, \mathbf{v}_p$, the corresponding normalized characteristic vectors. (The symmetric square root, $\Sigma^{1/2}$, is equal to $\sum_{j=1}^p \theta_j^{1/2}\mathbf{u}_j\mathbf{u}_j'$ where the θ_j's, $j = 1, \dots, p$ denote the characteristic roots of Σ and the $\mathbf{u}_j$'s are the corresponding normalized characteristic vectors.)

Letting

$$
\boldsymbol{\beta} = (\beta_1, \dots, \beta_p)' = P'\Sigma^{-1/2}(\boldsymbol{\mu} - \boldsymbol{\alpha}) \tag{8.3.1}
$$

and $\mathbf{Y} = (Y_1, \cdots, Y_p)' = \Sigma^{-1/2}(\mathbf{X} - \boldsymbol{\mu})$, and noting that $P'\mathbf{Y} \simeq \mathbf{Y} \sim S_p(\xi)$, one has

$$
Q = Q(\mathbf{Y}) = \sum_{i=1}^p \lambda_i(Y_i + \beta_i)^2.
$$

It should be noted that a similar representation of Q involving $r \leq p$ terms holds when Σ is a positive semidefinite matrix whose rank is r.

Letting $L^p(q)$ be the set of points $\mathbf{y}' = (y_1, \dots, y_p)$ such that $Q(\mathbf{y}) \leq q$ and $f(\mathbf{y})$ denote the density function of $\mathbf{Y}$, one has from Theorem 8.2.1

$$
\begin{aligned}
\Pr(Q(\mathbf{Y}) \leq q) &= \int_{L^p(q)} f(\mathbf{y})\,d\mathbf{y} \\
&= \int_{L^p(q)} \int_0^\infty w(t)\eta_{\mathbf{y}}(t^{-1}I)\,dt\,d\mathbf{y}
\end{aligned}
$$

$$= \int_0^\infty w(t) \int_{L^p(q)} \eta_{\mathbf{y}}(t^{-1}I) \, d\mathbf{y} \, dt$$

where I denotes the identity matrix.

Now, if one lets $\mathbf{Z} = t^{1/2}\mathbf{Y}$ and $L_1^p(q)$ represent the set of points $\mathbf{z} = (z_1, \ldots, z_p)'$ such that $\sum_{i=1}^p \lambda_i (z_i + t^{1/2}\beta_i)^2 \le tq$, then

$$\Pr(Q \le q) = \int_0^\infty w(t) \int_{L_1^p(q)} \eta_{\mathbf{z}}(I) \, d\mathbf{z} \, dt \qquad (8.3.2)$$

where

$$\int_{L_1^p(q)} \eta_{\mathbf{z}}(I) \, d\mathbf{z} = \Pr\left(\sum_{i=1}^p \lambda_i (Z_i + t^{1/2}\beta_i)^2 \le tq\right) \qquad (8.3.3)$$

and the Z_i's, $i = 1, \ldots, p$, are independently and identically distributed standard normal variables.

The integral appearing in (8.3.3) can be evaluated by means of Imhof's (1961) formula:

$$\Pr\left(\sum_{i=1}^p \lambda_i (Z_i + t^{1/2}\beta_i)^2 \le tq\right) = \frac{1}{2} - \frac{1}{\pi} \int_0^\infty \frac{\sin(\theta(u))}{u\,\rho(u)} \, du \qquad (8.3.4)$$

where

$$\theta(u) = \frac{1}{2} \sum_{j=1}^p \left\{ r_j \, \text{Arctan}(\lambda_j\, u) + \frac{t\beta_j^2 \lambda_j u}{1 + \lambda_j^2 u^2} \right\} - \frac{u\,t\,q}{2} \, ,$$

$$\rho(u) = \left\{ \prod_{j=1}^p (1 + \lambda_j^2 u^2)^{r_j/4} \right\} \exp\left\{ \frac{1}{2} \sum_{i=1}^p \frac{(t^{1/2}\beta_j \lambda_j u)^2}{(1 + \lambda_j^2 u^2)} \right\} \, ,$$

the λ_j's being the characteristic roots of $A\Sigma$ and the β_j's, as given in (8.3.1). Hence, with the above notation,

$$\Pr(Q \le q) = \frac{1}{2} - \frac{1}{\pi} \int_0^\infty w(t) \int_0^\infty \frac{\sin(\theta(u))}{u\,\rho(u)} \, du \, dt \qquad (8.3.5)$$

where $w(t)$ is as defined in Theorem 8.2.1. On differentiating the right-hand side of (8.3.5) with respect to q, one obtains the following representation of density function of Q:

$$f(q) = \frac{1}{2\pi} \int_0^\infty t\,w(t) \int_0^\infty \frac{\cos(\theta(u))}{\rho(u)} \, du \, dt \qquad (8.3.6)$$

where $\theta(u)$ and $\rho(u)$ are as defined in Imhof's formula. The ranges of integration can be made finite using for example the transformations $u^* = 1/(u+1)$ and $t^* = 1/(t+1)$ and numerical integration techniques may then be applied to either integral, should this be required for computational purposes.

It should be pointed out that the right-hand side of Equation (8.3.3) can also be expressed in terms of MacLaurin series, chi-square densities or series involving Laguerre polynomials; see, for example, Ruben (1962), Kotz, Johnson and Boyd (1967a,b), and Mathai and Provost (1992, Chapter 4). The main results of this section are summarized in the following theorem.

Theorem 8.3.1 *Let* $\mathbf{X} \sim C_p(\boldsymbol{\mu}, \Sigma, \xi)$, *A be a real symmetric matrix and* $\boldsymbol{\alpha}$ *be a p-dimensional real vector. Then, the distribution function and the density of the quadratic form* $Q = (\mathbf{X} - \boldsymbol{\alpha})'A(\mathbf{X} - \boldsymbol{\alpha})$ *can be respectively evaluated from the integral representations given in (8.3.5) and (8.3.6).*

Clearly the approach described in this section also applies to other functions of elliptically distributed vectors whose associated weighting functions can be determined.

8.4   MOMENTS AND APPROXIMATE DISTRIBUTION

First, Theorem 8.2.1 is extended to non-central elliptically contoured distributions. A representation of the moments of quadratic forms in elliptically contoured vectors is then obtained.

Theorem 8.4.1 *Let* $\mathbf{Z} \sim C_p(\boldsymbol{\mu}, \Sigma, \xi)$. *Then, under certain regularity conditions, the density of* $\mathbf{Z}$ *denoted by* $h(\mathbf{z})$ *can be represented as follows*

$$h(\mathbf{z}) = \int_0^\infty w(t)\, \eta_{\mathbf{z}}(\boldsymbol{\mu}, t^{-1}\Sigma)\, dt$$

where $\eta_{\mathbf{z}}(\boldsymbol{\mu}, t^{-1}\Sigma)$ *denotes the density function of a p-dimensional Gaussian random vector with mean* $\boldsymbol{\mu}$ *and covariance matrix* $t^{-1}\Sigma$, *and*

$$w(t) = (2\pi)^{p/2}\, |\Sigma|^{1/2}\, t^{-p/2}\, \mathcal{L}^{-1}\left(f(s)\right),$$

$\mathcal{L}^{-1}(f(s))$ *denoting the inverse Laplace transform of* $f(s)$ *with* $f(s) = h(\mathbf{z})$ *when* $s = (\mathbf{z} - \boldsymbol{\mu})'\Sigma^{-1}(\mathbf{z} - \boldsymbol{\mu})/2$.

The result follows from Theorem 8.2.1 by letting $\mathbf{X} = \mathbf{Z} - \boldsymbol{\mu}$. A representation of the density function of an important subclass of elliptically contoured distributions in terms of scale mixtures of normal variates is also discussed by Muirhead (1982).

Let $\mathbf{Z} \sim \mathcal{C}_p(\boldsymbol{\mu}, \Sigma, \xi)$ and $A = A'$; then according to Theorem 8.4.1, the moment-generating function of the non-central quadratic form $\mathbf{Z}'A\mathbf{Z}$ can be obtained as follows:

$$
M_{\mathbf{Z}'A\mathbf{Z}}(t) = \int_{-\infty}^{\infty} \cdots \int_{-\infty}^{\infty} e^{t\mathbf{z}'A\mathbf{z}} \int_{0}^{\infty} w(s)\, \eta_{\mathbf{z}}(\boldsymbol{\mu}, s^{-1}\,\Sigma)\, ds\, d\mathbf{z}
$$

$$
= \int_{0}^{\infty} w(s)\, M_{Q(s)}^{*}(t)\, ds \qquad (8.4.7)
$$

where

$$
M_{Q(s)}^{*}(t) = \int_{-\infty}^{\infty} \cdots \int_{-\infty}^{\infty} e^{t\mathbf{z}'A\mathbf{z}}\, \eta_{\mathbf{z}}(\boldsymbol{\mu}, s^{-1}\,\Sigma)\, d\mathbf{z}
$$

is the moment-generating function of the quadratic form $Q(s) = \mathbf{Y}'A\mathbf{Y}$ in which $\mathbf{Y} \sim N_p(\boldsymbol{\mu}, s^{-1}\Sigma)$.

The moments of $\mathbf{Z}'A\mathbf{Z}$ are obtained similarly. Whenever they exist, they are given by

$$
E(\mathbf{Z}A\mathbf{Z})^r = \int_{0}^{\infty} w(s)\, E(\mathbf{Y}A\mathbf{Y})^r\, ds \qquad (8.4.8)
$$

where $\mathbf{Y} \sim N_p(\boldsymbol{\mu}, s^{-1}\Sigma)$. The first four moments of $\mathbf{Y}'A\mathbf{Y}$ are

$$
E(\mathbf{Y}'A\mathbf{Y}) = \boldsymbol{\mu}'A\boldsymbol{\mu} + s^{-1}\mathrm{tr}(A\Sigma),
$$

$$
E(\mathbf{Y}'A\mathbf{Y})^2 = \{\boldsymbol{\mu}'A\boldsymbol{\mu} + s^{-1}\mathrm{tr}(A\Sigma)\}^2 + 2\{2s^{-1}\boldsymbol{\mu}A\Sigma A\boldsymbol{\mu} + s^{-2}\mathrm{tr}(A\Sigma)^2\},
$$

$$
\begin{aligned}
E(\mathbf{Y}'A\mathbf{Y})^3 &= \{\boldsymbol{\mu}'A\boldsymbol{\mu} + s^{-1}\mathrm{tr}(A\Sigma)\}^3 \\
&+ [6\{\boldsymbol{\mu}'A\boldsymbol{\mu} + s^{-1}\mathrm{tr}(A\Sigma)\}\{2s^{-1}\boldsymbol{\mu}'A\Sigma A\boldsymbol{\mu} + s^{-2}\mathrm{tr}(A\Sigma)^2\}] \\
&+ 8\{3s^{-2}\boldsymbol{\mu}'(A\Sigma)^2 A\boldsymbol{\mu} + s^{-3}\mathrm{tr}(A\Sigma)^3\},
\end{aligned}
$$

and

$$
\begin{aligned}
E(\mathbf{Y}'A\mathbf{Y})^4 &= 6\{2s^{-1}\boldsymbol{\mu}'A\Sigma A\boldsymbol{\mu} + s^{-2}\mathrm{tr}(A\Sigma)^2\} \\
&\times [\{\boldsymbol{\mu}'A\boldsymbol{\mu} + s^{-1}\mathrm{tr}(A\Sigma)\}^2 + 2\{2s^{-1}\boldsymbol{\mu}'A\Sigma A\boldsymbol{\mu} + s^{-2}\mathrm{tr}(A\Sigma)^2\}] \\
&+ 24\{\boldsymbol{\mu}'A\boldsymbol{\mu} + s^{-1}\mathrm{tr}(A\Sigma)\}\{3s^{-2}\boldsymbol{\mu}'(A\Sigma)^2 A\boldsymbol{\mu} + s^{-3}\mathrm{tr}(A\Sigma)^3\} \\
&+ \{\boldsymbol{\mu}'A\boldsymbol{\mu} + s^{-1}\mathrm{tr}(A\Sigma)\}[\{\boldsymbol{\mu}'A\boldsymbol{\mu} + s^{-1}\mathrm{tr}(A\Sigma)\}^3 \\
&+ 6\{\boldsymbol{\mu}'A\boldsymbol{\mu} + s^{-1}\mathrm{tr}(A\Sigma)\}\{2s^{-1}\boldsymbol{\mu}'A\Sigma A\boldsymbol{\mu} + s^{-2}\mathrm{tr}(A\Sigma)^2\} \\
&+ 8\{3s^{-2}\boldsymbol{\mu}'(A\Sigma)^2 A\boldsymbol{\mu} + s^{-3}\mathrm{tr}(A\Sigma)^3\}] \\
&+ 48\{4s^{-3}\boldsymbol{\mu}'(A\Sigma)^3 A\boldsymbol{\mu} + s^{-4}\mathrm{tr}(A\Sigma)^4\}.
\end{aligned}
$$

Note that in light of the conclusion of Section 3.2b.3 in Mathai and Provost (1992), the moments have the same representations when Σ is singular.

The first moments of $\mathbf{Z}'A\mathbf{Z}$ may be used to obtain approximate distributions such as those based on the Pearson or Johnson curve systems, the saddlepoint approximation, or on Edgeworth or Gram–Charlier expansions.

Furthermore, it can be seen that the technique used in (8.4.7) readily applies to other functions of elliptically contoured vectors whose counterparts for Gaussian vectors are known. The results derived in this section are summarized in the following theorem.

Theorem 8.4.2 *Let* $\mathbf{Z} \sim C_p(\mu, \Sigma, \xi)$ *and* $A = A'$; *then, whenever they exist, the moment-generating function and the moments of the quadratic form* $\mathbf{Z}'A\mathbf{Z}$ *can be evaluated from the integral representations given respectively in (8.4.7) and (8.4.8).*

The distributional results derived in Sections 8.2, 8.3 and 8.4 could be used for example for determining the distribution of the Mahalanobis distance when the vectors are elliptically contoured or that of the serial covariances associated with time series whose innovations are spherically distributed. The use of spherically distributed errors was discussed for instance by Jensen (1979) and Hwang and Chen (1986) in connection with some linear models, by Pázman (1988) for certain nonlinear models and by Krishnaiah and Lin (1986) and Basu and Das (1994) in connection with some time series models.

8.5   A NUMERICAL EXAMPLE

A generic numerical example is given in this section.

Let $\mathbf{X}$ be an elliptical vector with mean $\mu = (2, 4, -1, 3)'$ and scale parameter matrix Σ

$$\begin{pmatrix} 4 & -1 & 0.5 & -0.5 \\ -1 & 3 & 1 & -1 \\ 0.5 & 1 & 6 & 1 \\ -0.5 & -1 & 1 & 4 \end{pmatrix}.$$

The distribution function of the quadratic form $Q = (\mathbf{X} - \alpha)'A(\mathbf{X} - \alpha)$ where $\alpha = (0.5, 0.4, -0.6, 0.2)'$ and

$$A = \begin{pmatrix} 0.1 & -0.6 & 0.7 & -0.9 \\ -0.6 & 0.3 & 0.1 & 0.2 \\ 0.7 & 0.1 & 0.4 & 1 \\ -0.9 & 0.2 & 1 & 0.2 \end{pmatrix}$$

can be determined by making use of Theorem 8.3.1.

Exact and simulated values of the distribution function of Q were obtained for some selected points in the range of Q, first assuming that $\mathbf{X} \sim N_4(\boldsymbol{\mu}, \Sigma)$ and then that $\mathbf{X} \sim t_3(\boldsymbol{\mu}, \Sigma)$, a non-central t-vector with 3 degrees of freedom and scale parameter matrix Σ. The integrations were done numerically by making use of *Mathematica*, Version 3.0 with precision 5, and the simulations were carried out with 100,000 replications. The results are reported in Table 8.2.

TABLE 8.2 The distribution function of Q evaluated at selected points q

	Normal vector		t_3 vector	
q	Exact	Simulated	Exact	Simulated
-4.75	0.0208	0.0207	0.0499	0.5009
3	0.0778	0.0772	0.1002	0.1005
15	0.2532	0.2531	0.2482	0.2476
50	0.7393	0.7389	0.6854	0.6838
100	0.9674	0.9675	0.8925	0.8924
150	0.9970	0.9970	0.9468	0.9462

REFERENCES

Anderson, T. W. and Fang, K.-T. (1987). Cochran's theorem for elliptically contoured distributions, *Sankhya, Series A*, **49**, 305–315.

Basu, A. K. and Das, J. K. (1994). A Bayesian approach to Kalman filter for elliptically contoured distribution and its application in time series models, *Calcutta Statistical Association Bulletin*, **44**, 11–28.

Cambanis, S., Huang, S. and Simmons, G. (1981). On the theory of elliptically contoured distributions, *Journal of Multivariate Analysis*, **11**, 368–385.

Chmielewski, M. A. (1981). Elliptically symmetric distributions: A review and bibliography, *International Statistical Review*, **49**, 67–74.

Chu, K.-U. (1973). Estimation and decision for linear systems with elliptically random process, *IEEE Transaction on Automatic Control*, **18**, 499–505.

Devlin, S. J., Gnanadesikan, R. and Kettenring, J. R. (1976). Some multivariate applications of elliptical distributions. In *Essays in Probability and Statistics* (Ed., S. Ideka), pp. 365–393. Sinko Tsusho, Tokyo.

Fama, E. F. (1965). The behavior of stock-market prices, *Journal of Business*, **XXXVIII**, 34–105.

Fan, J.-Q (1986). Distributions of quadratic forms and non-central Cochran's theorem, *Acta Mathematicae Sinica (New Series)*, **2**, 185–198.

Fang, K.-T., Fan, J.-Q. and Xu, J.-L. (1987). The distributions of quadratic forms of random idempotent matrices with their applications, *Chinese Journal of Applied Probability and Statistics*, **3**, 289–297.

Fang, K.-T., Kotz, S. and Ng, K.-W. (1990). *Symmetric Multivariate and Related Distributions*, Chapman and Hall, London.

Fang, K.-T. and Zhang, Y.-T. (1990). *Generalized Multivariate Analysis*, Springer, New York and Science Press, Beijing.

Fraser, D. A. S. and Ng, K.-W. (1980). Multivariate regression analysis with spherical error, In *Multivariate Analysis 5* (Ed., P. R. Krishnaiah), pp. 369–386, North Holland, New York.

Girón, F. J. and Rojano, J. C. (1994). Bayesian Kalman filtering with elliptically contoured errors, *Biometrika*, **80**, 390–395.

Gradshteyn, I. S. and Ryzhik, I. M. (1980). *Table of Integrals, Series, and Products, Corrected and Enlarged Edition*, Academic Press Inc., New York.

Hsu, H. (1990). Non-central distributions of quadratic forms for elliptically contoured distributions, In *Statistical Inference in Elliptically Contoured and Related Distributions* (Eds., K.-T Fang and T. W. Anderson), pp. 97–102, Allerton Press Inc, New York.

Hwang, J.-T. and Chen, J. (1986). Improved confidence sets for the coefficients of a linear model with spherically symmetric errors, *Annals of Statistics*, **14**, 444–460.

Imhof, J. P. (1961). Computing the distribution of quadratic forms in normal variables, *Biometrika*, **48**, 419–426.

Jensen, D. R. (1979). Linear models without moments, *Biometrika*, **66**, 611–618.

King, M. L. (1980). Robust tests for spherical symmetry and their application to least squares regression, *American Statistician*, **8**, 1265–1271.

Kotz, S., Johnson, N. L. and Boyd, D. W. (1967a). Series representations of quadratic forms in normal variables. I: Central case, *Annals Mathematical Statistics*, **38**, 823–837.

Kotz, S., Johnson, N. L. and Boyd, D. W. (1967b). Series representations of quadratic forms in normal variables. II: Non-central case, *Annals Mathematical Statistics*, **38**, 838–848.

Krishnaiah, P. R. and Lin, J. (1986). Complex elliptically symmetric distributions, *Communication in Statististics, Theory and Methods*, **15**, 3693–3718.

Li, G. (1987). Moments of a random vector and its quadratic forms, *Journal of Statistics and Applied Probability*, **2**, 219–229.

Mandelbrot, B. (1963). The variation of certain speculative prices, *Journal of Business*, **XXXVI**, 394–419.

Mathai, A. M. and Provost, S. B. (1992). *Quadratic Forms in Random Variables, Theory and Applications*, Marcel Dekker Inc, New York.

Mathai, A. M., Provost, S. B. and Hayakawa, T. (1995). *Bilinear Forms and Zonal Polynomials*, Springer-Verlag, New York.

McGraw, D. K. and Wagner, J. F. (1968). Elliptically symmetric distributions, *IEEE Transaction on Information Theory*, **14**, 110–120.

Muirhead, R. J. (1982). *Aspects of Multivariate Statistical Theory*, John Wiley & Sons, New York.

Osiewalski, J. and Steel, M. F. J. (1993). Robust Bayesian inference in elliptical regression models, *Journal of Econometrics*, **57**, 345–363.

Pázman, A. (1988). Distribution of L.S. estimates in nonlinear models with spherically symmetrical error, In *Optimal Design and Analysis of Experiment* (Eds., Y. Dodge, V. Fedorov and H. Wynn), pp. 177–184, North-Holland/Elsevier, Amsterdam/New York.

Provost, S. P. and Rudiuk, E. (1992). The exact distribution of the ratio of two quadratic forms in noncentral normal variables, *Metron* **50**, 33–58.

Provost, S. P. and Rudiuk, E. (1994). The exact density of the ratio of two linear combinations of dependent chi-square random variables, *The Annals of the Institute of Statistical Mathematics*, **46**, 557–571.

Provost, S. P. and Rudiuk, E. (1995). The sampling distribution of the serial correlation coefficient, *The American Journal of Mathematical and Management Sciences*, **15**, 57–81.

Resnick, S. I. (1997). Heavy tail modeling and teletraffic data, *The Annals of Statistics*, **25**, 1805–1869.

Ruben, H. (1962). Probability content of regions under spherical normal distribution. IV: The distribution of homogeneous and nonhomogeneous quadratic functions in normal variables, *The Annals Mathematical Statistics*, **33**, 542–570.

Zellner, A. (1976). Bayesian and non-Bayesian analysis of the regression model with multivariate student-t error terms, *Journal of American Statistical Association*, **71**, 400–405.

CHAPTER 9

INVERSE NORMALIZING TRANSFORMATIONS AND AN EXTENDED NORMALIZING TRANSFORMATION

HAIM SHORE

Ben-Gurion University of the Negev, Beer-Sheva, Israel

Abstract: When available data are non-normal, a common practice is to normalize them by applying the Box-Cox power transformation. The general effectiveness of this transformation implies that an inverse normalizing transformation, viz. a power transformation of the standard normal quantile, may effectively deliver a general representation for many of the commonly applied theoretical statistical distributions.

Employing as a departure point the Box-Cox transformation, we develop in this paper several inverse normalizing transformations (INTs), and define criteria for their effectiveness. In terms of these criteria, the new transformations are shown to deliver good representations to differently shaped distributions having skewness values that range from zero to over 11. A new normalizing transformation, derived by conversion of a certain INT, turns out to be an extension of the classical Box-Cox transformation. The new normalizing transformation provides an appreciably better normalizing effect relative to the Box-Cox transformation. Some estimation procedures for the new INTs are developed.

Keywords and phrases: Box-Cox transformation, distribution fitting, inverse normalizing transformation

9.1 INTRODUCTION

The Box-Cox power transformation is widely applied to achieve a normalizing effect for non-normal data. The basis for this transformation, as articulated by Box and Cox (1964), is the empirical observation that a power transformation is equivalent to finding the "right" scale for given data. Furthermore, expressing data by their "right" scale can simultaneously achieve three objectives: Simplicity of structure of descriptive models, homogeneity of the error-variance and normalization [Box and Cox (1964, pp. 211, 213)]. In the years that have followed, since the introduction of the Box-Cox transformation, it has proved to be effective in achieving normality for a variety of published cases involving widely differing source distributions. Yet, the obvious implication of this general effectiveness, namely, the potential of the Box-Cox transformation to serve as a starting point for developing general representations for statistical distributions, has never been probed.

In this paper we use the Box-Cox transformation as a departure point to develop inverse normalizing transformations (INTs), namely: transformations that express the quantile of a non-negative r.v. X, in terms of the corresponding quantile of the standard normal variable Z. Efforts of this sort are not new in the literature. Perhaps the most well known of these are Cornish and Fisher (1937) and Fisher and Cornish (1960). The latter have developed series-expansions that describe the quantile of an asymptotically normally distributed r.v. in terms of a polynomial sum of the standard normal quantile, where the coefficients of the polynomial terms are functions of the cumulants of X. However, for a polynomial of degree k these expansions assume that all cumulants of X up to degree k are known; see, for example, Johnson, Kotz and Balakrishnan (1994). This assumption diminishes appreciably from the applicability of the Cornish-Fisher expansions.

In this paper we pursue a different approach, and examine various extended variations of the inverse Box-Cox transformation that may deliver good representations to distributions with widely differing shapes. In Section 9.2 we develop three new INTs, together with a new normalizing transformation. The latter is found to be an extension of the classical Box-Cox transformation, hence its name "The extended normalizing transformation" (ENT). In Section 9.3 we examine the goodness-of-fit obtained when the new INTs are fitted to differently shaped distributions. The effectiveness of the ENT is also numerically demonstrated. In Section 9.4 we develop estimation procedures to some of the new INTs. Finally, Section 9.5 discusses some implications of the new transformations.

9.2   DERIVATION OF THE TRANSFORMATIONS

The Box-Cox single-parameter transformation is defined for non-negative X by Box and Cox (1964):

$$X^{(\lambda)} = \left\{ \begin{array}{ll} (X^{\lambda} - 1)/\lambda, & \lambda \neq 0 \\ \log(X), & \lambda = 0. \end{array} \right. \tag{9.2.1}$$

Assume first that $\lambda \neq 0$. Since the Box-Cox transformation is generally known to be an effective procedure to achieve normality, let us assume that the transformed variable, $X^{(\lambda)}$, is normal with parameters u and v. Denoting the standard normal variate by Z, we may express the quantile-relationship between X and Z by:

$$(x^{\lambda} - 1)/\lambda = u + vz \tag{9.2.2}$$

or

$$x = (\lambda u + 1 + \lambda vz)^{(1/\lambda)} \tag{9.2.3}$$

where x and z are corresponding quantile values, namely: $F(x) = \Phi(z)$, where $F(\cdot)$ and $\Phi(\cdot)$ are the CDFs of X and Z, respectively.

Rewriting (9.2.3) so that the median of X (denote it by M) is preserved, we obtain:

$$x = (M)(1 + \lambda az)^{(1/\lambda)}, \qquad (\lambda a) > 0 \tag{9.2.4}$$

which has two parameters: λ and a. Eq. (9.2.4) expresses x as a power transformation of the standard normal quantile, however it does not preserve the domain of x since as z tends to be negative x will also become negative (for a non-even value of $1/\lambda$).

Next, assume that $\lambda = 0$. Then, similarly with the derivation of (9.2.4) we obtain:

$$x = (M) \exp[bz], \qquad b > 0 \tag{9.2.5}$$

where b is a parameter, to be determined.

Unlike (9.2.4), (9.2.5) has the desirable property that as z tends to negative infinity x tends to zero. Since we wish this property to be available also for (9.2.4), and since it would be desirable to have a single quantile function for x (irrespective of the value of λ), a reasonable merging of (9.2.4) and (9.2.5) would result in the following quantile-relationship:

$$\begin{aligned} x &= H(z; \alpha) = (M)(1 + az)^{b} \exp[cz] \\ &= (M) \exp[b \log(1 + az) + cz] \end{aligned} \tag{9.2.6}$$

where $H(z; \alpha)$ denotes the quantile-relationship between X and Z, and $\alpha = \{a, b, c\}$ is a vector of three parameters, to be determined. Note, that for $b = 1$, $c = 0$, X is normally distributed, while for $b = 0$ X follows a log-normal distribution.

The parameters in (9.2.6) should ensure that x is an increasing function of z. Differentiating x with respect to z, and requiring that $\partial x / \partial z > 0$, we obtain, assuming that $ab > 0$ and $(1 + az) > 0$:

$$[ab + c(1 + az)] > 0$$

or

$$-1/a < z < -(ab + c)/(ac), \qquad \text{for } c < 0 \qquad (9.2.7)$$

and

$$-1/a < z, \qquad \text{for } c > 0. \qquad (9.2.8)$$

Once the parameters $\{a, b, c\}$ are determined, (9.2.7) and (9.2.8) may be used to determine the valid domain of z.

Expression (9.2.6) may serve as a departure point to develop two further INTs. First, when (9.2.6) is fitted to differently shaped distributions we realize that the parameters a and c are very close in their absolute values, however they have opposite signs (refer to Table 9.1 in the next section). This suggests that setting: $c = -a$ in (9.2.6) may result in an INT that still provides good representation for a wide variety of distributions. The resulting transformation is:

$$
\begin{aligned}
x &= H(z; \alpha) = (M)(1 + az)^b \exp[-az] \\
&= (M) \exp[b \log(1 + az) - az].
\end{aligned}
\qquad (9.2.9)
$$

The constraint on the parameters is, from (9.2.7) and (9.2.8):

$$-1/a < z < (b - 1)/a. \qquad (9.2.10)$$

Note, that since (9.2.9) contains only two parameters (apart from M), these uniquely determine (and are determined) by the mean and the variance. Thus, matching of the mean and the variance of (9.2.9) with those of X may provide a convenient fitting procedure. This will be numerically demonstrated in Section 9.3.

A third INT may be derived from (9.2.9) if we note that the exponent therein contains two power transformations of $(1 + az)$, namely: the log term (which is equivalent to a zero power-transformation) and the linear term. Thus, it is plausible to investigate the adequacy of the following INT:

$$x = H(z; \alpha) = (M)\exp[bz^{(\lambda)}] \qquad (9.2.11)$$

where $z^{(\lambda)}$ is a Box-Cox transformation of $(1 + az)$, namely:

$$z^{(\lambda)} = [(1 + az)^{(\lambda)} - 1]/\lambda. \qquad (9.2.12)$$

A unique feature of (9.2.11) is that the quantile z appears only once. This implies that a normalizing transformation may be easily derived thereof. If (9.2.11) is found to yield good representation for a large variety of distributions, then a normalizing transformation based on (9.2.11) may prove to be extremely effective. In particular, unlike the Box-Cox transformation the new normalizing transformation will preserve the true range of the normal variate.

From (9.2.11) and (9.2.12), the associated normalizing transformation is:

$$z = B\{[1 + A\log(x/M)]^C - 1\}/C \qquad (9.2.13)$$

where $A \,(= \lambda/b)$, $B \,(= 1/(\lambda a))$ and $C \,(= 1/\lambda)$ are parameters, to be determined. It is interesting to note that (9.2.13) represents a double application of the original Box-Cox transformation: First we apply the log transformation $[\lambda = 0$ in Eq. (9.2.1)$]$ and then we apply a power transformation to a linear combination of the result $[\lambda \neq 0$ in Eq. (9.2.1)$]$. Also note that since:

$$\lim_{K \to 0}[1 + K\log(x)]^{(1/K)} = \exp[\log(x)] = x \qquad (9.2.14)$$

and A is close to zero (observe values in Table 9.3), we may write from (9.2.13), approximately:

$$(x/M) = (1 + az)^b \qquad (9.2.15)$$

or

$$z = D\{(x/M)^C - 1\}/C \qquad (9.2.16)$$

where $C = 1/b$, $D = 1/(ab)$. Thus, (9.2.13) is found to be an extension of the Box-Cox transformation, and will therefore be denoted the "extended normalizing transformation" (ENT).

The effectiveness of transformations (9.2.6), (9.2.9), (9.2.11), (9.2.13) and (9.2.16) will be examined in the next section.

9.3 NUMERICAL ASSESSMENT

To examine how well the new INTs may serve as general models for the quantile-relationship between X and Z, suitability criteria have to be established. These criteria will naturally be linked to the procedure used to determine the parameters' values, for example: If the fitting procedure is based on matching of moments, obviously the moments of the fitted transformation cannot be used to assess the goodness-of-fit. In this section, we use a set of quantile values of X and Z, which are incorporated in a nonlinear least-squares (NL-LS) procedure to determine the parameters of the examined transformation. Once the parameters are identified, we examine the suitability of the examined INTs by calculating numerically the mean, the variance and the skewness and kurtosis measures of the fitted transformation. Proximity between these values and the exact moments will provide the required criteria for the effectiveness of the transformations.

To numerically assess the new INTs, we have selected a test-set of nine distributions, with skewness values that range from 0.63 (Rayleigh) to over eleven (Weibull). For each distribution in the test-set, seventy-one quantile values of z: $-4.0(0.1)3.0$ have been selected, and the corresponding x values identified. These distributions are then fitted via the NL-LS procedure, and the resulting moments numerically calculated with numerical integration that extends from $P = 0.000003$ ($z = -4.5$) to $P = 0.999997$ ($z = 4.5$).

The results are shown for the three inverse normalizing transformations in Tables 9.1 [Eq. (9.2.6)], 9.2 [Eq. (9.2.9)] and 9.3 [Eq. 9.2.11)].

It is of interest to examine whether the constraint on the permissible values of z [Eqs. (9.2.7), (9.2.8) and (9.2.10)] comprise a real restriction on the use of the new INTs. For that end, we have calculated, for each distribution in the test-set, the related limits. These are given as Z_L and Z_U (lower limit and upper limit, respectively) in Tables 9.1 and 9.2. We realize that practically Eqs. (9.2.7), (9.2.8) and (9.2.10) do not constitute any real limitation for the application of transformations (9.2.6) and (9.2.9).

Examining the tables we realize that the last transformation (9.2.11) generally provides better fit than the other transformations. To appreciate the effectiveness of this transformation, we provide plots of (9.2.11) (in standardized values), together with the source (exact) quantile function, for some of the distributions in the test-set. These plots are given in Figure 9.1 (left-hand plots). Note, that due to the small deviations between the two graphs (exact and approximate) the latter are indistinguishable in Figure 9.1. Plots of the associated errors (approximate minus exact) are given as the right-hand plots in Figure 9.1.

Next we examine the effectiveness of the new normalizing transformation (9.2.13) in comparison to the Box-Cox transformation. To this end, we apply the ENT to the distributions in the tables (using the NL-LS routine to determine the parameters), and also to the Box-Cox transformation. For a fair comparison between the two normalizing transformations, we use the two-parameter version of the Box-Cox transformation [Box and Cox (1964, Eq. 2)], and add a scale parameter to obtain:

$$z = \begin{cases} b[(X+a)^\lambda - 1)/\lambda], & \lambda \neq 0 \\ b\log(X+a), & \lambda = 0. \end{cases} \qquad (9.3.17)$$

Thus, (9.2.13) and (9.3.17) share an equal number of parameters (3), one of which is a scale parameter. Also included in the comparison is the modified Box-Cox transformation, as given by (9.2.16).

The results of the comparison are assessed in terms of the proximity of the transformation's numerically calculated moments to the first four partial moments of the standard normal distribution. The reason for using partial moments, rather than complete moments, is that the first three moments of the standard normal variate characterize any symmetrically distributed standardized variable. Employing partial moments may thus provide a better test for normality. The results are given in Table 9.4. Unfortunately, we were unable to achieve satisfactory goodness-of-fit from applying the NL-LS routine to the three-parameter Box-Cox transformation (9.3.17). Therefore, only results from (9.2.13) and (9.2.16) are given. Examining the table we realize that the new normalizing transformations yield good normalizing effect though, as expected, (9.2.13) is slightly better than its derivative (9.2.16).

9.4   ESTIMATION

Assume that only sample data are available. In this case, either the formerly described fitting methodology may be applied to the sample data, or other approaches be adopted, like moment-matching or maximum-likelihood procedures.

Referring to the former approach, assume that we have n observations and denote by X_i the ith observation, and by $X_{(i)}$ the ith order statistic, namely: $X_{(1)} \leq X_{(2)} \leq \cdots \leq X_{(n)}$. Then, following the shifted, piecewise linear definition of the sample quantile function given by Parzen (1979), and also used by Grimshaw and Alt (1997), an estimator of the quantile function for a given value P is:

$$x = F^{-1}(P)$$
$$= n[(2i+1)/(2n) - P]X_{(i)} + n[P - (2i-1)/(2n)]X_{(i+1)},$$

$$(9.4.18)$$

for

$$(2i-1)/(2n) \le P \le (2i+1)/(2n).$$

The quantile estimator is left undefined for $P < 1/(2n)$ and for $P > [1 - 1/(2n)]$. Introducing for the ith order statistic: $P_i = i/n$, the sample quantile function becomes:

$$x_i = F^{-1}(P_i) = [X_{(i)} + X_{(i+1)}]/2 \qquad (9.4.19)$$

which is used as an estimator for the quantile value of x corresponding to $P = i/n$. The associated values of Z, $\{z_i\}$, may now be identified, and incorporated in the NL-LS procedure.

Referring next to moment-matching, we will develop here a fitting procedure for Eq. (9.2.11). A basic requirement for the fitting procedure will be that only the mean and the variance of X (or their sample estimates) need to be specified. Thus, the large standard errors associated with sample estimates of higher-order moments (like skewness and kurtosis) are to be avoided.

To develop the fitting routine, note first that transformation (9.2.11) contains three parameters. This implies that to apply regular moment-matching procedure the first three moments need to be specified, which violates the requirement formulated above. To circumvent this problem, let us define: $T = \log[X]$, and let μ_i denote the ith non-central moment of T (moment about zero). Then from (9.2.11) we obtain:

$$\mu_1(T) = \log(M) + bE\{[(1 + az)^\lambda - 1]/\lambda\}. \qquad (9.4.20)$$

Expanding the second term on the RHS (the expression following the expectation sign) into a Taylor series around zero, and taking expectation for the first five terms in the expansion, we obtain:

$$E\{[(1 + az)^\lambda - 1]/\lambda\}$$
$$= (1/2)a^2(\lambda - 1) + (1/8)a^4(\lambda - 1)(\lambda - 2)(\lambda - 3). \qquad (9.4.21)$$

Introducing from (9.4.21) into (9.4.20) we obtain an expression for b in terms of $\mu_1(T)$, M, a and λ. Introducing back into (9.2.11) we obtain a transformation with two parameters, a and λ, that may be identified by matching the transformation's mean and variance with those of X. Table 9.5 exhibits the transformation's parameters and the resulting skewness and kurtosis values. A high goodness-of-fit is achieved. Since only moments of first and second degrees are used, we expect the MSEs associated with the skewness and kurtosis measures, calculated from the fitted transformation, to be small. This will ensure that the fitted transformation represents well the underlying (presumably unknown) distribution of X. Further moment-matching procedures, and examination of the resulting MSEs, are given in Shore (1999).

Finally, maximum likelihood procedures may be developed for the various INTs. The resulting expressions are complex and will be introduced elsewhere.

9.5   CONCLUSIONS

In this paper we derive universal quantile relationships between a non-negative r.v. and the standard normal variable. Unlike previous derivations, which are based on polynomial expansions [like the Cornish-Fisher expansions; refer to Johnson, Kotz and Balakrishnan (1994) and Stuart and Ord (1987, pp. 232–233)], the present effort has an empirical basis that derives its validity from the universal effectiveness in achieving normality of the Box-Cox power transformation. Employing the inverse transformation as a departure point, we derive three INTs that are capable of expressing quantile-relationships between a non-normal r.v. and the standard normal variate for distributions with skewness values that range from 0 to at least 11.

The motivation for deriving $H(z; \alpha)$ is twofold. First, recognizing that in many applications the amount of data required to identify the correct distribution with satisfactory confidence is insufficient, the methodology developed here allows for universal solutions to stochastic optimization problems to be derived, irrespective of the source distributions. Furthermore, as shown elsewhere [Shore (1999)], the sampling uncertainty associated with fitting $H(z; \alpha)$ via two-moment matching procedures is considerably lower compared to alternative available approaches (like three- or four-moment matching).

A second reason for developing $H(z; \alpha)$ is the simplicity introduced by addressing various statistical analyses (like statistical process control and process capability analysis) in terms of a single r.v. (the standard normal variate). Thus, the complex task of identifying the correct distributions and apply appropriate distribution-specific parameter-estimation procedures is avoided.

An important implication of the new normalizing transformations pertains to the use of the Box-Cox transformation to achieve the stated three objectives [Box and Cox (1964)]: Simplicity of structure of descriptive models, homogeneity of variance and normality. While we have demonstrated in this paper that the new extended normalizing transformation and its derivative [Eqs. (9.2.13) and (9.2.16), respectively] achieve better normality relative to the three-parameter Box-Cox transformation, the other two objectives (simplicity of structure and homogeneity of variance) may in many applications be of higher priority. It remains to be seen, by the cumulative empirical evidence gathered in reported studies, whether the new normalizing transformations are also more effective with respect to these other objectives.

REFERENCES

Box, G. E. P. and Cox, D. R. (1964). An analysis of transformations, *Journal of the Royal Statistical Society, Series B*, **26**, 211–243.

Grimshaw, S. D. and Alt, F. B. (1997). Control charts for quantile function values, *Journal of Quality Technology*, **29**, 1–7.

Johnson, N. L., Kotz, S. and Balakrishnan, N. (1994). *Continuous Univariate Distributions, Vol. 1*, Second edition, John Wiley & Sons, New York.

Parzen, E. (1979). Nonparametric statistical data modeling, *Journal of the American Statistical Association*, **74**, 105–121.

Shore, H. (1999). Process capability analysis for non-normal populations based on inverse normalizing transformations, *Submitted for publication*.

Stuart, A. and Ord, J. K. (1987). *Kendall's Advanced Theory of Statistics, Vol. 1: Distribution Theory*, Griffin, London.

TABLE 9.1 Parameters values (9.2.6) and the resulting moments. The exact moments are the upper entries. Sk and Ku are the skewness and kurtosis measures

Dis.	a	b	c	Z_L	Z_U	Mean	Var.	Sk	Ku
Gamma	.09164	5.773	−.1436	−10.9	29.3	14.00	28.00	.7559	.8571
$\alpha = 7, \beta = 2$						14.00	27.99	.7555	.8447
Gamma	.1419	5.746	−.2122	−7.0	20.0	6.000	12.00	1.155	2.000
$\alpha = 3, \beta = 2$						6.002	11.98	1.155	1.977
Weibull	.2694	4.810	−.2839	−3.71	13.3	4.790	18.19	1.676	4.040
$\alpha = 1.125, \beta = 5$						4.785	18.15	1.675	3.982
Weibull	.2819	5.063	−.2833	−3.55	14.3	5.000	25.00	2.000	6.000
$\alpha = 1, \beta = 5$						4.992	24.96	1.994	5.877
Weibull	.3106	5.494	−.2668	−3.22	17.4	5.665	50.99	2.815	12.74
$\alpha = 0.8, \beta = 5$						5.654	50.94	2.793	12.31
Weibull	.3456	9.465	−.3657	−2.89	23.0	16.62	2723	11.35	287.6
$\alpha = 0.4, \beta = 5$						16.61	2677	10.43	203.7
Rayleigh	.2231	3.729	−.2624	−4.48	9.73	2.507	1.717	.6311	.2451
$\sigma = 2$						2.512	1.707	.6404	.2504
Extreme Value	.2876	.0415	.3405	−3.95	none	3.577	1.645	1.139	2.400
$\alpha = 3, \beta = 1$						3.576	1.683	1.099	2.208

TABLE 9.2 Parameters values (9.2.9) and the resulting moments. The exact moments are the upper entries. Sk and Ku are the skewness and kurtosis measures

Dis.	a	b	Z_L	Z_U	Mean	Var.	Sk	Ku
Gamma	.09965	4.859	−10.0	38.7	14.00	28.00	.7559	.8571
$\alpha = 7, \beta = 2$					14.00	27.93	.7578	.8609
Gamma	.1541	4.909	−6.49	25.4	6.000	12.00	1.155	2.000
$\alpha = 3, \beta = 2$					5.999	11.96	1.157	2.005
Gamma	.4861	6.742	−2.06	11.8	0.5000	1.000	4.000	24.00
$\alpha = 1/4, \beta = 2$					0.4379	0.8971	5.154	42.87
Weibull	.2733	4.714	−3.66	13.6	4.790	18.19	1.676	4.040
$\alpha = 1.125, \beta = 5$					4.784	18.16	1.675	3.982
Weibull	.2823	5.054	−3.54	14.4	5.000	25.00	2.000	6.000
$\alpha = 1, \beta = 5$					4.992	24.96	1.994	5.877
Weibull	.2992	5.799	−3.34	16.0	5.665	50.99	2.815	12.74
$\alpha = 0.8, \beta = 5$					5.655	50.89	2.794	12.29
Weibull	.3496	9.322	−2.86	23.8	16.62	2723	11.35	287.6
$\alpha = 0.4, \beta = 5$					16.62	2678	10.43	204.3
Rayleigh	.2323	3.446	−4.30	10.5	2.507	1.717	.6311	.2451
$\sigma = 2$					2.511	1.703	.6405	.2649
Extreme Value	.0149	24.94	−67.1	1607	3.577	1.645	1.139	2.400
$\alpha = 3, \beta = 1$					3.576	1.701	1.080	2.086

TABLE 9.3 Parameters values [Eqs. (9.2.11) and (9.2.12)] and the resulting moments. The exact moments are the upper entries. Sk and Ku are the skewness and kurtosis measures

Dis.	a	b	λ	Mean	Var.	Sk	Ku
Gamma	.05826	6.618	−1.170	14.00	28.00	.7559	.8571
$\alpha = 7, \beta = 2$				14.00	27.99	.7552	.8455
Gamma	.08962	6.740	−1.162	6.000	12.00	1.155	2.000
$\alpha = 3, \beta = 2$				6.001	11.98	1.154	1.977
Gamma	.4404	7.729	−.9113	0.5000	1.000	4.000	24.00
$\alpha = 1/4, \beta = 2$				0.4993	0.9959	3.960	23.00
Weibull	.1663	6.119	−1.087	4.790	18.19	1.676	4.040
$\alpha = 1.125, \beta = 5$				4.787	18.16	1.675	3.990
Weibull	.1783	6.430	−.9704	5.000	25.00	2.000	6.000
$\alpha = 1, \beta = 5$				4.995	24.97	1.995	5.897
Weibull	.2069	6.956	−.7494	5.665	50.99	2.815	12.74
$\alpha = 0.8, \beta = 5$				5.658	50.94	2.796	12.56
Weibull	.2444	11.85	−.5404	16.62	2723	11.35	287.6
$\alpha = 0.4, \beta = 5$				16.61	2679	10.45	205.2
Rayleigh	.1165	4.911	−1.881	2.507	1.717	.6311	.2451
$\sigma = 2$				2.510	1.711	.6363	.2458
ExtremeValue	.2500	1.421	0.9359	3.577	1.645	1.139	2.400
$\alpha = 3, \beta = 1$				3.575	1.697	1.082	2.116

TABLE 9.4   Parameters values for the normalizing transformations [(9.2.15), upper entries, and (9.2.16), lower entries], and the resulting first four upper partial moments. The corresponding upper partial moments of the standard normal variate are:

$$m_1 = 1/\sqrt{2\pi} = 0.3989; \quad m_2 = 1/2; \quad m_3 = \sqrt{2\pi} = 0.7979; \quad m_4 = 3/2$$

Dis.	A D	B	C	m_1	m_2	m_3	m_4
Gamma	−.1962	−13.20	−.6731	.3986	.4993	.7965	1.497
$\alpha = 7,\ \beta = 2$	2.619		.3148	.4006	.5021	.7985	.1491
Gamma	−.2230	−7.382	−.4383	.3975	.4972	.7930	1.493
$\alpha = 3,\ \beta = 2$	1.694		0.2943	.4024	.5035	.7957	1.471
Gamma	−.1976	−1.567	.1196	.3892	.4753	.7567	1.472
$\alpha = 1/4,\ \beta = 2$	.4000		.1258	.4315	.5280	.7660	1.238
Weibull	−.2973	−3.233	−.0872	.3919	.4858	.7740	1.469
$\alpha = 1.125,\ \beta = 5$	1.059		.2646	.4073	.5039	.7742	1.373
Weibull	−.2643	−3.233	−.0872	.3919	.4858	.7740	1.469
$\alpha = 1,\ \beta = 5$	.9417		.2352	.4073	.5039	.7742	1.373
Weibull	−.2114	−3.233	−.0872	.3919	.4858	.7740	1.469
$\alpha = 0.8,\ \beta = 5$	.7533		.1881	.4073	.5039	.7742	1.373
Weibull	−.1057	−3.233	−.0872	.3919	.4858	.7740	1.469
$\alpha = 0.4,\ \beta = 5$	.3767		.0941	.4073	.5039	.7742	1.373
Rayleigh	−.5286	−3.233	−.0872	.3919	.4858	.7740	1.469
$\sigma = 2$	1.883		.4703	.4073	.5039	.7742	1.373
ExtremeValue	.6000	4.700	1.100	.3943	.4947	.7961	1.515
$\alpha = 3,\ \beta = 1$	2.791		.0921	.3936	.4954	.8032	1.546

H. SHORE

TABLE 9.5 Parameters values* [Eqs. (9.2.11) and (9.2.12), two-moment fitting] and the resulting skewness (Sk) and kurtosis (Ku) measures. For these moments, the exact figures are the upper entries

Dis.	a	b	λ	Mean	Var.	Sk	Ku
Gamma	.06350	6.072	−1	14.00	28.00	.7559	.8571
$\alpha = 7,\ \beta = 2$				−	−	.7542	.8454
Gamma	.09779	6.181	−1	6.000	12.00	1.155	2.000
$\alpha = 3,\ \beta = 2$				−	−	1.152	1.971
Weibull	.1702	5.950	−1	4.790	18.19	1.676	4.040
$\alpha = 1.125,\ \beta = 5$				−	−	1.697	4.140
Weibull	.1670	6.700	−1.002	5.000	25.00	2.000	6.000
$\alpha = 1,\ \beta = 5$				−	−	2.022	6.106
Weibull	.1701	8.364	−1.001	5.665	50.99	2.815	12.74
$\alpha = 0.8,\ \beta = 5$				−	−	2.836	12.29
Weibull	.1703	16.71	−1	16.62	2723	11.35	287.6
$\alpha = 0.4,\ \beta = 5$				−	−	10.55	208.0
Rayleigh	.1701	3.348	−1	2.507	1.717	.6311	.2451
$\sigma = 2$				−	−	.6378	.2869
ExtremeValue	.00070	−496.7	−1	3.577	1.645	1.139	2.400
$\alpha = 3,\ \beta = 1$				−	−	1.124	2.302

* Unsatisfactory fit was achieved for the gamma with parameters: $\alpha = 1/4,\ \beta = 2$ (Sk= 4)

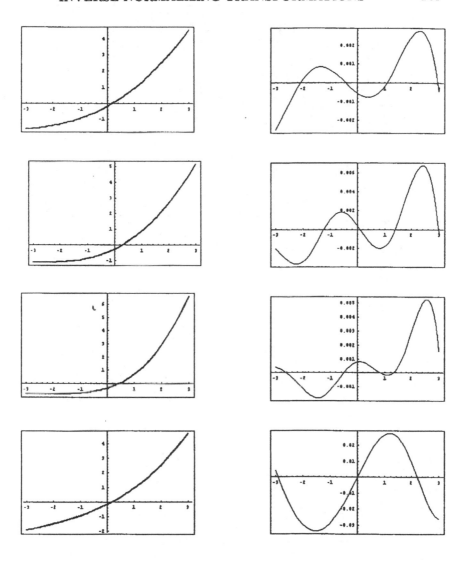

FIGURE 9.1 Plots of the quantile function (approximate and exact; left) and of the error function (approximate minus exact; right) for the (top to bottom):

Gamma(3,2), Weibull(1.125,5), Weibull(0.8,5) and ExtremeValue(3,1).

CHAPTER 10

CURVATURE: GAUSSIAN OR RIEMANN

WILLIAM CHEN

Internal Revenue Service, Washington, DC

Abstract: In this paper, we discuss the Gaussian curvature and discuss some of its basic properties. We also present some illustrative examples.

Keywords and phrases: Gaussian curvature, Riemannian curvature, Gauss equations

10.1 DEFINITION OF THE GAUSSIAN CURVATURE

We start with derivation of second fundamental form. It can be obtained by taking on the surface a curve C passing through a point p and considering the curvature vector of C at p. When t is the unit tangent vector of C then $\kappa = \frac{dt}{ds}$, curvature vector. We now decompose k into a component k_n, normal and a component k_g, tangential to the surface:

$$\frac{dt}{ds} = \kappa = \kappa_n + \kappa_g.$$

The vector κ_n is determined by C alone, not by any choice of the sense of t or N. The vector κ_g is called the tangential curvature vector or geodesic curvature vector.

From the equation $N \star t = 0$, we obtain by differentiation along C:

$$\frac{dt}{ds} \star N + t \star \frac{dN}{ds} = 0$$

$$\kappa_n \star N = -t \star \frac{dN}{ds} = -\frac{dX \star dN}{ds \star ds}$$

$$\kappa_n = -\frac{dx \star dN}{ds \star ds}.$$

Both N and X are surface functions of u and v [we may write $N(u,v)$, $X(u,v)$] which

$$
\begin{aligned}
dN &= N_u du + N_v dv; \\
dX &= X_u du + X_v dv; \\
dX \star dN &= (X_u du + X_v dv) \star (N_u du + N_v dv) \\
&= (X_u \star X_u) du^2 + (X_u N_v + X_v N_u) du dv + X_v N_v dv^2
\end{aligned}
$$

in turn depend on C. With

$$
\begin{aligned}
e &= -X_u \star N_u, \\
2f &= -(X_u \star N_v + X_v \star N_u), \\
g &= -X_v N_v,
\end{aligned}
$$

$$X_u \star N = 0, \qquad X_v \star N = 0,$$

$$
\begin{aligned}
X_{uu} \star N + X_u N_u &= 0, \\
X_{uv} \star N + X_u N_v &= 0, \\
X_{vu} \star N + X_v N_u &= 0, \\
X_{vv} \star N + X_v N_v &= 0
\end{aligned}
$$

differentiation gives

$$
\begin{aligned}
ds^2 = dx \star dx &= (X_u du + X_v dv)(X_u du + X_v dv) \\
&= E\,du^2 + 2F\,du\,dv + G\,dv^2
\end{aligned}
$$

where

$$
\begin{aligned}
E &= X_u \star X_u, \\
F &= X_u \star X_v, \\
G &= X_v \star X_v.
\end{aligned}
$$

These formula allow ready computation of e, f, g when the equation of the surface is given.

The distance of two points P and Q on a curve is found by integrating

$$
\begin{aligned}
2f &= -(X_u N_v + X_v N_u) = X_{uv} \star N + X_{vu} \star N = 2X_{uv} \star N, \\
e &= -X_u \star N_u = X_{uu} \star N, \\
f &= X_{uv} \star N, \\
g &= -X_v \star N_v = X_{vv} \star N,
\end{aligned}
$$

$ds^2 = dx \star dx$ along the curve and substituting for $dX(u, v)$. We find that

$$
e = \frac{(X_{uu} X_u X_v)}{\sqrt{EG - F^2}},
$$

$$
f = \frac{(X_{uv} X_u X_v)}{\sqrt{EG - F^2}},
$$

$$
g = \frac{(X_{vv} X_u X_v)}{\sqrt{EG - F^2}},
$$

where

$$
N = \frac{X_u \times X_v}{|X_u \times X_v|} = \frac{X_u \times X_v}{\sqrt{EG - F^2}},
$$

$$
(X_u \times X_v)(X_u \times X_v) = (X_u \star X_v)(X_u \star X_v) - (X_u \star X_v)^2
$$

$$
= EG - F^2,
$$

$$
|X_u \times X_v| = \sqrt{EG - F^2},
$$

$$
\kappa = -\frac{dX \star dN}{ds \star ds}
$$

$$
= \frac{e\, du^2 + 2f\, du\, dv + g\, dv^2}{E\, du^2 + 2F\, du\, dv + G\, dv^2}
$$

$$
= \frac{II}{I},
$$

where

$$
\begin{aligned}
II &= e\, du^2 + 2f\, du\, dv + g\, dv^2, \\
I &= E\, du^2 + 2F\, du\, dv + G\, dv^2,
\end{aligned}
$$

$$
\begin{aligned}
(Ek - e)(Gk - g) - (Fk - f)^2 &= 0, \\
(EG - F^2)k^2 - (eG - 2Ff + Eg)k + eg - f^2 &= 0.
\end{aligned}
$$

I is the first fundamental form, and II is the second fundamental form.

We see that the right-hand side depends only on u, v and $\frac{dv}{du}$. The coefficients e, f, g, E, F and G are constants at $P(u_0, v_0)$. So that kn is fully determined at $P(u_0, v_0)$ by the direction $\frac{dv}{du}$. All curves through P tangent to the same direction have therefore the same normal curvature.

We can now ask for the directions in which the normal curvature is a maximum or minimum:

$$(E + 2F\lambda + G\lambda^2)(f + g\lambda) - (e + 2f\lambda + g\lambda^2)(F + G\lambda) = 0,$$

$$\kappa = \frac{II}{I} = \frac{f + g\lambda}{F + G\lambda} = \frac{e + f\lambda}{E + F\lambda}$$

or

$$
\begin{aligned}
(E\kappa - e) + (F\kappa - f)\lambda &= 0, \\
(F\kappa - f) + (G\kappa - g)\lambda &= 0
\end{aligned}
$$

where $\lambda = \frac{du}{dv}$. The extreme values of κ can be

$$
\begin{vmatrix}
Ek - e & Fk - f \\
Fk - f & Gk - g
\end{vmatrix} = 0
$$

characterized by $\frac{d\kappa}{d\lambda} = 0$;

$$
\begin{aligned}
\kappa &= \frac{e\,du^2 + 2f\,du\,dv + g\,dv^2}{E\,du^2 + 2F\,du\,dv + G\,dv^2} \\
&= \frac{e + 2f\lambda + g\lambda^2}{E + 2F\lambda + G\lambda^2} \\
&= \kappa(\lambda)
\end{aligned}
$$

which can be simultaneously satisfied if and only if

$$
\begin{aligned}
M &= \frac{k_1 + k_2}{2} = \frac{Eg - 2Ff + eG}{2(EG - F^2)}, \\
K &= k_1 k_2 = \frac{eg - f^2}{EG - F^2}.
\end{aligned}
$$

This quadratic equation in κ has κ_1 and κ_2 as roots. Usually, we call them the principal curvatures κ_1 and κ_2.

We define M as the mean curvature and K as the Gaussian curvature (or total curvature).

10.2 EXAMPLES

Example 10.2.1 We find here the mean curvature and Gaussian curvature of a given monkey saddle parametrized surface:

$$X(u, v, u^3 - 3uv^2)$$
$$X_u(1, 0, 3u^2 - 3v^2)$$
$$X_v(0, 1, -6uv)$$
$$X_{uu}(0, 0, 6u)$$
$$X_{uv}(0, 0, -6v)$$
$$X_{vv}(0, 0, -6u).$$

Therefore,

$$E = 1 + (3u^2 - 3v^2)^2; \quad F = -6uv(3u^2 - 3v^2); \quad G = 1 + 36u^2v^2.$$

Furthermore, by inspection, a surface unit normal is

$$N = \frac{(-3u^2 + 3v^2, 6uv, 1)}{\sqrt{1 + 9u^4 + 18u^2v^2 + 9v^4}}$$

so that

$$e = \frac{6u}{\sqrt{1 + 9u^4 + 18u^2v^2 + 9v^4}},$$
$$f = \frac{-6v}{\sqrt{1 + 9u^4 + 18u^2v^2 + 9v^4}},$$
$$g = \frac{-6u}{\sqrt{1 + 9u^4 + 18u^2v^2 + 9v^4}}.$$

Therefore,

$$K = \frac{-36(u^2 + v^2)}{(1 + 9u^4 + 18u^2v^2 + 9v^4)^2},$$
$$H = \frac{-27u^5 + 54u^3v^2 + 81uv^4}{(1 + 9u^4 + 18u^2v^2 + 9v^4)^{3/2}}.$$

A glance at the expression for K shows that $(0,0,0)$ is a planar point of the monkey saddle, and that the other point is hyperbolic. Furthermore, the Gaussian curvature of the monkey saddle is invariant under all rotation about the z-axis, even though the monkey saddle itself does not have this property.

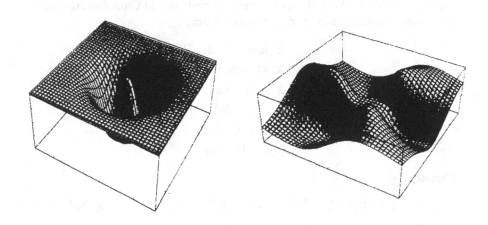

FIGURE 10.1

Example 10.2.2 We find here the mean curvature and Gaussian curvature of the torus parameterized by

$$X((a + b\cos v)\cos u, (a + b\cos v)\sin u, b\sin v)$$
$$X_u(-(a + b\cos v)\sin u, (a + b\cos v)\cos u, 0)$$
$$X_v(-b\sin v\cos u, -b\sin v\sin u, b\cos v)$$

and so $E = (a + b\cos v)^2$, $F = 0$, $G = b^2$. Furthermore,

$$X_{uu}(-(a + b\cos v)\cos u, -(a + b\cos v)\sin u, 0)$$
$$X_{uv}(b\sin v\sin u, -b\sin v\cos u, 0)$$
$$X_{vv}(-b\cos v\cos u, -b\cos v\sin u, b\sin v),$$

$$e = \frac{\begin{vmatrix} -(a + b\cos v)\cos u & -(a + b\cos v)\sin u & 0 \\ -(a + b\cos v)\sin u & (a + b\cos v)\cos u & 0 \\ -b\sin v\cos u & -b\sin v\sin u & b\cos v \end{vmatrix}}{b(a + b\cos v)}$$

$$= -\cos v(a + b\cos v),$$

$$f = \frac{\begin{vmatrix} b\sin v\sin u & -b\sin v\cos u & 0 \\ -(a+b\cos v)\sin u & (a+b\cos v)\cos u & 0 \\ -b\sin v\cos u & -b\sin v\sin u & b\cos v \end{vmatrix}}{b(a+b\cos v)} = 0,$$

$$g = \frac{\begin{vmatrix} -b\cos v\cos u & -b\cos v\sin u & -b\sin v \\ -(a+b\cos v)\sin u & (a+b\cos v)\cos u & 0 \\ -b\sin v\cos u & -b\sin v\sin u & b\cos v \end{vmatrix}}{b(a+b\cos v)} = -b.$$

Therefore,

$$K = \frac{\cos v}{b(a+b\cos v)} \;;\qquad H = \frac{a+2b\cos v}{2b(a+b\cos v)}\;.$$

Thus we see that the Gaussian curvature of the torus:

(a) $\kappa = 0$ along the curves given by $v = \pm\pi/2$. These are the parabolic points of the torus.

(b) $X(u,v)$: the set of hyperbolic points is $\pi/2 < v < 3\pi/2$,

(c) $X(u,v)$: the set of elliptic points is $-\pi/2 < v < \pi/2$.

In the following figure, the dark points are hyperbolic, the light points are elliptic, and the parabolic points have an intermediate color.

FIGURE 10.2

10.3 SOME BASIC PROPERTIES OF GAUSSIAN CURVATURE

Theorem 10.3.1 *A necessary and sufficient condition that a surface be developable is that the Gaussian curvature vanish.*

PROOF We have

$$
\begin{aligned}
eg - f^2 &= (X_u \star N_u)(X_v \star N_v) - (X_u \star N_v)(X_v \star N_u) \\
&= (X_u \times X_v)(N_u \times N_v) \\
&= (N N_u N_v)\sqrt{EG - F^2}
\end{aligned}
$$

which shows that $eg - f^2$ is identical with $(N N_u N_v) = 0$. This can happen either (a) when N_u or N_v vanishes, or (b) when N_u is collinear with N_v (N is perpendicular to Nu and Nv).

In case (a), N depends on only one parameter and the surface is the envelope of a family of ∞ planes, and hence is developable. In case (b) we take as one set of coordinate curves on the surface the asymptotic curves with equation $e\,du^2 + 2f\,du\,dv + g\,dv^2 = (\sqrt{e}\,du + \sqrt{g}\,dv)2 = 0$. If these curves are taken as the curves $v = $ constant in the new coordinate system, then $e = f = 0$ or $X_u N_u = X_v N_u = 0$, and hence, $N_u = 0$ which brings us back to case (a).   □

Theorem 10.3.2 *The differential equation of the developable surface $z = f(x, y)$ is $rt - s^2 = 0$, where*

$$
r = \frac{\partial^2 z}{\partial x^2} \; ; \; s = \frac{\partial^2 z}{\partial x \partial y} \; ; \; t = \frac{\partial^2 z}{\partial y^2} \; ; \; p = \frac{\partial z}{\partial x} \; ; \; q = \frac{\partial z}{\partial x} \; .
$$

PROOF We have

$$
\begin{aligned}
z &= f(x, y); \\
dz &= f_x dx + f_y dy; \\
p &= f_x; \qquad q = f_y; \\
ds^2 &= dx^2 + dy^2 + dz^2 \\
&= dx^2 + dy^2 + (f_x dx + f_y dy)^2 \\
&= (1 + p^2)dx^2 + 2pq\,dx\,dy + (1 + q^2)dy^2; \\
E &= 1 + p^2; \qquad F = pq; \qquad G = 1 + q^2; \\
EG - F^2 &= (1 + p^2)(1 + q^2) - (pq)^2 \\
&= 1 + p^2 + q^2
\end{aligned}
$$

$$
\int \begin{aligned}
&X:(x,y,z=f(x,y)); X_x:(1,0,f_x); X_y:(0,1,f_y); \\
&X_{xx}:(0,0,f_{xx}); X_{xy}:(0,0,f_{xy}); X_{yy}:(0,0,f_{yy});
\end{aligned}
$$

$$
e = \frac{(X_{xx}, X_x, X_y)}{\sqrt{EG - F^2}} = \frac{f_{xx}}{\sqrt{1 + p^2 + q^2}} \; ;
$$

$$f = \frac{(X_{xy}, X_x, X_y)}{\sqrt{EG - F^2}} = \frac{f_{xy}}{\sqrt{1 + p^2 + q^2}} \;;$$

$$g = \frac{(X_{yy}, X_x, X_y)}{\sqrt{EG - F^2}} = \frac{f_{yy}}{\sqrt{1 + p^2 + q^2}} \;;$$

$$eg - f^2 = f_{xx} \star f_{yy} - f_{yy} = rt - s^2 = 0. \qquad \square$$

Illustrative example. We now show that the surface

$$Z = a + bx + cy + \sum_{n=2}^{N} a_n (px + qy)^n,$$

where the coefficients of x and y are all constants, are developable:

$$\frac{\partial z}{\partial x} = b + \sum_{n=2}^{N} a_n np(px + qy)^{n-1},$$

$$\frac{\partial^2 z}{\partial x^2} = \sum_{n=2}^{N} a_n (n - 1)p^2(px + qy)^{n-2},$$

$$\frac{\partial^2 z}{\partial xy} = \sum_{n=2}^{N} a_n (n - 1)npq(px + qy)^{n-2},$$

$$\frac{\partial z}{\partial y} = c + \sum_{n=2}^{N} a_n nq(px + qy)^{n-1},$$

$$\frac{\partial^2 z}{\partial y^2} = \sum_{n=2}^{N} a_n n(n - 1)q^2(px + qy)^{n-2}.$$

PROOF We have

$$rt - s^2 = \frac{\partial^2 z}{\partial x^2} \frac{\partial^2 z}{\partial y^2} - \left[\frac{\partial^2 z}{\partial x \partial y}\right]^2$$

$$= \left[\sum_{n=2}^{N} a_n n(n - 1)p^2(px + qy)^{n-2}\right]$$

$$\times \left[\sum_{n=2}^{N} a_n n(n - 1)q^2(px + qy)^{n-2}\right]$$

$$- \left[\sum_{n=2}^{N} a_n (n - 1)npq(px + qy)^{n-2}\right]^2$$

$$= p^2 q^2 \left[\sum_{n=2}^{N} a_n n(n-1)(px+qy)^{n-2} \right]^2$$

$$- p^2 q^2 \left[\sum_{n=2}^{N} a_n n(n-1)px + qy)^{n-2} \right]^2$$

$$= 0. \qquad \qquad \Box$$

Theorem 10.3.3 *A sufficient small portion of a surface can be mapped isometrically into a plane if and only if it is a portion of a developable surface.*

Theorem 10.3.4 *All surfaces of the same constant Gaussian curvature are isometric.*

PROOF We distinguish among the three cases

$$K > 0, \qquad K = 0, \qquad K < 0.$$

Case 1. Since K satisfies the differential relation

$$K = - \frac{\partial^2 \sqrt{G}}{\sqrt{G} \partial u^2},$$

we find that $\sqrt{G}$ is of the form

$$\sqrt{G} = u c_1(v) + c_2(v).$$

We can impose on G the conditions

$$(\sqrt{G})_{u=0} = 0, \qquad \left(\frac{\partial \sqrt{G}}{\partial u} \right)_{u=0} = 1.$$

Hence, $\sqrt{G} = u$ and $ds^2 = du^2 + u^2 dv^2$.

This expression for ds^2 can be obtained for all surfaces with $K = 0$ by taking on it a geodesic polar coordinate system. All surfaces of zero curvature therefore are isometric. Taking $x = u \cos v$, $y = u \sin v$, we obtain ds^2 in the form $ds^2 = dx^2 + dy^2$, which shows that all developable surfaces can be isometrically mapped on a plane, the curvilinear coordinates corresponding to the rectangular Cartesian coordinates in the plane.

10.4   APPLICATIONS OF THE GAUSS EQUATIONS

Example 10.4.1 A correspondence can be established between the points of a catenoid and of a right helicoid such that at corresponding points the E, F, and G and, therefore, the Gaussian curvature are the same.

The Gaussian curvature of the catenoid is

$$K = \frac{f' f''}{u(1 + f'^2)^2} = \frac{-c^2}{c^4} .$$

Since $f' = \frac{c}{\sqrt{u^2 - c^2}}$, the Gaussian curvature of the right helicoid is

$$K_1 = \frac{-(f_1')^2}{[u_1^2 + (f_1')^2]^2} = \frac{-a^2}{(u_1^2 + a^2)^2} .$$

After substitution, we get $K = K_1$. This is a one-to-one correspondence as long as $-a \leq u \leq a$ and $0 \leq v \leq 2\pi$ for one full turn of the helix. It can be shown that one surface can actually pass into the other by a continuous bending. We show six different stages in this deformation. This can be demonstrated with a flexible piece of brass applied to a plaster model of a catenoid and bent so as to be applied to a model of a right helicoid.

The catenoid is

$$X \left(u \cos v, u \sin v, c \cosh^{-1} \frac{u}{c} \right) .$$

The right helicoid is

$$X(u_1 \cos v_1, u_1 \sin v_1, a v_1).$$

The first fundamental form of the catenoid is

$$ds^2 = \frac{u^2}{u^2 - c^2} \, du^2 + u^2 dv^2,$$

and that of the right helicoid is

$$ds_1^2 = du_1^2 + (u_1^2 + a^2) dv_1^2.$$

If we now write

$$a = c, \quad v = v_1, \quad u = \sqrt{u_1^2 + a^2} \text{ or } u_1 = \sqrt{u^2 - a^2}$$

then $ds_1^2 = ds^2$.

For $0 \leq v \leq 2\pi$ and $-a \leq u \leq +a$, we have thus established a one-to-one correspondence between the points on both surfaces such that at corresponding points the first fundamental forms are equal.

REFERENCES

Gray, A. (1993). *Modern Differential Geometry of Curves and Surfaces*, CRC Press, Boca Raton, Florida.

Kass, R. E. and Vos, P. W. (1997). *Geometrical Foundations of Asymptotic Inference*, John Wiley & Sons, New York.

Kreyszig, E. (1959). *Differential Geometry*, University of Toronto Press, Toronto.

Struik, D. J. (1961). *Lectures on Classical Differential Geometry*, Second Edition, Dover Publications, New York.

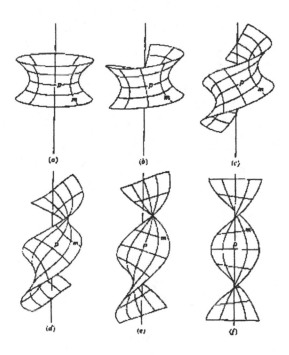

FIGURE 10.3

Part III
Inference

CHAPTER 11

CONVEX GEOMETRY, ASYMPTOTIC MINIMAXITY AND ESTIMATING FUNCTIONS

SCHULTZ CHAN   MALAY GHOSH

University of Florida, Gainesville, FL

Abstract: The paper generalizes the asymptotic minimaxity result of Huber (1964) to the multiparameter situation. The method of proof involves an application of convex geometry as well as the geometric interpretation of optimal estimating functions as orthogonal projections of score functions into appropriate linear subspaces.

Keywords and phrases: Convex geometry, asymptotic minimaxity, estimating functions, orthogonal projections, score functions

11.1 INTRODUCTION

Huber (1964), in his seminal paper, introduced a class of robust estimators whose members are intermediaries between the sample mean and the sample median. He then proved asymptotic minimaxity results related to these estimators. The main result in this context is that if within a convex set C of distribution functions, F_0 has the smallest Fisher information, and one calculates the asymptotic variance of the score function corresponding to F_0 with respect to every distribution in C, then the largest asymptotic variance results when F_0 is the true df. This explains the "max" part of asymptotic minimaxity. The "min" part is explained by the fact that this largest asymptotic variance is the least asymptotic variance within a class of competing functions ϕ (ψ in Huber's notation) involving both the parameter and the data when the asymptotic distribution of every such function ϕ is calculated under F_0. Huber (1964) found it convenient to consider the reciprocal of the asymptotic variance rather than the asymptotic variance itself. He also provided a necessary and sufficient condition for F_0 to have the least Fisher information when compared to other members of C. Also, Huber (1964) considered only the case when the parameter was real valued.

Our objective here is to extend the asymptotic minimaxity result of Huber (1964) to the multiparameter situation. More important, we tie Huber's result with the idea of finding an optimal estimating function.

Research on estimating function was pioneered by Godambe (1960), Kale (1962), and their associates. More recently, optimal estimating functions have been viewed as projections of score functions into appropriate linear subspaces [cf. Small and McLeish (1988, 1989, 1991, 1992, 1994)]. Chan and Ghosh (1998) have recently collected, extended and unified many optimal estimating function results via orthogonal projections. The present paper adopts the same geometric approach.

In addition to this geometric view of optimal estimating functions, we need to extend some convexity results of Huber (1964) to the multiparameter case. This convexity geometry is the second critical component in the derivation of our results.

Section 11.2 contains a simple result on convexity for matrix-valued functions. The multiparameter generalization of Huber's (1964) result is given in Section 11.3. Proof of a Lemma of a somewhat more technical nature is deferred to the Appendix.

11.2   A CONVEXITY RESULT

Let L be a linear space. A subset C of L is said to be convex if for every $x, y \in C$, $\lambda \in [0, 1]$,
$$\lambda x + (1 - \lambda)y \in C.$$
A function $f : C \to R$ is said to be convex if for any $x, y \in C$, $\lambda \in [0, 1]$,
$$f(\lambda x + (1 - \lambda)y) \le \lambda f(x) + (1 - \lambda)f(y).$$
A symmetric matrix-valued function $N : C \to M_{k \times k}$ (i.e., for any $x \in C$, $N(x)$ is a symmetric $k \times k$ matrix) is said to be convex, if for any $x, y \in C$, $\lambda \in [0, 1]$,
$$N(\lambda x + (1 - \lambda)y) \le \lambda N(x) + (1 - \lambda)N(y),$$
where for two $k \times k$ matrices A, B, $A \le B$ means that $B - A$ is nonnegative definite (n.n.d.). In the following, we only study properties of matrix valued convex functions.

For every $x, y \in C$, consider the function

$$N(\lambda; x, y) = N((1 - \lambda)x + \lambda y)$$

on $[0, 1]$. Then $N(\lambda; x, y)$ is a convex function on λ. The directional derivative of N at x in the direction of y is defined as

$$F_N(x; y) = \lim_{\lambda \to 0^+} \frac{N(\lambda; x, y) - N(0; x, y)}{\lambda}. \tag{11.2.1}$$

The existence of the limit is justified as follows: Since

$$\lambda_1 = \left(1 - \frac{\lambda_1}{\lambda_2}\right) 0 + \frac{\lambda_1}{\lambda_2} \lambda_2,$$

for $0 < \lambda_1 < \lambda_2 < 1$,

$$N(\lambda_1; x, y) \le \left(1 - \frac{\lambda_1}{\lambda_2}\right) N(0; x, y) + \frac{\lambda_1}{\lambda_2} N(\lambda_2; x, y).$$

This implies

$$\frac{N(\lambda_1; x, y) - N(0; x, y)}{\lambda_1} \le \frac{N(\lambda_2; x, y) - N(0; x, y)}{\lambda_2}, \tag{11.2.2}$$

that is $\frac{N(\lambda; x, y) - N(0; x, y)}{\lambda}$ is a nondecreasing function of λ in $(0, 1]$. Hence the limit in (11.2.1) is well defined.

From (11.2.1) and (11.2.2), for a convex function N,

$$F_N(x; y) \le N(1; x, y) - N(0; x, y) = N(y) - N(x),$$
$$\text{for all } x, y \in C. \tag{11.2.3}$$

The following result will be used repeatedly in the sequel.

Theorem 11.2.1 *Suppose that N is convex. Then for $x_0 \in C$, $N(x_0) \le N(y)$ for all $y \in C$ if and only if*

$$F_N(x_0; y) \ge 0 \tag{11.2.4}$$

for all $y \in C$.

PROOF Suppose that $N(x_0) \le N(y)$ for all $y \in C$. Then $\frac{N(\lambda; x_0, y) - N(0; x_0, y)}{\lambda}$ $(0 < \lambda \le 1)$ is non-negative definite. Hence

$$F_N(x_0; y) = \lim_{\lambda \to 0^+} \frac{N(\lambda; x_0, y) - N(0; x_0, y)}{\lambda}$$

is n.n.d., for all $y \in C$.

Conversely, if $F_N(x_0; y) \ge 0$ for all $y \in C$, then from (11.2.3),

$$N(y) - N(x_0) \ge F_N(x_0; y) \ge 0.$$

Thus $N(x_0) \le N(y)$ for all $y \in C$. □

11.3 ASYMPTOTIC MINIMAXITY

In this section, the famous asymptotic minimaxity result of Huber (1964) will be generalized. We begin with the sample space $\mathcal{X}$, a parameter space Θ which is an open subset of R^k. Let C be a convex set of distribution functions such that every $F \in C$ has an absolutely continuous density f such that

$$I(F) = E\left[\left(\frac{\partial \log f}{\partial \theta}\right)\left(\frac{\partial \log f}{\partial \theta}\right)^t \mid F, \theta\right] \qquad (11.3.5)$$

is positive definite. Consider functions $g : \mathcal{X} \times \Theta \to R^k$ satisfying the following four properties:

(i) $E(g \mid F, \theta) = 0$ for all $F \in C$ and $0 \in \Theta$;

(ii) $\frac{\partial g}{\partial \theta}$ exists for almost all $x \in \mathcal{X}$ and all $\theta \in \Theta$;

(iii) $E\left[\frac{\partial g}{\partial \theta} \mid F, \theta\right] = -E[g\, s^t \mid F, \theta]$, where $s = \frac{\partial \log f}{\partial \theta}$;

(iv) $E\left[\frac{\partial g}{\partial \theta} \mid F, \theta\right]$ is nonsingular.

A function g satisfying (i) is called an *unbiased estimating function* [cf. Godambe (1960)]. This unbiasedness criterion does not impose any restriction on g since if $E(g|\theta) \neq 0$, one can work with $g^* = g - E(g|\theta)$. A function g satisfying (i)–(iv) will be referred to as a regular unbiased estimating function.

Let L be the space of unbiased estimating functions with respect to C, i.e., every element of L is unbiased with respect to every distribution in C. Let L_0 be the subset of L which consists of all regular unbiased estimating functions in L.

Consider the function $K : L_0 \times C \to M_{k \times k}$ defined by

$$K(\phi, F) = E\left[\frac{\partial \phi}{\partial \theta} \mid F, \theta\right]^t (E[\phi\phi^t | F, \theta])^{-1} E\left[\frac{\partial \phi}{\partial \theta} \mid F, \theta\right], \qquad (11.3.6)$$

for all $\phi \in L_0$, $F \in C$. Note that when $k = 1$, then

$$K(\phi, F) = \frac{(\int_{\mathcal{X}} \phi f' dx)^2}{\int_{\mathcal{X}} \phi^2 f dx},$$

which is the expression given in Huber (1964).

For every $F \in C$, for any $g_1, g_2 \in L$, the inner product of g_1 and g_2 is defined by

$$\langle g_1, g_2 \rangle_F = E[g_1 g_2^t | F].$$

For any $g \in L$, $y_0 \in L_0$ is said to be an orthogonal projection of g into L_0 if

$$\langle g - y_0, y \rangle_F = 0$$

for all $y \in L_0$, i.e., $g - y_0$ is orthogonal to every element of L_0. Further, if the orthogonal projection exists, then it is unique. This result available in Small and McLeish (1994), is given more explicitly in Theorem 1 of Chan and Ghosh (1998).

In the following, for every $F \in C$, the orthogonal projection of the score function of F into the subspace L_0 with respect to the inner product $\langle \cdot, \cdot \rangle_F$ (if it exists) is denoted by ϕ_F.

Before proceeding further, we need the following lemma involving matrix convexity. Its proof is technical, and is deferred to the Appendix.

Lemma 11.3.1 *For any* $(M_1, M_2) \in M_{k \times k} \times M_{k \times k}^+$, *where* $M_{k \times k}^+$ *denotes the set of all $k \times k$ positive definite matrices, the matrix valued function* $M_1^t M_2^{-1} M_1$ *is convex in the sense that, for any* (M_1, M_2), $(M_3, M_4) \in M_{k \times k} \times M_{k \times k}^+$, $\lambda \in (0, 1)$,

$$J(\lambda) = [\lambda M_1 + (1 - \lambda) M_3]^t [\lambda M_2 + (1 - \lambda) M_4]^{-1} [\lambda M_1 + (1 - \lambda) M_3],$$

is convex in λ.

Next using the geometry of optimal estimating functions, a necessary and sufficient condition for the asymptotic minimaxity of estimating functions in the multiparameter case will be given. This result generalizes the asymptotic minimaxity result of Huber (1964).

Theorem 11.3.1 *Suppose the parameter space is multi-dimensional. Then* (ϕ_{F_0}, F_0) *is a saddle point of K, that is*

$$K(\phi, F_0) \leq K(\phi_{F_0}, F_0) \leq K(\phi_{F_0}, F),$$

for all $\phi \in L_0$, *and* $F \in C$, *if and only if*

$$\int_{\mathcal{X}} \left[\phi_{F_0} \left(\frac{\partial f}{\partial \theta} - \frac{\partial f_0}{\partial \theta} \right)^t + \left(\frac{\partial f}{\partial \theta} - \frac{\partial f_0}{\partial \theta} \right) (\phi_{F_0})^t - \phi_{F_0} \phi_{F_0}^t (f - f_0) \right] dx$$

$$(11.3.7)$$

is non-negative definite.

PROOF Let s_{F_0} denote the score function for F_0. Note that since ϕ_{F_0} is the orthogonal projection of s_{F_0} into L_0,

$$K(\phi, F_0) \leq K(\phi_{F_0}, F_0)$$

for all $\phi \in L_0$. This result follows from Theorem 5 of Chan and Ghosh (1996).

Also for any $F \in C$, consider the function

$$J_F : [0, 1] \to M_{k \times k},$$

given by

$$
\begin{aligned}
J_F(\lambda) = {} & \left(\int_{\mathcal{X}} \phi_{F_0} \left[(1-\lambda) \frac{\partial f_0}{\partial \theta} + \lambda \frac{\partial f}{\partial \theta} \right]^t dx \right)^t \\
& \times \left(\int_{\mathcal{X}} \phi_{F_0} \phi_{F_0}^t [(1-\lambda) f_0 + \lambda f] dx \right)^{-1} \\
& \times \left(\int_{\mathcal{X}} \phi_{F_0} \left[(1-\lambda) \frac{\partial f_0}{\partial \theta} + \lambda \frac{\partial f}{\partial \theta} \right]^t dx \right). \quad (11.3.8)
\end{aligned}
$$

From Lemma 11.3.1, J_F is convex, and by direct calculation,

$$
\begin{aligned}
J_F(0+) = {} & (M_1 - M_3)^t M_4^{-1} M_3 - M_3^t M_4^{-1} (M_2 - M_4) M_4^{-1} M_3 \\
& + M_3^t M_4^{-1} (M_1 - M_3),
\end{aligned}
$$

where

$$
\begin{aligned}
M_1 &= \int_{\mathcal{X}} \phi_{F_0} \left(\frac{\partial f^t}{\partial \theta} \right) dx, \quad M_3 = \int_{\mathcal{X}} \phi_{F_0} \left(\frac{\partial f_0^t}{\partial \theta} \right) dx, \\
M_2 &= \int_{\mathcal{X}} \phi_{F_0} \phi_{F_0}^t f \, dx, \quad M_4 = \int_{\mathcal{X}} \phi_{F_0} \phi_{F_0}^t f_0 \, dx.
\end{aligned}
$$

Since ϕ_{F_0} is the orthogonal projection of s_{F_0} into L_0 with respect to $\langle \cdot, \cdot \rangle_{F_0}$, $M_3 = M_4$. Hence,

$$
\begin{aligned}
J_F(0+) = {} & (M_1 - M_3)^t + (M_1 - M_3) - (M_2 - M_4) \\
= {} & \int_{\mathcal{X}} \left[\phi_{F_0} \left(\frac{\partial f}{\partial \theta} - \frac{\partial f_0}{\partial \theta} \right)^t + \left(\frac{\partial f}{\partial \theta} - \frac{\partial f_0}{\partial \theta} \right) (\phi_{F_0})^t \right. \\
& \left. - \phi_{F_0} \phi_{F_0}^t (f - f_0) \right] dx. \quad (11.3.9)
\end{aligned}
$$

Only if: Suppose that (ϕ_{F_0}, F_0) is a saddle point of K. Then for any $F \in C$, and every $\lambda \in (0, 1)$,

$$J_F(0) = K(\phi_{F_0}, F_0) \leq K(\phi_{F_0}, (1-\lambda) F_0 + \lambda F) = J_F(\lambda).$$

From the definition of $J_F(\lambda)$, $J_F'(0+)$ is non-negative definite. Hence,

$$\int_{\mathcal{X}} \left[\phi_{F_0} \left(\frac{\partial g}{\partial \theta} \right)^t + \frac{\partial g}{\partial \theta} (\phi_{F_0})^t - \phi_{F_0} \phi_{F_0}^t g \right] dx$$

is non-negative definite, where $g = f - f_0$.

If: Now suppose that

$$\int_{\mathcal{X}} \left[\phi_{F_0} \left(\frac{\partial g}{\partial \theta} \right)^t + \frac{\partial g}{\partial \theta} (\phi_{F_0})^t - \phi_{F_0} \phi_{F_0}^t g \right] dx$$

is non-negative definite, where $g = f - f_0$. Then from Theorem 11.2.1, J_F is a monotone function in $[0,1]$. Hence,

$$J_F(0) = K(\phi_{F_0}, F_0) \le J_F(1) = K(\phi_{F_0}, F_1).$$

Hence, (ϕ_{F_0}, F_1) is a saddle point of K. This completes the proof. $\square$

Corollary 11.3.1 *Assume that $F_0 \in C$ is such that $I(F_0) \le I(F)$ for all $F \in C$, and $s_{F_0} = \frac{\nabla f_0}{f_0} \in L_0$. Then (s_{F_0}, F_0) is a saddle point of K.*

PROOF For any $F \in C$, consider the function

$$
\begin{aligned}
Q_F(\lambda) &= I((1-\lambda)F_0 + \lambda F) \\
&= \int_{\mathcal{X}} \frac{\partial(f_0 + \lambda(f_0 - f))}{\partial \theta} \left(\frac{\partial(f_0 + \lambda(f_0 - f))}{\partial \theta} \right)^t \frac{1}{f_0 + \lambda(f - f_0)} dx.
\end{aligned}
$$

Then by Lemma 11.3.1, Q_F is convex, and attains its minimum at $\lambda = 0$. Thus, since $s(F_0) = \phi_{F_0}$,

$$Q_F'(0+) = \int_{\mathcal{X}} \left[\phi_{F_0} \left(\frac{\partial g}{\partial \theta} \right)^t + \frac{\partial g}{\partial \theta} (\phi_{F_0})^t - \phi_{F_0} \phi_{F_0}^t g \right] dx \qquad (11.3.10)$$

is non-negative definite. The above equality follows from the Lebesgue dominated convergence theorem and the facts that

$$\frac{1}{\lambda} \left[\frac{\frac{\partial f_\lambda}{\partial \theta} \left(\frac{\partial f_\lambda}{\partial \theta} \right)^t}{f_\lambda} - \frac{\frac{\partial f_0}{\partial \theta} \left(\frac{\partial f_0}{\partial \theta} \right)^t}{f_0} \right]$$

$$\to \frac{\partial f_0}{\partial \theta} \frac{1}{f_0} \left(\frac{\partial g}{\partial \theta} \right)^t + \frac{\partial g}{\partial \theta} \left(\frac{\partial f_0}{\partial \theta} \right)^t \frac{1}{f_0} - \frac{\frac{\partial f_0}{\partial \theta} \left(\frac{\partial f_0}{\partial \theta} \right)^t}{(f_0)^2} g,$$

and

$$\frac{1}{\lambda}\left[\frac{\frac{\partial f_\lambda}{\partial \theta}\left(\frac{\partial f_\lambda}{\partial \theta}\right)^t}{f_\lambda} - \frac{\frac{\partial f_0}{\partial \theta}\left(\frac{\partial f_0}{\partial \theta}\right)^t}{f_0}\right] \leq \frac{\frac{\partial f}{\partial \theta}\left(\frac{\partial f}{\partial \theta}\right)^t}{f} - \frac{\frac{\partial f_0}{\partial \theta}\left(\frac{\partial f_0}{\partial \theta}\right)^t}{f_0}.$$

APPENDIX

PROOF OF LEMMA 11.3.1 By straightforward calculation, and using repeatedly the relation

$$\frac{dM^{-1}}{d\lambda} = -M^{-1}\frac{dM}{d\lambda}M^{-1},$$

one gets

$$\begin{aligned}
\frac{dJ(\lambda)}{d\lambda} &= (M_1 - M_3)^t[\lambda M_2 + (1-\lambda)M_4]^{-1}[\lambda M_1 + (1-\lambda)M_3]\\
&\quad -[\lambda M_1 + (1-\lambda)M_3]^t[\lambda M_2 + (1-\lambda)M_4]^{-1}\\
&\quad \times (M_2 - M_4)[\lambda M_2 + (1-\lambda)M_4]^{-1}[\lambda M_1 + (1-\lambda)M_3]\\
&\quad +[\lambda M_1 + (1-\lambda)M_3]^t[\lambda M_2 + (1-\lambda)M_4]^{-1}(M_1 - M_3),
\end{aligned}$$

(11.3.11)

and

$$\begin{aligned}
\frac{d^2 J(\lambda)}{d\lambda^2} &= 2\{(M_1 - M_3)^t[\lambda M_2 + (1-\lambda)M_4]^{-1}(M_1 - M_3)\\
&\quad \times [\lambda M_1 + (1-\lambda)M_3]^t[\lambda M_2 + (1-\lambda)M_4]^{-1}(M_2 - M_4)\\
&\quad \times [\lambda M_2 + (1-\lambda)M_4]^{-1}\\
&\quad \times (M_2 - M_4)[\lambda M_2 + (1-\lambda)M_4]^{-1}\\
&\quad \times [\lambda M_1 + (1-\lambda)M_3]\\
&\quad -(M_1 - M_3)^t[\lambda M_2 + (1-\lambda)M_4]^{-1}(M_2 - M_4)\\
&\quad \times [\lambda M_2 + (1-\lambda)M_4]^{-1}\\
&\quad \times [\lambda M_1 + (1-\lambda)M_3]\\
&\quad \times [\lambda M_1 + (1-\lambda)M_3]^t[\lambda M_2 + (1-\lambda)M_4]^{-1}(M_2 - M_4)\\
&\quad \times [\lambda M_2 + (1-\lambda)M_4]^{-1}(M_1 - M_3)\}\\
&= 2(AA^t + B^tB - AB - B^tA^t) = 2(A - B^t)(A - B^t)^t \geq 0,
\end{aligned}$$

(11.3.12)

where

$$A = (M_1 - M_3)^t[\lambda M_2 + (1-\lambda)M_4]^{-1/2},$$

$$B = [\lambda M_2 + (1-\lambda)M_4]^{-1/2}(M_2 - M_4)[\lambda M_2 + (1-\lambda)M_4]^{-1}[\lambda M_1 + (1-\lambda)M_3].$$

This completes the proof of the Lemma.   □

Acknowledgement Research partially supported by NSF Grant Numbers SBR-9423996 and SBR-9810968.

REFERENCES

Chan, S. and Ghosh, M. (1998). Orthogonal projections and geometry of estimating functions, *Journal of Statistical Planning and Inference*, **67**, 227–245.

Godambe, V. P. (1960). An optimum property of a regular maximum likelihood estimation, *Annals of Mathematical Statistics*, **31**, 1208–1212.

Huber, P. J. (1964). Robust estimation of a location parameter, *Annals of Mathematical Statistics*, **35**, 73–101.

Kale, B. K. (1962). An extension of Cramer-Rao inequality for statistical estimation functions, *Skandinavisk Aktuarietidskrift*, **45**, 60–89.

McLeish, D. L. and Small, C. (1992). A projected likelihood function for semiparametric models, *Biometrika*, **79**, 93–102.

Small, C. and McLeish, D. L. (1988). Generalization of ancillarity, completeness and sufficiency in an inference function space, *Annals of Statistics*, **16**, 534–551.

Small, C. and McLeish, D. L. (1989). Projection as a method for increasing sensitivity and eliminating nuisance parameters, *Biometrika*, **76**, 693–703.

Small, C. and McLeish, D. L. (1991). Geometrical aspects of efficiency criteria for spaces of estimating functions, In *Estimating Functions* (Ed., V. P. Godambe), pp. 267–276, Oxford University Press, England, Oxford.

Small, C. and McLeish, D. L. (1994). *Hilbert Space Methods in Probability and Statistical Inference*, John Wiley & Sons, New York.

CHAPTER 12

NONNORMAL FILTERING VIA ESTIMATING FUNCTIONS

A. THAVANESWARAN

University of Manitoba, Winnipeg, Manitoba, Canada

M. E. THOMPSON

University of Waterloo, Waterloo, Ontario, Canada

Abstract: A result of Godambe (1999) on optimal combination of estimating functions for discrete time stochastic processes, extended to non-orthogonal estimating functions, is applied to nonnormal filtering problems, in which the posterior mode of the signal distribution is efficient. The extensions so obtained may be applicable in a wider context than the standard notions based upon the conditional mean. Recursive formulas are implied, and their properties are considered.

Keywords and phrases: Censored correlated data, estimating functions, recursive filtering, nonnormal filtering, posterior mode, biostatistical time series

12.1 INTRODUCTION

The state space approach is a useful tool for nonstationary time series [Kitagawa (1987)]. Recently, efforts have been made to extend the ordinary Kalman filter methodology in various directions [Thavaneswaran and Thompson (1986, 1988), Thompson and Thavaneswaran (1999), Kitagawa (1987) and Naik-Nimbalkar and Rajarshi (1995)]. The purpose of this paper is to explore further the implications for filtering of a recent optimality result of Godambe (1999). The main application is to filtering for discrete time series models.

Much filtering theory, both in the stationary process context and otherwise, is based on ad hoc application of minimum mean squares or minimum dispersion methods. This has the advantage of avoiding specific distributional assumptions. However, the standard results can sometimes be improved and the general estimating function theory which is now available provides a clear focus on particular classes of estimating functions and, within these, leads to optimal filters such that the information associated with the combined estimating function is large. This theory is not yet widely used in filtering and it deserves attention both for its concentration on information issues and for the focus it gives on the role of the choice of a family of estimating functions from which to choose an optimal one. Applications of orthogonal combination of estimating functions for prediction problems are given in Thavaneswaran and Heyde (1999).

The following basic results on combining estimating equations are implicit in the paper of Godambe (1999), and references therein. Consider a probability space (Ω, F, P), on which ξ and θ are jointly distributed random variables, and θ is real valued. In this context an estimating function for θ is a real valued function $g(\xi, \theta)$ or $g(\theta)$, and it is unbiased if $Eg(\xi, \theta)$ or $Eg(\theta)$ is zero. Define $Eff(g)$, the efficiency or "information" associated with g, by

$$Eff(g) = \left(E\frac{\partial g}{\partial \theta}\right)^2 / Eg^2 . \tag{12.1.1}$$

Note that if $\hat{\theta}$ satisfies $g(\xi, \hat{\theta}) = 0$ or $g(\hat{\theta}) = 0$, $[Eff(g)]^{-1}$ approximates the mean squared error $E(\hat{\theta} - \theta)^2$.

The first result gives the optimal combination of a pair of estimating functions unbiased with respect to the conditional distribution of ξ given θ and the marginal (prior) distribution of θ, respectively.

Theorem 12.1.1 *Consider estimating functions $h_1(\xi, \theta)$ and $h_2(\theta)$ such that $E(h_1(\xi, \theta) \mid \theta) = 0$ and $Eh_2(\theta) = 0$. Let h take the form*

$$h(\xi, \theta) = c_1 h_1(\xi, \theta) + c_2 h_2(\theta). \tag{12.1.2}$$

Then

(i) $Eff(h)$ is maximized when

$$c_1 = \left(E\frac{\partial h_1}{\partial \theta}\right) / Eh_1^2 , \quad c_2 = \left(E\frac{\partial h_2}{\partial \theta}\right) / Eh_2^2$$

(ii) When c_1 and c_2 are chosen as in (i),

$$E(h^2) = Eff(h_1) + Eff(h_2)$$

and

$$Eff(h) = Eff(h_1) + Eff(h_2) \ .$$

The second result affirms the optimality of the "posterior score" among functions of the form (12.1.2).

Theorem 12.1.2 *Among all linear combinations*

$$h(\xi, \theta) = c_1 h_1(\xi, \theta) + c_2 h_2(\theta)$$

of estimating function pairs $h_1(\xi, \theta)$ and $h_2(\theta)$ such that $E(h_1(\xi, \theta) \mid \theta) = 0$ and $Eh_2(\theta) = 0$, the efficiency $Eff(h)$ is maximized when

$$h_1(\xi, \theta) = \frac{\partial \ln f(\xi \mid \theta)}{\partial \theta} \ ,$$

$$h_2(\theta) = \frac{\partial \ln \lambda(\theta)}{\partial \theta}$$

$c_1 = c_2 = 1.$

In that case

$$Eff(h) = E(I(\theta)) + I_\lambda$$

where

$$I(\theta) = -E\left(\frac{\partial^2 \ln f(\xi \mid \theta)}{\partial \theta^2}\mid\theta\right)$$

$$I_\lambda = -E\frac{\partial^2 \ln \lambda(\theta)}{\partial \theta^2} = E\left(\frac{\partial \ln \lambda(\theta)}{\partial \theta}\right)^2 \ .$$

In this paper, we will illustrate the ideas of combining the estimating functions in a number of linear non-Gaussian process filtering problems. In Section 12.2, the combination of estimating functions approach is extended to non-orthogonal cases. In Section 12.3 it is indicated how the approach might be extended to cases of more general (non-normal) state space models.

12.2 LINEAR AND NONLINEAR FILTERS

The linear filter is optimal for Gaussian processes and (in the minimum mean square error or MMSE sense) for linear time series models with martingale difference innovations [Hannan and Heyde (1972)]. When one leaves this context it becomes important to assess possible improvements in efficiency associated with various extensions. In this section we give an example where the MMSE *linear* filter is the same as the optimal *linear* filter (in the sense of the estimating function criterion) and then show that the optimal (nonlinear) filter has approximately the same efficiency as the MMSE (nonlinear) filter, the posterior mean.

Let us consider the question of obtaining a filtered estimate of θ given $\xi = \theta + \eta$, where θ and η are independent with means zero and known variances σ_θ^2 and σ_η^2 respectively. The best linear filter of θ in the minimum mean square sense is $\alpha\xi$, where

$$\alpha = \frac{\sigma_\theta^2}{\sigma_\theta^2 + \sigma_\eta^2}$$

and the variance of the filtering error is

$$P = \frac{\sigma_\theta^2 \sigma_\eta^2}{\sigma_\theta^2 + \sigma_\eta^2}.$$

It can be shown that the optimal linear filter in the estimating function sense can be obtained as a root of the combined [Heyde (1987) and Godambe (1999)] linear estimating functions for θ ,

$$g_L^*(\xi, \theta) = \frac{\theta - E\theta}{\sigma_\theta^2} - \frac{\xi - \theta}{\sigma_\eta^2}. \tag{12.2.3}$$

The root is

$$\theta^* = \frac{\sigma_\theta^2}{\sigma_\theta^2 + \sigma_\eta^2}\xi,$$

and the information associated with the estimating function (12.2.3) is given by

$$Eff(g_L^*) = \frac{1}{\sigma_\theta^2} + \frac{1}{\sigma_\eta^2} = P^{-1}.$$

Thus the MMSE linear filter and the optimal linear filter coincide.

The best filter (not necessarily linear) for θ in the mean square sense is $E[\theta|\xi]$. Suppose now that η is normal and that θ is nonnormal with density $\lambda(.)$. Under differentiability conditions it can easily be shown that

$$E[\theta|\xi] = \xi + \sigma_\eta^2 \frac{\partial \ln \lambda(\theta)}{\partial \theta} \Big|_{\theta=\xi} + o(\sigma_\eta^2), \tag{12.2.4}$$

$$E[[\theta - E[\theta|\xi]]^2|\xi] \le E[[\theta - \alpha\xi]^2|\xi], \tag{12.2.5}$$

and the difference is given by

$$(\alpha\xi - E[\theta|\xi])^2 = ((1 - \alpha)\xi + \frac{\partial \ln \lambda(\theta)}{\partial \theta} \Big|_{\theta=\xi} \sigma_\eta^2 + o(\sigma_\eta^2))^2, \tag{12.2.6}$$

Thus, when the density of θ is known the filter $E[\theta \mid \xi]$, which has smaller mean square error than the best linear filter, can be computed, and approximated by (12.2.4). In the sense of Godambe (1999) the optimal estimating function for θ is

$$g^*(\xi, \theta) = -\frac{\partial \ln \lambda(\theta)}{\partial \theta} - \frac{\xi - \theta}{\sigma_\eta^2}, \qquad (12.2.7)$$

which gives the equation for the posterior mode. The information associated with g^* is given by

$$Eff(g^*) = I_\lambda + \frac{1}{\sigma_\eta^2}$$

where $I_\lambda = E[\frac{\partial \ln \lambda(\theta)}{\partial \theta}]^2$; and this information is maximal.

Since the equation for the approximate minimum mean square error filter (12.2.4) is close to the Equation (12.2.7), we can argue that for small σ_η^2 the minimum mean square error filter has approximately the same information and the same efficiency as the posterior mode.

In a slightly more refined approximation, when η is normally distributed,

$$E[\theta \mid \xi] \simeq \xi + \sigma_\eta^2 \frac{\lambda'(\xi)}{\lambda(\xi)} \left[1 + \frac{\lambda'''(\xi)}{2\lambda'(\xi)}\sigma_\eta^2 - \frac{\lambda''(\xi)}{2\lambda(\xi)}\sigma_\eta^2\right]$$

while the posterior mode [solution of (12.2.7)] is

$$\hat{\theta} \simeq \xi + \sigma_\eta^2 \frac{\lambda'(\xi)}{\lambda(\xi)} \left[1 + \frac{\lambda''(\xi)}{2\lambda(\xi)}\sigma_\eta^2 - \frac{[\lambda'(\xi)]^2}{\lambda^2(\xi)}\sigma_\eta^2\right] .$$

Thus the two estimates differ when σ_η^2 is not very small, and the equation for $E[\theta \mid \xi]$ is suboptimal under Godambe's criterion.

Note It follows from Godambe (1999) that the information associated with the posterior score function (12.2.7) is maximal. However the corresponding estimate need not have minimum mean square error. Moreover if, for example, ξ has a Cauchy distribution then the MMSE filter of θ is not defined but the information associated with the posterior score is well defined.

12.2.1   Optimal Combination Extension

The result of Godambe (1999) uses combinations of orthogonal estimating functions. We now give some combination results when the components need not be orthogonal, and apply them in the normal case. Again consider a probability space $(\Omega, \mathcal{F}, \mathcal{P})$, on which ξ and θ are jointly distributed and θ is real valued.

Let $g_1(\xi, \theta)$, $g_2(\xi, \theta)$ be fixed unbiased estimating functions having finite and positive variances, and such that the expectations of $\partial g_1/\partial \theta$ and $\partial g_2/\partial \theta$ are finite, with $E[\partial g_1/\partial \theta] \neq 0$.

The following theorem can be used to obtain the filtering equations in a wider context. It is reproduced from Thompson and Thavaneswaran (1999).

Theorem 12.2.1 *In the class of all unbiased estimating functions*

$$g = g_1 + cg_2,$$

(i) the one which minimizes Var g *is given by*

$$g^* = g_1 + C^* g_2$$

where

$$C^* = -\mathrm{Cov}(g_1, g_2)/\mathrm{Var}\ g_2$$

and

(ii) the one which maximizes $Eff(g)$ *is given by*

$$g^0 = g_1 + C^0 g_2$$

where

$$C^0 = \frac{E\left[\frac{\partial g_2}{\partial \theta}\right] \mathrm{Var}\ g_1 - E\left[\frac{\partial g_1}{\partial \theta}\right]\ \mathrm{Cov}(g_1, g_2)}{E\left[\frac{\partial g_1}{\partial \theta}\right] \mathrm{Var}\ g_2 - E\left[\frac{\partial g_2}{\partial \theta}\right]\ \mathrm{Cov}(g_1, g_2)}.$$

PROOF The proof follows by evaluating the expressions and optimizing w.r.t. C.   □

Notes

(i) The two criteria are equivalent if $E\left[\frac{\partial g_2}{\partial \theta}\right] = 0$, in which case $C^* = C^0 = -\mathrm{Cov}\ (g_1, g_2)/\mathrm{Var}\ g_2$.

(ii) When g_1^0 and g_2^0 are orthogonal and information unbiased estimating functions as in Godambe (1999), then $C^0 = 1$, and the optimal combined estimating function is

$$g^0 = g_1^0 + g_2^0 .$$

We now show how to apply the results above to the updating of a filter in a state space model.

12.3 APPLICATIONS TO STATE SPACE MODELS

12.3.1 Linear State Space Models

Again assuming a probability space $(\Omega, \mathcal{F}, \mathcal{P})$, consider the following state space model in discrete time:

$$
\begin{aligned}
\theta_{t+1} &= a(t)\theta_t + c(t) + b(t)u_{t+1} \\
\xi_{t+1} &= A(t)\theta_{t+1} + B(t)v_{t+1}
\end{aligned}
$$

where $\{\theta_t\}$ is an unobserved sequence of random variables, $\{\xi_t\}$ is an observed sequence of random variables and $\{u_t\}, \{v_t\}$ are independent sequences of independent variables having mean 0 and variance σ_u^2, σ_v^2 respectively. The functions $a(t), b(t), c(t), A(t)$ and $B(t)$ are $\mathcal{F}_t^\xi$-measurable, i.e. functions of the observations up to time t.

Let $\tilde{\theta}_t = E(\theta_t|\mathcal{F}_t^\xi)$, $\gamma_t = \text{Var}(\theta_t|\mathcal{F}_t^\xi)$.

Since it is in some ways natural to think of the innovations as elementary estimating function components, consider combinations of

$$
\begin{aligned}
g_1 &= \theta_{t+1} - a(t)\tilde{\theta}_t - c(t), \\
g_2 &= \xi_{t+1} - A(t)a(t)\tilde{\theta}_t - A(t)c(t),
\end{aligned}
$$

where θ_{t+1} plays the role of θ in Section 12.2. It is easy to show that $E(g_1|\mathcal{F}_t^\xi) = 0$ and $E(g_2|\mathcal{F}_t^\xi) = 0$. The 'optimal' combination will be

$$
\theta_{t+1} - a(t)\tilde{\theta}_t - \frac{\text{Cov}(g_1, g_2|\mathcal{F}_t^\xi)}{\text{Var}(g_2|\mathcal{F}_t^\xi)}(\xi_{t+1} - A(t)a(t)\tilde{\theta}_t - A(t)c(t)) . \quad (12.3.8)
$$

Now $g_2 = \xi_{t+1} - A(t)\theta_{t+1} + A(t)g_1$, and $E\left((\xi_{t+1} - A(t)\theta_{t+1})g_1|\mathcal{F}_t^\xi\right) = 0$. Thus, $\text{Cov}(g_1, g_2|\mathcal{F}_t^\xi) = A(t)\,\text{Var}(g_1 > |\mathcal{F}_t^\xi)$. This gives as 'optimal' estimate of θ_{t+1}

$$
\hat{\theta}_{t+1} = a(t)\tilde{\theta}_t + c(t) + \frac{A(t)[a^2(t)\gamma_t + b^2(t)\sigma_u^2](\xi_{t+1} - A(t)a(t)\tilde{\theta}_t - A(t)c(t))}{B^2(t)\sigma_v^2 + A^2(t)[a^2(t)\gamma_t + b^2(t)\sigma_u^2]},
$$

$$(12.3.9)$$

which agrees with the one based on the combination of the orthogonal estimating functions $\theta_{t+1} - a(t)\tilde{\theta}_t - c(t)$ and $\xi_{t+1} - A(t)\theta_{t+1}$ as in Godambe (1999): Setting

$$
\frac{\theta_{t+1} - a(t)\tilde{\theta}_t - c(t)}{a^2(t)\gamma_t + b^2(t)\sigma_u^2} - A(t)\frac{(\xi_{t+1} - A(t)\theta_{t+1})}{B^2(t)\sigma_v^2} = 0 \quad (12.3.10)
$$

gives the same result as (12.3.9).

The point estimation would be regarded as **recursive** if $\widehat{\theta}_{t+1} = \tilde{\theta}_{t+1} = E(\theta_{t+1}|\mathcal{F}_{t+1}^{\xi})$. We cannot conclude this in general, though we conclude that $\hat{\theta}_{t+1}$ is $\mathcal{F}_{t+1}^{\xi}$ - measurable and

$$E(\hat{\theta}_{t+1}|\mathcal{F}_t^{\xi}) = E(\theta_{t+1}|\mathcal{F}_t^{\xi}) . \tag{12.3.11}$$

12.3.2   Generalized Nonnormal Filtering

Here we assume a more general form for the state process (so that θ_{t+1} is non-normal), and

$$\xi_{t+1} = \theta_{t+1} + \eta_{t+1}$$

where $\{\eta_{t+1}\}$ is a sequence of mean zero normal random variables, $\{\xi_{t+1}\}$ is an observed sequence of variables and $\{(\theta_{t+1} \mid \theta_t)\}$ has density $\lambda_{t+1}(\ )$.

It can be shown that if θ_t were known the optimal filter could be obtained as the solution of the combined [Godambe (1999)] estimating functions for θ_{t+1},

$$\frac{\xi_{t+1} - \theta_{t+1}}{\sigma_\eta^2} + \frac{\partial \ln \lambda_{t+1}(\theta_{t+1} \mid \theta_t)}{\partial \theta_{t+1}} = 0 . \tag{12.3.12}$$

An estimate of θ_{t+1} can be obtained by solving Equation (12.3.12) in terms of ξ_{t+1} and θ_t and plugging in an estimate of θ_t. Fahrmeir (1994) has studied properties of such recursions in the biostatistical context.

12.3.3   Robust Estimation Filtering Equations

Assume that instead of a normal distribution, η_{t+1} of the previous section has a general zero mean distribution with density $f(\eta)$. For example, η_{t+1} might have a heavy-tailed distribution such as the Laplace distribution or the Cauchy distribution. We now have no hope of obtaining a simple recursive relation for the posterior mean. However, we can take Godambe's formulation as a starting point and investigate a combination of orthogonal estimating functions. By replacing the first component of the combined estimating function (12.3.12) by the score function corresponding to f, we obtain a filtering equation

$$\frac{\partial f(\xi_{t+1} - \theta_{t+1})}{\partial \theta_{t+1}} + \frac{\partial \ln \lambda_{t+1}(\theta_{t+1} \mid \theta_t)}{\partial \theta_{t+1}} = 0 . \tag{12.3.13}$$

In particular, if f is the Laplace distribution, we obtain the "Least Absolute Deviation" filtering equation

$$(I_{(\xi_{t+1}-\theta_{t+1})\geq 0} - I_{(\xi_{t+1}-\theta_{t+1})<0}) + \frac{\partial \ln \lambda_{t+1}(\theta_{t+1} \mid \theta_t)}{\partial \theta_{t+1}} = 0 . \tag{12.3.14}$$

In either case, the estimate of θ_{t+1} can be obtained by solving the equation in terms of ξ_{t+1} and θ_t and plugging in the estimate of θ_t.

When the distribution of η_{t+1} is symmetric, both combined estimating equations (12.3.12) and (12.3.14) are unbiased regardless of the form of f. For example, when f is the Cauchy distribution both are unbiased, and we would expect (12.3.14) to be more efficient than (12.3.12). Equation (12.3.13) is unbiased conditional on θ_t, and it obviously defines a recursion, in the sense of defining an estimate of θ_{t+1} in terms of ξ_{t+1} and an estimate of θ_t. The properties of such recursions and their modifications need careful study.

12.3.4   Censored Autocorrelated Data

Now we give an interesting example of filtering with censored autocorrelated data from a biostatistical time series. Here the original process is imperfectly observed not because of additive noise but because of censoring.

Censored observations may arise naturally in time series if there is an upper or lower limit of detection – for example when one is monitoring levels of an airborne contaminant or recording daily bioassays of hormone levels in a patient. Regression analysis with autoregressive errors when some observations are left censored had been studied in Zeger and Brookmeyer (1986). The case below is similar, but with right censored observations.

Let θ_{t+1} satisfy

$$\theta_{t+1} = a\theta_t + u_{t+1}$$

where the u_{t+1}'s are independent and identically distributed Gaussian variates with mean zero and variance σ^2. We observe possibly right censored observations, and suppose censoring of θ_{t+1} happens rarely (with high probability at most two consecutive observations are censored), and happens whenever $\theta_{t+1} > L_{t+1}$. We consider the following four cases:

$A : \theta_t \leq L_t,\ \theta_{t+1} \leq L_{t+1}$ (both observed)

$B : \theta_t \leq L_t,\ \theta_{t+1} > L_{t+1}$

$C : \theta_t > L_t,\ \theta_{t+1} \leq L_{t+1}$

$D : \theta_t > L_t,\ \theta_{t+1} > L_{t+1}$

where L_t's are the known limits of detection. In cases B, D an elementary estimating function would be

$$\theta_{t+1} - E(\theta_{t+1} \mid \theta_{t+1} > L_{t+1}, \theta_t, \cdots) = \theta_{t+1} - a\theta_t - \frac{\phi(a\theta_t - L_{t+1})}{\Phi(a\theta_t - L_{t+1})},$$
$$(12.3.15)$$

while in cases A, C an elementary estimating function would be

$$\theta_{t+1} - E(\theta_{t+1} \mid \theta_{t+1} \leq L_{t+1}, \theta_t, \cdots) = \theta_{t+1} - a\theta_t + \frac{\phi(a\theta_t - L_{t+1})}{1 - \Phi(a\theta_t - L_{t+1})} \; ;$$

$$(12.3.16)$$

here ϕ and Φ are, respectively, the density and distribution functions of the standard normal law.

Assuming a value of a, we can estimate the unobserved θ_{t+1} in case B through setting (12.3.15) to 0, then the unobserved θ_{t+1} in case D by setting (12.3.15) to 0 and using the (case B) estimate of θ_t. We can then obtain a new estimate of a by setting the sum of all terms of type (12.3.15) or (12.3.16) equal to 0; then obtain new estimates of the unobserved θ_{t+1}'s, and so on.

It is of interest to note that even though the error distribution is normal, the filtered estimate of the censored observation, in this case the posterior mean, becomes a nonlinear function of the observations.

REFERENCES

Fahrmeir, L. (1994). Dynamic modelling and penalized likelihood estimation for discrete time survival data, *Biometrika*, **81**, 317–330.

Godambe, V. P. (1960). An optimum property of regular maximum likelihood equation, *Annals of Mathematical Statistics*, **31**, 1208–1211.

Godambe, V. P. (1999). Linear Bayes and optimal estimation, *Annals of the Institute of Statistical Mathematics*, **51**, 201–215.

Hannan, E. J. and Heyde, C. C. (1972). On limit theorems for quadratic functions of discrete time series, *Annals of Mathematical Statistics*, **43**, 2058–2066.

Heyde, C. C. (1987). On combining quasi-likelihood estimating functions, *Stochastic Processes and Their Applications*, **25**, 281–287.

Kitigawa, G. (1987). Non-Gaussian state space modelling of nonstationary time series (with discussion), *Journal of the American Statistical Association*, **82**, 1032–1063.

Lipčer, R. and Shiryayev, A. (1978). *Statistics of Random Processes, Vol. II, Applications*, Springer-Verlag, Heidelberg.

Naik-Nimbalkar, U. V. and Rajarshi, M. B. (1995). Filtering and smoothing via estimating functions, *Journal of the American Statistical Association*, **90**, 301–306.

Thavaneswaran, A. and Heyde, C. C. (1999). Prediction via estimating functions, *Journal of Statistical Planning and Inference*, **77**, 89–101.

Thavaneswaran, A. and Thompson, M. E. (1986). Optimal estimation for semimartingales, *Journal of Applied Probability*, **23**, 409–417.

Thavaneswaran, A. and Thompson, M. E. (1988). A criterion for filtering in semimartingale models, *Stochastic Processes and Their Applications*, **28**, 259–265.

Thompson, M. E. and Thavaneswaran, A. (1999). Filtering via estimating functions, *Applied Mathematics Letters*, **12**, 61–67.

Zeger, S. L. and Brookmeyer, R. (1986). Regression analysis with censored autocorrelated data, *Journal of the American Statistical Association*, **81**, 722–729.

CHAPTER 13

RECENT DEVELOPMENTS IN CONDITIONAL-FREQUENTIST SEQUENTIAL TESTING

BENZION BOUKAI

Indiana University–Purdue University, Indianapolis, IN

Abstract: Recent developments on a new sequential testing procedure which unifies conditional frequentist and Bayesian approaches to testing are presented. The new testing procedure has simultaneously valid Bayesian and Conditional Frequentist interpretations, which greatly improve interpretability of results. It is also considerably easier to use than the conventional sequential tests and it reports error probabilities which are independent of the stopping rule employed.

Keywords and phrases: Bayes factor, likelihood ratio, composite hypothesis, SPRT, sequential test, conditional test, error probabilities

13.1 INTRODUCTION

Conditional frequentist tests of a precise hypothesis versus a composite alternative have recently been developed in Berger, Boukai and Wang (1997), and have been shown to be equivalent to conventional Bayes tests in that the reported frequentist error probabilities equal the posterior probabilities of the hypotheses. These recent development were lead by Berger, Brown and Wolpert (1994) who advocated a 'common' conditional frequentist viewpoint. They considered, in fixed sample and in sequential settings, the testing of simple versus simple hypotheses and presented a

185

conditional frequentist testing method that can be made exactly equivalent to the Bayesian testing method. This equivalence was made possible by proposing a particular conditioning strategy which allows an agreement between the two approaches.

There are several surprising aspects of this result; not only that both the Bayesian and the Conditional Frequentist might have the same decision rule for rejecting or accepting the null hypothesis, but also that they will both report the same conditional error probabilities upon rejecting or accepting. That is, the reported conditional error probabilities are the same as the posterior probabilities of the relevant hypotheses.

The new conditional testing method was generalized to testing simple null hypothesis against a composite alternative by Berger, Boukai and Wang (1997), in the fixed sample size settings, and by Berger, Boukai and Wang (1999), in its sequential version. They demonstrated that even when testing a simple hypothesis against a composite alternative, the new testing procedure allows for a valid Bayesian interpretation as well as for a valid conditional frequentist interpretations. The appeal of such a testing procedure is evident. The new 'common' conditional approach for testing does not seem to comprise of an artificial compromise between the Bayesian and the frequentist approaches, but rather appears to indicate that there is a testing method that is simultaneously frequentist and Bayesian. This approach was further generalized in the fixed sample setting, by Dass and Berger (1999), to the case where the null and the alternative hypotheses are both composite and have a related invariance structure.

From a frequentist viewpoint, sequential testing of composite hypotheses is typically viewed to be quite difficult. Wald (1947) considered a generalization of the SPRT by utilizing some a weight function, (i.e. some sort of prior distribution, on composite hypotheses). The primary difficulties prevail in the computation of error probabilities and in the related matter of choosing a suitable stopping rule. There are also inherent deficiencies in unconditional frequentist testing, particularly when testing precise hypotheses. Most notably is that the reported error probabilities do not depend on the evidentiary strength of the observed data. Thus, for an $\alpha = 0.05$ level test, one reports the same error probability upon rejection whether the data is just at the rejection boundary or well within the rejection region. For the SPRT, this criticism only applies when there can be substantial 'overshoot' of the stopping boundary. The traditional remedy to this problem is often to report a p-value or attained significance level, [e.g. Siegmund (1985)]. However, the p-value is not perceived as a true frequentist error measure; see Berger, Boukai and Wang (1997) for discussion and earlier references.

In this article we review some of the developments regarding the new conditional test as were obtained in Berger, Boukai, and Wang (1999) for the sequential settings. Apart from the dual interpretability as a valid conditional frequentist and as a Bayesian testing procedure, which rectify much of the deficiencies mentioned above, the new sequential test appears to be much more easier to use than the conventional sequential tests. Among the interesting properties of the new sequential tests is the lack of dependence of the reported error probabilities on the stopping rule employed.

13.2   THE SETUP

Let $X_1, X_2, \ldots,$ be a sequence of observable random variables and for each $n = 1, 2, \ldots,$ write $\boldsymbol{X}_n = (X_1, X_2, \ldots, X_n)$ and let $\mathcal{F}_n = \sigma\{X_1, X_2, \cdots, X_n\}$ denote the corresponding sigma-algebra. In addition, let $\mathcal{F}_\infty$ denote the smallest sigma-algebra containing all the $\mathcal{F}_n$. Here, n represents time and $\mathcal{F}_n$ represents the data available at time n. Let N be a proper stopping time (adaptive to $\{\mathcal{F}_n\}$) and let $\mathcal{F}_N$ denote the collection of all events $D \in \mathcal{F}_\infty$ determined prior to N, i.e. $\mathcal{F}_N \equiv \sigma\{ D \in \mathcal{F}_\infty; \ \{N = n\} \cap D \in \mathcal{F}_n, \forall n \geq 1\}$. Given $\theta \in \Theta$, we let P_θ denote the unique probability measure on $\mathcal{F}_\infty$ under which, $\boldsymbol{X}_n$ has joint probability density function (p.d.f.) $f_n(\boldsymbol{x}_n | \theta)$, $\boldsymbol{x}_n = (x_1, x_2, \ldots, x_n)$, for each $n = 1, 2, \ldots$.

Consider the problem of sequentially testing a simple hypothesis versus a composite alternative as given by

$$H_0 : \ \theta = \theta_0 \quad \text{versus} \quad H_1 : \ \theta \in \Theta_1, \tag{13.2.1}$$

for some given $\theta_0 \in \Theta$, $\theta_0 \notin \Theta_1 \subset \Theta$. Often we will take Θ_1 to be $\Theta_1 = \{\theta \in \Theta : \ \theta \neq \theta_0\}$.

In the Bayesian framework, one usually specifies the *prior probabilities*, π_0 for H_0 being true and $1 - \pi_0$ for H_1 being true and then proceeds to construct the posterior probability (given the data) of H_0 being true. When no specific prior probabilities of the hypotheses are available, it is intuitively appealing to choose $\pi_0 = 1/2$. We will use this default choice in the remainder of the paper. Thus, we assume the default prior probability of $\pi_0 = 1/2$ for the simple hypothesis $H_0 : \theta = \theta_0$ while assigning to Θ_1 the prior density $(1 - \pi_0)\pi(\theta) \equiv \pi(\theta)/2$, where $\pi(\cdot)$ is a proper p.d.f. over Θ_1 with respect to Lebesgue measure.

For each fixed n, the marginal p.d.fs of $\boldsymbol{x}_n$ under H_0 and H_1 in (13.2.1) are given by $m_{0,n}(\boldsymbol{x}_n) = f_n(\boldsymbol{x}_n | \theta_0)$ and

$$m_{1,n}(\boldsymbol{x}_n) = \int_{\Theta_1} f_n(\boldsymbol{x}_n | \theta) \pi(\theta) d\theta, \tag{13.2.2}$$

respectively. Clearly, with a given prior p.d.f. $\pi(\cdot)$, the marginal p.d.fs, $m_{0,n}(\cdot)$ and $m_{1,n}(\cdot)$ of $\boldsymbol{x}_n$ are completely specified. For a Bayesian, the sequential test of the hypotheses (13.2.1) can be reduced to a sequential test of the "simple" versus "simple" hypotheses

$$H_0' : \boldsymbol{X}_n \sim m_{0,n}(\boldsymbol{x}_n) \quad \forall n \geq 1, \quad vs, \quad H_1' : \boldsymbol{X}_n \sim m_{1,n}(\boldsymbol{x}_n) \quad \forall n \geq 1,$$
(13.2.3)

which my be carried-on based on the corresponding likelihood ratio

$$B_n = \frac{m_{0,n}(\boldsymbol{x}_n)}{m_{1,n}(\boldsymbol{x}_n)}.$$
(13.2.4)

B_n is also the Bayes factor in favor of H_0, which is often viewed as the odds of H_0 to H_1 (of H_0' to H_1') arising from the data. If the stopping rule, N, of the sequential experiment is indeed, a proper stopping time, so that $P_\theta(N < \infty) = 1$, then B_N in (13.2.4) is well defined, (in fact, B_N is $\mathcal{F}_N$ measurable).

Let P_i denote probability under H_i', $i = 0, 1$, in (13.2.3). That is, for any $\mathcal{D} \in \mathcal{F}_N$, $P_0(\mathcal{D}) = P_{\theta_0}(\mathcal{D})$, while

$$P_1(\mathcal{D}) = \int_{\Theta_1} P_\theta(\mathcal{D})\pi(\theta)d\theta.$$
(13.2.5)

For $i = 0, 1$, let $F_i(\cdot)$ denote the distribution function of B_N: $F_i(b) \equiv P_i(B_N \leq b)$, $b \in \mathbb{R}$. Wherever they exist, we write F_i^{-1} for the inverse function of F_i, $i = 0, 1$, and write

$$\psi(s) = F_0^{-1}\{1 - F_1(s)\}, \qquad \psi^{-1}(s) = F_1^{-1}\{1 - F_0(s)\}.$$
(13.2.6)

These functions will be seen important in the development.

13.3   THE 'CONVENTIONAL' APPROACHES

The classical Frequentist approach to sequentially testing the hypotheses (13.2.1) or (13.2.3) is to construct, based on the stopping rule N at hand, appropriate *rejection* and *acceptance* regions and to report the corresponding (though fixed) error probabilities. The incorrect rejection of the null hypothesis, the *Type I error*, has a probability α', and the incorrect acceptance of the null hypothesis, the *Type II error*, has a probability β'. However, when dealing with a composite hypotheses, as H_1 in (13.2.1), these frequentist's measures of respective errors are considered as functions of the parameter θ. Let

$$\alpha(\theta) = P_\theta(Rejecting \ H_0) \quad and \quad \beta(\theta) = P_\theta(Accepting \ H_0). \ (13.3.7)$$

These functions, $\alpha(\theta)$ and $\beta(\theta)$ $\theta \in \Theta$, are known as the power function and operating characteristic functions of the test. In most cases however, the *rejection* and *acceptance* regions of the sequential test are closely related to the stopping boundaries of N and any frequentist evaluation of these error probabilities requires an accurate account of the effect of the particular stopping rule employed.

Example 1 Consider the testing problem of the "simple" hypotheses (13.2.3), and recall that the SPRT with boundaries $0 < R < A < \infty$ (usually $R < 1 < A$) continues sampling as long as $R < B_n < A$; it stops at the first n (if any) for which either $B_n \leq R$ or $B_n \geq A$, and it rejects H_0' if $B_n \leq R$ and accepts H_0' if $B_n \geq A$. That is, the SPRT of (13.2.3) is based on the stopping time

$$N = \inf\{ \, n \geq 1; \ B_n \notin (R, A)\}, \qquad (13.3.8)$$

and rejects H_0' if and only if $B_N \leq R$. The test's boundaries R and A are constructed so that to achieve the desired, error probabilities, α' and β', for the Type I and the Type II errors, respectively. That is, R and A are related to α' and β' by $\alpha' = P_0(B_N \leq R) = F_0(R)$ and $\beta' = P_1(B_N \geq A) = 1 - F_1(A)$.

It is a remarkable property of the SPRT that (assuming again that N is a proper stopping time) if one neglects the boundaries' "overshoot" then the respective error probabilities can simply be approximates as $\alpha' \approx R(A-1)/(A-R)$ and $\beta' \approx (1-R)/(A-R)$, (which entails approximations to R and A). Following Wald's (1947) proposal, the SPRT of H_0' versus H_1' can also be employed as a sequential test of H_0 versus H_1 in (13.2.1).

Example 1 (continued) With the stopping rule (13.3.8), the extension of the SPRT to a sequential test of the simple versus composite hypothesis (13.2.1), can be written as;

$$\begin{cases} if \ B_N \leq R, & \text{Reject } H_0 \text{ and report error probability } \alpha(\theta_0); \\ if \ B_N \geq A, & \text{Accept } H_0 \text{ and report error probability } \beta(\theta), \end{cases} \quad (13.3.9)$$

where by (13.3.7), the reported error probabilities now become

$$\alpha(\theta_0) = P_{\theta_0}(B_N \leq R), \qquad and \qquad \beta(\theta) = P_\theta(B_N \geq A). \quad (13.3.10)$$

The relation between the error probabilities $(\alpha(\theta), \beta(\theta))$ in (13.3.10) and (α', β') of Example 1 is evident. Clearly, $\alpha' \equiv \alpha(\theta_0)$ and it is straightforward to realize that by (13.2.5),

$$\beta' = \int_{\Theta_1} \beta(\theta)\pi(\theta)d\theta. \qquad (13.3.11)$$

That is, for the frequentist, β' is the "average" probability of the Type II error, averaged with respect to the weight function (the prior p.d.f.) $\pi(\cdot)$. Clearly, the frequentist who utilizes the test (13.3.9), may report either (α', β') or $(\alpha(\theta_0), \beta(\theta))$. In either case, these error probabilities as reported by the frequentist are data–independent and hence, are constants over the rejection and the acceptance regions, respectively, [see also discussion in Berger, Boukai and Wang (1999)].

To allow the reporting of data–dependent error probabilities, the conditional frequentist considers some statistic $S(\boldsymbol{X}_N)$, where larger values of $S(\boldsymbol{X}_N)$ are indicative of data with greater evidentiary strength, for or against H_0, and then reports the error probabilities conditional on $S(\boldsymbol{X}_N) = s$, where s denotes the observed value of $S(\boldsymbol{x}_N)$. In similarity to (13.3.7), these conditional error probabilities (abbreviated here, CEP), can be obtained at any value of $\theta \in \Theta$ as; $\alpha(s|\theta) = P_\theta(Rejecting \;\; H_0|S(\boldsymbol{X}_N) = s)$ and similarly, $\beta(s|\theta) = P_\theta(Accepting \;\; H_0|S(\boldsymbol{X}_N) = s)$.

Example 1 (continued) For the conditional frequentist, the sequential test of H_0 versus H_1, as is based on the stopping rule N (13.3.8), becomes

$$\begin{cases} if \; B_N \leq R, & \text{Reject } H_0 \text{ and report CEP } \alpha(s|\theta_0); \\ if \; B_N \geq A, & \text{Accept } H_0 \text{ and report CEP } \beta(s|\theta), \end{cases}$$

where s is the observed value of $S(\boldsymbol{X}_N)$. Here the reported conditional error probabilities are obtained from (13.3.9) as, $\alpha(s|\theta_0) = P_{\theta_0}(B_N \leq R|S(\boldsymbol{X}_N) = s)$ and $\beta(s|\theta) = P_\theta(B_N \geq A|S(\boldsymbol{X}_N) = s)$.

It is not hard to realize [c.f. Berger (1985)] that even in the most favorable case of i.i.d. observations, accurate evaluation of the error probabilities, either conditional or unconditional, are not easy to come by. Clearly the choice of the conditioning statistics greatly affects the reported CEP $\alpha(s|\theta_0)$ and $\beta(s|\theta)$. And unfortunately, optimal choices rarely exist [see Kiefer (1977) and Brown (1978) also see discussion in Berger, Boukai, and Wang (1997, 1999)].

In contrast, the Bayesian need not search for a "good" choice of conditioning statistics since he/she utilizes the most extreme form of conditioning, namely conditioning on the given data. It can be verified that, upon stopping at $N = n$, the posterior probability of H_0 being true, given the data $\boldsymbol{x}_n$, is

$$\alpha^*(B_n) \equiv \Pr(H_0|\boldsymbol{x}_n) = B_n/(1 + B_n), \qquad (13.3.12)$$

whereas the posterior probability of H_1 being true is

$$\beta^*(B_n) \equiv \Pr(H_1|\boldsymbol{x}_n) = 1/(1 + B_n). \qquad (13.3.13)$$

These posterior probabilities of H_0 and H_1, $\alpha^*(B_N)$ and $\beta^*(B_N)$, respectively, are very simple to compute and they provide data–dependent measures of the evidentiary strength in favor or against H_0. Typically under, say, '0-1' loss, one then accepts or rejects H_0 according to whether its posterior probability is greater than or less than 1/2, and the reported error probability is just the posterior probability of the rejected hypothesis. Note that however, that posterior probabilities are unaffected by the choice of the stopping rule being employed.

Example 1 (continued) The Bayesian's version of the sequential test (13.3.9) for H_0 versus H_1, as is based on the stopping time N (13.3.8), (with $R < 1 < A$), can be written as

$$\begin{cases} if \ B_N \leq R, & \text{Reject } H_0 \text{ and report posterior probability } \alpha^*(B_N); \\ if \ B_N \geq A, & \text{Accept } H_0 \text{ and report posterior probability } \beta^*(B_N). \end{cases}$$

Here, the reported posterior probabilities (of H_0 and H_1, respectively) are $\alpha^*(B_N) = B_N/(1 + B_N)$ and $\beta^*(B_N) = 1/(1 + B_N)$.

13.4   THE NEW CONDITIONAL SEQUENTIAL TEST

Recall the basic setup of Section 13.2 and consider the problem of sequentially testing a simple hypothesis H_0 against a composite alternative H_1 as stated in (13.2.1).

   Let N be any stopping time of the sequential experiment and let ψ and ψ^{-1} be as are defined in (13.2.6). The only requirements we impose on the sequential experiment is that,

Condition I *The range of B_N is of the form $\mathcal{B} = (R_L, R_U] \cup [A_L, A_U)$, where $R_U \leq 1 \leq A_L$, R_L could be zero and A_U could be infinity. Furthermore, assume that ψ exists on $(R_L, R_U]$ and ψ^{-1} exists on $[A_L, A_U)$.*

Since we are dealing with continuous densities, this condition will be satisfied by commonly encountered stopping rules. These include (13.3.8), the standard 'open-ended' SPRT stopping rule. If the B_n can range from zero to infinity, it is easy to see that $(R_L, R_U] = (0, R]$ and $[A_L, A_U) = [A, \infty)$, and the remaining part of Condition I can be easily verified. Other variants include the truncated at m version of (13.3.8), namely, $N_m = \min(N, m)$. Since the range of B_m must include 1, so must that of B_N; hence $R_U = 1 = A_L$ and $R_L = 0$ and $A_U = C_m$ for some constant $C_m > 1$. See Berger, Boukai and Wang (1999) for other examples.

Corresponding to the constants R_U and A_L defined above, let

$$r = \min\{R_U, \psi^{-1}(A_L)\}, \qquad a = \max\{A_L, \psi(R_U)\}. \qquad (13.4.14)$$

These constants, r and a, define the decision boundaries for a modified Bayes sequential test, T^*, as follows:

$$\begin{cases} \text{if } B_N \leq r, & \text{reject } H_0, \text{ and report CEP} \\ & \qquad \alpha^*(B_N) = B_N/(1+B_N); \\ \text{if } r < B_N < a, & \text{make no decision,} \qquad\qquad (13.4.15) \\ \text{if } B_N \geq a, & \text{accept } H_0, \text{ and report CEP} \\ & \qquad \beta^*(B_N) = 1/(1+B_N). \end{cases}$$

Recall that [see (13.3.12)–(13.3.13)], given $\{N = n\}$ and x_n, $\alpha^*(B_n) \equiv B_n/(1+B_n)$ and $\beta^*(B_n) \equiv 1/(1+B_n)$, are the posterior probabilities of H_0 and H_1, respectively. This new sequential test T^* is similar to the one presented in Berger, Brown and Wolpert (1994) and it was shown in Berger, Boukai and Wang (1999) to define a conditional (frequentist) sequential test. This was achieved by considering, for the conditional frequentist, the conditioning statistics

$$S(B_N) = \min(B_N, \psi^{-1}(B_N)). \qquad (13.4.16)$$

as defined over the domain $\mathcal{B}_S = \{b : b \in \mathcal{B};\ 0 \leq S(b) \leq r\}$. (The complement of $\mathcal{B}_S$ is the no-decision region.) Since T^* rejects H_0 if $B_N \leq r$ and accepts H_0 if $B_N \geq a$, it follows immediately that the conditional error probabilities in this case become $\alpha(s|\theta_0) = P_{\theta_0}(B_N \leq r | S(B_N) = s)$ and $\beta(s|\theta) = P_\theta(B_N \geq a | S(B_N) = s)$, $\theta \in \Theta$.

Theorem 13.4.1 [Berger, Boukai and Wang (1999)] *For the sequential test T^* in (13.4.15) of the simple versus composite hypotheses (2.1) and the conditioning statistic $S(B_N)$ given in (13.4.16), it holds that $\alpha(s|\theta_0) = \alpha^*(B_N)$ and*

$$\int_{\Theta_1} \beta(s|\theta)\pi(\theta|s)d\theta = \beta^*(B_N),$$

where $\pi(\theta|s)$ denotes the posterior p.d.f. of θ conditional on H_1 being true and on $\{S(B_N) = s\}$.

Clearly, by this Theorem, the conditional Type I error probability and the posterior probability of H_0 are equal. However, the situation for Type II error is a bit complicated because the frequentist probability of Type II error necessarily depends on the unknown θ, while the posterior probability of H_1, is necessarily a fixed number (with respect to θ). The relationship between the posterior probability of H_1, $\beta^*(B_N) = Pr(H_1|\boldsymbol{x}_N)$, and the conditional frequentist Type II error probability, $\beta(s|\theta) = P_\theta(B_N \geq a|S(B_N) = s)$, is however, quite natural. It is similar to the relation (13.3.11) between β' and $\beta(\theta)$ discussed in Example 1. Similarly, the posterior probability of H_1 can be interpreted as the weighted conditional probability of the Type II, with a weight function being the posterior probability of θ given $S(B_N) = s$. Also, for those who wish to report weighted power, this procedure provides a reasonable choice of weight function.

Also observe that this conditional test is an exact frequentist test in the sense that it does not involve any type of approximation such as ignoring 'overshoot'. The conditional error probabilities are available explicitly here and incorporate the overshoot in the error statement; the greater the overshoot, the smaller the stated error. The inclusion of the particular *no-decision* region in the sequential test is seen to allow that duality of interpretations. This *no-decision* region disappears if the stopping rule is chosen so that $F_1(R_U) + F_0(A_L) = 1$. This can virtually always be achieved, if desired. (In the case of the SPRT for instance, with natural "standard" rule N as in (13.3.8), we have with this choice that $r = R \equiv R_U$ and $a = A \equiv A_L$ and the no-decision region disappears, then for each R a solution of A is a straightforward one.)

13.5 AN APPLICATION

We illustrate the application of the new conditional sequential testing procedure to the two-sided normal testing problem. Let $X_1, X_2, \cdots$, be a sequence of i.i.d. $N(\theta, \sigma^2)$ random variables and consider the sequential testing of $H_0 : \theta = \theta_0$ against $H_1 : \theta \neq \theta_0$, in the presence of the unknown σ^2. A conventional Bayesian approach in this case, [see Berger, Boukai and Wang (1997, 1999)], is to assume a hierarchical prior structure defined as follows. For the first-stage prior distribution of θ, take $\pi_1(\theta|\sigma^2, \xi)$ as the $\mathcal{N}(\theta_0, \xi\sigma^2)$ density. For the second-stage prior of (σ^2, ξ), take $\pi_2(\sigma^2, \xi) = \sigma^{-2}g(\xi)d\sigma^2 d\xi$, where $g(\cdot)$ is some proper prior density for $\xi > 0$. Straightforward computation yields, for $n \geq 2$,

$$B_n = (n - 1 + y_n)^{-n/2} \times \left[\int_0^\infty \frac{(1 + n\xi)^{(n-1)/2}}{\{(n-1)(1+n\xi) + y_n\}^{n/2}} g(\xi)d\xi \right]^{-1},$$

where $y_n = n(\bar{x} - \theta_0)^2/S_n^2$ and S_n^2 is the usual sample variance. For this testing problem, Berger, Boukai and Wang (1997) recommended the prior

$$g(\xi) = (2\pi)^{-\frac{1}{2}} \xi^{-\frac{3}{2}} \exp\{-\frac{1}{2\xi}\}, \qquad \xi > 0.$$

For some predetermined stopping boundaries R and A, $R < 1 < A$, consider the truncated at m stopping time N_m discussed in Section 13.3. It can be verified from (13.4.14) that the test T^* applies here with $r = R_U = A_L = 1$, and with a satisfying the equation $F_0(a) = 1 - F_1(1)$. Thus the resulting conditional sequential test is given by,

$$\begin{cases} if \ B_N \leq 1, & \text{reject } H_0, \text{ and report CEP } \alpha^*(B_N); \\ if \ 1 < B_N < a, & \text{make no decision}, \\ if \ B_N \geq a, & \text{accept } H_0, \text{ and report CEP } \beta^*(B_N). \end{cases}$$

Table 13.1 presents values of a as were determined by $M = 10^4$ simulation runs for selected choices of boundaries R and A and various choices of the truncation value m. In addition, other unconditional quantities are also presented in the table. These include the unconditional probabilities of Type I and Type II errors, $\alpha' = P_0(B_N < 1)$ and $\beta' = P_1(B_N > a)$, respectively; the expected stopping times $E_0(N)$, $E_1(N)$, under H_0' and H_1', respectively; and the corresponding probabilities of 'no decision', $p_0 = P_0(1 < B_N < a)$ and $p_1 = P_1(1 < B_N < a)$.

TABLE 13.1 Truncated two-sided normal sequential testing with $R = 0.1$ and unknown σ^2

A	m	a	α'	β'	p_0	p_1	$E_0(N)$	$E_1(N)$
8	50	3.670	0.048	0.116	0.129	0.059	44.4	16.6
	100	6.050	0.042	0.103	0.094	0.035	56.2	21.9
	200	8.014	0.038	0.104	0.084	0.017	62.7	25.9
	300	8.016	0.035	0.104	0.083	0.013	64.9	27.0
9	50	3.640	0.049	0.117	0.128	0.059	48.0	17.3
	100	4.540	0.045	0.097	0.091	0.036	67.3	23.6
	200	9.008	0.040	0.094	0.073	0.019	77.7	28.5
	300	9.015	0.037	0.094	0.071	0.012	80.8	30.4
10	50	3.630	0.048	0.117	0.129	0.060	48.1	17.2
	100	4.010	0.046	0.095	0.089	0.039	77.6	24.8
	200	7.850	0.042	0.086	0.065	0.021	92.7	31.3
	300	10.008	0.040	0.086	0.061	0.013	97.9	34.8

Example 2 Consider the case of $\theta_0 = 0$ in (13.2.1) and the data set in Table 6.3 of Armitage (1975). The data were presented as the difference in

time to recovery between paired patients who were administered different hypotensive agents. Armitage used a type of approximate sequential t-test truncated at $m = 62$. Suppose $R = 0.1$ and $A = 9$ are chosen; intuitively, these would correspond to desiring to reject H_0 when the odds are 10 to 1 in favor of H_1, while desiring to accept H_0 when the odds are 9 to 1 in its favor. These, together with $m = 62$, define the stopping boundaries, and are shown in Figure 13.1 together with the data, graphed as B_n versus n. Computation then yields $a = 3.72$; the resulting *decision boundaries* are also shown in Figure 13.1.

For the given data, the stopping boundary $A = 9$ would have been reached with $n = 52$ observations; indeed, $B_{52} = 9.017$. The conclusion of the test would then be to accept H_0 (H_0') and report *conditional error probability* $\beta^*(B_{52}) = 1/(1 + B_{52}) = 0.100$.

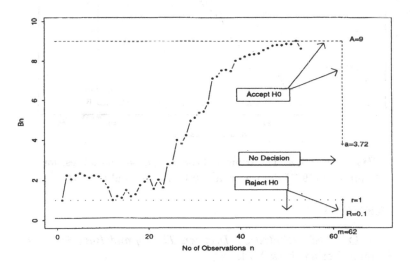

FIGURE 13.1 The truncated conditional sequential test for Armitage's (1975) data with $R = 0.1$, $A = 9$, $m = 62$, and $N = 52$

Example 2 (continued) Note that the experiment reported by Armitage actually was actually stopped at the 53rd observation. As further illustration, suppose that $m = 53$ had been the predetermined truncation time for the sequential test but now with $A = 10$ and $R = 0.1$. For these boundaries, computation yields $a = 3.66$. This situation is depicted in Figure 13.2, and the experiment would not have stopped until reaching the truncation time $m = 53$. Now, since $B_{53} = 8.613 > a = 3.66$, one would accept H_0, reporting conditional error probability $\beta^*(B_{53}) = 1/(1 + B_{53}) = 0.104$. For design purposes, one might be interested in pre-experimental properties of these tests, including their unconditional error probabilities and expected stopping times. For the situation of Figure 13.1, we have $\alpha' = 0.048$ and $\beta' = 0.109$, and the expected stopping times under H_0' and H_1' are 55.7 and 19.4, respectively. The corresponding numbers for the situation of Figure 13.2 are 0.041, 0.115, 51.0, and 17.6.

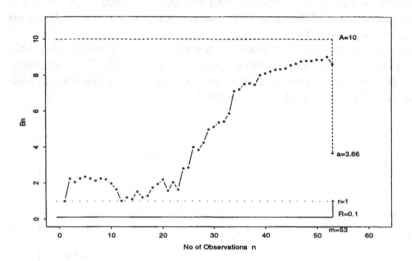

FIGURE 13.2 The truncated conditional sequential test for Armitage's (1975) data with $R = 0.1$, $A = 10$, and $m = 53$

REFERENCES

Berger, J. O. (1985). *Statistical Decision Theory and Bayesian Analysis*, Springer-Verlag, New York.

Berger, J. O., Boukai, B. and Wang, Y. (1997). Unified frequentist and Bayesian testing of a precise hypothesis, *Statistical Sciences*, **12**, 133–160.

Berger, J. O., Boukai, B. and Wang, Y. (1999). On a simultaneous Bayesian-Frequentist sequential testing of nested hypotheses, *Biometrika*, **86**, 79–92.

Berger, J. O., Brown, L. D. and Wolpert, R. L. (1994). A unified conditional frequentist and Bayesian test for fixed and sequential simple hypothesis testing, *Annals of Statistics*, **22**, 1787–1807.

Brown, L. D. (1978). A contribution to Kiefer's theory of conditional confidence procedures, *Annals of Statistics*, **6**, 59–71.

Dass, C. S. and Berger, J. O. (1999). Unified Bayesian and conditional testing of composite hypotheses, Working Paper Series 98-43, Institute of Statistics and Decision Sciences, Duke University, Durham, North Carolina.

Kiefer, J. (1977). Conditional confidence statements and confidence estimators (with discussion), *Journal of the American Statistical Association*, **72**, 789–827.

Siegmoud, D. (1985). *Sequential Analysis: Test and Confidence Intervals*, Springer-Verlag, New York.

Wald, A. (1947). *Sequential Analysis*, John Wiley & Sons, New York.

CHAPTER 14

SOME REMARKS ON GENERALIZATIONS OF THE LIKELIHOOD FUNCTION AND THE LIKELIHOOD PRINCIPLE

TAPAN K. NAYAK   SUBRATA KUNDU

George Washington University, Washington, DC

Abstract: This paper discusses the sufficiency principle (SP), the weak conditionality principle (WCP), the likelihood function (LF), and the likelihood principle (LP) for a general statistical inference problem. It is argued that a general statistical problem can be regarded as a prediction problem by treating the quantity (z) of inferential interest as the realized but unobserved value of a random vector Z. The LF is defined as the density of the data given z and the unknown fixed parameters (θ) of the model, considered as a function of z and θ. The SP and WCP are modified such that they are equivalent to the LP based on the proposed LF.

Keywords and phrases: Evidence, prediction, sufficiency, weak conditionality

14.1 INTRODUCTION

This paper deals with generalizations of the likelihood function (LF) and the likelihood principle (LP). For convenience of exposition we briefly review the LF and LP for standard parametric inference problems before discussing the generalizations and related issues. Let Y be a random observable taking values in a sample space according to a probability distribution belonging to a given family $P = \{f(y|\theta), \theta \in \Theta\}$. Here, θ is an unknown parameter vector; the observed value of Y, denoted by y, constitute the data, and the goal is to make inferences about θ or some function(s) of it. The experiment is formalized as $E = \{Y, \theta, P\}$.

Definition 14.1.1 In the above context, given the data y,

$$l(\theta|y) \propto f(y|\theta), \tag{14.1.1}$$

considered as a function of θ, is called the likelihood function. Actually, the likelihood function is defined upto a constant multiple, and two LFs are equivalent if they are proportional to each other.

The likelihood principle says that in making inferences or decisions about θ based on y, the LF $l(\theta|y)$ contains all relevant experimental information about θ provided by y and E. We shall denote this information by $Ev(E, y; \theta)$. Let $E_1 = \{Y_1, \theta, P_1\}$ and $E_2 = \{Y_2, \theta, P_2\}$ be two experiments where $P_1 = \{f_1(y_1|\theta), \theta \in \Theta\}$ and $P_2 = \{f_2(y_2|\theta), \theta \in \Theta\}$ are the families of distributions of Y_1 and Y_2, respectively, and θ is a common parameter vector. Then, the LP can be stated as follows [cf., Berger and Wolpert (1988, p. 26)].

Formal LP Let E_1 and E_2 be two experiments as stated above with outcomes y_1 and y_2 such that $f_1(y_1|\theta)$ and $f_2(y_2|\theta)$ are proportional to each other, as functions of θ. Then, $Ev(E_1, y_1; \theta) = Ev(E_2, y_2; \theta)$, i.e., the experimental evidence about θ when y_1 is observed in experiment E_1, and y_2 is observed in experiment E_2, respectively, are identical.

There have been considerable debate about validity and relevance of the LP. We refer to Berger and Wolpert (1988), for an excellent discussion of the LP, its consequences, main controversies about it, and many relevant references. There are some measure-theoretic difficulties, but we shall not concentrate on them. Actually, this paper also suffers from those difficulties in the continuous case. Our discussions are valid for discrete distributions, and provide intuitive guides in the continuous case. We focus mainly on the validity and interpretations of generalizations of the LP when the problem involves unobserved random variables as in prediction problems, and falls outside the parametric framework considered earlier. Such issues were raised and discussed by Hill (1987), Bayarri, De Groot, and Kadane (1988), and Bjørnstad (1996).

One criticism stems from the role and meaning of θ. While Berger and Wolpert (1988, p. 4) stated that "θ will be understood to contain *all* unknown features of the probability distribution" (of Y), Bayarri, De Groot, and Kadane (1988) argued that what should be regarded as Y and what should be regarded as θ is not clear when the problem involves unobserved random variables (W) either as a part of the model or for prediction. They examined three possibilities: (i) exclude W in the LF definition, i.e., $LF \propto f(y|\theta)$, (ii) include W in Y, i.e., $LF \propto f(y, w|\theta)$, and (iii) include W in θ, i.e., $LF \propto f(y|w, \theta)$, and concluded that none of them is fully satisfactory. While the LF and many of its variations, such as, marginal, conditional, profile, and integrated likelihood functions have been used extensively for statistical inferences, Bayarri, De Groot, and Kadane (1988) argued that no general definition of the LF can be given and hence the LP and likelihood methods are questionable.

Another criticism of the LP is that Berger and Wolpert (1988), Birnbaum (1962) and most other authors do not define $Ev(E, y; \theta)$ but, allow it to be any concept of evidence. For example, Bjørnstad (1996) stated that "it can be a report of the experimental results, the inferences made, or the methods used, or a collection of different measures of evidence." LeCam (1988) questioned the existence of $Ev(E, y; \theta)$, and Hill (1987) argued that as $Ev(E, y; \theta)$ is not defined and what constitute θ is unclear, the LP as stated above is stronger than can be justified, and he proposed a restricted LP.

In this paper we address, in part, the question: "What is the likelihood function?" raised by Bayarri, De Groot, and Kadane (1988). First, we feel that for this question to be meaningful it should be stated clearly what is wanted in such a definition. Otherwise, one can suggest any definition as an answer to the question. Indeed, a definition is useful when it is given in a well defined context for clear purposes. It appears from the literature that usually one or more of the following are wanted in a definition of the LF.

(A) The LF should be such that good inferences can be made based solely on it, i.e., it should be a good summary of all available and relevant information. Various predictive likelihoods [e.g., Lauritzen (1974); Hinkley (1979); Butler (1986)] attempt to achieve this goal; see Bjørnstad (1990) for an informative review.

(B) The LF should lead to equivalency of the LP, and sufficiency and weak conditionality principles (suitably modified), to generalize Birnbaum's (1962) seminal result that generated much attention for the LP.

(C) The LF should extracts all *experimental information* that is relevant for the inference problem i.e., once the LF is given one would not need to know any part of (E, y) for making inferences.

The difference between the first and third goals is that in the later case the LF is required to extract only all experimental information, whereas in the former case it is expected to capture all experimental information as well as other relevant information such as prior information. As our goal is to generalize the LP we shall define the LF for achieving goals (B) and (C).

We agree with Butler (1987) and Bjørnstad (1996) that the LF and the LP should be defined in a well defined context. Further, it is desirable to consider a context that is as broad as possible, preferably including *all* statistical inference problems. Recently, Bjørnstad (1996) attempted to provide such a discussion; he gave a general definition of the LF and proved that it achieves goal (B). We, however, find some parts of that discussion unclear. We shall show that his description of a general statistical problem and hence his generalizations of various principles are ambiguous. In Section 14.2, we argue that a general statistical problem can be regarded as a prediction problem. The description of the problem is then completed by specifying a family of joint distributions of the data and the quantities of interest. The sufficiency, and weak conditionality principles are modified suitably in Section 14.3. The general LP is presented in Section 14.4. We argue that our LF contains all experimental information, and generalize Birnbaum's (1962) result by proving that the general LP is equivalent to the generalized sufficiency and weak conditionality principles.

14.2   A GENERAL FRAMEWORK

Since the goal is to provide a general discussion of the LF and the LP, which applies to *all* statistical inference problems, it is first necessary to consider a general framework covering all statistical problems. We think this task is rather ambitious because we are not aware of every statistical problem that mankind has faced (or imagined) in the past or will face in future. Thus, while our set up described below covers most known problems, we shall refrain from claiming that it covers *all* problems.

As the goal of a general statistical analysis is to make inferences about some unknown quantity z (may be vector valued) from the data y, we think a reasonable approach to describe a general statistical problem is to describe y, and z, and state what we know and/or assume about their nature. For statistical analysis it is common to regard the data (y) as the observed value of some random observable Y. The quantities of inferential interest may be fixed unknowns or the realized but unobserved value of a random vector Z, or a combination of both. For mathematical generality z can always be regarded as the realized value of Z. Since a random vector is described completely by its probability distribution, our knowledge and/or assumptions about all statistical aspects of (Y, Z) can be expressed by specifying a family $P = \{f(y, z|\theta), \theta \in \Theta\}$ of the joint distributions of Y and Z and postulating that the true distribution belongs to P. Here, θ just indexes the family P and it is natural to require P to be identifiable with respect to θ. The parameter θ may or may not have simple physical interpretations. Thus, a general statistical problem can be described as (Y, Z, P). Intuitively, specification of a probability model seems necessary for relating y to z and hence making inferences about z from y. We define the likelihood function for a general problem as

$$l(z, \theta|y) \propto f(y|z, \theta), \qquad (14.2.2)$$

considered as a function of z and θ. This is a natural modification of (14.1.1), and was also suggested by Bayarri and De Groot (1988).

We should note that while our description of a general statistical problem specifies all probabilistic properties of Y and Z and captures the mathematical part of the model, it does not specify the loss function relevant to the problem. As our main goal is to generalize the LF and the LP, which do not concern the loss function, we do not include the loss function in our statement of the problem. Accepting Y, Z and P as the starting point, it can be seen that our framework covers standard estimation and prediction problems under frequentist, Bayes, and empirical Bayes models. In standard parametric problems Z is a function of θ and $f(z|\theta)$ is degenerate. In Bayesian models, P contains a single distribution $f(y, \theta)$, which can be factored into two components, $f(\theta)$, the prior distribution of θ, and $f(y|\theta)$. Further, $f(y|\theta)$ considered as a function of θ is called the likelihood function, and it is said to capture all experimental information. Although frequentist, Bayes, and empirical Bayes problems can be brought under a common model, we think it is important to recognize their differences for proper interpretation of the results.

We formalize the experiment in the general case as $E = \{Y, (Z, \theta), P\}$. The experiment produces y and z from a distribution in P, or equivalently a value of z according to $f(z|\theta)$, and then a value of y from $f(y|z, \theta)$ for some $\theta \in \Theta$. It is important to note that the statistician observes only y, and z remains unobserved. So, the statistician performs only a part of the experiment. Letting $E^n = \{Z, \theta, P^z\}$, and $E^s = \{Y, (z, \theta), P^*)\}$, where $P^z = \{f(z|\theta), \theta \in \Theta\}$, $P^* = \{f(y|z, \theta), z \in \mathcal{Z}, \theta \in \Theta\}$, and $\mathcal{Z}$ is the sample space of Z, it seems reasonable to regard E as consisting of E^n and E^s of which the statistician performs only E^s. Conceptually, the nature first selects a value of θ and performs E^n to obtain a value of z, and then the statistician performs E^s to obtain a value of y. Our likelihood function in (14.2.2) captures the information provided by the outcome of the statistician's experiment, i.e., the statistician's post-experimental information. Clearly, post-experimental information makes sense for the statistician's experiment, but not for the nature's experiment as its outcome remains unknown to us. Our break up of the experiment and information is consistent with parametric Bayesian analysis where the likelihood function in (14.1.1) captures the statistician's post-experimental information, and the prior distribution of θ captures our pre-experimental knowledge about the nature's experiment. We shall use $Ev(E, y; z, \theta)$ to denote the information about z and θ, when y is observed in the experiment E.

To describe the general statistical problem, Bjørnstad (1996) started with the data $Y = y$ and some unobservable variables W, which may consist of variables to be predicted and/or parameters in a Bayes or empirical Bayes model. The unknown fixed parameters characterizing the distribution of (Y, W) are denoted by θ. He then formalized the model as $P = \{f(y, w|\theta), \theta \in \Theta\}$, and the experiment as the triple $E = \{Y, (W, \theta), P\}$. Finally, he regarded the quantities of interest for statistical analysis to be a function of w and possibly y, $z = z_y = z(w, y)$, and may be θ or some part of θ.

As it has been noted in Nayak (1999), Bjørnstad's framework is not well defined because what constitute W is unclear. Since any function of W can also be regarded as a function of W and W_+, where W_+ are some additional random variables, the problem can also be stated in terms of Y, W, W_+, θ, β, and z_*, where β are some additional parameters necessary for complete specification of the joint distribution of Y, W, and W_+, and $z_*(w, w_+, y) = z(w, y)$ expresses the quantity of interest as a function of y, w, and w_+. Thus, his set up is not well defined and a statistical problem can be given many different formal descriptions. Our formulation avoids this ambiguity by describing the problem using the joint distribution of Y and Z. It may be convenient to consider some unobserved variables W for specifying (and motivating) the model, as in random effects, and measurement error models, but it may not be necessary to go beyond the family of joint distributions of Y and Z after it has been obtained.

14.3 SUFFICIENCY AND WEAK CONDITIONALITY

14.3.1 The Sufficiency Principle

As we regard a general statistical problem to be a prediction problem, we take sufficiency to mean predictive sufficiency. Thus, a statistic T is said to be sufficient if

$$f(y|t, z, \theta) = f(y|t) \quad \text{for all } z \text{ and } \theta, \quad (14.3.3)$$

i.e., T is sufficient in the usual sense for the conditional model $f(y|z,\theta)$. This is also equivalent to $f(y|t,\theta) = f(y|t)$ and $f(z|y,\theta) = f(z|t,\theta)$. Further, T is sufficient if and only if there exists two nonnegative function g and h such that the factorization $f(y|z,\theta) = g(t,z,\theta)h(y)$ holds. In most applications, this provides an easy method for finding a minimal sufficient statistic T. Prediction sufficiency and its many properties have been studied by Skibinsky (1967), Takeuchi and Akahira (1975), Torgersen (1977), Dawid (1979) and others.

Using (14.3.3) it is possible to construct Y^* only from the knowledge of T such that (Y^*, Z) and (Y, Z) have the same joint distributions. Further, when z is scalar, for any given predictor $k(Y)$, there exists a predictor based only on T, viz., $k_1(T) = E[(Y)|T]$, which is at least as good as $k(Y)$ in terms of the mean squared error, i.e., $E_\theta[k_1(T) - Z]^2 \le E_\theta[k(Y) - Z]^2$ for all θ. The equality holds if and only if $k(Y)$ is a function of Y only through T with probability 1. A (predictive) sufficient statistic T is also Bayes sufficient i.e., for any prior distribution of θ, $f(z, \theta|y) = f(z, \theta|t)$ with $t = T(y)$. Thus, in many ways $t = T(y)$ contains all information in the data y. The general sufficiency principle is described as follows.

General SP Let $Ev(E, y; z, \theta)$ denote the experimental evidence about z, and θ supplied by (E, y). Let T be a (predictive) sufficient statistic and y_1 and y_2 be such that $T(y_1) = T(y_2)$. Then, $Ev(E, y_1; z, \theta) = Ev(E, y_2; z, \theta)$.

Our general SP is the same as Bjørnstad's (1996) sufficiency principle for prediction (SPP). He, however, defined general SP differently, which is ambiguous as discussed below. Recall that his framework contains Y, W, θ and $z = z(y, w)$. Letting $\mathcal{Y}$ denote the sample space of Y, and $Y_z = \{y \in \mathcal{Y}|z(w, y) = z$ for some $w\}$, he defined T to be regular sufficient for (z, θ) if $f(y|t, z, \theta) = f(y|t)$ for all $y \in Y_z$. Then he described the general SP as follows.

Bjørnstad's SP Let $T(Y)$ be regular sufficient for (z, θ) and assume (a) $z(w, y_1) = z(w, y_2)$ $(= z)$ for all w, and (b) $T(y_1) = T(y_2)$. Then, $Ev(E, y_1; z, \theta) = Ev(E, y_2; z, \theta)$.

Thus, he requires the additional condition in (a) to be satisfied. We find this unnecessary, and it creates additional complications as it is not clear what is to be included in W. One undesirable consequence of requiring condition (a) is illustrated in the following example.

Example 14.3.1 Let $X_1, \cdots, X_n$ be iid $N(\theta, 1), Y = (X_1, \cdots, X_n), W \sim N(0, 1), Z = X_n + W$, where Y, W are independent. Bjørnstad (1996) used this example for a different purpose. It can be seen easily that $T(Y) = (\bar{X}, X_n)$ is minimal predictive sufficient, and the SP (both our, and Bjørnstad's) implies that $Ev(E, y; z, \theta)$ depends on y only through $(\bar{x}, x_n)$.

The same problem can also be stated as: $Z = X_n - X_1 + W_*$ where Y is as before, and W_* is an unobservable random variable with $f(w_*|y, \theta)$ being $N(x_1, 1)$. Here also, $T(Y) = (\bar{X}, X_n)$ is minimal predictive sufficient, and our SP implies that $Ev(E, y; z, \theta)$ depends on y only through $(\bar{x}, x_n)$, which is intuitively sensible. However, we cannot reach this conclusion from Bjørnstad's SP as $T(y_1) = T(y_2)$ does not imply that condition (a) is satisfied. His SP implies that $Ev(E, y; z, \theta)$ depends on y only through $(\bar{x}, x_n)$ and x_1. Thus, different representations of Z leads to different data reductions based on Bjørnstad's SP. Our SP does not have this drawback.

14.3.2   Weak Conditionality

The weak conditionality principle (WCP) due to Basu (1975) in the standard parametric case, is a modification of Birnbaum's (1962) conditionality principle, and it can be described as follows. Let $E_1 = \{Y_1, \theta, P_1\}$ and $E_2 = \{Y_2, \theta, P_2\}$ be two experiments with common θ, and $P_j = \{f_j(y_j|\theta), \theta \in \Theta\}, j = 1, 2$. Let E^* be a mixture of E_1 and E_2 with mixture probability of each experiment being 0.5. That is, in E^*, first $J = 1$ or 2 is observed, each having probability 0.5, and experiment E_J is then performed. The outcome of E^* can be represented as (j, y_j), and E^* can be formalized as $E^* = \{Y^*, \theta, P^*\}$ where, $Y^* = (J, Y_J)$ and $P^* = \{f^*(y^*|\theta) = f^*(j, y_j|\theta) = (1/2)f_j(y_j|\theta), \theta \in \Theta\}$. Then, the WCP says that $Ev(E^*, (j, y_j); \theta) = Ev(E_j, y_j; \theta)$.

To generalize the WCP let $E_1 = \{Y_1, (Z, \theta), P_1\}$ and $E_2 = \{Y_2, (Z, \theta), P_2\}$ be two experiments with *common* (Z, θ), and $P_j = \{f_j(y_j, z|\theta); \theta \in \Theta\}, j = 1, 2$. Since a complete description of (Z, θ) is given by Θ and $\{f(z|\theta), \theta \in \Theta\}$, we interpret "*common* (Z, θ)" as Θ and $\{f(z|\theta), \theta \in \Theta\}$ being common to the two experiments. Otherwise, it would seem that the experiments are informative about z, creating more confusion. Alternatively, one may require θ as well as z, the realized but unobserved value of Z, to be the same for two experiments. We think this requirement is unnecessarily stringent. Also, this is equivalent to treating z as an unknown fixed parameter, which leads to the standard parametric case. It should be noted, however, that if z is treaded as a fixed unknown parameter, $\{f(y|z, \theta)\}$ need not be identifiable with respect to (z, θ). With our broader interpretation of "common (Z, θ)", the general WCP is presented below.

General WCP Let E_1 and E_2 be two experiments with common (Z, θ), and let E^* be the mixture of E_1 and E_2 with equal probabilities. Formally, $E^* = \{(J, Y_J), (Z, \theta), P^*\}$, where $P^* = \{f^*((j, y_j), z|\theta), \theta \in \Theta\}$, and $f^*((j, y_j), z|\theta) = (1/2)f_j(y_j, z|\theta)$. Then,

$$Ev(E^*, (j, y_j); z, \theta) = Ev(E_j, y_j; z, \theta).$$

To generalize the WCP Bjørnstad (1996) considered two experiments $E_j = \{Y_j, (W, \theta), P_j\}, P_j = \{f_j(y_j, w|\theta), \theta \in \Theta\}, j = 1, 2$, with identical (W, θ) and the same Z function $Z(W, \cdot)$. He did not elaborate much on what he meant by "identical (W, θ)", and "the same Z function; $Z(W, \cdot)$", except that his WCP "deals with two experiments about the *same* unknown physical quantities θ, W". Here, one implicit restriction is that Y_1 and Y_2 must be of same dimension, otherwise the Z function cannot be the same in the two experiments. Also, unlike us he did not assume (p. 800) that the distributions of Z given θ are the same in the two experiments.

14.4   THE LIKELIHOOD PRINCIPLE

We defined the likelihood function as $l(z, \theta|y) \propto f(y|z, \theta)$, considered as a function of z and θ. This definition of the LF depends on the inferential goal, which Bayarri, De Groot and Kadane (1988) did not consider, and it includes the fixed unknown parameters θ used for specifying the joint distributions of Y and Z. Also, (14.2.2) is a natural modification of the parametric LF in (14.1.1) as both are based on the density of the observed data conditional on the unknowns relevant to the problems. As noted earlier, Bayarri and De Groot (1988) also suggested (14.2.2) as the general LF. Taking (14.2.2) as the general LF we next state the general LP, and generalize Birnbaum's (1962) theorem.

General LP   Let $E_i = \{Y_i, (Z, \theta), P_i\}, P_i = \{f_i(y_i, z|\theta), \theta \in \Theta\}, i = 1, 2$, be two experiments with common (z, θ), i.e., with common Θ and $\{f(z|\theta), \theta \in \Theta\}$. Suppose y_1, y_2 are two outcomes of E_1 and E_2, respectively, such that $l_1(z, \theta|y_1) \propto l_2(z, \theta|y_2)$, where $l_i(z, \theta|y_i) \propto f_i(y_i|z, \theta)$. Then, $Ev(E_1, y_1; z, \theta) = Ev(E_2, y_2; z, \theta)$.

Theorem 14.4.1   The general LP follows from the general sufficiency and weak conditionality principles. Conversely, the SP and the WCP follow from the LP.

PROOF   (SP, WCP $\Rightarrow$ LP). Let E_1, E_2, y_1, y_2 be as defined in the general LP and suppose

$$f_1(y_1|z, \theta) = cf_2(y_2|z, \theta) \text{ for all } z \text{ and } \theta, \qquad (14.4.4)$$

where c is a constant independent of θ and z. Let E^* be the mixture of E_1 and E_2, each with probability 0.5, as in the WCP. Then, from the general WCP

$$Ev(E^*, (j, y_j); z, \theta) = Ev(E_j, y_j; z, \theta), \quad j = 1, 2.$$

We shall now complete the proof by showing that the general SP and (14.4.4) implies

$$Ev(E^*, (1, y_1); z, \theta) = Ev(E^*, (2, y_2); z, \theta). \tag{14.4.5}$$

Note that in the mixture experiment E^*,

$$f^*((j, y_j), z|\theta) = (1/2)f_j(y_j, z|\theta). \tag{14.4.6}$$

Since $f(z|\theta)$ is the same in the two experiments, (14.4.4) and (14.4.6) imply that

$$f^*((1, y_1)|z, \theta)/f^*((2, y_2)|z, \theta) = c \quad \text{for all } z \text{ and } \theta. \tag{14.4.7}$$

Now, (14.4.7) implies that $(1, y_1)$ and $(2, y_2)$ are in the same minimal sufficient partition set of E^*, and hence (14.4.5) follows from the general SP.

To prove LP $\Rightarrow$ SP, let T be a sufficient statistics and $T(y_1) = T(y_2)$. Then, (14.4.4) holds from the definition of sufficiency, and then the LP implies that $Ev(E, y_1; z, \theta) = Ev(E, y_2, z, \theta)$.

To prove LP $\Rightarrow$ WCP, let one experiment be E^* and the other be E_j. Then, from (14.4.6) and our definition of the LF,

$$\begin{aligned}
l^*(z, \theta|(j, y_j)) &\propto f^*((j, y_j)|z, \theta) \\
&= f^*((j, y_j), z|\theta)/f(z|\theta) \\
&= (1/2)f_j(y_j, z|\theta)/f(z|\theta) \\
&= (1/2)f_j(y_j|z, \theta) \\
&\propto l_j(z, \theta|y_j),
\end{aligned}$$

and it follows from the LP that $Ev(E^*, (j, y_j); z, \theta) = Ev(E_j, y_j; z, \theta), j = 1, 2.$ □

As in the standard parametric case, the general LP implies that $Ev(E, y; z, \theta)$, the experimental information about z and θ when y is observed in the experiment E, depends on (E, y) only through the likelihood function $l(z, \theta|y)$. The general LP does not say that the LF contains all available information relevant for the inference problem. It only says that the LF contains all information relevant to the inference problem provided by the observed outcome of the experiment actually performed by the statistician, i.e., the observed outcome of E^s. There may be other relevant information, namely, the loss function, the prior distribution of θ, and $f(z|\theta)$. We do not consider the prior distribution of θ, and $f(z|\theta)$ as parts of the experimental information. To us, they contain our pre-experimental knowledge about the nature's experiment E^n. As we stated earlier, there is no post-experimental information from the nature's experiment E^n, whose outcome remains unknown to us. The LP also does not say how to use the LF for making inferences or, how to combine the LF with other relevant information.

Bjørnstad (1996), Butler (1987), and Berger and Wolpert (1988) have not favored taking (14.2.2) as the general definition of the LF. They prefer to take

$$l_0(z, \theta|y) \propto f(y, z|\theta) \qquad (14.4.8)$$

as the general LF (as a function of z and θ). One criticism of (14.2.2) is that if Y and Z are independent given θ, then (14.2.2) does not involve z (and we lose the information y carries about z through θ). We think this criticism is not valid as the LP does not say that the LF contains all relevant information, and does not forbid combining the LF with other information such as $f(z|\theta)$, and the prior distribution of θ. Indeed, (14.4.8) can be obtained by combining (14.2.2) with $f(z|\theta)$, and is better suited for achieving goal (A). But in our view, it contains more than the experimental information. It may also be noted that the choice of the LF for achieving goal (A) is not unanimous. Other predictive likelihoods have also been proposed in the literature, and we refer to Bjørnstad (1990) for an excellent review and further references.

It should be noted that Theorem 14.4.1 can also be proved taking (14.4.8) as the LF definition. This is because (14.4.4) is equivalent to

$$f_1(y, z|\theta) = cf_2(y_2, z|\theta) \qquad \text{for all } z \text{ and } \theta$$

as $f(z|\theta)$ is the same in the two experiments. Thus, both (14.2.2) and (14.4.8) achieve the goal of generalizing Birnbaum's theorem, but their interpretations are different. In our opinion, (14.4.8) contains more than sample information and hence we like to take (14.2.2) as the general definition of the LF. We may emphasize the following points about our definition of the LF. (1) It depends on the specific objective or inferential aim of the statistical investigation, which Bayarri, De Groot, and Kadane (1988) seemed to overlook but Bjørnstad (1996) attempted to take into account. (2) It is designed to capture only the information provided by the observed outcome of the experiment actually performed by the statistician. (3) It includes the model parameters θ even when they are not of inferential interest. Thus, it captures all post-experimental information on z and θ.

Bayarri and De Groot (1988) gave some additional arguments for taking (14.2.2) as the LF. It may also be noted that if in the WCP *common* (Z, θ) is interpreted as z and θ being common to the two experiments, i.e., if z is treated as an unknown fixed parameter, Theorem 14.4.1 holds with (14.2.2) as the LF.

14.5 DISCUSSION

In this paper, we viewed a general statistical inference problem as a prediction problem, and in that set up we discussed the SP, WCP, LF, and the LP. We think a well defined context is necessary for discussing such basic principles. Our context as well as the starting point is Y, Z, and $P = \{f(y, z|\theta), \theta \in \Theta\}$, and the SP, WCP, LF, and the LP come into play after this starting point is given. Thus, we did not address what Mallows (1998) called "The Zeroth Problem", namely, how to decide what the relevant data (y) and the quantities of inferential interest (z) are, and how they relate to the purpose of the statistical study. Clearly, one should use good judgment and care in choosing Y, Z, and especially, P.

It may be noted that according to our formulation, a Bayesian problem does not contain nuisance parameters. For example, if Y consist of iid observations $X_1, \cdots, X_n$ from $N(\mu, \sigma)$ and the goal is to estimate μ, ordinarily one takes a prior distribution of $\theta = (\mu, \sigma)$ and calculates the posterior distribution, $f(\mu, \sigma|y)$, of (μ, σ), and then integrates out σ to obtain $f(\mu|y)$, which is of main interest. But, the same posterior distribution of μ is obtained starting from the model $f(\mu) = \int f(\mu, \sigma)d\sigma$ and $f(y|\mu) = [\int f(y|\mu, \sigma)f(\mu, \sigma)d\sigma]/f(\mu)$ for (Y, μ). However, one may choose to report the calculations for the first approach for better disclosure of the arguments and judgments used in the analysis, and for allowing others to make inferences about σ. The point is that while formulating a problem (Bayesian or non-Bayesian) we should think if there are some quantities that are not of interest to us, but are quite likely to be of interest to others and hence should be included in Z. It may be noted that a non-Bayesian problem may contain nuisance parameters (in θ) even when Z contains the quantities that are of interest to us only.

It has been more than 36 years since Birnbaum's paper was published, but the LP has still remained controversial, at least judged by how often it is violated in practice. This may be due to the radical consequences of the LP, but it may also be due to lack of a clear definition of "evidence" (or information). While it is not necessary to define $Ev(E, y; \theta)$ for proving Birnbaum's theorem, it is important to have a good practical interpretation of $Ev(E, y; \theta)$ for convincing others that it applies to their analyses. Also, the use of the word *all* as in *all information* and *all unknowns*, without adequate qualifications may have contributed to the scepticism about the LP. Indeed, this was the source of the concerns of Hill (1987), and Bayarri, De Groot, and Kadane (1988). This is not surprising when we think about the difficulties the word "*all*" caused in set theory. The LP still occupies a somewhat strange position in statistical science. Mainly the Bayesian methods do not violate it, but there how the joint density of Y, Z, and θ is factored into likelihood, prior, and other components appears to be unimportant.

REFERENCES

Basu, D. (1975). Statistical Information and Likelihood (with discussion), *Sankhyā, Series A*, **37**, 1–71.

Bayarri, M. J. and De Groot, M. H. (1988). Discussion: Auxiliary parameters and simple likelihood functions, In *The Likelihood Principle* Second Edition, by Berger and Wolpert, Institute of Mathematical Statistics, Hayward, California. 160.3–160.7.

Bayarri, M. J., De Groot, M. H. and Kadane, J. B. (1987). What is the likelihood function? (with discussion), In *Statistical Decision Theory and Related Topics IV*, Vol. 1 (Eds., S. S. Gupta and J. O. Berger), Springer-Verlag, New York.

Berger, J. O. and Wolpert, R. L. (1988). *The Likelihood Principle* Second Edition, Institute of Mathematical Statistics, Hayward, California.

Bjørnstad, J. F. (1990). Predictive likelihood : A review (with discussion), *Statistical Science*, **5**, 242–265.

Bjørnstad, J. F. (1996). On the generalization of the likelihood function and the likelihood principle, *Journal of American Statistical Association*, **91**, 791–806.

Butler, R. W. (1986). Predictive likelihood inference with applications (with discussion), *Journal of the Royal Statistical Society, Series B*, **48**, 1–38.

Butler, R. W. (1987). A likely answer to 'What is the likelihood function?', In *Statistical Decision Theory and Related Topics IV*, Vol. 1 (Eds., S. S. Gupta and J. O. Berger), Springer-Verlag, New York.

Birnbaum, A. (1962). On the foundations of statistical inference (with discussion), *Journal of American Statistical Association*, **57**, 269–306.

Dawid, A. P. (1979). Conditional independence in statistical theory (with discussion), *Journal of the Royal Statistical Society, Series B*, **41**, 1–31.

Hill, B. M. (1987). On the validity of the likelihood principle, *The American Statistician*, **41**, 95–100.

Hinkley, D. V. (1979). Predictive likelihood, *Annals of Statistics*, **7**, 718–728.

Lauritzen, S. L. (1974). Sufficiency, prediction and extreme models, *Scandinavian Journal of Statistics*, **1**, 128–134.

LeCam, L (1988). Discussion, In *The Likelihood Principle*, Second Edition, by Berger and Wolpert, Institute of Mathematical Statistics, Hayward, California. 182–185.2.

Mallows, C (1998). The zeroth problem, *The American Statistician*, **52** 1–9.

Nayak, T. K. (1999). On best unbiased prediction and its relationships to unbiased estimation, *Journal of Statistical Planning and Inference* (to appear).

Skibinsky, (1967). Adequate subfields and sufficiency, *Annals of Mathematical Statistics*, **38**, 155–161.

Takeuchi, K. and Akahira, M. (1975). Characterization of prediction sufficiency (adequacy) in terms of risk functions, *The Annals of Statistics*, **3**, 1018–1024.

Torgersen, E. N. (1977). Prediction sufficiency when loss function does not depend on the unknown parameter, *The Annals of Statistics*, **5**, 155–163.

CHAPTER 15

CUSUM PROCEDURES FOR DETECTING CHANGES IN THE TAIL PROBABILITY OF A NORMAL DISTRIBUTION

RASUL A. KHAN

Cleveland State University, Cleveland, OH

Abstract: Let $x_1, x_2, \ldots$ be independent random variables having normal distribution with unknown parameters, and consider the problem of detecting changes in the tail probability $p = P(x_1 > x_0)$, where x_0 is a known constant. To monitor changes in p we consider some cusum procedures including a Shewhart control chart and its generalization based on a suitable integer-valued random walk. Both of these procedures are based on noncentral t-statistics under group sampling although a special case of the generalized version is also applicable without group sampling. Some cusum procedures based on noncentral t-statistics under group sampling are also considered, and the average run lengths of these procedures are evaluated and/or estimated by simulations to compare their relative merits.

Keywords and phrases: Normal, tail probability, Shewhart chart, cusum, noncentral t-statistics, average run length (ARL), approximation, simulation

15.1 INTRODUCTION

Control charts and cusum (cumulative sum) procedures are often used for monitoring shifts in the mean of a normal or nearly normal distribution. In practice one can expedite these procedures under group sampling of

small or moderate sample sizes with fairly good results and minimum se-
quentialization. Moreover, group sampling is a practical convenience and
very natural in some problems. In what follows we consider the problem
of monitoring changes in the tail probability of a normal distribution. Let
$x_1, x_2, \ldots$ be independent $N(\mu, \sigma^2)$ random variables with unknown mean μ
and unknown variance σ^2. Let $p = P(x_1 > x_0) = \Phi((\mu - x_0)/\sigma)$, where x_0
is a known constant and $\Phi(.)$ is the standard normal distribution function.
The process $x_1, x_2, \ldots$ is said to be in control if $\mu \geq x_0 + a\sigma$, or equivalently, $p \geq \beta = \Phi(a)$ for a given $\beta(0 < \beta < 1)$. With no loss of generality and
without any further comment in the sequel we let $x_0 = 0$ for one can always
replace x_i by $x_i - x_0$. Thus letting $\theta = \mu/\sigma$ and $\omega = \{(\mu, \sigma) : \theta \geq a\}$, it is
assumed that $x_1, \ldots, x_{t-1}$ are independent $N(\mu, \sigma^2)$ when $(\mu, \sigma) \in \omega$, while
$x_t, x_{t+1}, \ldots$ are independent $N(\mu, \sigma^2)$ with $(\mu, \sigma) \in \bar{\omega}$, where $t\,(t \geq 1)$ is an
unknown change-point. Such a change in distribution at an unknown time
point is particularly important in some industrial applications such as reliability control and it is of interest to detect a change in p or equivalently
in θ as quickly as possible. The problem is interesting and some detection
procedures are explored.

The object of this note is to study some cusum procedures and provide evaluations and/or simulated estimates of the average run lengths
(ARLs). In Section 15.2 we begin with Shewhart chart and its generalization based on a suitable integer-valued random walk dependent on non-
central t-statistics under group sampling although the generalized version
is also applicable without group sampling. In Section 15.3 we consider some
cusum procedures based on noncentral t-statistics under group sampling,
and a modified cusum without group sampling. Exact evaluations and/or
approximate or simulated estimates of the ARLs are given. Simulations
are reported in Section 15.4 and the relative merits of these procedures are
discussed.

15.2   A SHEWHART CHART AND A CUSUM SCHEME

In order to motivate a Shewhart control chart we consider a random sample
$(x_1, x_2, \ldots, x_m)$ of size $m \geq 2$ from a $N(\mu, \sigma^2)$ distribution and define $\bar{x} = (x_1 + \ldots + x_m)/m$ and $s^2 = \sum_{i=1}^{m}(x_i - \bar{x})^2/(m-1)$. Letting $\theta = \mu/\sigma$ and c
as a constant we have

$$P(\bar{x} \geq c(s/\sqrt{m}))$$
$$= P\{\frac{\sqrt{m}(\bar{x} - \mu)}{\sigma} + \theta\sqrt{m} \geq c(s/\sigma)\}$$
$$= P\{t[m-1, \theta\sqrt{m}] \geq c\} \geq P\{t[m-1, \theta_0\sqrt{m}] \geq c\}, \theta \geq \theta_0,$$

where θ_0 is the control value and $t[q, \delta]$ denotes a noncentral t random variable with q degrees of freedom and noncentrality parameter δ. Let $c = t_\gamma[m - 1, \theta_0\sqrt{m}]$ denote a γ-fractile such that

$$P\{t[m - 1, \theta_0\sqrt{m}] \leq t_\gamma[m - 1, \theta_0\sqrt{m}]\} = \gamma.$$

Thus

$$P\{\bar{x} \geq \frac{s}{\sqrt{m}} t_\gamma[m - 1, \theta_0\sqrt{m}]\} \geq 1 \quad \gamma, \; (\mu, \sigma) \in \omega.$$

Now suppose that independent random samples $\{x_{ij}, j = 1, 2, ..., m, i \geq 1\}$ are taken at certain intervals of time and $(\bar{x}_i, s_i^2)$ are computed. Define the stopping time

$$\nu = \inf\{i \geq 1 : \bar{x}_i < t_\gamma[m - 1, \theta_0\sqrt{m}] (s_i/\sqrt{m})\}, \tag{15.2.1}$$

and a corrective action is taken as soon as ν terminates. Thus the statistics $\sqrt{m}\,\bar{x}_i/s_i$ are plotted on a control chart with a boundary line $t_\gamma[m - 1, \theta_0\sqrt{m}]$, and a corrective action is signaled whenever $\sqrt{m}\,\bar{x}_i/s_i$ falls below the boundary line for the first time. It is a simple-minded procedure and its average run length (ARL) $mE\nu$ can be easily computed. To this end, we clearly have

$$P(\nu > n) = \alpha^n(\theta), \tag{15.2.2}$$

where $\alpha(\theta) = P\{t[m - 1, \theta\sqrt{m}] \geq t_\gamma[m - 1, \theta_0\sqrt{m}]\}$. Hence the ARL is given by

$$\text{ARL}(\theta) = m\,E_\theta\nu = m \sum_{n=0}^{\infty} P(\nu > n) = m/(1 - \alpha(\theta)). \tag{15.2.3}$$

Since $\alpha(\theta) \geq \alpha(\theta_0) = 1 - \gamma$ when $\theta \geq \theta_0$, we have $\text{ARL}(\theta) \geq \text{ARL}(\theta_0) = m/\gamma$ for $\theta \geq \theta_0$. By choosing m and γ we can achieve $\text{ARL}(\theta_0)$ at any desired level to satisfy $\text{ARL}(\theta_0) \geq A_0$ for any preassigned A_0. This simple control procedure based on noncentral t-statistics is quite practical and completely analogous to the standard Shewhart control chart had σ been known.

We will now define a closely related cusum procedure based on a suitable integer-valued random walk. This may be considered a generalization of the preceding Shewhart control scheme. As before, $(\bar{x}_i, s_i^2)$ are respectively the mean and the variance of the ith random sample of size m, and likewise $\theta = \mu/\sigma$ and $\theta \geq \theta_0$ are the control values. For $k \geq 0$ and $i = 1, 2, ...$ let

$$\begin{aligned} \xi_i &= -1 \quad \text{if } \bar{x}_i \geq k(s_i/\sqrt{m}) \\ &= +1 \quad \text{if } \bar{x}_i < k(s_i/\sqrt{m}). \end{aligned}$$

Also, let $p = p(\theta) = P(\xi_i = 1)$ and $q(\theta) = 1 - p(\theta)$. Note that

$$p(\theta) = P(t[m-1, \theta\sqrt{m}] \le k), \qquad (15.2.4)$$

where $t[q, \delta]$ is a noncentral t random variable with q $d.f.$ and noncentrality parameter δ. Set $S_0 = 0$, $S_n = \xi_1 + ... + \xi_n$, $n \ge 1$. Then $\{S_n, n \ge 0\}$ is a random walk on the integers and it is natural to choose a positive integer h as the boundary of the ensuing cusum procedure. Consequently, we define a cusum stopping time by

$$N = \inf\{n \ge 1: S_n - min_{0 \le i \le n} S_i = h\}, \qquad (15.2.5)$$

and a corrective action is taken at the termination of N. An alternative representation of N is obtained by letting $W_0 = 0$, $W_n = \max(0, W_{n-1} + \xi_n)$, $n \ge 1$, and noting that

$$N = \inf\{n \ge 1: W_n = h\}. \qquad (15.2.6)$$

A little reflection makes it clear that N is indeed the Shewhart scheme ν defined by (15.2.1) when $h = 1$ and $k = t_\gamma[m-1, \theta_0\sqrt{m}]$. For higher values of h, N is much more stringent. For example, $h = 2$ makes termination possible only when two consecutive samples produce $\xi_i = 1$, $\xi_{i+1} = 1$, etc. It should be remarked that this procedure is applicable even if $k = 0$. This is due to the fact that $p = P(\xi_i = 1)$ still depends on θ only. In fact, if $k = 0$ we have $p = p(\theta) = 1 - \Phi(\theta\sqrt{m})$, $q = \Phi(\theta\sqrt{m})$, and one can use N with a suitable h. In case $k = 0$, the procedure is also applicable even if $m = 1$ so that no group sampling is required.

The procedure N defined by (15.2.6) has been studied by Khan (1984) and the associated ARL is given by

$$\begin{aligned}
\text{ARL} = EN &= \frac{h}{(p-q)} - \frac{q}{(p-q)^2}\left(1 - (q/p)^h\right), \quad p \ne q \\
&= h(h+1)/2p, \quad p = q.
\end{aligned} \qquad (15.2.7)$$

For $k = t_\gamma[m-1, \theta_0\sqrt{m}]$ we note from (15.2.2) and (15.2.4) that $p(\theta) = 1 - \alpha(\theta)$ and $q = \alpha(\theta)$. Moreover, if $h = 1$, (15.2.7) becomes $EN = 1/p(\theta)$ so that $\text{ARL}(\theta) = m\,EN$ agrees with (15.2.3). Some other properties of N can be found in Khan (1995).

15.3 NONCENTRAL t-STATISTICS BASED CUSUM PROCEDURES

Let $\{x_{ij}, i = 1, 2, ..., m, j \ge 1\}$ be random samples of size $m \ge 4$ taken at regular intervals of time, and let $y_i = \bar{x}_i/s_i$ where $\bar{x}_i$ and s_i^2 are the ith sample mean and variance as defined in Section 15.2. In what follows we define two intuitive cusum procedures based on the sequence $y_1, y_2, \ldots$ with the hope that the size based cusum procedures are more efficient than those discussed in Section 15.2. Clearly, the distribution of y_i's depends on $\theta = \mu/\sigma$, and like the preceding section, the control values are $\theta \geq \theta_0$, where θ_0 is known and fixed. Let $d_i = y_i - E_{\theta_0}y_1 + k$, where k ($k \geq 0$) is a reference constant, and set $S_0 = 0$, $S_n = d_1 + d_2 + \cdots + d_n$, $n \geq 1$. For any $h > 0$ we define the cusum procedure by

$$T = \inf\{n \geq 1: \ max_{0 \leq i \leq n} \ S_i - S_n \geq h\},$$

and a corrective action is signaled as soon as T terminates. Let $d_i^* = -d_i$, $S_0^* = 0$, $S_n^* = d_1^* + \ldots + d_n^*$, $W_0 = 0$, $W_n = \max(0, W_{n-1} + d_n^*)$, $n \geq 1$. Clearly, T can be written as

$$
\begin{aligned}
T &= \inf\{n \geq 1: S_n^* - min_{0 \leq i \leq n} \ S_i^* \geq h\} \\
&= \inf\{n \geq 1: W_n \geq h\}. \tag{15.3.8}
\end{aligned}
$$

The ARL(θ) (average run length) is defined as $E_\theta T$ under the assumption that $y_1, y_2, \ldots$ are iid. We note that $y_1 = t[\nu, \theta\sqrt{m}]/\sqrt{m}$, $\nu = m - 1$, and it is easily verified [cf. Resnikoff and Lieberman (1957)] that the mean $\beta(\theta)$ and the variance $\sigma^2(\theta)$ of y_1 are given by

$$\beta(\theta) = \theta\sqrt{\frac{\nu}{2}} \frac{\Gamma(\frac{\nu-1}{2})}{\Gamma(\frac{\nu}{2})}, \text{ and } \sigma^2(\theta) = \frac{\nu}{(\nu-2)(\nu+1)}(1 + m\theta^2) - \beta^2(\theta).$$

Thus $E_\theta d_i = \eta(\theta) = \beta(\theta - \theta_0) + k$, and $var(d_i) = \sigma^2(\theta)$. It is also known [cf. Ghosh (1970, p. 22)] that y_i's are approximately normal $N(\theta, (1+\theta^2/2)/m)$ for large m. Clearly, $E_\theta d_i^* = -\eta(\theta)$ and $var(d_i^*) = \sigma^2(\theta)$, and that d_i^*'s can be suitably approximated by normal random variables for large m. Since it is desirable to keep the group sample size m at moderate levels, the resulting ARL cannot be approximated by the usual known results based on normal theory unless m is fairly large. Some alternative asymptotics are needed to supplement the normal approximation for $E_\theta T$. However, before discussing any approximation we define another intuitive cusum procedure. To this end, we define

$$\bar{x}(\alpha, n) = \frac{\sum_{i=\alpha}^{n} \sum_{j=1}^{m} x_{ij}}{(n - \alpha + 1)m}, \quad s^2(\alpha, n) = \frac{\sum_{i=\alpha}^{n} \sum_{j=1}^{m}(x_{ij} - \bar{x}(\alpha, n))^2}{((n - \alpha + 1)m - 1)},$$

and note that

$$\bar{x}_i = \bar{x}(i, i) = (\sum_{j=1}^{m} x_{ij})/m, \text{ and } s_i^2 = s^2(i, i) = \sum_{j=1}^{m}(x_{ij} - \bar{x}_i)^2/(m - 1).$$

Set $\omega_0 = 0$, and define

$$\omega_1 = \max\{0, -\frac{\overline{x}(1,1)}{s(1,1)} + (\theta_0 - k)\},$$

$$\omega_2 = \max\{0, -\frac{\overline{x}(1,2)}{s(1,2)} + (\theta_0 - k)\} \quad \text{if } \omega_1 > 0$$

$$= \max\{0, -\frac{\overline{x}(2,2)}{s(2,2)} + (\theta_0 - k)\} \quad \text{if } \omega_1 = 0,$$

and in general, for $i = 1, 2, ..., n-1$,

$$\omega_n = \max\{0, -\frac{\overline{x}(i,n)}{s(i,n)} + \theta_0 - k\} \text{ if } \omega_j = 0, \ j \le i-1, \ \omega_i > 0,$$

$$\omega_{i+1} > 0, \ldots, \omega_{n-1} > 0,$$

$$= \max\{0, -\frac{\overline{x}(n,n)}{s(n,n)} + \theta_0 - k\} \text{ if } \omega_{n-1} = 0.$$

Thus we continue to combine samples to improve our estimate $\hat{\theta}$ as long as $-\hat{\theta} + (\theta_0 - k) > 0$, and the renewal occurs as soon as $-\hat{\theta} + (\theta_0 - k) \le 0$. The new intuitive cusum procedure is defined by

$$T_1 = \inf\{n \ge 1 : n\omega_n \ge h\}. \tag{15.3.9}$$

A little reflection shows that T_1 is equivalent to repeated application (after each renewal) of the stopping time M until exit occurs at h where M is defined by

$$M = \inf\left\{n \ge 1 : \sum_{i=1}^{n}(\frac{-\overline{x}_i}{s(1,n)} + \theta_0 - k) \le 0 \quad \text{or}\right.$$

$$\left.\sum_{i=1}^{n}(\frac{-\overline{x}_i}{s(1,n)} + \theta_0 - k) \ge h\right\}, \tag{15.3.10}$$

where

$$s^2(1,n) = \sum_{i=1}^{n}\sum_{j=1}^{m}(x_{ij} - \overline{x})^2/(nm-1),$$

$$\overline{x} = \overline{x}(1,n) = \sum_{i=1}^{n}\sum_{j=1}^{m}x_{ij}/nm = \sum_{i=1}^{n}\overline{x}_i/n.$$

The following general result provides the alternative asymptotics which can supplement the normal approximations for the associated ARLs of the preceding cusum procedures.

Theorem 15.3.1 Let $\xi_1, \xi_2, \ldots$ be iid random variables with mean μ and variance σ^2, and set $S_0 = 0$, $S_n = \xi_1 + \cdots + \xi_n$, $n \geq 1$. Define the cusum stopping time $\tau = \tau(h) = \inf\{n \geq 1 : S_n - \min_{0 \leq i \leq n} S_i \geq h\}$. Then (as $h \to \infty$):

(a) $\tau/h \overset{a.s.}{\longrightarrow} \mu^{-1}$ and $\mu E\tau/h \longrightarrow 1$ if $\mu > 0$

(b) $\sigma^2 E\tau/h^2 \longrightarrow 1$ if $\mu = 0$,

(c) $(\tau - (h/\mu))/\sigma(h/\mu^3)^{\frac{1}{2}} \overset{\mathcal{L}}{\longrightarrow} N(0,1)$ if $\mu > 0$.

PROOF The asymptotics (a) and (b) are known [cf. Khan (1979) and Robbins (1976)]. Since $\tau/h \overset{a.s.}{\longrightarrow} \mu^{-1}$, Renyi-Anscome Theorem [cf. Gut (1988, p. 15)] gives

$$\frac{S_\tau - \mu\tau}{\sigma\sqrt{\tau}} \overset{\mathcal{L}}{\longrightarrow} N(0,1) \text{ as } h \to \infty. \tag{15.3.11}$$

Clearly, from the definition of τ we have

$$\frac{h - \mu\tau}{\sigma\sqrt{\tau}} + (\sigma\sqrt{\tau})^{-1} \min_{0 \leq i \leq \tau} S_i$$
$$\leq \frac{S_\tau - \mu\tau}{\sigma\sqrt{\tau}} \leq \frac{h - \mu\tau}{\sigma\sqrt{\tau}} + \frac{\xi_\tau}{\sigma\sqrt{\tau}} + (\sigma\sqrt{\tau})^{-1} \min_{0 \leq i \leq \tau - 1} S_i. \tag{15.3.12}$$

Since $\mu > 0$, $\min(0, S_1, \ldots, S_n) = \max(0, -S_1, \ldots, -S_n)$ has a limit distribution, and it can be shown that $(\min_{0 \leq i \leq n} S_i)/\sqrt{n} \overset{a.s.}{\longrightarrow} 0$ as $n \to \infty$. Moreover, $\xi_\tau/\sigma\sqrt{\tau} \overset{P}{\longrightarrow} 0$ as $h \to \infty$ by Lemma 2.6 of Gut (1988). Hence (c) follows from (15.3.11) and (15.3.12). Thus Theorem 5.1 of Gut (1988, p. 85) continues to hold if $\upsilon(h) = \inf\{n \geq 1 : S_n \geq h\}$ is replaced by $\tau(h)$. $\square$

Remark 15.3.1 It is known [cf. Khan (1995)] that under appropriate conditions, τ has an asymptotic exponential distribution with a suitable parameter if $\mu < 0$. The required conditions are related to the moment generating function of ξ_1 which obviously cannot be applied to the noncentral t-distribution.

Clearly, (a) can be used to approximate the ARLs of (15.3.8) and (15.3.9) when $\theta < \theta_0 - k$, while (c) can be used to estimate the standard deviations of T and T_1. If $k = 0$, then (b) can be used to approximate the ARL under θ_0 (or under $\theta_0 - k$ even if $k \neq 0$) without any normality, etc. Thus if $\theta < \theta_0 - k$, then $\text{ARL}(T,\theta) \approx \frac{h}{\eta(\theta)} \approx \frac{h}{(\theta_0-\theta-k)}$ if m is also large. Now recalling (15.3.10) let $y_n = \bar{x}_i/s(1,n)$, and note that $y_n = t[\nu, \theta\sqrt{m}]/\sqrt{m}$, $\nu = nm - 1$. It then follows that

$$E_\theta\, y_n = \theta\sqrt{\frac{\nu}{2}}\,\frac{\Gamma\left(\frac{\nu-1}{2}\right)}{\Gamma\left(\frac{\nu}{2}\right)} \text{ and } E_\theta\, y_n^2 = \frac{\nu}{(\nu-2)\,m}\,(1+m\,\theta^2).$$

Consequently, by Stirling's formula we see that $E_\theta\, y_n \to \theta$ and $var(y_n) \to 1/m$ as $n \to \infty$. Hence it follows from (15.3.9), (15.3.10) and (a) that

$$\text{ARL}(T_1,\theta) \approx \frac{h}{(\theta_0 - \theta - k)} \quad \text{if } \theta < \theta_0 - k. \qquad (15.3.13)$$

Such approximations seem to work even for moderate h and m. However, the difficulty is under θ_0 if we do use a suitable reference constant k. Then there is no choice but to use some normal approximation at least under θ_0. Let $-\eta(\theta)$ and $\sigma^2(\theta)$ be the mean and variance of the random variables of our cusum procedures, then the normal approximation for $\text{ARL}(\theta)$ is given by

$$\text{ARL}(\theta) \approx \frac{\sigma^2(\theta)}{2\eta^2(\theta)}\left(\exp\left(\frac{2h\eta(\theta)}{\sigma^2(\theta)}\right) - 1\right) - \frac{h}{\eta(\theta)}. \qquad (15.3.14)$$

The uses of these approximations are discussed in the next section on simulation.

Finally, we now consider a fully sequential cusum procedure without group sampling. Let $x_1, x_2, \ldots$ be independent $N(\mu, \sigma^2)$ random variables, and the process is said to in control if $\theta = \mu/\sigma \geq \theta_0$, and out of control if $\theta < \theta_0$. Let $\bar{x}_n$ and s_n^2 $(n \geq 2)$ be the usual sample mean and variance based on $(x_1, \ldots, x_n)$. For a suitable positive reference constant k define

$$\omega_0 = 0, \ \omega_n = \max\left\{0, -\frac{\bar{x}_n}{s_n} + \theta_0 - k\right\}, \qquad n = 2, 3, \ldots,$$

and renewal occurs whenever $\omega_n = 0$. The associated cusum procedure is defined by

$$T_0 = T_0(h) = \inf\{n \geq 2 : n\,\omega_n \geq h\}. \qquad (15.3.15)$$

Of course, whenever renewal occurs, one has to take at least two observations to obtain new ω_n's. Obviously, T_0 is equivalent to repeated application of a stopping time M_0 until termination at h, where M_0 is defined by

$$M_0 = \inf\{n \geq 2 : \sum_{i=1}^{n}(-\frac{x_i}{s_n} + \theta_0 - k) \leq 0 \text{ or } \sum_{i=1}^{n}(-\frac{x_i}{s_n} + \theta_0 - k) \geq h\}.$$

Given the nature of T_0 and M_0, no suitable approximation of the ARL seems possible. Therefore, only the simulated estimates of the ARL are given in the next section.

15.4 SIMULATIONS

Exact evaluations and simulations of the ARL of the Shewhart chart and its generalized version under group sampling use the values of the noncentrality parameter $\delta = \theta\sqrt{m}$ from the table of noncentral t-distribution of Resnikoff and Lieberman (1957). Table 15.1 gives the exact and simulated ARL for the Shewhart control chart. Table 15.2 gives the exact and simulated ARL of the generalized version N defined by (15.2.5). Table 15.3 gives the approximate and simulated ARL of T defined by (15.3.8). The approximation is based on (15.3.14) with a modified $\sigma^2(\theta)$. Recall that $\eta(\theta)$ and $\sigma^2(\theta)$ are the mean and variance of d_i which is a linear function of a noncentral t-statistic. For small or moderate m, (15.3.14) gives poor approximation. Surprisingly, the estimates get improved by using the variance $(\sigma_c^2 + \sigma^2(\theta))/2$ where $\sigma_c^2 = \sigma^2(0)$ is the variance of the central t-statistic. Of course, one can use the normal approximation for the t-statistic itself with mean θ and variance $(1+\theta^2/2)/m$ if m is fairly large. Table 15.4 gives the estimated ARL of T_1 defined by (15.3.9). Here we use (15.3.13) to approximate the ARL if $\theta < \theta_0$ while (15.3.14) is used under θ_0 with $\eta(\theta) = k$ and $\sigma^2(\theta_0) = 2/m$. As noted earlier, $var(\overline{x}_i/s(1,n)) \approx 1/m$ for large n (which is certainly the case under θ_0), and thus $\sigma^2(\theta_0) \approx 1/m$. However, $\sigma^2(\theta_0) \approx 2/m$ seems to compensate the variation caused by $s(1,n)$. Still it provides only a rough estimate of ARL(θ_0). The simulated estimates of the ARL for (15.3.15) is also included in Table 15.4.

TABLE 15.1 ARL of Shewhart chart

θ	$m = 5,\ \gamma = .0038,\ t_\gamma = 2$		$m = 10, \gamma = .0032,\ t_\gamma = 3.75$	
	Theoretical	Simulated	Theoretical	Simulated
2.326348	263.16	264.64	312.50	311.73
1.959964	47.17	46.34	34.25	34.38
1.750686	20.62	20.91	12.66	12.50
1.514102	9.26	9.24	5.15	5.15
1.281552	4.84	4.82	2.66	2.62
1.036433	2.82	2.85	1.65	1.64
.674490	1.62	1.65	1.14	1.14

TABLE 15.2 ARL of generalized cusum N

| | $m=5, k=2, h=2$ | | $m=1, k=0, h=1$ | | $m=1, k=0, h=2$ | |
θ	Exact	Simulated	Exact	Simulated	Exact	Simulated
2.326348	69515.23	—	100	100.47	10100	10458.24
1.959964	2272.16	2274.99	40	40.47	1640	1644.11
1.750686	445.74	447.28	25	25.32	650	655.16
1.514102	94.99	95.49	15.39	15.20	252.07	253.56
1.281552	28.27	28.92	10	10.12	110	110.86
1.036433	10.76	10.72	6.67	6.84	51.11	50.79
.674490	4.24	4.13	4	4.05	20	20.50

In the remaining Tables 15.3 and 15.4, the choice of k is determined by $\theta_0 - k = (\theta_0 + \theta_1)/2$, and thus $k = (\theta_0 - \theta_1)/2 = .183192$ is used in both tabes. This is in line with the usual practice, and with this choice of k, the approximate and simulated ARL are given. Only simulated estimates of ARL are given for T_0.

TABLE 15.3 ARL of cusum T

| | $m=5, h=10.5$ | | $m=10, h=4$ | |
θ	Approximate	Simulated	Approximate	Simulated
2.326348	264.27	281.39	334.19	317.01
1.959964	34.35	33.20	16.26	17.18
1.750686	19.16	19.79	8.60	9.59
1.514102	12.60	13.41	5.57	6.44
1.281552	9.40	10.26	4.13	4.87
1.036433	7.41	8.15	3.24	3.88
.674490	5.63	6.23	2.46	3.06

TABLE 15.4 ARL of T_1 and T_0

| | $T_1, m=5, h=4.3$ | | $T_1, m=10, h=2.5$ | | $T_0, h=1.5$ |
θ	Approximate	Simulated	Approximate	Simulated	Simulated
2.326348	276.58	295.27	273.89	310.19	112.26
1.959964	23.47	23.64	13.65	14.46	37.74
1.750686	10.96	12.45	6.37	7.53	20.84
1.514102	6.84	8.18	3.97	4.87	11.89
1.281552	4.99	6.02	2.90	3.66	8.35
1.036433	3.89	4.69	2.26	2.92	6.10
.674490	2.93	3.59	1.70	2.23	4.53

The relative merits of these procedures can now be compared. Comparing Tables 15.1–15.3 we see that the cusum procedure of Section 15.2.3 based on noncentral t-statistics is more efficient than that of Shewhart or its generalized version. However, the Shewhart chart or the generalized cusum have the advantage of the availability of their exact ARL. Nevertheless, procedures (15.3.8) and (15.3.9) do indeed perform better, and (15.3.9) appears to be some what more efficient than (15.3.8). The cusum procedure N with $k = 0$ and $m = 1$, and the cusum procedure (15.3.15) are indeed useful without group sampling. However, some suitable approximations of the ARL associated with T_0 in (15.3.15) is an interesting open problem.

REFERENCES

Ghosh, B. K. (1970). *Sequential Tests of Statistical Hypotheses*, Addison-Wesley, Cambridge, MA.

Gut, A. (1988). *Stopped Random Walks*, Springer-Verlag, New York.

Khan, R. A. (1978). Wald's approximations to the average run length in cusum procedures, *Journal of Statistical Planning and Inference*, **2**, 63–77.

Khan, R, A. (1979). Some first passage problems related to cusum procedures, *Stochastic Processes and their Applications*, **9**, 207–215.

Khan, R. A. (1984). On cumulative sum procedures and the SPRT with applications, *Journal of the Royal Statistical Society, Series B*, **46**, 79–85.

Khan, R. A. (1995). Detecting changes in probabilities of a multi-component process, *Sequential Analysis*, **14**, 375–388.

Page, E. S. (1954). Continuous inspection schemes, *Biometrika*, **41**, 100–115.

Resnikoff, G. J. and Lieberman, G. J. (1957). Tables of the noncentral t-distribution, *Stanford University Press*, Stanford, CA.

Reynolds, M. R. (1975). Approximations to the average run length in cumulative sum control charts, *Technometrics*, **17**, 65–71.

Robbins, N. B. (1976). Convergence of some expected first passage times, *Annals of Probability*, **4**, 1027–1029.

REFERENCES

CHAPTER 16

DETECTING CHANGES IN THE VON MISES DISTRIBUTION

KAUSHIK GHOSH

George Washington University, Washington, DC

Abstract: Tests for detecting changes in the concentration parameter and/or the mean direction when data are from von Mises distribution are developed. Critical values are obtained through simulations. Finally, the tests are compared with respect to their powers using simulations.

Keywords and phrases: Change-point, directional data, von Mises distribution, generalized likelihood ratio test, power

16.1 INTRODUCTION

Suppose we have a set of independent and identically distributed measurements on 2-dimensional directions, say α_1, α_2, ..., α_n. These measurements, called angular or circular data, can be represented as points on the circumference of a circle with unit radius. They may be wind directions, the vanishing angles at the horizon for a group of birds or the times of arrival at a hospital emergency room where the 24 hour cycle is represented as a circle [see Mardia (1972), Fisher (1993) for more on circular data]. In Ghosh, Jammalamadaka and Vasudaven (1999), tests to detect presence of change-points in the preferred direction were proposed and studied. In this article, we cover other aspects by considering detection of changes in the concentration (variability) and/or the preferred direction.

Formally, let α_1, ..., α_n be angular measurements measured in a time-ordered or space-ordered sequence . Assume that for some unknown (but

225

fixed) k, $(1 \leq k \leq n)$, α_1, ..., $\alpha_k \sim F_1$ and α_{k+1}, ..., $\alpha_n \sim F_2 (\neq F_1)$. Here, k is called the change-point of the data. If $k = n$, there are no observations from F_2, meaning all the observations are from the same population and hence, there is no change-point. We are interested in testing for the presence of a change-point. Hence, we are testing $H_0 : k = n$ vs. $H_1 : 1 \leq k \leq n - 1$. For concreteness and simplicity, we assume that the two populations F_1 and F_2 are von Mises (sometimes also known as the Circular Normal) with concentration parameters κ_1 and κ_2 and mean directions μ_1 and μ_2 respectively.

We say that a random angle α has von Mises distribution with mean direction $\mu \in [-\pi, \pi)$ and concentration parameter $\kappa > 0$ (i.e. $vM(\mu, \kappa)$) if it has the probability density:

$$f(\alpha) = \frac{1}{2\pi I_0(\kappa)} e^{\kappa \cos(\alpha - \mu)}, \quad -\pi \leq \alpha < \pi.$$

von Mises distribution is in many ways analogous to the normal distribution on the line and plays a central role in modeling circular data. Many real-life circular data can be modeled by such a distribution.

A change in distribution at k implies that α_1, ..., $\alpha_k \sim vM(\mu_1, \kappa_1)$ and α_{k+1}, ..., $\alpha_n \sim vM(\mu_2, \kappa_2)$. Here, we have a 4-dimensional parameter $\theta = (\mu_1, \mu_2, \kappa_1, \kappa_2)$ with the parameter space $\Omega = [-\pi, \pi) \times [-\pi, \pi) \times (0, \infty) \times (0, \infty)$.

While there has been an extensive study of change-point problems for the linear data case, only a few have dealt with circular data. Lombard (1986) proposed nonparametric tests for detecting change-points in angular data. Csörgő and Horváth (1996) give the asymptotic distributions of the statistic proposed by Lombard. Ghosh et. al. (1999) deals with detecting changes in mean direction in the parametric set up of the von Mises distribution. To the author's knowledge, no other work is currently available that deals with changes in directions.

This article is organized as follows: Section 16.2 deals with the derivation of tests when there is a change in κ (possibly along with a change in μ). We use the generalized likelihood ratio method to derive tests for H_0 vs. H_1. The exact critical values of the test statistics are obtained through simulations after conditioning on the length the resultant of the observed data.

An alternative method, with a Bayesian flavor, assumes that the change-point is equally likely to be at any one of the intermediate points. Hence, using a discrete uniform prior over the possible change-point values, we get an alternate statistic. If we have further information about the possible point of change, we can incorporate that into an appropriate prior on k

and derive the corresponding Bayes procedure.

The results on the critical values of some of the test statistics, obtained through simulations, are presented in Section 16.3. They are provided as nomograms from which one can read the 5% values. The author may be contacted for the code, if other values are of interest. Finally, in Section 16.4, we compare the powers of the two procedures for various alternatives.

In what follows, we will use $I_0(\cdot)$ to denote the modified Bessel function of order zero, $I_1(\cdot)$ to denote the modified Bessel function of order one and $A(\cdot)$ to denote the ratio $\frac{I_1(\cdot)}{I_0(\cdot)}$. We will also use the notation $\Psi(\cdot)$ for the following function:

$$\Psi(x) = xA^{-1}(x) - \log I_0(A^{-1}(x)). \tag{16.1.1}$$

It is worth noting here that $I_1(\cdot)$ is the derivative of $I_0(\cdot)$.

16.2   THE TESTS

16.2.1   Change in κ, μ Fixed and Known

Let $\mu_1 = \mu_2 = \mu$ (say). If the change is at k, the likelihood of the data is

$$\frac{1}{[2\pi I_0(\kappa_1)]^k} \frac{1}{[2\pi I_0(\kappa_2)]^{n-k}} e^{\kappa_1 \sum_{i=1}^{k} \cos(\alpha_i - \mu) + \kappa_2 \sum_{i=k+1}^{n} \cos(\alpha_i - \mu)},$$

which is maximized for

$$\hat{\kappa}_1 = A^{-1}\left(\frac{\sum_{i=1}^{k} \cos(\alpha_i - \mu)}{k}\right) \quad \text{and} \quad \hat{\kappa}_2 = A^{-1}\left(\frac{\sum_{i=k+1}^{n} \cos(\alpha_i - \mu)}{n-k}\right).$$

Under H_0, the likelihood is

$$\frac{1}{[2\pi I_0(\kappa)]^n} e^{\kappa \sum_{i=1}^{n} \cos(\alpha_i - \mu)},$$

which is maximized for

$$\hat{\kappa} = A^{-1}\left(\frac{\sum_{i=1}^{n} \cos(\alpha_i - \mu)}{n}\right).$$

If Λ_k is the likelihood ratio when change is at k, we reject the null hypothesis of no change for change at k if

$$-\log \Lambda_k = k\Psi\left(\frac{C_1}{k}\right) + (n-k)\Psi\left(\frac{C_2}{n-k}\right) - n\Psi\left(\frac{C}{n}\right)$$

is too large. Here, $C_1 = \sum_{i=1}^{k} \cos(\alpha_i - \mu)$, $C_2 = \sum_{i=k+1}^{n} \cos(\alpha_i - \mu)$, $C = \sum_{i=1}^{n} \cos(\alpha_i - \mu)$ and $\Psi(\cdot)$ is as in (16.1.1). Without loss of generality, we can assume that $\mu = 0$.

Since k is unknown, we use the "sup" and "avg" methods as in Ghosh, Jammalamadaka and Vasudaven (1999) to get the following statistics for testing H_0 against H_1:

$$T_1 = \sup_{k \in \{1, \, \ldots, \, n-1\}} \left[k\Psi\left(\frac{C_1}{k}\right) + (n-k)\Psi\left(\frac{C_2}{n-k}\right) - n\Psi\left(\frac{C}{n}\right) \right]$$

$$(16.2.2)$$

and

$$T_2 = \frac{1}{n-1} \sum_{k=1}^{n-1} \left[k\Psi\left(\frac{C_1}{k}\right) + (n-k)\Psi\left(\frac{C_2}{n-k}\right) - n\Psi\left(\frac{C}{n}\right) \right]. \quad (16.2.3)$$

When H_0 is true, the distributions of these statistics depend on the unknown common concentration parameter κ. To make them free of κ, one could, as we do, condition on the overall sum of cosines C (sufficient statistic for κ). Although exact distributions are very difficult to find, we can resort to simulations to obtain the cut-offs of these conditional tests.

16.2.2   Change in κ, μ Fixed but Unknown

In this case, the maximizer of the likelihood assuming change at k is given by the following three equations:

$$\hat{\kappa}_1 \sum_{i=1}^{k} \sin(\alpha_i - \hat{\mu}) + \hat{\kappa}_2 \sum_{i=k+1}^{n} \sin(\alpha_i - \hat{\mu}) = 0,$$

$$A(\hat{\kappa}_1) = \frac{\sum_{i=1}^{k} \cos(\alpha_i - \hat{\mu})}{k} \quad \text{and} \quad A(\hat{\kappa}_2) = \frac{\sum_{i=k+1}^{n} \cos(\alpha_i - \hat{\mu})}{n-k}.$$

Under H_0, the likelihood is maximized at

$$\hat{\mu} = \overline{\alpha} \quad \text{and} \quad \hat{\kappa} = A^{-1}\left(\frac{R}{n}\right)$$

where $(\overline{\alpha}, R)$ are the direction and length respectively of the overall resultant.

Hence, if Λ_k is the likelihood ratio, we have

$$-\log \Lambda_k = k\Psi\left(\frac{C_1}{k}\right) + (n-k)\Psi\left(\frac{C_2}{n-k}\right) - n\Psi\left(\frac{R}{n}\right)$$

where $C_1 = \sum_{i=1}^{k}\cos(\alpha_i-\hat{\mu})$, $C_2 = \sum_{i=k+1}^{n}\cos(\alpha_i-\hat{\mu})$, $C = \sum_{i=1}^{n}\cos(\alpha_i-\hat{\mu})$ and $\Psi(\cdot)$ is as in (16.1.1). The two statistics of interest for testing H_0 against H_1 are then

$$T_1 = \sup_{k\in\{1,\,\ldots,\,n-1\}}\left[k\Psi\left(\frac{C_1}{k}\right) + (n-k)\Psi\left(\frac{C_2}{n-k}\right) - n\Psi\left(\frac{R}{n}\right)\right]$$
$$(16.2.4)$$

and

$$T_2 = \frac{1}{n-1}\sum_{k=1}^{n-1}\left[k\Psi\left(\frac{C_1}{k}\right) + (n-k)\Psi\left(\frac{C_2}{n-k}\right) - n\Psi\left(\frac{R}{n}\right)\right]. \quad (16.2.5)$$

As before, to make the null distributions of the tests free of the unknown parameters, we condition on the overall resultant R. Exact distributions being difficult to obtain, we need to resort to simulations to get the cut-offs of these conditional tests.

16.2.3   Change in μ or κ or Both

Assuming that the change-point is at k, the likelihood of the observed data is maximized for

$$\hat{\mu}_1 = \overline{\alpha}_{1k}, \ \hat{\mu}_2 = \overline{\alpha}_{2k}, \ \hat{\kappa}_1 = A^{-1}\left(\frac{R_{1k}}{k}\right) \ \text{ and } \ \hat{\kappa}_2 = A^{-1}\left(\frac{R_{2k}}{n-k}\right).$$

Here, $(\overline{\alpha}_{1k},\ R_{1k})$ denote the direction and length respectively of the resultant of the unit vectors given by $\alpha_1,\ \ldots,\ \alpha_k$ while $(\overline{\alpha}_{2k},\ R_{2k})$ denote those for $\alpha_{k+1},\ \ldots,\ \alpha_n$.

Under H_0 (no change-point), $\mu_1 = \mu_2 (\equiv \mu,\ \text{say})$ and $\kappa_1 = \kappa_2 (\equiv \kappa,\ \text{say})$. The likelihood is then maximized for

$$\hat{\mu} = \overline{\alpha} \ \text{ and } \ \hat{\kappa} = A^{-1}\left(\frac{R}{n}\right)$$

where $(\overline{\alpha},\ R)$ are the direction and length respectively of the resultant of the n unit vectors given by $\alpha_1,\ \ldots,\ \alpha_n$.

Thus, assuming that the change-point is at k, the likelihood ratio becomes

$$\Lambda_k = \frac{[I_0(\hat{\kappa}_1)]^k [I_0(\hat{\kappa}_2)]^{n-k}}{[I_0(\hat{\kappa})]^n}$$

$$\times \, e^{\hat{\kappa} \sum_{i=1}^{n} \cos(\alpha_i - \hat{\mu}) - \hat{\kappa}_1 \sum_{i=1}^{k} \cos(\alpha_i - \hat{\mu}_1) - \hat{\kappa}_2 \sum_{i=k+1}^{n} \cos(\alpha_i - \hat{\mu}_2)}.$$

After some algebra, we have

$$-\log \Lambda_k = \frac{k}{n} \Psi\left(\frac{R_1}{k}\right) + \frac{n-k}{n} \Psi\left(\frac{R_2}{n-k}\right) - \Psi\left(\frac{R}{n}\right).$$

As before, k being unknown, the "sup" and "avg" methods give rise to the statistics:

$$T_1 = \sup_{k \in \{2, \, \ldots, \, n-2\}} \left[\frac{k}{n} \Psi\left(\frac{R_1}{k}\right) + \frac{n-k}{n} \Psi\left(\frac{R_2}{n-k}\right) - \Psi\left(\frac{R}{n}\right) \right]$$

$$\text{(16.2.6)}$$

and

$$T_2 = \frac{1}{n-3} \sum_{k=2}^{n-2} \left[\frac{k}{n} \Psi\left(\frac{R_1}{k}\right) + \frac{n-k}{n} \Psi\left(\frac{R_2}{n-k}\right) - \Psi\left(\frac{R}{n}\right) \right].$$

$$\text{(16.2.7)}$$

Note that unlike the previous cases, k runs from 2 through $n-2$. This is because, the concentration parameters being unknown, we need at least two observations to estimate them.

The test statistics so obtained are functions of the resultant lengths of the angles, whose distributions under the null hypothesis depend on the common (unknown) concentration parameter κ. This can be made free of κ upon conditioning by the overall resultant length R, since the conditional distribution of (R_1, R_2) given R is independent of κ [see Watson and Williams (1956)]. Results of these conditional simulations are presented in the next section.

16.3   SIMULATION RESULTS

All the tests proposed in the previous section have no simple known distributional form. Thus, to obtain their cut-off values, we need to do Monte-Carlo simulations. For illustration and simplicity, we present here simulation results for the last case (subsection 16.2.3) only. In particular, the results are based on simulations of the following two statistics:

$$T_1^* = \sup_{k \in \{2, \, \ldots, \, n-2\}} \left\{ \frac{k}{n} \Psi \left(\frac{R_1}{k} \right) + \frac{n-k}{n} \Psi \left(\frac{R_2}{n-k} \right) \right\} \Big| R = r$$

(16.3.8)

and

$$T_2^* = \frac{1}{n-3} \sum_{k=2}^{n-2} \left\{ \frac{k}{n} \Psi \left(\frac{R_1}{k} \right) + \frac{n-k}{n} \Psi \left(\frac{R_2}{n-k} \right) \right\} \Big| R = r.$$

(16.3.9)

All the codes were written in the C language with calls to the IMSL/C/STAT library for the random number generators. In particular, we made extensive use of the routine imsls_f_random_von_mises to generate all the von Mises random deviates.

Since κ is unknown, we sample from a conditional von Mises distribution, the conditioning event being the given length r of the resultant R. Each (r, n) combination results in a different distribution and we considered $n = 10(2)20(5)30(10)50$ and $\frac{r}{n} = 0.05(0.05)0.95$. Since the conditional sampling discards a lot of the random numbers for not meeting the conditioning criterion, sampling procedure was sometimes computer intensive — especially in the very extreme $\frac{r}{n}$ values. Apart from using imsls_f_random_von_mises, we used imsls_f_random_binomial to draw the conditional samples. The results of 100,000 simulations appear as nomograms in Figures 16.1 and 16.2.

A striking feature as observed on close examination of these nomograms is the decreasing trend for a fixed $\frac{r}{n}$ as n increases. In comparison, the nomograms when we were testing for changes only in mean had an increasing trend. This can be explained by the convexity of the function $\Psi(\cdot)$.

The other notable feature is that the curves all become close to horizontal as n increases. This asymptotic behavior is pronounced after a considerable sample is taken, say $n = 50$ or more. This can be explained using the same argument as used to explain the horizontal nature of nomograms in the test for change in mean.

16.4 POWER COMPARISONS

To examine the behavior of the tests proposed in (16.3.8) and (16.3.9) under various alternatives, we did a Monte-Carlo study of their power curves. For each of the alternatives under consideration, we ran 500 simulations to get

the powers. the cut-of for each test was based on 100 simulations. Figures
16.3–16.6 summarize the effects of changes in n, Δ, κ and k on the power
curves. In Figure 16.3, we chose $\Delta = \frac{\pi}{2}$, $\kappa_1 = 1.5$, $\kappa_2 = 4.5$. We varied
n through the values 5(5)10(10)40(20)80 choosing $k = \frac{n}{2}$ in each case. As
expected, the power increases with sample size increase; he effect being
more pronounced for the "sup" statistic where the power for $n = 30$ is close
to 1. In comparison, the "avg" statistic reaches a maximum power of about
65% for n as large as 80.

In Figure 16.4, we examine the effect of change in difference in means
$\Delta = |\mu_1 - \mu_2|$. We took $n = 20$, $\kappa_1 = 1$, $\kappa_2 = 4$, $k = 10$ and keeping these
fixed, we varied Δ over the range $\frac{\pi}{20}\left(\frac{\pi}{20}\right)\frac{\pi}{2}$. Again, power increases with
increase in the difference of the means and the "sup" statistic performs
better than the "avg".

Figure 16.5 depicts the effect of changing one concentration parameter
on the power of the test. Here, we have taken $n = 20$, $\Delta = \frac{\pi}{2}$, $\kappa_1 = 2$ and
$k = 10$. Under this set up, the powers of the two tests obtained by varying
κ_2 over the range 0.5(0.5)4.5 are obtained. Clearly, the "sup" statistic is
the uniform winner.

Finally, Figure 16.6 demonstrates the effect of the position of change-
point k on the powers of the two tests, when everything else remains the
same. In particular, we chose $n = 30$, $\Delta = \frac{\pi}{2}$, $\kappa_1 = 1$, $\kappa_2 = 4$ and
$k = 1(2)15$. as before, "sup" is the clear winner, with maximum power
close to 100% for k close to $\frac{n}{2}$ while the "avg" gets max power close to
65%.

A notable feature of the study of the power curves is that in all these
cases, the "sup" statistic is doing better, unlike the tests for change in
mean only in which "avg" seems to be doing better in most cases. Also,
as expected, both the tests are symmetric in k (i.e., power at k is same as
power at $n - k$) and is also symmetric in κ_1 and κ_2.'

Summarizing, we would suggest using the "sup" statistic in almost all
cases, if we are looking for changes in mean and/or concentration. A bonus
is that we can also get the estimate of the location of the change-point in
this case — this is the value of k at which the "sup" is attained.

16.5   AN EXAMPLE

As an illustration of the proposed method of change-point detection, consider the following data taken from Fisher (1993):

108.2	19.7	2.5	346.2	38.9	57.9	306.1	358.7	312.3	30.3
144.9	22.2	189.1	304.4	267.5	167.5	289.7	301.8	322.2	244.0
290.4	184.3	234.7	252.4	265.6	328.0	312.6	6.6	219.4	296.4
326.5	297.9	226.2	172.8	247.9	127.5	337.3	331.0	256.5	352.6
310.6	218.1	351.9	105.0	343.2	329.6	358.2	344.4	104.4	79.1
90.9	24.8	84.3	279.7	282.7	81.8	133.1	340.7	10.7	308.4
282.7	81.8	133.1	340.7	10.7	308.4				

These are the measurements of azimuth of flares when they start burning, collected by Dr. F. Lombard and later analyzed in Lombard (1986). We will use the first 30 observations to detect for presence of a change-point, assuming there is at most one change (AMOC).

The values of T_1^* and T_2^* calculated from this data set are 0.459 and 0.291 respectively, with $n = 30$ and $\frac{r}{n} = 0.392$. Comparing with the 5% cut-off values in Figures 16.1 and 16.2, we see both the "sup" and "avg" tests are significant (since the observed values are bigger than the tabulated values). This suggests the presence of a change-point in the distribution of the azimuth direction of the (first 30) burning flares.

This conforms with the findings of Lombard (1986) in which the presence of a change-point in this data set is shown using non-parametric techniques.

Acknowledgements The author is grateful to Professor S. Rao Jammalamadaka for his suggestions and valuable comments during the course of this research and to the anonymous referee for the comments that improved the presentation of this chapter.

REFERENCES

Csörgő, M. and Horváth, L. (1996). A note on the change-point problem for angular data, *Statistics and Probability Letters*, **27**, 61—65.

Fisher, N. I. (1993). *The Statistical Analysis of Circular Data*, Cambridge University Press, Cambridge.

Ghosh, K., Jammalamadaka, S. R. and Vasudaven, M. (1999). Change-point problems for the von Mises distribution, *Journal of Applied Statistics*, **26**, 415–426.

Lombard, F. (1986). The change-point problem for angular data: A non-parametric approach, *Technometrics*, **28**, 391–397.

Mardia, K. V. (1972). *Statistics of Directional Data*, Second Edition, Academic Press, London.

Watson, G. S. and Williams, E. J. (1956). On the construction of significance tests on the circle and the sphere, *Biometrika*, **43**, 344–352.

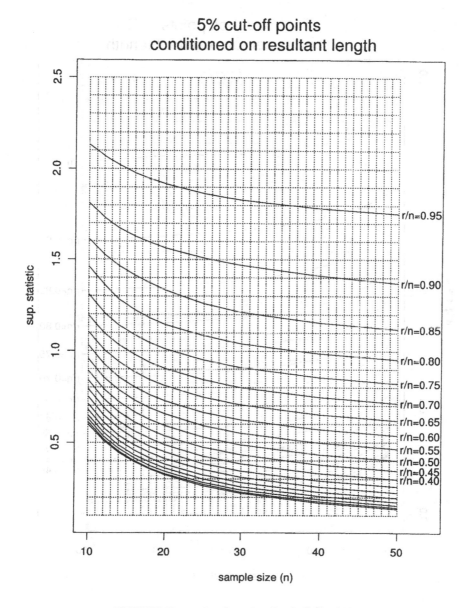

FIGURE 16.1 Cut-offs of the sup statistic

5% cut-off points
conditioned on resultant length

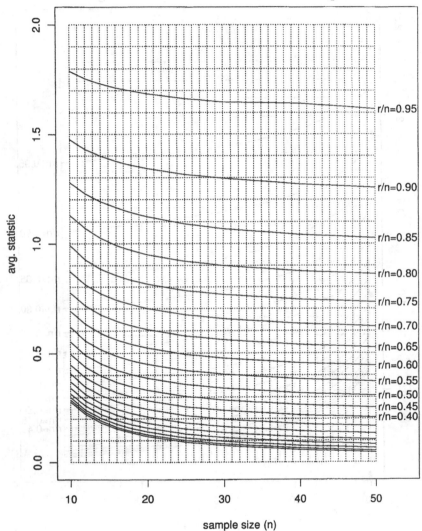

FIGURE 16.2 Cut-offs of the avg statistic

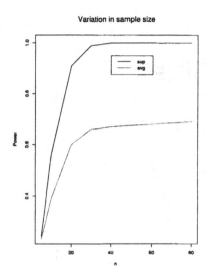

FIGURE 16.3 Effect of n on power

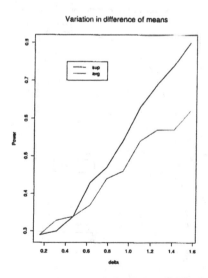

FIGURE 16.4 Effect of Δ on power

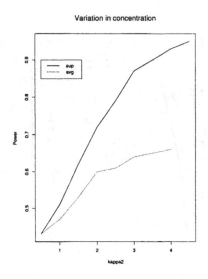

FIGURE 16.5 Effect of κ on power

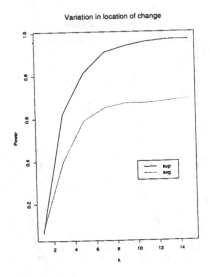

FIGURE 16.6 Effect of k on power

CHAPTER 17

ONE-WAY RANDOM EFFECTS MODEL WITH A COVARIATE: NONEGATIVE ESTIMATORS

PODURI S. R. S. RAO

University of Rochester, Rochester, NY

Abstract: For the balanced case of the one-way random effects model with a covariate, the Analysis of Covariance Estimator (ANCOVA) is adjusted for nonnegativeness. The Minimum Norm Quadratic Estimator (MINQE), which is nonnegative, is also considered and a modification is suggested to reduce its bias. B ased on the principle of the Minimum Variance Quadratic Unbiased Estimation (MIVQUE), a nonnegative estimator for the variance component which utilizes its prior information is developed. The sensitivity of the above MIVQUE and MINQE type estimators to the a priori values are examined. The relative merits of all the estimators are compared.

Keywords and phrases: Random effects model, analysis of covariance, minimum norm quadratic estimator, minimum variance quadratic estimator

17.1 INTRODUCTION

The balanced one-way random effects model with a covariate is

$$y_{ij} = \mu + \alpha_i + \beta(x_{ij} - \bar{x}) + \varepsilon_{ij} \tag{17.1.1}$$

for $i = (1, \ldots, k)$ and $j = (1, \ldots, m)$. In this model, y_{ij} represents m observations from each of k groups or treatments and x_{ij} the covariate, with its mean $\bar{x} = \sum x_{ij}/n$ where $n = km$ is the total sample size. The

239

overall mean and the common slope for the k groups are represented by μ and β respectively. The residuals ε_{ij} in each group are assumed to have mean zero and variance σ^2. They are assumed to be uncorrelated within each group and among the groups. The random effect α_i is assumed to be independent of ε_{ij} with mean zero and variance σ_α^2.

Cochran (1946) and Anderson and Bancroft (1952) consider the AN-COVA procedure for the estimation of σ^2 and σ_α^2. For both the balanced and unbalanced cases of the above model, the Minimum Variance Quadratic Estimation (MIVQUE) proposed by C. R. Rao (1971, 1972) and some of its modifications were examined by P. S. R. S. Rao and Miyawaki (1989) and P. S. R. S. Rao (1977).

The purpose of this article is to suggest nonnegative estimators for σ_α^2 based on the ANCOVA and MIVQUE procedures and examine their relative merits. In Section 17.2, we present the ANCOVA estimator as presented in P. S. R. S. Rao (1977) and derive a nonnegative estimator, ANCOVA(N). A nonnegative estimator derived from the Minimum Norm Quadratic Estimation (MINQE) procedure considered by P. S. R. S. Rao and Chaubey (1978) is presented in Section 17.3. We have also made an adjustment to this estimator to reduce its bias and considered the MINQE*. In Section 17.4, we derive a nonnegative estimator MIVQUE(N) based on the MIVQUE procedure. This estimator follows the procedure of PROPE (proportional priors) considered by P. S. R. S. Rao (1997) for the one-way random effects model. It depends on the prior values of σ^2 and σ_α^2. We have also considered special cases of this estimator based on the relative magnitudes of these prior values. We note that for the balanced case of the model, the MIVQUE does not depend on these prior values and it coincides with the ANCOVA estimator. In Section 17.5, we examine the biases and variances of the different estimators.

17.2 ANCOVA ESTIMATOR AND ITS MODIFICATION

17.2.1 Ancova Estimator

Let

$$T_{xy} = \sum\sum(x_{ij} - \bar{x})(y_{ij} - \bar{y}), \quad B_{xy} = m\sum(\bar{x}_i - \bar{x})(\bar{y}_i - \bar{y})$$

and

$$W_{xy} = \sum\sum(x_{ij} - \bar{x})(y_{ij} - \bar{y})$$

where $(\bar{x}_i, \bar{y}_i)$ and $(\bar{x}, \bar{y})$ are the group means and the overall means respectively. Similarly, let T_{xx}, B_{xx} and W_{xx} denote the total, between and within sums of squares for the covariate.

Let

$$T_{y.x} = \sum\sum[(y_{ij} - \bar{y}) - b_1(x_{ij} - \bar{x})]^2,$$

where $b_1 = T_{xy}/T_{xx}$. Similarly, let

$$W_{y.x} = \sum\sum[(y_{ij} - \bar{y}) - b_2(x_{ij} - \bar{x})]^2,$$

where $b_2 = W_{xy}/W_{xx}$. An unbiased estimator of σ^2 is $\hat{\sigma}^2 = W_{y.x}/(n-k-1)$. Since

$$E(T_{y.x} - W_{y.x}) = [(k-2) + W_{xx}/T_{xx}]m\sigma_\alpha^2 + (k-1)\sigma^2,$$

an unbiased estimator of σ_α^2 is given by

$$\sigma_\alpha^2 = (T_{y.x} - W_{y.x})/A - (k-1)W_{y.x}/(n-k-1)A, \qquad (17.2.2)$$

where $A = [(k-2) + W_{xx}/T_{xx}]m$. Note that $(W_{xx}/T_{xx}) = W_{xx}/(W_{xx} + B_{xx}) = 1/(1+R)$, where $R = B_{xx}/W_{xx} = [(k-1)/(n-k)]F_{k-1,n-k}$; $F_{k-1,n-k}$ is the F-distribution with $(k-1)$ and $(n-k)$ d.f.

The variance of the estimator in (17.2.2) is

$$V(\hat{\sigma}_\alpha^2) = 2E^2/(k-1)A^2 + 2(k-1)^2\sigma^4/(n-k-1)A^2. \qquad (17.2.3)$$

17.2.2 Adjustment for Nonnegativeness

From (17.2.2), a nonnegative estimator, ANCOVA(N), for σ_α^2 is

$$\hat{\sigma}_{\alpha N}^2 = (T_{y.x} - W_{y.x})/A. \qquad (17.2.4)$$

Denoting the expectation of $(T_{y.x} - W_{y.x})$ by E, the expectation and bias of this estimator are E/A and $(k-1)\sigma^2/A$ respectively. The variance of this estimator is $2E^2/(k-1)A^2$. Its Mean Square Error is

$$MSE(\hat{\sigma}_{\alpha N}^2) = 2E^2/(k-1)A^2 + (k-1)^2\sigma^4/A^2. \qquad (17.2.5)$$

From (17.2.3) and (17.2.5),

$$MSE(\hat{\sigma}_{\alpha N}^2) - V(\hat{\sigma}_\alpha^2) = (k-1)^2\sigma^4(n-k-3)/(n-k-1)A^2. \quad (17.2.6)$$

This difference can be large if F is significant.

17.3   THE MINQE AND A MODIFICATION

Let $(\gamma_\alpha^2, \gamma^2)$ denote the a priori values of $(\sigma_\alpha^2, \sigma^2)$, and let $W = 1/(\sigma_\alpha^2 + \sigma^2/m)$. Following P. S. R. S. Rao and Chaubey (1978) and P.S.R.S Rao (1997), the MINQE for σ_α^2 is given by

$$\hat{\sigma}_\alpha^2 = \gamma_\alpha^4 (W^2/k) \sum [(\bar{y}_i - \bar{y} - b(\bar{x}_i - \bar{x})]^2, \qquad (17.3.7)$$

where $b = (B_{xy} + W_{xy})/(B_{xx} + W_{xx})$. The expectation of the summation on the right hand side of (17.3.7) is

$$
\begin{aligned}
N = &\ (k-1)\nu + (R/m)[W^2 R\nu/m + (\gamma_\alpha^2 + \gamma^2)/\gamma^4](WR/m + 1/\gamma^2)^2 \\
&\ -2W\nu R/(WR + m\gamma^2) ,
\end{aligned}
$$

where $\nu = \sigma_\alpha^2 + \sigma^2/m$. With this expression, from (17.3.7), $E(\hat{\sigma}_\alpha^2) = \gamma_\alpha^4 W^2 N/k$.

The above estimator is biased. The modified estimator MINQE*, obtained by replacing k in the denominator of (17.3.7) by $(k-1)$, will have smaller bias.

17.4   AN ESTIMATOR DERIVED FROM THE MIVQUE PROCEDURE

Following the procedure in P. S. R. S. Rao (1997), we derive a nonnegative estimator for σ_α^2, MIVQUE(N), as follows. Let E denote the expectation of $\sum [(\bar{y}_i - \bar{y}) - b(\bar{x}_i - \bar{x})]^2$, with $9\sigma_\alpha^2, \sigma^2)$ replaced by $(c\sigma_\alpha^2, c\sigma^2)$, where c is a constant. From this approach,

$$
\begin{aligned}
E = &\ (k-1)(\gamma_\alpha^2 + \gamma^2/m) + (B_{xx}/mD^2)[WB_{xx}/m(\gamma_\alpha^2 + \gamma^2)W_{xx}/\gamma^4] \\
&\ -2B_{xx}/mD,
\end{aligned}
$$

where $D = WB_{xx}/m + W_{xx}/\gamma^2$.

Now, a nonnegative estimator for σ_α^2 is

$$\hat{\sigma}_\alpha^2 = (\gamma_\alpha^2/E) \sum [(\bar{y}_i - \bar{y}) - b(\bar{x}_i - \bar{x})]^2. \qquad (17.4.8)$$

The expectation of this estimator is given by $E(\hat{\sigma}_\alpha^2) = \gamma_\alpha^2(N/E)$, where N is the same as presented in Section 17.3.

17.4.1 Special Cases of the Estimator

We consider the cases where $r = (\gamma_\alpha^2/\gamma^2)$ is large or small. When r is large, the estimator in (17.4.8) becomes

$$\hat{\sigma}_\alpha^2 = \gamma_\alpha^2 W \frac{\sum[(\bar{y}_i - \bar{y}) - b(\bar{x}_i - \bar{x})]^2}{(k-1)\gamma_\alpha^2 + \gamma^2/m) + (\gamma_\alpha^2 + \gamma^2)R/m} \ . \tag{17.4.9}$$

In this expression, $b = W_{xy}/W_{xx}$. The expectation of this estimator is

$$E(\hat{\sigma}_\alpha^2) = \gamma_\alpha^2 \frac{[(k-1)\nu + (\sigma_\alpha^2 + \sigma^2)R/m}{(k-1)gamma_\alpha^2 + \gamma^2/m) + (\gamma_\alpha^2 + \gamma^2)R/m} \ . \tag{17.4.10}$$

When r is small, the estimator in (17.4.8) becomes

$$\hat{\sigma}^2\alpha = \frac{\gamma_\alpha^2}{\gamma_\alpha^2 + \gamma^2/m} \frac{\sum[(\bar{y}_i - \bar{y}) - b(\bar{x}_i - \bar{x})]^2}{k-2} \ . \tag{17.4.11}$$

In this expression, $b = B_{xy}/B_{xx}$. The expectation and variance of this estimator are

$$E(\hat{\sigma}_\alpha^2) = \gamma_\alpha^2 \nu/(\gamma_\alpha^2 + \gamma^2/m) \tag{17.4.12}$$

and

$$V(\hat{\sigma}_\alpha^2) = [2/(k-2)][\gamma_\alpha^2/(\gamma_\alpha^2 + \gamma^2/m)]^2 \nu^2. \tag{17.4.13}$$

17.5 COMPARISON OF THE ESTIMATORS

To examine the estimators for their sensitivity to the prior values and to compare their biases, in Tables 17.1 and 17.2, we present the expected values of the estimators for $(\sigma_\alpha^2, \sigma^2) = (10, 1)$ when $(\gamma_\alpha^2, \gamma^2) = (10, 1)$, (20,2) and (5,1). The expected values for $k = 3$, $m = 10$ and $n = 30$ are presented in Table 17.1 and for $k = 6$, $m = 11$ and $n = 66$ are presented in Table 17.2. For these two cases, $F_{2,27}(.05)$ and $F_{5,60}(.05)$ are equal to 5.5 and 2.37 respectively, and we have computed the expectations with these values. We have considered these significant points, since the model in (17.1.1) should be considered only when the effect of the covariate is significant.

TABLE 17.1 Expected values of the
estimators when $(\sigma_\alpha^2, \sigma^2) = (10, 1)$; $k = 3$,
$m = 10$ and $n = 30$

$(\gamma_\alpha^2, \gamma^2)$	(10,1)	(20,2)	(5,1)
ANOVA	10	10	10
ANOVA(N)	10.12	10.12	10.12
MINQE	6.98	7.42	6.57
MINQE*	10.28	10.73	9.80
MIVQUE(N)	10	10.22	9.79
r large	10	10	9.88
r small	10	10	9.90

TABLE 17.2 Expected values of the
estimators when $(\sigma_\alpha^2, \sigma^2) = (10, 1)$; $k = 6$,
$m = 11$ and $n = 66$

$(\gamma_\alpha^2, \gamma^2)$	(10,1)	(20,2)	(5,1)
ANOVA	10	10	10
ANOVA(N)	10.09	10.09	10.09
MINQE	8.42	8.62	8.15
MINQE*	10.07	10.27	9.77
MIVQUE(N)	10	10.04	9.89
r large	10	10	9.91
r small	10	10	9.91

As can be seen from Tables 17.1 and 17.2, MINQE* and MIVQUE(N)
are not sensitive to the relative magnitudes of the prior values of the variance components. The biases of these estimators are also negligible.

From (17.2.3), the variance of $\hat{\sigma}_\alpha^2$ of the ANOVA procedure for all the
three cases in Table 17.1 is equal to 102.35. For all the three cases of
Table 17.2, it is equal to 40.76. From the expressions in (17.2.4), (17.3.7)
and (17.4.8), we can expect the variances of the ANOVA(N), MINQE*
and MIVQUE(N) to be close to these variances, provided the effect of the
covariate is significant. We are investing the biases and variances of all the
estimators described in Sections 17.2, 17.3andrefs4 through simulations.

REFERENCES

Anderson, R. L. and Bancroft, T. A. (1952). *Statistical Theory in Research*, McGraw Hill, New York.

Cochran, W. G. (1946). Analysis of covariance, Paper presented at the Institute of Mathematical Statistics Meetings, Princeton University.

Rao, C. R. (1971). Minimum variance quadratic unbiased estimation of variance components, *Journal of Multivariate Analysis*, **1**, 445–456.

Rao, C. R. (1972). Estimation of variance and covariance components in linear models, *Journal of the American Statistical Association*, **67**, 112–115.

Rao, P. S. R. S. (1997). *Variance Components Estimation: Mixed models, Methodologies and Applications*, Chapman and Hall, London, New York.

Rao, P. S. R. S. and Chaubey, Y. P.(1978). Three modifications of the principle of the MINQUE, *Communications in Statistics—Theory and Methods*, **7**, 767–778.

Rao, P. S. R. S. and Miyawaki, N. (1989). Estimation procedures for one-way variance components model, *Bulletin of the International Statistical Institute*, **2**, 243–244.

CHAPTER 18

ON A TWO-STAGE PROCEDURE WITH HIGHER THAN SECOND-ORDER APPROXIMATIONS

NITIS MUKHOPADHYAY

University of Connecticut, Storrs, CT

Abstract: We reconsider the general Stein-type two-stage methodology of Mukhopadhyay and Duggan (1999) which had incorporated *partial information* in the form of a known and positive lower bound for the otherwise *unknown nuisance parameter*. This revised methodology was shown to enjoy customary second-order properties and expansions for functions of the associated stopping variable, under appropriate conditions. In this paper, new machineries are provided which help one to obtain the *third-order* and *higher than third-order expansions* of the analogous functions of the associated stopping variable, under appropriate conditions. These general techniques are then applied in a variety of estimation as well as selection and ranking problems.

Keywords and phrases: General procedure, third-order expansions, higher-order expansions, exact consistency, confidence regions, point estimation, selection and ranking

18.1 INTRODUCTION

In his classic papers, Stein (1945, 1949) developed two-stage procedures for constructing a confidence interval J for estimating the unknown mean μ in a $N(\mu, \sigma^2)$ population when $\sigma \in R^+$ is completely unknown, in such a way that J had the fixed-width $2d$ and

$$P_{\mu,\sigma}\{\mu \in J\} \geq 1 - \alpha \text{ whatever be } \mu \in R, \sigma \in R^+. \qquad (18.1.1)$$

Here, the two numbers $d(> 0)$ and $0 < \cdot \alpha < 1$ are preassigned before data collection. The novelty of this two-stage procedure came out loud and clear, particularly because no fixed-sample procedure could solve this problem [Dantzig (1940)] in the first place.

But the Stein two-stage procedure had one important poor characteristic of significant *over-sampling*, even asymptotically. In order to reduce over-sampling, other types of two-stage, purely sequential and multi-stage estimation strategies have been developed in the literature. Extensive reviews of this literature can be found in Ghosh, Mukhopadhyay and Sen (1997). Let us simply add that the three-stage, purely sequential, and accelerated sequential estimation strategies allow one to claim *second-order* approximations of the associated confidence coefficient and the expected sample size, among other characteristics. However, when one of these three-stage, purely sequential, and accelerated sequential procedures is implemented, we then fail to conclude the *exact consistency* property described by (18.1.1).

Mukhopadhyay and Duggan (1997) had developed a modification of Stein's two-stage fixed-width confidence interval estimation procedure for μ in a $N(\mu, \sigma^2)$ population, when the experimenter has some additional prior knowledge to justify an assumption that $\sigma > \sigma_L$ and $\sigma_L(> 0)$ is known. The corresponding two-stage procedure still enjoyed the exact consistency property (18.1.1) *and* the methodology also had attractive *second-order* characteristics. The key ideas from this paper were then extended and synthesized in a latter article by the same authors (1999). Such general techniques eventually provided second-order properties of the Mukhopadhyay-Duggan type two-stage sampling strategies in the case of a series of other difficult and interesting problems in reliability, multivariate analysis, regression analysis as well as selection and ranking.

In the meantime, Mukhopadhyay (1999) has successfully expanded both the lower and upper bounds of $E_{\mu,\sigma}(N)$ up to the order $O(d^6)$, and those for the coverage probability up to the order $o(d^4)$, for the two-stage procedure of Mukhopadhyay and Duggan (1997). The approximations for the coverage probability were then further sharpened by providing the remainder term up to the order $O(d^6)$. These sharper rates of convergence are referred to as the *third-order approximations and beyond* via double sampling.

The readers will easily locate references to the available approximations for $E_{\mu,\sigma}(N)$ and the coverage probability respectively up to the orders $o(1)$ and $o(d^2)$ in the context of the existing multi-stage and sequential fixed-width confidence interval estimation methodologies in the literature. These go by the name, *second-order expansions*. Refer to Woodroofe (1977, 1982), Ghosh and Mukhopadhyay (1981), and Ghosh *et al.* (1997), among other sources. Mukhopadhyay's (1999) orders of approximation are obviously much sharper and these are shown to hold for an appropriately modified Stein type two-stage sampling design when the unknown parameter σ exceeds σ_L and $\sigma_L(> 0)$ is known!

The objective of the present paper is two-fold. First, we go back to the general formulation of Mukhopadhyay and Duggan (1999) and derive specific *third-* and *higher-order expansions* for the expected values of functions of the stopping variable, N. These higher-order expansions constitute a substantial leap from the available results of Mukhopadhyay (1999). In that earlier paper, we took the first step in this direction in a special case. The general methodology and the main results (Theorems 18.2.1–18.2.3) are explained in Section 18.2. The Section 18.3 builds the essential tools and then provides the proofs of the main results. The second objective of this paper is to substantiate that these general machineries can be immediately applied in a variety of interesting problems in statistical inference. Some examples are included in Section 18.4 from the areas of reliability, multivariate estimation, regression analysis and multiple comparisons.

The breadth of the techniques proposed in the Sections 18.2 and 18.3 should be clear from the wide variety of applications given in Section 18.4. We hope that this synthesis will provide some impetus to researchers to investigate higher than second-order expansions for other multi-stage strategies and/or other challenging problems.

18.2 GENERAL FORMULATION AND MAIN RESULTS

Let us revisit the general two-stage procedure studied by Mukhopadhyay and Duggan (1999). In many problems in estimation and multiple decision theory, the expression of the so called "optimal" fixed sample size looks like:

$$n_0^* = qh^{*-1}\theta^\tau \text{ where } q, h^*, \tau \text{ are known positive numbers, but } \theta(> 0)$$

$$\text{is an unknown nuisance parameter.} \tag{18.2.2}$$

We *assume* that $\theta > \theta_L(> 0)$ where θ_L is known.

The explicit roles of q, h^*, τ and θ will become clear from the specific applications given in the Section 18.4. In this section and in Section 18.3, the asymptotic analysis are carried out assuming that $h^* \to 0$. Throughout the text, $[u]^*$ and $I(.)$ will respectively stand for the largest integer $< u$ and the indicator function of $(.)$.

Let $m(\geq m_0)$ be the initial sample size where

$$m \equiv m(h^*) = \max\{m_0, [qh^{*-1}\theta_L^\tau]^* + 1\}, \qquad (18.2.3)$$

$m_0(\geq 1)$ being a fixed integer. Based on the pilot sample of size m, suppose that one considers a statistic $U(m)$ such that $P\{U(m) > 0\} = 1$, $E[U(m)] = \theta$, and

$p_m U(m)/\theta$ is distributed as $\chi^2_{p_m}$ where p_m is a positive integer of the form $c_1 m + c_2$ with positive integer c_1 and integer c_2.   (18.2.4)

Let q^* be a positive number such that

$$q^* \equiv q_m^* = q + a_1 m^{-1} + O(m^{-2}) \qquad (18.2.5)$$

with some real number a_1. Mukhopadhyay and Duggan (1999) then defined a positive integer valued random variable as follows:

$$N \equiv N(h^*) = \max\{m, [q_m^* h^{*-1} U^\tau(m)]^* + 1\}. \qquad (18.2.6)$$

In various applications, as we will see in Section 18.4, one starts with m random samples in the first stage and thereby obtains $U(m)$ which leads to the determination of N, an estimator of n_0^*. If $N = m$, one does not take any more observations at the second stage. But, if $N > m$, then one samples the difference $(N - m)$ at the second stage. In either situation, one proceeds with the appropriate inference procedures given the nature of a particular application which depends on *all* N observations gathered from the population.

Under the assumptions (18.2.4) and (18.2.5), Mukhopadhyay and Duggan (1999) proved the following results among others:

$$P(N = m) = O(\eta^{\frac{1}{2}p_m}) \text{ where } \eta = (\theta_L/\theta) \exp\{1 - (\theta_L/\theta)\}$$
$$\text{is a positive proper fraction;} \qquad (18.2.7)$$

$$n_0^{*-\frac{1}{2}}(N - n_0^*) \xrightarrow{\mathcal{L}} N(0, \sigma_0^2) \text{ as } h^* \to 0 \text{ where}$$
$$\sigma_0^2 = 2c_1^{-1}\tau^2(\theta/\theta_L)^\tau; \qquad (18.2.8)$$

$$n_0^{*-1}(N - n_0^*)^2 \text{ is uniformly integrable for } 0 < h^* < h_0^*$$
$$\text{with sufficiently small } h_0^*. \qquad (18.2.9)$$

Mukhopadhyay and Duggan's (1999) central piece of the results can be summarized as follows:

Suppose that $g : R^+ \to R^+$ and let us denote $g^{(s)}(x) = d^s g(x)/dx^s$, $s = 1, 2$. Let $g(x)$ be a twice differentiable function such that

  (i) $g^{(2)}(x)$ is continuous at $x = 1$;
  (ii) $\left|g^{(2)}(x)\right| \leq \Sigma_{i=1}^r p_i x^{-u_i}$ for all $x \in R^+$, where p_i's and u_i's
        are fixed but non-negative numbers;

$$(18.2.10)$$

Then, under the assumptions (18.2.4), (18.2.5) and (18.2.10), the following expanded forms of the lower and upper bounds for $E[g(N/n_0^*)]$ were obtained up to the *second-order approximations* as $h^* \to 0$:

$$g(1) \; + \; n_0^{*-1}\left\{\psi g^{(1)}(1) + \frac{1}{2}\sigma_0^2 g^{(2)}(1)\right\} + o(n_0^{*-1}) \leq E[g(N/n_0^*)] \leq g(1)$$

$$+ \; n_0^{*-1}\left\{(\psi + 1)g^{(1)}(1) + \frac{1}{2}\sigma_0^2 g^{(2)}(1)\right\} + o(n_0^{*-1}) \text{ if } g^{(1)}(1) > 0;$$

$$(18.2.11)$$

$$g(1) \; + \; n_0^{*-1}\left\{(\psi + 1)g^{(1)}(1) + \frac{1}{2}\sigma_0^2 g^{(2)}(1)\right\} + o(n_0^{*-1})$$

$$\leq E[g(N/n_0^*)] \leq g(1)$$

$$+ \; n_0^{*-1}\left\{\psi g^{(1)}(1) + \frac{1}{2}\sigma_0^2 g^{(2)}(1)\right\} + o(n_0^{*-1}) \text{ if } g^{(1)}(1) < 0;$$

$$(18.2.12)$$

where $\psi = \{\tau(\tau - 1)c_1^{-1} + a_1\}(\theta/\theta_L)^\tau$ and recall the expression of σ_0^2 from (18.2.8). Mukhopadhyay and Duggan (1999) also showed the following *second-order* expansion of $E(N - n_0^*)$:

$$\psi + o(h^{*1/2}) \leq E(N - n_0^*) \leq \psi + 1 + o(h^{*1/2}). \qquad (18.2.13)$$

In the derivations of (18.2.11) and (18.2.12), the expansion from (18.2.13) was a crucial ingredient.

   The expressions which are analogous to both the lower and upper bounds in (18.2.11) and (18.2.12), but expanded up to the order $o(n_0^{*-2})$ or $O(n_0^{*-3})$, are respectively called the *third-order* and *higher than third-order approximations* for $E[g(N/n_0^*)]$.

   The aim of the present paper is to derive approximations in (18.2.11)–(18.2.13) which are significantly sharper. In order to achieve this, we need more accurate expansions of all the intermediate ingredients used in the original work of Mukhopadhyay and Duggan (1999). Also, we need a more accurate expansion of q_m^* than what was assumed in (18.2.5). These kinds of expansions are customarily known as the *Cornish-Fisher expansions*. When q_m^* stands for the appropriate percentage points of the Student's t or F distribution, we can indeed include more explicit terms on the rhs of (18.2.5) involving m^{-2}, m^{-3} and so on. One may refer to Johnson and Kotz (1970, pp. 84 and 102).

In order to achieve higher than second-order approximations in (18.2.11)–(18.2.13), we assume the following expansion of q_m^*/q instead of what was assumed earlier in (18.2.5). Let us now suppose that

$$q_m^* q^{-1} = 1 + a_1 m^{-1} + a_2 m^{-2} + a_3 m^{-3} + O(m^{-4}). \qquad (18.2.14)$$

Now, the three main results are successively given below.

Theorem 18.2.1 *For m and N respectively defined in (18.2.3) and (18.2.6), with the assumptions (18.2.4) and (18.2.14), we have as $h^* \to 0$:*

$$\psi + \{a_2 + a_1 b_1(1) + b_2(1)\}(\theta/\theta_L)^{2\tau} n_0^{*-1} + \{a_3 + a_2 b_1(1) + a_1 b_2(1)$$

$$+ b_3(1)\}(\theta/\theta_L)^{3\tau} n_0^{*-2} + O(n_0^{*-3})$$

$$\leq E(N) - n_0^* \leq \psi + 1 + \{a_2 + a_1 b_1(1) + b_2(1)\}(\theta/\theta_L)^{2\tau} n_0^{*-1}$$

$$+ \{a_3 + a_2 b_1(1) + a_1 b_2(1) + b_3(1)\}(\theta/\theta_L)^{3\tau} n_0^{*-2} + O(n_0^{*-3})$$

where $\psi = \{a_1 + b_1(1)\}(\theta/\theta_L)^\tau$. The expressions for a_i and $b_i(1), i = 1, 2, 3$, respectively appear in (18.2.14) and Lemma 18.3.1.

The expression for ψ given by the Theorem 18.2.1 certainly agrees with that given by (18.2.13) which was obtained earlier by Mukhopadhyay and Duggan (1999). But, we emphasize how strong the new expansion of $E(N) - n_0^*$ is when compared with that in (18.2.13). The new remainder term is $O(h^{*3})$ instead of $o(h^{*1/2})$!

Suppose again that $g : R^+ \to R^+$ and let us recall the notation, $g^{(s)}(x) = d^s g(x)/dx^s$, $s = 1, \ldots, 6$. For sharper approximations of $E[g(N/n_0^*)]$, now we have to demand more of the function $g(.)$ compared with what Mukhopadhyay and Duggan (1999) had required in (18.2.10).

In the Theorem 18.2.2, we require that $g(x)$ satisfies the following conditions:

(i) $g^{(4)}(x)$ is continuous at $x = 1$;

(ii) $\left|g^{(4)}(x)\right| \leq \Sigma_{i=1}^u p_i x^{-s_i}$ for all $x \in R^+$, where p_i's and s_i's are fixed but non-negative numbers.

$$(18.2.15)$$

In the Theorem 18.2.3, we require that $g(x)$ instead satisfies the following conditions:

(i) $g^{(6)}(x)$ is continuous at $x = 1$;
(ii) $\left| g^{(6)}(x) \right| \le \Sigma_{i=1}^{t} p_i x^{-v_i}$ for all $x \in R^+$, where p_i's and v_i's are fixed but non-negative numbers;

$$(18.2.16)$$

Theorem 18.2.2 *For m and N respectively defined in (18.2.3) and (18.2.6), with the assumptions (18.2.4) and (18.2.14), and with $g(.)$ satisfying (18.2.15), we have as $h^* \to 0$:*

$$E[g(N/n_0^*)]$$

$$= g(1) + n_0^{*-1} g^{(1)}(1) E(N - n_0^*) + \frac{1}{2} n_0^{*-1} g^{(2)}(1) E \left[\frac{(N - n_0^*)^2}{n_0^*} \right]$$

$$+ \frac{1}{6} g^{(3)}(1) E \left[\frac{(N - n_0^*)^3}{n_0^{*3}} \right] + \frac{1}{24} n_0^{*-2} g^{(4)}(1) E \left[\frac{(N - n_0^*)^4}{n_0^{*2}} \right] + o(n_0^{*-2}).$$

Theorem 18.2.3 *For m and N respectively defined in (18.2.3) and (18.2.6), with the assumptions (18.2.4) and (18.2.14), and with $g(.)$ satisfying (18.2.16), we have as $h^* \to 0$:*

$$E[g(N/n_0^*)]$$

$$= g(1) + n_0^{*-1} g^{(1)}(1) E(N - n_0^*) + \frac{1}{2} n_0^{*-1} g^{(2)}(1) E \left[\frac{(N - n_0^*)^2}{n_0^*} \right]$$

$$+ \frac{1}{6} g^{(3)}(1) E \left[\frac{(N - n_0^*)^3}{n_0^{*3}} \right] + \frac{1}{24} n_0^{*-2} g^{(4)}(1) E \left[\frac{(N - n_0^*)^4}{n_0^{*2}} \right]$$

$$+ O(n_0^{*-3}).$$

The major difference between the two Theorems 18.2.2 and 18.2.3 can be seen from the rates of convergence of the two remainder terms. In the proof of the Theorem 18.2.2, we directly exploit the Theorem 18.2.1, the Lemmas 18.3.2, 18.3.3 and Lemma 18.3.7, part (i), plus the condition (18.2.15). Theorem 18.2.3 is stronger than the Theorem 18.2.2 and hence we need stronger machineries to deal with the Theorem 18.2.3. In the proof of the Theorem 18.2.3, we exploit the Theorem 18.2.1, the Lemmas 18.3.2–18.3.5, Lemma 18.3.7, part (ii), plus the condition (18.2.16). One may, however, note that the Lemma 18.3.6 is essential in the proof of the Lemma 18.3.7.

In order to apply the Theorems 18.2.2 and 18.2.3, one needs to consider several situations separately involving the signs of $g^{(s)}(1)$, $s = 1, \ldots, 4$.

Now, with some specific $g(.)$ function, one will be able to write down the lower and upper bounds of $E[g(N/n_0^*)]$ up to the order $o(n_0^{*-2})$ or $O(n_0^{*-3})$, by combining the available bounds from the Theorem 18.2.1 and Lemmas 18.3.2–18.3.6. In order to get the correct upper and lower bounds, naturally one has to be particularly mindful of the signs of the derivatives $g^{(s)}(1)$, $s = 1, 2, 3, 4$. The first example in Section 18.4 shows some details in order to clarify the technicalities.

Proofs of the three theorems, particularly the details needed in the verifications of the Theorems 18.2.2 and 18.2.3, are involved and technical in nature. These are deferred to Section 18.3. In order to appreciate the usefulness of these results, the reader may first decide to skip to Section 18.4 and have a look at the wide range of applications before considering all the details given in Section 18.3. Two remarks follow.

Remark 18.2.1 In the higher-order bounds obtained from Theorems 18.2.1–18.2.3, one may consider the coefficients of terms looking like $\theta^{\kappa_1} n_0^{*-1}$ $\theta^{\kappa_2} n_0^{*-2}$ or $\theta^{\kappa_3} n_0^{*-3}$ where κ_1, κ_2 and κ_3 are some fixed appropriate numbers. These coefficients are all known and fixed real numbers. In the fixed-width confidence interval problem for a normal mean, the Remark 3.2 in Mukhopadhyay (1999) gave explicit comparisons of his third- and higher-order bounds for $E_{\mu,\sigma}(N - n_0^*)$ with the corresponding second-order bounds from Mukhopadhyay and Duggan (1997). The third- and higher-order bounds provided better approximations.

Remark 18.2.2 In a three-stage procedure, one customarily estimates a *fraction* of n_0^* in the second stage with the help of pilot samples and takes these observations in the first and second stage combined. The remaining observations are taken in the third stage. A fairly broad class of three-stage procedures are known to have second-order properties. One may refer to Chapter 6 in Ghosh, Mukhopadhyay and Sen (1997) for a review. From (18.2.3), observe that $m \approx qh^{*-1}\theta_L^\tau = (\theta_L/\theta)^\tau n_0^* = \rho n_0^*$, say, where $0 < \rho < 1$ is unknown. It may be tempting to view our first-stage sampling as equivalent to what is customarily carried out in the second stage of a three-stage design. But, the big difference is that we have not assumed any knowledge about the number $\rho \in (0, 1)$. In a three-stage sampling strategy, the experimenter works with an *arbitrary, but fixed* $\rho \in (0, 1)$ which is chosen along with the other design constants. How to make an appropriate choice of ρ is not really a simple matter. In the present two-stage methodology, however, we are on safe grounds because we know for sure that $m < n_0^*$.

18.3   PROOFS OF THE MAIN RESULTS

Under the assumption of (18.2.14), we can show that the following expansions hold:

$$
\begin{aligned}
q_m^{*2}q^{-2} &= 1 + 2a_1 m^{-1} + (a_1^2 + 2a_2)m^{-2} + 2(a_3 + a_1 a_2)m^{-3} + O(m^{-4}); \\
q_m^{*3}q^{-3} &= 1 + 3a_1 m^{-1} + 3(a_1^2 + a_2)m^{-2} + (3a_3 + 6a_1 a_2 + a_1^3)m^{-3} \\
&\quad + O(m^{-4}); \\
q_m^{*4}q^{-4} &= 1 + 4a_1 m^{-1} + 2(3a_1^2 + 2a_2)m^{-2} + 4(a_3 + 3a_1 a_2 + a_1^3)m^{-3} \\
&\quad + O(m^{-4}); \\
q_m^{*5}q^{-5} &= 1 + 5a_1 m^{-1} + 5(2a_1^2 + a_2)m^{-2} + 5(a_3 + 4a_1 a_2 + 2a_1^3)m^{-3} \\
&\quad + O(m^{-4}); \\
q_m^{*6}q^{-6} &= 1 + 6a_1 m^{-1} + 3(5a_1^2 + 2a_2)m^{-2} \\
&\quad + 2(3a_3 + 15a_1 a_2 + 10a_1^3)m^{-3} + O(m^{-4}).
\end{aligned}
\tag{18.3.17}
$$

Along the lines of the arguments used in Mukhopadhyay and Duggan (1999), we find it easier first to work with the random variable

$$
T \equiv T(h^*) = \max\left\{m, q_m^* h^{*-1} U^\tau(m)\right\}
\tag{18.3.18}
$$

instead of directly working with the stopping variable N from (2.5). The results obtained for T are not difficult to transcribe into analogous results for N in view of the following relationship:

$$
T \le N \le T + 1.
\tag{18.3.19}
$$

Next, observe that in order to expand various moments of N, we must at first expand appropriate higher moments of $U(m)$. With any fixed $s > 0$, one has

$$
E[U^{s\tau}(m)] = \{2\theta/p_m\}^{s\tau}\Gamma(\tfrac{1}{2}p_m + s\tau)\{\Gamma(\tfrac{1}{2}p_m)\}^{-1}.
\tag{18.3.20}
$$

Lemma 18.3.1 *One can expand $E[U^{s\tau}(m)]$ for large m and obtain:*

$$
E[U^{s\tau}(m)] = \theta^{s\tau}\left\{1 + b_1 m^{-1} + b_2 m^{-2} + b_3 m^{-3}\right\} + O(m^{-4})
$$

where

$$
\begin{aligned}
b_1 \equiv b_1(s) &= s\tau(s\tau - 1)c_1^{-1}, \\
b_2 \equiv b_2(s) &= s\tau c_1^{-2}\{-(s\tau - 1)c_2 - \frac{5}{3}s^2\tau^2 + \frac{1}{2}s^3\tau^3 + \frac{3}{2}s\tau - \frac{1}{3}\},
\end{aligned}
$$

$$b_3 \equiv b_3(s) = s\tau c_1^{-3}[(s\tau - 1)c_2^2 + c_2\{s\tau(2s\tau - 1)(s\tau + 3)$$
$$-8s^2\tau^2(\frac{1}{3} + \frac{1}{8}s\tau) + \frac{2}{3}\}$$
$$-\frac{1}{6}s^2\tau^2(2s\tau - 1)(2s\tau - 3)(s\tau + 5)$$
$$+4s^3\tau^3(2s\tau - 1)(\frac{1}{3} + \frac{1}{8}s\tau)$$
$$-8s^3\tau^3(\frac{1}{4} + \frac{1}{6}s\tau + \frac{1}{48}s^2\tau^2)].$$

The proof of Lemma 18.3.1 is particularly long and tedious. One needs to expand each individual term involved in (18.3.20) very carefully and let the painstaking algebra take care of the rest. The details are left out for brevity.

In the proofs of the bits and pieces which follow, we repeatedly combine terms such as $m^r P(N = m)$ for different values of r, and replace all of them with one generic expression, namely, $O(m^{-k})$ or $O(n_0^{*-k})$ where $k(> 0)$ can be chosen sufficiently large in view of (18.2.7).

18.3.1 Proof of Theorem 18.2.1

Recall the random variable T from (18.3.18) and write the basic inequality

$$q_m^* h^{*-1} U^\tau(m) \leq T \leq mI(N = m) + q_m^* h^{*-1} U^\tau(m). \qquad (18.3.21)$$

The result (18.2.7) also holds for the variable T. Thus, combining (18.2.14) and Lemma 18.3.1 with $s = 1$, one has

$$E(T/n_0^*)$$
$$= q_m^* q^{-1} E[U^\tau(m)\theta^{-\tau}] + O(m^{-k})$$
$$= \{1 + a_1 m^{-1} + a_2 m^{-2} + a_3 m^{-3} + O(m^{-4})\}\{1 + b_1(1)m^{-1}$$
$$+b_2(1)m^{-2} + b_3(1)m^{-3} + O(m^{-4})\} + O(m^{-k})$$
$$= 1 + \{a_1 + b_1(1)\}m^{-1} + \{a_2 + a_1 b_1(1) + b_2(1)\}m^{-2}$$
$$+\{a_3 + a_2 b_1(1) + a_1 b_2(1) + b_3(1)\}m^{-3} + O(m^{-k}). \qquad (18.3.22)$$

Hence, we obtain

$$E(T - n_0^*)$$
$$= \{a_1 + b_1(1)\}n_0^* m^{-1} + \{a_2 + a_1 b_1(1) + b_2(1)\}n_0^* m^{-2}$$
$$+\{a_3 + a_2 b_1(1) + a_1 b_2(1) + b_3(1)\}n_0^* m^{-3} + O(m^{-3}).$$
$$(18.3.23)$$

Now, without any loss of generality, suppose that we make h^* approach zero in such a way that $q\theta_L^\tau/h^*$ remains an integer $(> m)$. Now, we can rewrite (18.3.23) as

$$
\begin{aligned}
E(T &- n_0^*) \\
&= \{a_1 + b_1(1)\}(\theta/\theta_L)^\tau + \{a_2 + a_1 b_1(1) + b_2(1)\}(\theta/\theta_L)^{2\tau} n_0^{*-1} \\
&\quad + \{a_3 + a_2 b_1(1) + a_1 b_2(1) + b_3(1)\}(\theta/\theta_L)^{3\tau} n_0^{*-2} + O(n_0^{*-3}).
\end{aligned}
$$

$$(18.3.24)$$

The desired result finally follows from (18.3.24) in view of the relationship between T and N given by (18.3.19). □

18.3.2 Auxiliary Lemmas

Before we can set out to prove Theorems 18.2.2 and 18.2.3, we need to verify a series of crucial lemmas. We first take care of these.

Lemma 18.3.2 *For the stopping variable N from (18.2.6), we have as $h^* \to 0$:*

$$
\begin{aligned}
k_{2,1}(\theta/\theta_L)^\tau &- 2 + k_{2,2}(\theta/\theta_L)^{2\tau} n_0^{*-1} + k_{2,3}(\theta/\theta_L)^{3\tau} n_0^{*-2} + O(n_0^{*-3}) \\
&\leq E\{(N - n_0^*)^2/n_0^*\} \\
&\leq k_{2,1}(\theta/\theta_L)^\tau + 2 + \{k_{2,2}(\theta/\theta_L)^{2\tau} + 2(a_1 + b_1(1))(\theta/\theta_L)^{2\tau}\} n_0^{*-1} \\
&\quad + \{k_{2,3}(\theta/\theta_L)^{3\tau} + 2(a_2 + a_1 b_1(1) + b_2(1))(\theta/\theta_L)^{2\tau}\} n_0^{*-2} \\
&\quad + O(n_0^{*-3})
\end{aligned}
$$

where the coefficients $k_{2,i}$, $i = 1, 2, 3$, come from (18.3.27).

PROOF We combine (18.2.14) and Lemma 18.3.1 with $s = 2$ to write

$$
\begin{aligned}
E(T^2/n_0^{*2}) \\
&= q_m^{*2} q^{-2} E[U^{2\tau}(m)\theta^{-2\tau}] + O(m^{-k}) \\
&= \{1 + 2a_1 m^{-1} + (a_1^2 + 2a_2)m^{-2} + 2(a_3 + a_1 a_2)m^{-3} \\
&\quad + O(m^{-4})\}\{1 + b_1(2)m^{-1} + b_2(2)m^{-2} + b_3(2)m^{-3} \\
&\quad + O(m^{-4})\} + O(m^{-k}) \\
&= 1 + \{2a_1 + b_1(2)\}m^{-1} + \{a_1^2 + 2a_2 + 2a_1 b_1(2) \\
&\quad + b_2(2)\}m^{-2} + \{2(a_3 + a_1 a_2) + (a_1^2 + 2a_2)b_1(2) \\
&\quad + 2a_1 b_2(2) + b_3(2)\}m^{-3} + O(m^{-4}).
\end{aligned}
$$

$$(18.3.25)$$

Now, we recall the expression of $E[T/n_0^*]$ from (18.3.22) and combine this with (18.3.25) to obtain

$$E\{(T - n_0^*)^2/n_0^{*2}\} = k_{2,1}m^{-1} + k_{2,2}m^{-2} + k_{2,3}m^{-3} + O(m^{-4}),$$

$$(18.3.26)$$

where we denote

$$
\begin{aligned}
k_{2,1} &= b_1(2) - 2b_1(1), \\
k_{2,2} &= (a_1^2 + 2a_2) + 2a_1b_1(2) + b_2(2) - 2(a_2 + a_1b_1(1) + b_2(1)), \\
k_{2,3} &= 2(a_3 + a_1a_2) + (a_1^2 + 2a_2)b_1(2) + 2a_1b_2(2) + b_3(2) \\
&\quad -2(a_3 + a_2b_1(1) + a_1b_2(1) + b_3(1)).
\end{aligned}
$$

$$(18.3.27)$$

Next, utilizing (18.3.19), one can verify that

$$E\{(T - n_0^*)^2/n_0^*\} - 2 \;\leq\; E\{(N - n_0^*)^2/n_0^*\} \leq E\{(T - n_0^*)^2/n_0^*\}$$
$$+2E\{T/n_0^*\} + n_0^{*-1}. \qquad (18.3.28)$$

The desired result then follows by combining (18.3.22), (18.3.26) and (18.3.28).   □

Remark 18.3.1 From (18.3.26), it is clear that

$$\lim_{h^* \to 0} E\{(T - n_0^*)^2/n_0^*\} = k_{2,1}(\theta/\theta_L)^\tau.$$

But, we have $k_{2,1} = b_1(2) - 2b_1(1) = 2\tau^2 c_1^{-1}$. Hence,

$$\lim_{h^* \to 0} E\{(T - n_0^*)^2/n_0^*\} = 2\tau^2 c_1^{-1}(\theta/\theta_L)^\tau,$$

which coincides with the variance of the limiting distribution of $(T - n_0^*)/n_0^{*1/2}$. The equation (18.3.28) already implies the uniform integrability of $(T - n_0^*)^2/n_0^*$ and thus the actual limiting value of its expectation makes good intuitive sense in view of the distributional convergence given in (18.2.8). We have actually verified more than what the uniform integrability property of $(T - n_0^*)^2/n_0^*$ alone would allow one to conclude in this situation.

Lemma 18.3.3 *For the stopping variable N from (2.5), we have as $h^* \to 0$:*

$$-k_{3,1}n_0^{*-1} + k_{3,2}n_0^{*-2} + O(n_0^{*-3})$$

$$\leq E\left\{\frac{(N - n_0^*)^3}{n_0^{*3}}\right\} \leq k_{3,1}n_0^{*-1} + k_{3,3}n_0^{*-2} + O(n_0^{*-3})$$

where the coefficients $k_{3,i}$, $i = 1, 2, 3$, come from (18.3.32).

PROOF Using arguments as before, we can show that

$$E(T^3/n_0^{*3}) = 1 + \{3a_1 + b_1(3)\}m^{-1}$$
$$+\{3(a_1^2 + a_2) + 3a_1b_1(3) + b_2(3)\}m^{-2}$$
$$+O(m^{-3}). \qquad (18.3.29)$$

Then, we combine (18.3.22), (18.3.25) and (18.3.29) to obtain

$$E\{(T - n_0^*)^3/n_0^{*3}\} = \{3a_1b_1(3) + b_2(3) - 6a_1b_1(2)$$
$$-3b_2(2) + 3a_1b_1(1) + 3b_2(1)\}m^{-2}$$
$$+O(m^{-3}). \qquad (18.3.30)$$

Next, we use (18.3.19) to get the following bounds:

$$n_0^{*-3}(T - n_0^*)^3 - 3n_0^{*-2}(2T + 1)$$
$$\leq n_0^{*-3}(N - n_0^*)^3$$
$$\leq n_0^{*-3}(T - n_0^*)^3 + 3n_0^{*-2}\{(T - n_0^*)^2/n_0^*\} + 6n_0^{*-2}T$$
$$+n_0^{*-3}(3T + 1). \qquad (18.3.31)$$

Hence, after combining (18.3.24), (18.3.26) and (18.3.30), the result follows with

$$k_{3,1} = 6,$$
$$k_{3,2} = -6(a_1 + b_1(1))(\theta/\theta_L)^\tau + \{3a_1b_1(3) + b_2(3) - 3b_2(2)$$
$$-6a_1b_1(2) + 3a_1b_1(1) + 3b_2(1)\}(\theta/\theta_L)^{2\tau},$$
$$k_{3,3} = 3 + 6(a_1 + b_1(1))(\theta/\theta_L)^\tau + 3k_{2,1}(\theta/\theta_L)^\tau$$
$$+\{3a_1b_1(3) + b_2(3) - 6a_1b_1(2) - 3b_2(2)$$
$$+3a_1b_1(1) + 3b_2(1)\}(\theta/\theta_L)^{2\tau}, \qquad (18.3.32)$$

with $k_{2,1}$ defined earlier in (18.3.27). $\qquad \square$

Lemma 18.3.4 *For the stopping variable N from (18.2.6), we have as $h^* \to 0$:*

$$E\{(N - n_0^*)^4/n_0^{*2}\} = \{b_2(4) - 4b_2(3) + 6b_2(2) - 4b_2(1)\}(\theta/\theta_L)^{2\tau} + O(n_0^{*-1})$$

where the coefficients $b_2(s)$, $s = 1, 2, 3, 4$, come from Lemma 18.3.1.

PROOF Using arguments as before, we can show that

$$E(T^4/n_0^{*4})$$
$$= 1 + \{4a_1 + b_1(4)\}m^{-1} + \{2(3a_1^2 + 2a_2) + 4a_1b_1(4) + b_2(4)\}m^{-2}$$
$$+O(m^{-3}). \qquad (18.3.33)$$

Then, we combine (18.3.22), (18.3.25), (18.3.29) and (18.3.33) to obtain

$$
\begin{aligned}
E\{(T - n_0^*)^4/n_0^{*4}\} \\
= \ & \{b_2(4) - 4b_2(3) + 6b_2(2) - 4b_2(1)\}(\theta/\theta_L)^{2\tau} n_0^{*-2} \\
& + O(n_0^{*-3})
\end{aligned}
\tag{18.3.34}
$$

Now, one may proceed along the lines of proof of the Lemma 2.5 in Mukhopadhyay (1999) and complete this proof.   □

Remark 18.3.2 From (18.3.34), it is clear that

$$
\lim_{h^* \to 0} E\{(T - n_0^*)^4/n_0^{*2}\} = \{b_2(4) - 4b_2(3) + 6b_2(2) - 4b_2(1)\}(\theta/\theta_L)^{2\tau}.
$$

But, we have $b_2(4) - 4b_2(3) + 6b_2(2) - 4b_2(1) = 12\tau^4 c_1^{-2}$. Hence,

$$
\lim_{h^* \to 0} E\{(T - n_0^*)^2/n_0^*\} = 12\tau^4 c_1^{-2}(\theta/\theta_L)^{2\tau}
$$

which coincides with the fourth central moment of the limiting distribution of $(T - n_0^*)/n_0^{*1/2}$. The equation (18.3.34) already implies the uniform integrability of $(T - n_0^*)^4/n_0^{*2}$ and thus the actual limiting value of its expectation makes good intuitive sense in view of the distributional convergence given in (18.2.8). We have, however, verified more than what the uniform integrability property of $(T - n_0^*)^4/n_0^{*2}$ alone would allow one to conclude in this situation.

Lemma 18.3.5 *For the stopping variable N from (18.2.6), we have as $h^* \to 0$:*

$$
E\{(N - n_0^*)^5/n_0^{*5}\} = O(n_0^{*-3}).
$$

PROOF Using arguments as before, we can show that

$$
\begin{aligned}
E(T^5/n_0^{*5}) = \ & 1 + \{5a_1 + b_1(5)\}m^{-1} + \{5(2a_1^2 + a_2) \\
& + 5a_1 b_1(5) + b_2(5)\}m^{-2} + O(m^{-3}).
\end{aligned}
\tag{18.3.35}
$$

Then, we combine (18.3.22), (18.3.25), (18.3.29), (18.3.33) and (18.3.35) to obtain

$$
E\{(T - n_0^*)^5/n_0^{*5}\} = O(m^{-3}).
\tag{18.3.36}
$$

Now, one may again use (18.3.19) to write down the appropriate lower and upper bounds for $E\{(N - n_0^*)^5/n_0^{*5}\}$ along the lines of what was done at the end of the Lemmas 18.3.2–18.3.3, and thus complete this proof.   □

Lemma 18.3.6 *For the stopping variable N from (18.2.6), we have the uniform integrability of $(N - n_0^*)^6/n_0^{*3}$ for $0 < h^* < h_0$ for sufficiently small h_0.*

PROOF In view of (18.3.19), it will suffice to verify the desired result when N is replaced by T. First let us examine the behavior of $P(T \geq 2n_0^*)$. It will obviously converge to zero as $h^* \to 0$, but the question is at what rate? Now, let us write

$$P(T \geq 2n_0^*) \leq P\left\{U(m) \geq (2qq_m^{*-1})^{1/\tau}\theta\right\}$$

$$= P\left\{U(m) - \theta \geq \theta[(2qq_m^{*-1})^{1/\tau} - 1]\right\}. \quad (18.3.37)$$

Since $\lim_{h^* \to 0} qq_m^{*-1} = 1$, for sufficiently small $0 < h^*(< h_0)$, we can make $\theta[(2qq_m^{*-1})^{1/\tau} - 1] \geq \varepsilon$ for some $\varepsilon(> 0)$. Now, from (18.3.37) we can claim that

$$
\begin{aligned}
P(T \geq 2n_0^*) &\leq P\{|U(m) - \theta| \geq \varepsilon\} \\
&= O(p_m^{-k}) \text{ for arbitrarily large } k(> 0), \\
&\qquad \text{using the Markov inequality} \\
&= O(n_0^{*-k}). \quad (18.3.38)
\end{aligned}
$$

Let us denote $t = \left|(T - n_0^*)/n_0^{*1/2}\right|$. Hence, we have

$$
\begin{aligned}
\int_{[T \geq 2n_0^*]} t^6 dP &= E\left[(T - n_0^*)^6 n_0^{*-3} I(T \geq 2n_0^*)\right] \\
&= n_0^{*3} E\left[(Tn_0^{*-1} - 1)^6 I(T \geq 2n_0^*)\right] \\
&\leq n_0^{*3} E^{1/2}\left[(Tn_0^{*-1} - 1)^{12}\right] P^{1/2}(T \geq 2n_0^*), \\
&\qquad \text{by the Cauchy-Schwartz inequality} \\
&= n_0^{*3} O(1) O(n_0^{*-k/2}), \text{ using (18.3.38) and the fact} \\
&\qquad \text{that } Tn_0^{*-1} \text{ have all moments finite} \\
&\to 0 \text{ as } h^* \to 0, \text{ since } k \text{ is large.} \quad (18.3.39)
\end{aligned}
$$

Also, we note that

$$
\begin{aligned}
\int_{[T = m]} t^6 dP &= E\left[(m - n_0^*)^6 n_0^{*-3} I(T = m)\right] \\
&= O(n_0^{*3}) P(T = m) \\
&\leq O(n_0^{*3}) O(\eta^{\frac{1}{2}p_m}), \text{ by using (18.2.7)} \\
&\to 0 \text{ as } h^* \to 0. \quad (18.3.40)
\end{aligned}
$$

Next, in the light of (18.3.39) and (18.3.40) and the Theorem 1.5 of Woodroofe (1982), it will now suffice to show that

$$P\{m < T < 2n_0^* \cap |t| > x\} \leq H(x) \text{ where } H(x) \text{ is}$$

such that $x^5 H(x)$ is Lebesgue integrable for sufficiently.

large x and n_0^*.   (18.3.41)

We first consider the set $\{T > m \cap t < -x\}$. When $x \geq [1-(\theta_L/\theta)^\tau]n_0^{*1/2}$, we have

$$P\{T > m \cap t < -x\} = P\{T > m \cap T < n_0^* - xn_0^{*1/2}\} = 0. \quad (18.3.42)$$

When $0 < x < [1 - (\theta_L/\theta)^\tau]n_0^{*1/2}$, for large enough m, we can write

$$
\begin{aligned}
&P\{T > m \cap t < -x\} \\
&= P\{T > m \cap T < n_0^* - xn_0^{*1/2}\} \\
&\leq P\{U(m) - \theta < \theta[(q/q_m^*)^{1/\tau}(1 - xn_0^{*-1/2})^{1/\tau} - 1]\} \\
&\leq P\{U(m) - \theta < -\theta x \tau^{-1} n_0^{*-1/2}\} \\
&\leq P\{|U(m) - \theta|^8 > (\theta x \tau^{-1} n_0^{*-1/2})^8\} \\
&\leq c^*/x^8, \text{ for some constant } c^*(> 0), \text{ independent}
\end{aligned}
$$

of x, using the Markov inequality.   (18.3.43)

Next, we consider the set $\{T < 2n_0^* \cap t > x\}$. When $x \geq n_0^{*1/2}$, we have

$$P\{T < 2n_0^* \cap t > x\} = P\{T < 2n_0^* \cap T > n_0^* + xn_0^{*1/2}\} = 0. \quad (18.3.44)$$

Then, we simply need to show that for $0 < x < n_0^{*1/2}$, we can conclude:

$$P\{T < 2n_0^* \cap t > x\} \leq c^{**}/x^8, \text{ for some constant}$$

$c^{**}(> 0)$, independent of x.   (18.3.45)

The claim made in (18.3.45) can be justified along the lines of the arguments used in (18.3.44). Some details are left out for brevity. The proof is now complete.   □

Lemma 18.3.7 *For the stopping variable N defined in (18.2.6), we have as $h^* \to 0$:*

(i) $E\left[n_0^{*-2}(N - n_0^*)^4 g^{(4)}(W)\right] = 12\tau^4 c_1^{-2}(\theta/\theta_L)^{2\tau} g^{(4)}(1) + o(1)$
       if $g(.)$ satisfies (2.14);

(ii) $E\left[n_0^{*-3}(N - n_0^*)^6 g^{(6)}(W)\right] = 120\tau^6 c_1^{-3}(\theta/\theta_L)^{3\tau} g^{(6)}(1) + o(1)$
       if $g(.)$ satisfies (2.15);

where W is a random variable between Nn_0^{*-1} and 1.

PROOF We sketch a proof of part (i) only, since the proof of part (ii) is very similar. From (18.2.15), observe that

$$\left| g^{(4)}(x) \right| \leq \Sigma_{i=1}^{u} p_i x^{-s_i} \qquad (18.3.46)$$

where p_i's and s_i's are non-negative numbers. On the set $\{N = m\}$, we have $W > N/n_0^* = m/n_0^*$. Hence, using (18.2.7) and (18.3.46), we get

$$\left| E\left[n_0^{*-2}(N - n_0^*)^4 g^{(4)}(W) I(N = m) \right] \right|$$

$$\leq \Sigma_{i=1}^{u} p_i (n_0^*/m)^{s_i} \frac{(m - n_0^*)^4}{n_0^{*2}} P(N = m)$$

$$= O(1)O(m^2)O(\eta^{\frac{1}{2}p_m})$$

$$= o(1). \qquad (18.3.47)$$

Next, on the set $\{N > m\}$, we have $W > \sigma_L^2 \sigma^{-2}$. Also, $n_0^{*-2}(N - n_0^*)^4$ is uniformly integrable, and hence utilizing (18.2.15) again, we obtain

$$\left| n_0^{*-2}(N - n_0^*)^4 g^{(4)}(W) I(N > m) \right|$$

$$\leq n_0^{*-2}(N - n_0^*)^4 \left\{ \Sigma_{i=1}^{u} p_i (\sigma_L^2 \sigma^{-2})^{-s_i} \right\},$$

so that $n_0^{*-2}(N - n_0^*)^4 g^{(4)}(W) I(N > m)$ is then uniformly integrable. Thus, in view of Lemma 18.3.4 and the fact that $g^{(4)}(x)$ is continuous at $x = 1$, we obtain

$$E\left[n_0^{*-2}(N - n_0^*)^4 g^{(4)}(W) I(N > m) \right]$$

$$= \{b_2(4) - 4b_2(3) + 6b_2(2) - 4b_2(1)\}(\theta/\theta_L)^{2\tau} + o(1). \qquad (18.3.48)$$

Now, from the Remark 18.3.1 we recall that $b_2(4) - 4b_2(3) + 6b_2(2) - 4b_2(1) = 12\tau^4(\theta/\theta_L)^{2\tau}$, and hence (18.3.47) and (18.3.48) together would complete the proof of part (i).

To prove part (ii), simply recall the condition (18.2.16) to bound $\left| g^{(6)}(x) \right|$ and the fact that $g^{(6)}(x)$ is assumed continuous at $x = 1$. One will also need the Lemma 18.3.6 and the distributional convergence from (18.2.8). Further details are left out. □

18.3.3 Proof of Theorem 18.2.2

Using the Taylor expansion, with some suitable random variable W between 1 and Nn_0^{*-1}, we obtain

$$
g(N/n_0^*) = g(1) + n_0^{*-1}(N - n_0^*)g^{(1)}(1) + \frac{1}{2}n_0^{*-1}\frac{(N - n_0^*)^2}{n_0^*}g^{(2)}(1)
$$

$$
+ \frac{1}{6}\frac{(N - n_0^*)^3}{n_0^{*3}}g^{(3)}(1) + \frac{1}{24}n_0^{*-2}\frac{(N - n_0^*)^4}{n_0^{*2}}g^{(4)}(W).
$$

$$(18.3.49)$$

Then, by taking the expectations throughout (18.3.49), we obtain

$$
E[g(N/n_0^*)]
$$

$$
= g(1) + n_0^{*-1}E[N - n_0^*]g^{(1)}(1) + \frac{1}{2}n_0^{*-1}E\left[\frac{(N - n_0^*)^2}{n_0^*}\right]g^{(2)}(1)
$$

$$
+ \frac{1}{6}E\left[\frac{(N - n_0^*)^3}{n_0^{*3}}\right]g^{(3)}(1) + \frac{1}{24}n_0^{*-2}E\left[\frac{(N - n_0^*)^4}{n_0^{*2}}g^{(4)}(W)\right].
$$

$$(18.3.50)$$

Now, we can bound $E[N - n_0^*]$ using the terms from the Theorem 18.2.1 up to the order $o(n_0^{*-1})$. Next, we can bound $E[(N - n_0^*)^2/n_0^*]$ using the terms from the Lemma 18.3.2 up to the order $o(n_0^{*-1})$ also. Then, we bound $E[(N - n_0^*)^3/n_0^{*3}]$ using the terms from the Lemma 18.3.3 up to the order $O(n_0^{*-3})$. At the end, $E[\{(N-n_0^*)^4/n_0^{*2}\}g^{(4)}(W)]$ can be bounded up to the order $o(1)$, using the terms from the Lemma 18.3.7, part (i). The desired result will then follow from (18.3.50). Some details are left out. $\square$

18.3.4 Proof of Theorem 18.2.3

Using the Taylor expansion, with some suitable random variable W between 1 and Nn_0^{*-1}, we obtain

$$
g(N/n_0^*)
$$

$$
= g(1) + n_0^{*-1}(N - n_0^*)g^{(1)}(1) + \frac{1}{2}n_0^{*-1}\frac{(N - n_0^*)^2}{n_0^*}g^{(2)}(1)
$$

$$
+ \frac{1}{6}\frac{(N - n_0^*)^3}{n_0^{*3}}g^{(3)}(1) + \frac{1}{24}n_0^{*-2}\frac{(N - n_0^*)^4}{n_0^{*2}}g^{(4)}(1)
$$

$$
+ \frac{1}{120}\frac{(N - n_0^*)^5}{n_0^{*5}}g^{(5)}(1) + \frac{1}{720}n_0^{*-3}\frac{(N - n_0^*)^6}{n_0^{*3}}g^{(6)}(W).
$$

$$(18.3.51)$$

Then, by taking the expectations throughout (18.3.51), we obtain

$$
\begin{aligned}
E[g(N/n_0^*)] & \\
= \ & g(1) + n_0^{*-1} E[N - n_0^*]g^{(1)}(1) + \frac{1}{2}n_0^{*-1}E\left[\frac{(N - n_0^*)^2}{n_0^*}\right]g^{(2)}(1) \\
& + \frac{1}{6}E\left[\frac{(N - n_0^*)^3}{n_0^{*3}}\right]g^{(3)}(1) + \frac{1}{24}n_0^{*-2}E\left[\frac{(N - n_0^*)^4}{n_0^{*2}}\right]g^{(4)}(1) \\
& + \frac{1}{120}E\left[\frac{(N - n_0^*)^5}{n_0^{*5}}\right]g^{(5)}(1) + \frac{1}{720}n_0^{*-3}E\left[\frac{(N - n_0^*)^6}{n_0^{*3}}g^{(6)}(W)\right].
\end{aligned}
$$

$$(18.3.52)$$

Then, by taking the expectations throughout (18.3.52), we obtain the desired result in view of the Lemma 18.3.7, part (ii). In this expansion, we go up to the remainder term of order $O(n_0^{*-3})$ which requires us to invoke the Lemmas 18.3.2–18.3.5 and Theorem 18.2.1. The details are left out. $\square$

Remark 18.3.3 The proofs of the Theorems 18.2.2 and 18.2.3 may seem obvious to some readers. But, one must also realize that this appearance is deceiving. The major hard work lies in building the essential machineries before one is ready to prove Theorems 18.2.2 and 18.2.3.

18.4 APPLICATIONS OF THE MAIN RESULTS

Let us briefly go back to the construction of a fixed-width confidence interval for the mean of a normal population when the variance σ^2 is also unknown. In the case when one has the available information that $\sigma > \sigma_L$ where σ_L is known and positive, we had mentioned the paper of Mukhopadhyay and Duggan (1997) which appropriately modified the two-stage sampling design of Stein (1945, 1949) and obtained the associated second-order properties. Mukhopadhyay (1999) derived the *third-* and *higher-order expansions* of both the lower and upper bounds for the corresponding $E_{\mu,\sigma}[N - n_0^*]$ and the confidence coefficient for the normal problem alone. It should be apparent that those expansions given in Mukhopadhyay (1999) would follow from the present Theorems 18.2.1–18.2.3. But, instead of supplying any more details on the normal problem, let us go forward and examine a few other interesting problems where the general results obtained here can be directly applied.

18.4.1 Negative Exponential Location Estimation

Let $X_1, X_2, \ldots$ be i.i.d. random variables with the probability density function:

$$f(x; \mu, \sigma) = \sigma^{-1} \exp\{-(x - \mu)/\sigma\} I(x > \mu), \qquad (18.4.53)$$

where $-\infty < \mu < \infty$, $0 < \sigma < \infty$ are two unknown parameters. This distribution has been used widely in reliability as well as survival analyses where the minimum guarantee time μ is of course assumed positive. One may refer to Basu (1971, 1991) and Balakrishnan and Basu (1995), among other sources.

We will, however, continue to assume that $\mu \in R$ in order to avoid additional complexities. Sequential and multistage estimation problems for μ and σ were reviewed in Basu (1991) and Mukhopadhyay (1988, 1995).

Having recorded $X_1, \ldots, X_n$ we estimate μ, σ respectively by $T_n = \min\{X_1, \ldots, X_n\}$ and $S_n = (n - 1)^{-1} \Sigma_{i=1}^n (X_i - T_n)$, with $n \geq 2$. Now suppose that given two *preassigned numbers* $d(> 0)$ and $0 < \alpha < 1$, we wish to construct a confidence interval I for μ such that the length of I is d and $P_{\mu,\sigma}\{\mu \in I\} \geq 1 - \alpha$. Here, the scale parameter σ is treated as a nuisance parameter. A two-stage procedure was originally proposed by Ghurye (1958), but it does not have the (first-order) efficiency property in the sense of Chow and Robbins (1965) and Ghosh and Mukhopadhyay (1981).

Let us consider $I_n = [T_n - d, T_n]$ as the confidence interval for μ. Now, $P_{\mu,\sigma}\{\mu \in I\} \geq 1 - \alpha$ provided that n is the smallest integer $\geq a\sigma/d = C$, say, where $a = \log(1/\alpha)$.

Here, C plays the role of n_0^* from (2.1) with $q = a$, $\theta = \sigma$, $\tau = 1$ and $h^* = d$. But, let us suppose that $\sigma > \sigma_L$ where $\sigma_L(> 0)$ is available from prior knowledge and the nature of the practical applications on hand. With $\theta_L = \sigma_L$ and $m_0 \geq 2$, one then defines m as in (18.2.3) and N as in (18.2.6) with $U(m) = S_m$, and implements the two-stage sampling design, with q_m^* being the upper $100\alpha\%$ point of the F distribution with degrees of freedom 2 and $2(m - 1)$. This is the modified version of Ghurye's (1958) two-stage sampling design when one has the prior knowledge that $\sigma > \sigma_L$ where $\sigma_L(> 0)$ is known.

Based on all the observations $X_1, \ldots, X_N$, we then propose the fixed-width confidence interval $I_N = [T_N - d, T_N]$ for μ. The asymptotic analysis is carried out when d converges to zero.

Since $I(N = n)$ and T_n are independent for all $n \geq m$, we have

$$P_{\mu,\sigma}\{\mu \in I_N\} = E_{\mu,\sigma}[1 - \exp(-Nd/\sigma)] = E_{\mu,\sigma}[g(N/C)] \quad (18.4.54)$$

where $g(x) = 1 - \exp(-ax)$, $x > 0$.

The condition (18.2.4) is satisfied with $p_m = 2(m-1)$, that is $c_1 = 2$, $c_2 = -2$. It is also easy to see that $\alpha = \{1 + q_m^*(m-1)^{-1}\}^{-(m-1)}$ and hence

$$q_m^* = (m-1)\left[\alpha^{-1/(m-1)} - 1\right]. \tag{18.4.55}$$

Now, we need to expand q_m^* up to the order $O(m^{-4})$ as in (18.2.14) and we proceed as follows. First, we write

$$\log\left\{\alpha^{-1/(m-1)}\right\} = am^{-1}(1 - m^{-1})^{-1}$$
$$= a\left[m^{-1} + m^{-2} + m^{-3} + O(m^{-4})\right], \tag{18.4.56}$$

so that from (18.4.55) and (18.4.56) one has

$$q_m^* = (m-1)[\exp\{a(m^{-1} + m^{-2} + m^{-3} + O(m^{-4})\} - 1]$$
$$= a + (\tfrac{1}{2}a^2)m^{-1} + (\tfrac{1}{2}a^2 + \tfrac{1}{6}a^3)m^{-2} + (a^2 + \tfrac{1}{3}a^3 + \tfrac{1}{24}a^4)m^{-3}$$
$$+ O(m^{-4}). \tag{18.4.57}$$

From (18.4.57) we readily see that the expansion of q_m^*/q with $q = a$ satisfies (18.2.14) where

$$a_1 = \tfrac{1}{2}a, \; a_2 = \tfrac{1}{2}a + \tfrac{1}{6}a^2 \text{ and } a_3 = a + \tfrac{1}{3}a^2 + \tfrac{1}{24}a^3. \tag{18.4.58}$$

From Ghurye (1958) and Mukhopadhyay (1988), it follows that for all fixed μ, σ, d and α,

$$P_{\mu,\sigma}\{\mu \in I_N\} \geq 1 - \alpha. \quad [Exact \; Consistency \; Property] \tag{18.4.59}$$

Next, with $s = \tau = 1$ and $c_1 = 2, c_2 = -2$, we get the expressions of $b_i(1)$, $i = 1, 2, 3$. In this special situation, we obtain

$$b_1(1) = 0, b_2(1) = 0 \text{ and } b_3(1) = \tfrac{3}{4}. \tag{18.4.60}$$

Now, one evaluates the expression of ψ given in Theorem 18.2.1 and obtains $\psi = \tfrac{1}{2}a\sigma\sigma_L^{-1}$. Thus, using Theorem 18.2.1, we immediately claim that $E_{\mu,\sigma}(N - C)$ lies between

$$\tfrac{1}{2}a\sigma\sigma_L^{-1} + (\tfrac{1}{2}a + \tfrac{1}{6}a^2)(\sigma\sigma_L^{-1})^2 C^{-1}$$
$$+ (a + \tfrac{1}{3}a^2 + \tfrac{1}{24}a^3 + \frac{3}{4})(\sigma\sigma_L^{-1})^3 C^{-2} + O(d^3) \quad \text{and}$$

$$(\tfrac{1}{2}a\sigma\sigma_L^{-1} + 1) + (\tfrac{1}{2}a + \tfrac{1}{6}a^2)(\sigma\sigma_L^{-1})^2 C^{-1}$$
$$+ (a + \tfrac{1}{3}a^2 + \tfrac{1}{24}a^3 + \frac{3}{4})(\sigma\sigma_L^{-1})^3 C^{-2} + O(d^3) \tag{18.4.61}$$

where $C = a\sigma/d$, the optimal fixed sample size with $a = \log(1/\alpha)$.

These bounds for $E_{\mu,\sigma}(N-C)$ are certainly much sharper than the ones obtained earlier in Mukhopadhyay and Duggan (1999) which were given up to the order $o(d^{1/2})$.

Next, we consider the higher order expansions for the coverage probability. From (18.4.54), recall that $P_{\mu,\sigma}\{\mu \in I_N\} = E_{\mu,\sigma}[g(N/C)]$ where $g(x) = 1 - \exp(-ax)$, $x > 0$. Observe that $g^{(s)}(x) = (-1)^{s+1}a^s e^{-ax}$ and hence we know the signs for all these derivatives at $x = 1$ for $s = 1, ..., 6$. Also, one obviously can conclude that $\left|g^{(4)}(x)\right| \leq a^4$ and $\left|g^{(6)}(x)\right| \leq a^6$ for all $x > 0$. In other words, one can verify the conditions in (18.2.15) and (18.2.16) as needed. Thus, we can right away use the Theorems 18.2.2 and 18.2.3 to write down the upper and lower bounds for $P_{\mu,\sigma}\{\mu \in I_N\} - (1-\alpha)$ up to the order $o(d^2)$ or $O(d^3)$, as desired. Mukhopadhyay and Duggan (1999) obtained bounds for $P_{\mu,\sigma}\{\mu \in I_N\} - (1 - \alpha)$ up to the order $o(d)$. The details are left out for brevity.

Remark 18.4.1 In the next few applications, we do not show much of the details for the sake of brevity. In each case, we first explain the nature of the problem and then show generally how the underlying structure can be identified with the one we have developed in the Sections 18.2 and 18.3.

18.4.2 Multivariate Normal Mean Vector Estimation

Consider $\mathbf{X_1}, \mathbf{X_2}, ...,$ a sequence of independent $N_p(\mu, \sigma^2\mathbf{H})$ random variables where $\mu \in \mathbf{R^p}$, $\sigma \in R^+$ are two unknown parameters, but H is a $p \times p$ positive definite (p.d.). matrix. Here, σ^2 is the nuisance parameter. First, we discuss the *fixed-size confidence region* problem and then the *minimum risk point estimation* problem for the mean vector μ. In practice, let us suppose, however, that $\sigma > \sigma_L$ where $\sigma_L(> 0)$ is known.

Fixed-size confidence region

Having recorded $\mathbf{X_1}, ..., .\mathbf{X_n}$, we estimate μ and σ^2 respectively by $\overline{\mathbf{X}}_n = n^{-1}\Sigma_{i=1}^n \mathbf{X_i}$ and $S_n^2 = (np-p)^{-1}\Sigma_{i=1}^n(\mathbf{X_i} - \overline{\mathbf{X}}_n)'\mathbf{H^{-1}}(\mathbf{X_i} - \overline{\mathbf{X}}_n)$ with $n \geq 2$. Given $d(> 0)$ and $0 < \alpha < 1$ we consider the *fixed-size ellipsoidal confidence region*

$$\mathcal{R}_\backslash = \left\{\omega \in \mathbf{R^p} : (\overline{\mathbf{X}}_n - \omega)'\mathbf{H^{-1}}(\overline{\mathbf{X}}_n - \omega) \leq \mathbf{d^2}\right\} \qquad (18.4.62)$$

for μ, and we require that $P_{\mu,\sigma}\{\mu \in \mathcal{R}_\backslash\} \geq 1 - \alpha$ which holds if n is the smallest integer $\geq a\sigma^2/d^2 = C$, say. Here, $F(a) = 1 - \alpha$ with $F(x) = P(\chi_p^2 \leq x)$, $x > 0$.

Here again, C plays the role of n_0^* with $q = a$, $h^* = d^2$, $\theta = \sigma^2$ and $\tau = 1$. Mukhopadhyay and Al-Mousawi (1986) proposed a two-stage procedure which had the *exact consistency property* but it was not first-order efficient. Mukhopadhyay and Al-Mousawi (1986) also had developed other multistage procedures. Nagao (1996) came up with the second-order·properties associated with his sequential procedure when the dispersion matrix has some special structure, including the one considered in Mukhopadhyay and Al-Mousawi (1986).

With $\theta_L = \sigma_L^2$ and $m_0 \geq 2$, we then define m as in (18.2.3) and N as in (18.2.6) with $U(m) = S_m^2$, and implement the two-stage sampling design with $q_m^* = pb_m$ where b_m is the upper $100\alpha\%$ point of the F distribution with degrees of freedom p and $p(m-1)$. The condition (18.2.4) is satisfied with $p_m = pm - p$, that is with $c_1 = p = -c_2$. Utilizing the results from Scheffe and Tukey (1944), we can express b_m, and hence $q_m^* a^{-1}$, satisfying (18.2.14).

Since $I(N = n)$ and $\overline{X}_n$ are independent for all $n \geq m$, we have

$$P_{\mu,\sigma}\{\mu \in \mathcal{R}_N\} = E_{\mu,\sigma}\left[F(Nd^2/\sigma^2)\right] = E_{\mu,\sigma}\left[g(N/C)\right] \qquad (18.4.63)$$

with $g(x) = F(ax)$, $x > 0$ while the confidence region $\mathcal{R}_N$ corresponds to (18.4.62) based on $X_1, ..., X_N$.

From Theorem 2 in Mukhopadhyay and Al-Mousawi (1986) it follows that

$$P_{\mu,\sigma}\{\mu \in \mathcal{R}_N\} \geq \infty - \alpha \qquad [Exact\ Consistency\ Property], \qquad (18.4.64)$$

for all fixed $\mu, \sigma, \mathbf{d}$ and α. Using Theorem 18.2.1, we can immediately bound $E_{\mu,\sigma}[N - C]$ up to the order $O(d^6)$ as $d \to 0$.

Then, one may look at the function $g(x) = F(ax)$, $x > 0$, and obtain the successive derivatives $g^{(s)}(x)$, $s = 1, ..., 6$. Their behavior will depend on the dimension p also. But, once (18.2.15) and (18.2.16) are verified, the Theorems 18.2.2 and 18.2.3 will provide both the lower and upper bounds for $P_{\mu,\sigma}\{\mu \in \mathcal{R}_N\} - (1 - \alpha)$ up to the order $o(d^4)$ or $O(d^6)$, as the case may be. Further details are omitted.

Minimum risk point estimation

Let $\mathbf{X}$'s be i.i.d. $N_p(\mu, \sigma^2\mathbf{H})$ as before. Suppose that the loss function in estimating μ by $\overline{X}_n$ is taken to be

$$L_n = A\{(\overline{X}_n - \mu)'\mathbf{H}^{-1}(\overline{X}_n - \mu)\}^{r/2} + cn^t \qquad (18.4.65)$$

where A, c, r and t are known positive numbers. The type of loss function given by (18.4.65) was adopted by Wang (1980). The situation when $r = 2$

and $t = 1$ corresponds to the customary scenario of squared error loss plus linear cost of sampling.

Now, the risk associated with (18.4.65) is given by $R_n = E_{\mu,\sigma}(L_n) = B\sigma^r n^{-\frac{1}{2}r} + cn^t$ with $B = 2^{\frac{1}{2}r} A\Gamma(\frac{1}{2}(p+r))/\Gamma(\frac{1}{2}p)$, whereas this risk is *minimized* (approximately) if $n = n^* = \{K^* c^{-1}\}^{2/(2t+r)} \sigma^{2r/(2t+r)}$ with $K^* = \frac{1}{2}rBt^{-1}$. The corresponding *minimum risk* is given by $R_{n^*} = B\sigma^r n^{*-\frac{1}{2}r} + cn^{*t} = cn^{*t}\{1 + BK^{*-1}\} = c(1 + 2tr^{-1})n^{*t}$, and our goal is to achieve this minimum risk approximately.

Note that n^* plays the role of n_0^* where $q = K^{*2/(2t+r)}$, $h^* = c^{2/(2t+r)}$, $\theta = \sigma^2$ and $\tau = r/(2t+r)$. With $\theta_L = \sigma_L^2$ and $m_0 \geq 2$, we then define m as in (18.2.3) and N as in (18.2.6) where $U(m) = S_m^2$, the same as in Section 18.4.2, and $q_m^* = q$. So, the condition (18.2.14) applies in this case with $a_1 = a_2 = a_3 = 0$. The condition (18.2.4) is satisfied with $c_1 = -c_2 = p$.

After implementing the two-stage procedure (18.2.6), we propose to estimate μ by $\overline{\mathbf{X}}_N$. Using Theorem 18.2.1, we can immediately bound $E_{\mu,\sigma}(N - n^*)$ up to the order $O(n^{*-3})$ as $c \to 0$.

Since $I(N = n)$ and $\overline{\mathbf{X}}_n$ are independent for all $n \geq m$, we have $E_{\mu,\sigma}(L_N) = B\sigma^r E_{\mu,\sigma}(N^{-\frac{1}{2}r}) + cE_{\mu,\sigma}(N^t)$, and hence the

$$
\begin{aligned}
\text{Regret} \quad &= \quad E_{\mu,\sigma}(L_N) - R_{n^*} \\
&= \quad c\{2tr^{-1}[E_{\mu,\sigma}((N/n^*)^{-r/2}) - 1] + [E_{\mu,\sigma}((N/n^*)^t) - 1]\}n^{*t}.
\end{aligned}
$$

(18.4.66)

Then, with the function $g(x) = x^{-r/2}$, we can bound $E_{\mu,\sigma}((N/n^*)^{-r/2})$ and then with another function $h(x) = x^t$, we can bound $E_{\mu,\sigma}((N/n^*)^t)$ up to the desired order by utilizing the Theorems 18.2.2 and 18.2.3. Further details are omitted.

18.4.3 Linear Regression Parameters Estimation

Consider the linear regression model with normally distributed errors. We write

$$Y_i = \mathbf{x}_i'\beta + \varepsilon_i, \quad i = 1, 2, \dots \tag{18.4.67}$$

where ε_i's are i.i.d. $N(0, \sigma^2)$, β is an unknown $p \times 1$ vector of parameters, and $\mathbf{x}_i$'s are known vectors. Let us denote $\mathbf{Y}_n' = (Y_1, \dots, Y_n)$ and $X_n' = (\mathbf{x}_1, \dots, \mathbf{x}_n)$ and assume that the model is of full rank, that is the rank of the $p \times p$ matrix $X_n' X_n$ is $p(< n)$. Also, we assume that the nuisance parameter $\sigma(> 0)$ is unknown.

Having recorded $(\mathbf{x_i}, \mathbf{Y_i})$, $i = 1, ..., n$, we estimate β by the least squares estimator $\widehat{\beta}_n = (X'_n X_n)^{-1} X'_n \mathbf{Y_n}$ and use the loss function

$$L_n = A(\widehat{\beta}_n - \beta)'(\mathbf{n^{-1} X'_n X_n})(\widehat{\beta}_n - \beta) \qquad (18.4.68)$$

with $A(> 0)$ known. Our goal is to make the risk $\leq W$ where $W(> 0)$ is preassigned.

Hence, the sample size n has to be the smallest integer $\geq Ap\sigma^2/W = n^*$, say, which corresponds to n_0^* with $q = Ap$, $h^* = W$, $\tau = 1$ and $\theta = \sigma^2$. Let us assume that $\sigma > \sigma_L$ where $\sigma_L(> 0)$ is known. Let $S_m^2 = (m - p)^{-1}\{(\mathbf{Y_m} - \mathbf{X_m}\widehat{\beta}_m)'(\mathbf{Y_m} - \mathbf{X_m}\widehat{\beta}_m)\}$, the mean square error, where m is given by (18.2.3) with $\theta_L = \sigma_L^2$ and $m_0 > p+2$. We implement the two-stage procedure (18.2.6) with $U(m) = S_m^2$.

With $q_m = \frac{1}{2}(m - 1)[\Gamma\{\frac{1}{2}(m - p - 2)\}/\Gamma\{\frac{1}{2}(m - p)\}]$, let us define $q_m^* = Apq_m$. One can verify that for all fixed β, σ^2, A and W, the associated risk

$$E_{\beta,\sigma}(L_N) = Ap\sigma^2 E_{\beta,\sigma}(N^{-1}) \leq W \qquad [\textit{Risk Efficiency Property}]. \qquad (18.4.69)$$

The condition (18.2.4) holds with $p_m = m - p$, that is with $c_1 = 1$, $c_2 = -p$. This structure can again be identified with that developed in Sections 18.2 and 18.3. Hence, one will be able to obtain higher order bounds for both $E_{\beta,\sigma}(N - n^*)$ and $E_{\beta,\sigma}(L_N/W)$. Further details are omitted.

In this setup, minimum risk point estimation problem for β or the fixed-size confidence region problem for β could also be easily introduced under similar sort of two-stage sampling schemes when $\sigma > \sigma_L$ with known $\sigma_L(> 0)$. In order to review such procedures when $\sigma(> 0)$ is completely unknown, one should refer to Ghosh, Mukhopadhyay and Sen (1997), Mukhopadhyay and Abid (1986), and Mukhopadhyay (1991). Finster's (1983, 1985) papers are also relevant. We omit the details for brevity.

18.4.4   Multiple Decision Theory

The tools developed in Sections 18.2 and 18.3 will now be applied for two interesting selection and ranking problems. The emphasis still lies in achieving higher-order characteristics for the two-stage methodologies developed for such problems in Mukhopadhyay and Duggan (1999).

Selecting the best normal population

Consider independent populations $\pi_1, ..., \pi_k$ with $k \geq 2$, and let $X_{i1}, ..., X_{in}, ...$ be i.i.d. $N(\mu_i, \sigma^2)$ random variables from π_i, with $\mu_i \in R$ and $\sigma \in R^+$, $i = 1, ..., k$. Let us write $\overline{X}_{in} = n^{-1}\Sigma_{j=1}^n X_{ij}$, $U_{in}^2 = (n -$

$1)^{-1}\Sigma_{j=1}^{n}(X_{ij} - \overline{X}_{in})^2$, and $U_n = k^{-1}\Sigma_{i=1}^{k}U_{in}^2$ for $n \geq 2$, $i = 1, ..., k$. We assume that all the parameters are unknown whereas σ^2 is considered the nuisance parameter.

Let us denote $\mu' = (\mu_1, ..., \mu_k)$ and write $\mu_{[1]} \leq \cdots \leq \mu_{[k-1]} \leq \mu_{[k]}$ for the ordered μ values. Here and elsewhere, "CS" would stand for "Correct Selection". We adopt Bechhofer's (1954) *indifference zone formulation*. So, let there be two preassigned numbers $\delta^*(> 0)$ and $P^* \in (k^{-1}, 1)$, and our goal is to select the population associated with $\mu_{[k]}$, referred to as the *best population*, so that $P_{\mu,\sigma}(\text{CS}) \geq P^*$ whenever $\mu \in \Omega(\delta^*) = \{\mu : \mu_{[k]} - \mu_{[k-1]} \geq \delta^*\}$, the *preference zone*. The parameter subspace $\Omega^c(\delta^*)$ is referred to as the *indifference zone*.

Define $C = a^2\sigma^2/\delta^{*2}$ where "a" satisfies the integral equation:

$$\int_{-\infty}^{\infty} \Phi^{k-1}(y + a)\phi(y)dy = P^*,$$

$\phi(\cdot)$ being the standard normal density and $\Phi(x) = \int_{-\infty}^{x} \phi(y)dy$, $x \in R$.

If σ^2 were known, then C could be interpreted as the optimal fixed sample size required from each π in conjunction with the selection of the population giving rise to the largest sample mean. When $\sigma^2(> 0)$ is completely unknown, a two-stage procedure was developed by Bechhofer, Dunnett and Sobel (1954). For a review of other multistage sampling techniques for this problem, refer to Chapter 3 in Mukhopadhyay and Solanky (1994).

Let us, however, assume that $\sigma > \sigma_L$ where $\sigma_L(> 0)$ is known in advance. Now, C plays the role of n_0^* with $q = a^2$, $\theta = \sigma^2$, $\tau = 1$ and $h^* = \delta^{*2}$. One then defines m as in (18.2.3), with $m_0 \geq 2$, $\theta_L = \sigma_L^2$, and considers N as in (18.2.6) with $U(m) = U_m$ and $q_m^* = q$. We then implement the two-stage methodology and select the population associated with $\max_{1 \leq i \leq k} \overline{X}_{iN}$ based on the observations $\{X_{i1}, ..., X_{iN}\}$, $i = 1, ..., k$.

Since $I(N = n)$ and $(\overline{X}_{1n}, ..., \overline{X}_{kn})$ are independent for all $n \geq m$, we have [from Theorem 3.2.1 in Mukhopadhyay and Solanky (1994)]:

$$\inf_{\mu \in \Omega(\delta^*)} P_{\mu,\sigma}(\text{CS}) = E_\sigma[\int_{-\infty}^{\infty} \{\Phi(y + N^{1/2}\delta^*\sigma^{-1})\}^{k-1}\phi(y)dy]$$

$$= E_\sigma[g(N/C)] \qquad (18.4.70)$$

where $g(x) = f(x^{\frac{1}{2}})$ and $f(x) = \int_{-\infty}^{\infty} \Phi^{k-1}(y + ax)\phi(y)dy$, $x > 0$. The condition (18.2.4) is satisfied with $p_m = k(m - 1)$, that is $c_1 = -c_2 = k$. Obviously (18.2.14) holds with $a_1 = a_2 = a_3 = 0$.

By taking the derivatives of $g(x)$ inside the integral, one can successively determine $g^{(s)}(x)$, $s = 1, ..., 6$ and thereby one can also show that the conditions (18.2.15) and (18.2.16) would be satisfied.

This structure can again be identified with that developed in Sections 18.2 and 18.3. Hence, one will be able to obtain higher order bounds for both $E_\sigma(N - C)$ and $\underset{\mu \in \Omega(\delta^*)}{Inf} P_{\mu,\sigma}(CS)$. Further details are omitted.

Selecting the best negative exponential population

Consider independent populations $\pi_1, ..., \pi_k$ with $k \geq 2$, and let $X_{i1}, ..., X_{in}, ...$ be i.i.d. random variables obtained from π_i having the negative exponential probability density function $f(x; \mu_i, \sigma)$, defined via (4.1), with $\mu_i \in R$, $\sigma \in R^+$, $i = 1, ..., k$. Let us write $T_{in} = \min\{X_{i1}, ..., X_{in}\}$, $U_{in} = (n - 1)^{-1}\Sigma_{j=1}^n(X_{ij} - T_{in})$, and $U_n = k^{-1}\Sigma_{i=1}^k U_{in}$ for $n \geq 2$, $i = 1, ..., k$. We assume that all the parameters are unknown whereas σ is considered the nuisance parameter. Let us denote as before μ, $\mu_{[\cdot]}$, and pursue the *indifference zone formulation* again, given two preassigned numbers $\delta^*(> 0)$ and $P^* \in (k^{-1}, 1)$. We define the *preference zone* $\Omega(\delta^*)$ as before and the problem is to select the population associated with $\mu_{[k]}$, referred to as the *best population*, in such a way that $P_{\mu,\sigma}(CS) \geq P^*$ whenever $\mu \in \Omega(\delta^*)$. Let $C = a\sigma/\delta^*$ where "a" is obtained by solving the equation, $\int_0^\infty \{1 - \exp(-z - a)\}^{k-1} \exp(-z)dz = P^*$.

If σ were known, then C could be interpreted as the optimal fixed sample size required from each π in conjunction with the selection of the population associated with the largest sample minimum order statistics among the corresponding T_{iC}'s. We tacitly assume that C is an integer. When $\sigma^2(> 0)$ is completely unknown, Desu, Narula and Villarreal (1977) developed a two-stage procedure for this selection problem. For a review of other multistage sampling techniques in this problem, refer to Chapter 4 in Mukhopadhyay and Solanky (1994), and Panchapakesan (1995).

Let us, however, assume that $\sigma > \sigma_L$ where $\sigma_L(> 0)$ is known in advance. Now, C plays the role of n_0^* where $q = a$, $\theta = \sigma$, $\tau = 1$ and $h^* = \delta^*$. One then defines m as in (18.2.3), with $m_0 \geq 2$, $\theta_L = \sigma_L$. Then, consider N as in (18.2.6) with $U(m) = U_m$ and $q_m^* = q$. We then implement the two-stage methodology and select the population associated with $\max_{1 \leq i \leq k} T_{iN}$ based on the observations $\{X_{i1}, ..., X_{iN}\}$, $i = 1, ..., k$.

Since $I(N = n)$ and $(T_{1n}, ..., T_{kn})$ are independent for all $n \geq m$, we have [from Theorem 4.2.1 in Mukhopadhyay and Solanky (1994)]:

$$\underset{\mu \in \Omega(\delta^*)}{\inf} P_{\mu,\sigma}(CS) = E_\sigma[\int_0^\infty \{1 - \exp(-z - N\delta^*\sigma^{-1})\}^{k-1} \exp(-z)dz]$$

$$= E_\sigma[g(N/C)] \qquad (18.4.71)$$

where $g(x) = \int_0^\infty \{1 - \exp(-z - ax)\}^{k-1} \exp(-z)dz$, $x > 0$. The expression in (18.4.71) can be further simplified as follows:

$$\inf_{\mu \in \Omega(\delta^*)} P_{\mu,\sigma}(\text{CS}) = \Sigma_{u=0}^{k-1} b(k,u) E_\sigma[g_u(N/C)] \tag{18.4.72}$$

where $b(k,u) = \binom{k-1}{u}(-1)^u(u+1)^{-1}$ and $g_u(x) = \exp(-uax)$, $x > 0$, $u = 0, 1, ..., k-1$. The condition (18.2.4) is satisfied with $p_m = 2k(m-1)$, that is with $c_1 = -c_2 = 2k$. Obviously (18.2.14) holds with $c_1 = c_2 = c_3 = 0$.

Using (18.4.72), one can successively determine $g^{(s)}(x)$, $s = 1, ..., 6$ and thereby one can also show that the conditions (18.2.15) and (18.2.16) would be satisfied.

This structure can again be identified with that developed in Sections 18.2 and 18.3. Hence, one will be able to obtain higher order bounds for both $E_\sigma(N - C)$ and $\underset{\mu \in \Omega(\delta^*)}{Inf} P_{\mu,\sigma}(\text{CS})$. Further details are omitted.

18.5   CONCLUDING THOUGHTS

After going through the techniques used in the proofs of the three Theorems 18.2.1–18.2.3, it becomes obvious in principle that both the lower and upper bounds for $E[g(N/n_0^*)]$ could be expanded up to the order $O(n_0^{*-k})$ for any $k(> 0)$. One will certainly need more terms in (18.2.14) and conditions like those in (18.2.16) for appropriately chosen higher derivatives of $g(.)$. Of course, any such sharpening of the rate of convergence will need more efforts, but the direction of the approach should be quite apparent. For brevity alone, we have gone only up to the order $O(n_0^{*-3})$.

Another point is to be noted here. In several applications, including the normal problem (Section 1) of Stein, as well as the problems such as those mentioned in the Sections 18.4.1, 18.4.2, 18.4.3, and 18.4.4, the relevant $g(.)$ function is frequently monotone in nature. Then, depending on the problem on hand, we may be satisfied with either the lower *or* upper bound for $E[g(N/n_0^*)]$. In view of (18.3.19), in the monotone case, we may then simply use $E[g(T/n_0^*)]$ or $E[g(\{T + 1\}/n_0^*)]$ as the appropriate lower *or* upper bound for $E[g(N/n_0^*)]$. Then, one will simply write down the appropriate expansions up to the order $O(n_0^{*-3})$ for $E[g(T/n_0^*)]$ or $E[g(\{T + 1\}/n_0^*)]$, as the case may be. Here, one will directly exploit (18.3.22), (18.3.26), (18.3.30), (18.3.34), (18.3.36) and Lemmas 18.3.6 and 18.3.7. By this time, such calculations may be considered fairly routine. So, we leave out the details.

Acknowledgement The referee's queries prompted the Remarks 18.2.1 and 18.2.2 of which Remark 18.2.2 is more substantial. Thanks to the referee.

18.6 REFERENCES

Balakrishnan, N. and Basu, A. P. (Eds.) (1995). *The Exponential Distributions: Theory, Methods and Applications*, Gordon and Breach Publishers, Amsterdam.

Basu, A. P. (1971). On a sequential rule for estimating the location parameter of an exponential distribution, *Naval Research Logistic Quarterly*, **18**, 329–337.

Basu, A. P. (1991). Sequential methods in reliability and life testing, In *Handbook of Sequential Analysis* (Eds., B. K. Ghosh and P. K. Sen), pp. 581–592, Marcel Dekker, New York.

Bechhofer, R.E. (1954). A single-sample multiple decision procedure for ranking means of normal populations with known variances. *Annals of Mathematical Statistics*, **25**, 16–39.

Bechhofer, R. E., Dunnett, C. W. and Sobel, M. (1954). A two-sample multiple decision procedure for ranking means of normal populations with a common unknown variance, *Biometrika,* **41**, 170–176.

Chow, Y. S. and Robbins, H. (1965). On the asymptotic theory of fixed width sequential confidence intervals for the mean, *Annals of Mathematical Statistics*, **36**, 457-462.

Dantzig, G. B. (1940). On the non-existence of tests of Student's hypothesis having power functions independent of σ, *Annals of Mathematical Statistics*, **11**, 186–192.

Desu, M. M., Narula, S. C. and Villarreal, B. (1977). A two-stage procedure for selecting the best of k exponential distributions, *Communications in Statistics, Series A,* **6**, 1231–1243.

Finster, M. (1983). A frequentist approach to sequential estimation in the general linear model, *Journal of the American Statistical Association*, **78**, 403–407.

Finster, M. (1985). Estimation in the general linear model when the accuracy is specified before data collection, *Annals of Statistics*, **13**, 663–675.

Ghosh, M. and Mukhopadhyay, N. (1981). Consistency and asymptotic efficiency of two-stage and sequential estimation procedures, *Sankhyā, Series A,* **43**, 220–227.

Ghosh, M., Mukhopadhyay, N. and Sen, P. K. (1997). *Sequential Estimation*, John Wiley & Sons, New York.

Ghurye, S. G. (1958). Note on sufficient statistics and two-stage procedures, *Annals of Mathematical Statistics*, **29**, 155–166.

Johnson, N. L. and Kotz, S. (1970). *Continuous Univariate Distributions-2*, John Wiley & Sons, New York.

Mukhopadhyay, N. (1988). Sequential estimation problems for negative exponential populations. *Communications in Statistics—Theory and Methods*, **17**, 2471–2506.

Mukhopadhyay, N. (1991). Parametric sequential point estimation, In *Handbook of Sequential Analysis* (Eds., B. K. Ghosh and P. K. Sen), pp. 245–267, Marcel Dekker, New York.

Mukhopadhyay, N. (1995). Two-stage and multi-stage estimation, In *The Exponential Distribution: Theory, Methods and Applications* (Eds., N. Balakrishnan and A. P. Basu), pp. 429–452. Gordon and Breach, Amsterdam.

Mukhopadhyay, N. (1999). Higher than second-order approximations via two-stage sampling, *Sankhyā, Series A*, **61**, 254–269.

Mukhopadhyay, N. and Abid, A. D. (1986). On fixed-size confidence regions for the regression parameters, *Metron*, **44**, 297–306.

Mukhopadhyay, N. and Al-Mousawi, J. S. (1986). Fixed-size confidence regions for the mean vector of a multinormal distribution, *Sequential Analysis*, **5**, 139–168.

Mukhopadhyay, N. and Duggan, W. T. (1997). Can a two-stage procedure enjoy second-order properties? *Sankhyā, Series A*, **59**, 435–448.

Mukhopadhyay, N. and Duggan, W. T. (1999). On a two-stage procedure having second-order properties with applications, *Annals of the Institute of Statistical Mathematics*, **51**, in press.

Mukhopadhyay, N. and Solanky, T. K. S. (1994). *Multistage Selection and Ranking Procedures*, Marcel Dekker, New York.

Nagao, H. (1996). On fixed width confidence regions for multivariate normal mean when the covariance matrix has some structure, *Sequential Analysis*, **15**, 37–46. Correction: *Sequential Analysis*, **17**, 125–126.

Panchapakesan, S. (1995). Selection and ranking procedures, In *The Exponential Distribution: Theory, Methods and Applications* (Eds., N. Balakrishnan and A. P. Basu), pp. 259–278. Gordon and Breach, Amsterdam.

Scheffe, H. and Tukey, J. W. (1944). A formula for sample sizes for population tolerance limits, *Annals of Mathematical Statistics*, **15**, 217.

Stein, C. (1945). A two sample test for a linear hypothesis whose power is independent of the variance, *Annals of Mathematical Statistics*, **16**, 243–258.

Stein, C. (1949). Some problems in sequential estimation (abstract), *Econometrica*, **17**, 77–78.

Wang, Y.H. (1980). Sequential estimation of the mean of a multinormal population, *Journal of the American Statistical Association*, **75**, 977–983.

Woodroofe, M. (1977). Second order approximations for sequential point and interval estimation, *Annals of Statistics*, **5**, 984–995.

Woodroofe, M. (1982). *Nonlinear Renewal Theory in Sequential Analysis*, CBMS Lecture Notes No. 39, SIAM Publications, Philadelphia.

CHAPTER 19

BOUNDED RISK POINT ESTIMATION OF A LINEAR FUNCTION OF K MULTINORMAL MEAN VECTORS WHEN COVARIANCE MATRICES ARE UNKNOWN

MAKOTO AOSHIMA

University of Tsukuba, Ibaraki, Japan

YOSHIKAZU TAKADA

Kumamoto University, Kumamoto, Japan

Abstract: The problem of constructing an estimator with a risk bounded by a preassigned number is considered for a linear function of mean vectors of k multinormal distributions when covariance matrices are fully unknown. We provide a new two–stage procedure which does improve that in Aoshima (1998a). The procedure is shown to be asymptotically efficient.

Keywords and phrases: Asymptotic efficiency, bounded risk, exact consistency, two–stage procedure

19.1 INTRODUCTION

Suppose that there exist k independent p–variate normal populations π_i : $N_p(\boldsymbol{\mu}_i, \boldsymbol{\Sigma}_i)$, $i = 1, ..., k$, where all the parameters are unknown. Let $\boldsymbol{\mu} =$

279

$\sum_{i=1}^{k} b_i \boldsymbol{\mu}_i$, where b_i's are known and nonzero scalars. Let $\boldsymbol{X}_{i1}, ..., \boldsymbol{X}_{in_i}$, be random sample vectors of size n_i from π_i, $i = 1, ..., k$, and let

$$T_{\mathbf{n}} = \sum_{i=1}^{k} b_i \overline{\boldsymbol{X}}_{i(n_i)},$$

where $\boldsymbol{n} = (n_1, ..., n_k)$ and $\overline{\boldsymbol{X}}_{i(n_i)} = \sum_{j=1}^{n_i} \boldsymbol{X}_{ij}/n_i$, $i = 1, ..., k$. For fixed $\boldsymbol{n}$ suppose that the loss incurred in estimating $\boldsymbol{\mu}$ by $T_{\mathbf{n}}$ is given by

$$L_{\mathbf{n}} = \|T_{\mathbf{n}} - \boldsymbol{\mu}\|^2. \tag{19.1.1}$$

The problem is to construct an estimator $T_{\mathbf{n}}$ of $\boldsymbol{\mu}$ such that the risk associated with (19.1.1) is bounded by some preassigned number W (> 0), i.e.

$$E(L_{\mathbf{n}}) \leq W \tag{19.1.2}$$

for all $(\boldsymbol{\mu}_i, \boldsymbol{\Sigma}_i)$, $i = 1, ..., k$.

If $\boldsymbol{\Sigma}_i, i = 1, ..., k$, were known, the optimal–fixed sample size would be given by

$$n_i^{\star} = \frac{1}{W} |b_i| \sqrt{tr(\boldsymbol{\Sigma}_i)} \sum_{i=1}^{k} |b_i| \sqrt{tr(\boldsymbol{\Sigma}_i)} \quad (i = 1, ..., k), \tag{19.1.3}$$

as in Aoshima (1998a). That gives $n^{\star} = \sum_{i=1}^{k} n_i^{\star}$, the minimum total sample size required to satisfy the requirement (19.1.2). In the present problem with $\boldsymbol{\Sigma}_i$'s unknown, there does not exist any fixed–sample size procedure to achieve the requirement (19.1.2) [see Takada (1988)]. Aoshima (1998a) gave a solution by using a two–stage procedure as follows: First take the initial samples of size m ($> \max(3, p)$) from π_i's, and compute the sample covariance matrices $\boldsymbol{S}_{im}$, $i = 1, ..., k$, as

$$\boldsymbol{S}_{im} = \sum_{j=1}^{m} \left(\boldsymbol{X}_{ij} - \overline{\boldsymbol{X}}_{i(m)} \right) \left(\boldsymbol{X}_{ij} - \overline{\boldsymbol{X}}_{i(m)} \right)' / \nu, \tag{19.1.4}$$

where $\nu = m - 1$. Let ℓ_{im} denote the maximum latent root of $\boldsymbol{S}_{im}$ for each $i = 1, ..., k$. Let $B = E\left(\nu / \min_{1 \leq i \leq k} \chi_{\nu(i)}^2 \right)$, where $\chi_{\nu(i)}^2$, $i = 1, ..., k$, are i.i.d. chi–square random variables with ν d.f. The sample size of the two–stage procedure is defined by

$$N_i = \max \left\{ m, \left[\frac{pB}{W} |b_i| \sqrt{\ell_{im}} \sum_{i=1}^{k} |b_i| \sqrt{\ell_{im}} \right] + 1 \right\} \quad (i = 1, ..., k). \tag{19.1.5}$$

Let $N = (N_1, ..., N_k)$. Secondly, take the additional samples of size $N_i - m$ from π_i, $i = 1, ..., k$, and construct $T_N = \sum_{i=1}^{k} b_i \overline{X}_{i(N_i)}$ for estimating μ. Then, the estimator T_N satisfies the requirement (19.1.2). Aoshima (1998a) also showed that the two–stage procedure based on (19.1.5) is more efficient than the Ghosh, Mukhopadhyay and Sen (1997)–type two–stage procedure asymptotically.

In this paper, we provide a new two–stage procedure satisfying the requirement (19.1.2) exactly for general Σ_i, $i = 1, ..., k$, for which the sample size is smaller than that in (19.1.5) w.p.1 for every π_i. The stopping rule proposed here does mimic the expression (19.1.3). The procedure is shown not only to improve Aoshima's (1998a) two–stage procedure in terms of the sample size but also to be asymptotically efficient, along with some other asymptotic properties when $W \to 0$.

19.2   TWO–STAGE PROCEDURE

We shall consider the estimation of n_i^* $(i = 1, ..., k)$ which is obtained by replacing Σ_i in (19.1.3) with S_{im} and denote

$$R_i = \max \left\{ m, \ \left[\frac{1}{z} |b_i| \sqrt{tr(S_{im})} \sum_{i=1}^{k} |b_i| \sqrt{tr(S_{im})} \right] + 1 \right\} \quad (i = 1, ..., k).$$

(19.2.6)

Let $R = (R_1, ..., R_k)$, where $z = z(m, k, W)$ is a positive constant such that T_R satisfies the requirement (19.1.2). When we adopt R instead of N based on (19.1.5) in the two–stage procedure, the determination of the constant z becomes very crucial.

We propose a two–stage procedure based on (19.2.6) as follows: First, the intial samples of size m (> 3) are taken from π_i's and S_{im}, $i = 1, ..., k$, are computed as (19.1.4). Let $z = W(\nu - 2)/\nu$. The sample size of the two–stage procedure is defined by (19.2.6) for each π_i. The additional samples of size $R_i - m$ from π_i, $i = 1, ..., k$, are taken and $T_R = \sum_{i=1}^{k} b_i \overline{X}_{i(R_i)}$ is constructed.

The following theorem verifies that the estimator T_R is a solution to the problem.

Theorem 19.2.1 *The two–stage procedure based on (19.2.6) with $z = W(\nu -2)/\nu$ satisfies the requirement (19.1.2) for all $(\mu_i, \Sigma_i), i = 1, ..., k$.*

PROOF By noting that $I(R = n)$ is independent of T_n for every fixed n, we have that

$$E\left(L_{\mathbf{R}}\right) = E\left(\sum_{i=1}^{k} \frac{b_i^2}{R_i} tr\left(\mathbf{\Sigma}_i\right)\right)$$

$$\leq zE\left(\frac{\sum_{i=1}^{k} |b_i|\sqrt{tr\left(\mathbf{\Sigma}_i\right)/V_{im}}}{\sum_{i=1}^{k} |b_i|\sqrt{tr\left(\mathbf{\Sigma}_i\right)V_{im}}}\right), \tag{19.2.7}$$

where $V_{im} = tr\left(\mathbf{S}_{im}\right)/tr\left(\mathbf{\Sigma}_i\right)$, $i = 1,...,k$. Let

$$c_i = \frac{|b_i|\sqrt{tr\left(\mathbf{\Sigma}_i\right)V_{im}}}{\sum_{i=1}^{k} |b_i|\sqrt{tr\left(\mathbf{\Sigma}_i\right)V_{im}}}, \quad i = 1,...,k.$$

Then, we write

$$E\left(L_{\mathbf{R}}\right) \leq z \sum_{i=1}^{k} E\left(c_i V_{im}^{-1}\right). \tag{19.2.8}$$

But, note that

$$E\left(c_i V_{im}^{-1}\right) = Cov(c_i, V_{im}^{-1}) + E(c_i)E\left(V_{im}^{-1}\right)$$

$$\leq E(c_i)E\left(V_{im}^{-1}\right)$$

$$\leq \frac{\nu}{\nu-2}E(c_i), \tag{19.2.9}$$

since $Cov\left(c_i, V_{im}^{-1}\right) \leq 0$, and $E\left(V_{im}^{-1}\right) \leq \nu/(\nu-2)$ from Kubokawa (1990). Combining (19.2.9) with (19.2.8), we conclude the result since $\sum_{i=1}^{k} c_i = 1$ w.p.1. $\qquad\square$

We observe that $R_i \leq N_i$ $(i = 1,...,k)$ w.p.1 since $tr(\mathbf{S}_{im}) \leq p\ell_{im}$ w.p.1 and $\nu/(\nu-2) \leq B$, where the equality holds only when $(p,k) = (1,1)$. Also, note that the constant z does not depend on k contrary to B. From the Table in Aoshima (1998a), we observe that the larger (p,k) are, the more efficient the two–stage procedure based on (19.2.6) with $z = W(\nu-2)/\nu$ is than based on (19.1.5). Hence, we conclude that the estimator $\mathbf{T_R}$ improves upon $\mathbf{T_N}$.

19.3   ASYMPTOTIC PROPERTIES

Under the condition that $m = m(W)$:

$$m(W) \to \infty, \quad Wm(W) \to 0 \quad \text{as} \quad W \to 0, \tag{19.3.10}$$

we study asymptotic properties of the two–stage procedure proposed in Section 19.2. First we have the following first–order asymptotics, especially in which part (b) says that the proposed procedure is asymptotically efficient.

Theorem 19.3.1 *For the two–stage procedure based on (19.2.6) with $z = W(\nu - 2)/\nu$, we have as $W \to 0$:*

(a) $R_i/n_i^* \to 1$ $(i = 1, ..., k)$ *in probability;*

(b) $E\left(\sum_{i=1}^k R_i/n^\star\right) \to 1;$

(c) $E(L_{\mathbf{R}}/W) \to 1$ *for all* $(\boldsymbol{\mu}_i, \boldsymbol{\Sigma}_i)$, $i = 1, ..., k$,

where n_i^* $(i = 1, ..., k)$ is the same as in (19.1.3) and $n^\star = \sum_{i=1}^k n_i^*$.

PROOF   Let us write the basic inequality from (19.2.6) that

$$\frac{1}{z}|b_i|\sqrt{tr\,(S_{im})} \sum_{i=1}^k |b_i|\sqrt{tr\,(S_{im})} \le R_i$$

$$\le m + \frac{1}{z}|b_i|\sqrt{tr\,(S_{im})} \sum_{i=1}^k |b_i|\sqrt{tr\,(S_{im})} \quad (19.3.11)$$

for each π_i. From (19.3.11) we have

$$\frac{\nu\sqrt{tr\,(S_{im})} \sum_{i=1}^k |b_i|\sqrt{tr\,(S_{im})}}{(\nu - 2)\sqrt{tr\,(\Sigma_i)} \sum_{i=1}^k |b_i|\sqrt{tr\,(\Sigma_i)}} \le \frac{R_i}{n_i^*}$$

$$\le \frac{m}{n_i^*} + \frac{\nu\sqrt{tr\,(S_{im})} \sum_{i=1}^k |b_i|\sqrt{tr\,(S_{im})}}{(\nu - 2)\sqrt{tr\,(\Sigma_i)} \sum_{i=1}^k |b_i|\sqrt{tr\,(\Sigma_i)}}$$

for $i = 1, ..., k$. Then part (a) follows since $tr\,(S_{im}) \to tr\,(\Sigma_i)$ w.p.1 as $W \to 0$ and $mn_i^{*-1} \to 0$ as $W \to 0$ under (19.3.10).

From (19.3.11) we also have

$$\frac{1}{z}\left(\sum_{i=1}^k |b_i|\sqrt{tr\,(S_{im})}\right)^2 \le \sum_{i=1}^k R_i \le km + \frac{1}{z}\left(\sum_{i=1}^k |b_i|\sqrt{tr\,(S_{im})}\right)^2,$$

$$(19.3.12)$$

which implies that

$$\frac{\nu}{\nu - 2}\left(\frac{\sum_{i=1}^k |b_i|\sqrt{tr\,(S_{im})}}{\sum_{i=1}^k |b_i|\sqrt{tr\,(\Sigma_i)}}\right)^2 \le \frac{\sum_{i=1}^k R_i}{n^\star}$$

$$\le \frac{km}{n^\star} + \frac{\nu}{\nu - 2}\left(\frac{\sum_{i=1}^k |b_i|\sqrt{tr\,(S_{im})}}{\sum_{i=1}^k |b_i|\sqrt{tr\,(\Sigma_i)}}\right)^2. \quad (19.3.13)$$

Taking expectations throughout (19.3.13), we obtain part (b) when $W \to 0$.
We write

$$L_{\mathbf{R}}/W = \sum_{i=1}^{k} \frac{n_i^{\star} |b_i| \sqrt{tr(\Sigma_i)}}{R_i \sum_{i=1}^{k} |b_i| \sqrt{tr(\Sigma_i)}} \ .$$

From part (a) we have $L_{\mathbf{R}}/W \to 1$ in probability as $W \to 0$. So, in order to have part (c), it is enough to show that $\{L_{\mathbf{R}}/W\}$ is uniformly integrable for small $W > 0$. From (19.2.7) and with $z = W(\nu - 2)/\nu$, we note that

$$L_{\mathbf{R}}/W \leq \frac{\nu - 2}{\nu} \frac{\sum_{i=1}^{k} |b_i| \sqrt{tr(\Sigma_i)/V_{im}}}{\sum_{i=1}^{k} |b_i| \sqrt{tr(\Sigma_i) V_{im}}}$$

$$\leq \frac{\nu - 2}{\nu} \max_{1 \leq i \leq k} V_{im}^{-1}$$

$$\leq \frac{\nu - 2}{\nu} \sum_{i=1}^{k} V_{im}^{-1}.$$

Then the following Lemma 19.3.1 shows that $\{L_{\mathbf{R}}/W\}$ is uniformly integrable. $\square$

Lemma 19.3.1 Let $V_{im} = tr(S_{im})/tr(\Sigma_i)$, $i = 1, ..., k$. Then, if $m \geq m_0 > 5$, $\{V_{im}^{-1} : m \geq m_0\}$ is uniformly integrable for each $i = 1, ..., k$.

PROOF Let $\lambda_{i(j)}$, $j = 1, ..., p$, denote the latent roots of Σ_i for $i = 1, ..., k$. We can write from Aoshima (1998a) that

$$V_{im} = \frac{1}{\nu} \sum_{j=1}^{p} \tau_{i(j)} \chi_{\nu(j)}^2, \quad i = 1, ..., k, \tag{19.3.14}$$

where $\tau_{i(j)} = \lambda_{i(j)}/\sum_{j=1}^{p} \lambda_{i(j)}$, $j = 1, ..., p$, and $\chi_{\nu(j)}^2$, $j = 1, ..., p$, denote i.i.d. chi–square random variables with $\nu = m - 1$ d.f. Note that $\sum_{j=1}^{p} \tau_{i(j)} = 1$ and $\tau_{i(j)} \in (0, 1)$ for $j = 1, ..., p$. Then we have

$$E\left(V_{im}^{-2}\right) = \nu^2 E\left\{ \left(\sum_{j=1}^{p} \tau_{i(j)} \chi_{\nu(j)}^2 \right)^{-2} \right\}$$

$$\leq \nu^2 \max_{1 \leq j \leq p} E\left\{ \left(\chi_{\nu(j)}^2 \right)^{-2} \right\}$$

$$= \frac{\nu^2}{(\nu - 2)(\nu - 4)}, \tag{19.3.15}$$

where the inequality (19.3.15) holds in a way similar to Kubokawa (1990), by using the convexity of $f(x) = x^{-2}$, $x > 0$. Hence, we obtain

$$\sup_{m \geq m_0} E\left(V_{im}^{-2}\right) < \infty,$$

if $m \geq m_0 > 5$. The proof is complete. □

For the regret of the two–stage procedure based on (19.2.6), we have the following theorem.

Theorem 19.3.2 *Under the condition (19.3.10), the asymptotic regret of (19.2.6) with $z = W(\nu - 2)/\nu$ is infinity, that is,*

$$\lim_{W \to 0} \left\{ E\left(\sum_{i=1}^{k} R_i\right) - n^\star \right\} = \infty,$$

where $n^\star = \left(\sum_{i=1}^{k} |b_i| \sqrt{tr\left(\Sigma_i\right)}\right)^2 / W$.

PROOF From the left hand side of the inequality (19.3.12), we have

$$E\left(\sum_{i=1}^{k} R_i\right) > \frac{1}{z} E\left\{ \left(\sum_{i=1}^{k} |b_i| \sqrt{tr\left(S_{im}\right)}\right)^2 \right\}$$

$$= \frac{1}{z} \left\{ \sum_{i=1}^{k} b_i^2 \, tr\left(\Sigma_i\right) + \right.$$

$$\left. \sum_{i \neq j} |b_i||b_j| E\left(\sqrt{tr\left(S_{im}\right)}\right) E\left(\sqrt{tr\left(S_{jm}\right)}\right) \right\}.$$

$$(19.3.16)$$

From (19.3.14), it is easy to show that

$$E\left(V_{im}^{1/2}\right) = 1 - \frac{1}{4\nu} \sum_{j=1}^{p} \tau_{i(j)}^2 + O\left(\nu^{-2}\right),$$

that is

$$E\left(\sqrt{tr\left(S_{im}\right)}\right) = \sqrt{tr\left(\Sigma_i\right)} \left\{ 1 - \frac{1}{4\nu} \frac{tr\left(\Sigma_i^2\right)}{\left(tr\left(\Sigma_i\right)\right)^2} + O\left(\nu^{-2}\right) \right\}. \quad (19.3.17)$$

Combining (19.3.17) with (19.3.16) and noting that $z^{-1} = W^{-1}(1 + 2/\nu + O(\nu^{-2}))$,

$$
E\left(\sum_{i=1}^{k} R_i\right) \geq W^{-1}\left(1 + \frac{2}{\nu} + O(\nu^{-2})\right)\left\{\left(\sum_{i=1}^{k} |b_i|\sqrt{tr(\Sigma_i)}\right)^2\right.
$$

$$
\left. - \frac{1}{2\nu}\sum_{i\neq j} |b_i||b_j|\sqrt{tr(\Sigma_i)}\frac{tr(\Sigma_j^2)}{(tr(\Sigma_j))^{3/2}} + O(\nu^{-2})\right\}
$$

$$
\geq n^* + W^{-1}\left\{\frac{2}{\nu}\left(\sum_{i=1}^{k} |b_i|\sqrt{tr(\Sigma_i)}\right)^2\right.
$$

$$
\left. - \frac{1}{2\nu}\sum_{i\neq j} |b_i||b_j|\sqrt{tr(\Sigma_i)}\sqrt{tr(\Sigma_j)} + O(\nu^{-2})\right\}
$$

$$
= n^* + W^{-1} \times
$$

$$
\left\{\frac{1}{2\nu}\left(3\left(\sum_{i=1}^{k} |b_i|\sqrt{tr(\Sigma_i)}\right)^2 + \sum_{i=1}^{k} b_i^2 tr(\Sigma_i)\right) + O(\nu^{-2})\right\}.
$$

$$
(19.3.18)
$$

The inequality (19.3.18) holds because of $tr(\Sigma_j^2) \leq (tr(\Sigma_j))^2$. The proof is complete.   □

Remark A suitable modification of the condition (19.3.10) would enable us to study the second–order asymptotic properties of (19.2.6) along with its bounded regret. Aoshima (1998b) discussed it for the case when $k = 1$ by using the techniques given in Mukhopadhyay (1999) and Aoshima and Mukhopadhyay (1999).

Acknowledgements The research of the first author was supported by Grant-in-Aid for Encouragement of Young Scientists, the Ministry of Education, Science, Sports and Culture, Japan, under Contract Number 721-8035-09780215.

REFERENCES

Aoshima, M. (1998a). Bounded risk point estimation of a linear function of k multinormal mean vectors, *Journal of the Japan Statistical Society*, **28**, 153–161.

Aoshima, M. (1998b). Second–order properties of improved two–stage procedures for a multivariate normal distribution, submitted.

Aoshima, M. and Mukhopadhyay, N. (1999). Second–order properties of a two–stage fixed–size confidence region when the covariance matrix has a structure, *Communications in Statistics—Theory and Methods (Sugiura Festschrift Volume)*, **28**, 839–855.

Ghosh, M., Mukhopadhyay, N. and Sen, P. K. (1997). *Sequential Estimation,* John Wiley & Sons, New York.

Kubokawa, T. (1990). Two–stage estimators with bounded risk in a growth curve model, *Journal of the Japan Statistical Society*, **20**, 77–87.

Mukhopadhyay, N. (1999). Second–order properties of a two–stage fixed–size confidence region for the mean vector of a multivariate normal distribution, *Journal of Multivariate Analysis*, **68**, 250–263.

Takada, Y. (1988). Two–stage procedures for a multivariate normal distribution, *Kumamoto Journal of Mathematics*, **1**, 1–8.

CHAPTER 20

THE ELUSIVE AND ILLUSORY MULTIVARIATE NORMALITY

GOVIND S. MUDHOLKAR

University of Rochester, Rochester, NY

DEO KUMAR SRIVASTAVA

St. Jude Children's Research Hospital, Memphis, TN

Abstract: The assumption of multivariate normality is intrinsic in most statistical methods used for analyzing multidimensional data. The normality assumption is, in general, dubious even in the univariate case. In multivariate setting the assumption is harder to justify since it implies joint normality in addition to marginal normality. Unlike the assumption of univariate normality, tests for which are a part of most of the software packages, the multivariate normality is rarely verified in practice because of paucity of simple and readily available procedures. In this paper we first survey the literature on testing multivariate normality and then variously examine the appropriateness of normality assumptions for some well known data sets, such as Rao's Cork data, Fisher's Iris data, used for illustrative purposes in such leading multivariate analysis monographs as Seber (1984), Anderson (1984) and Mardia (1994). It is seen that, under such scrutiny, the support for multivariate normality for these data is at best equivocal. Hence, in general, such an assumption may be inappropriate and may essentially be illusory. The results emphasize the need for robust multivariate methods.

Keywords and phrases: Multivariate analysis, normal theory, robust alternatives

20.1 INTRODUCTION

Since the earliest days of multivariate analysis, beginning with the introduction of Mahalanobis' D^2 or equivalently Hotelling's T^2 tests, Fisher's discriminant analysis, Rao's classification methods included in the monographs such as Kendall (1961), Rao (1962) and Anderson (1984), multivariate normality has been the canonical underlying assumption. Although there exists considerable literature on nonnormal, principally nonparametric, methods for analyzing multivariate data normal theory methods remain the mainstay of the practice. However, as reflected in Geary's (1947) provocative comment, "*Normality is a myth; there never was, and never will be, a normal distribution*", the assumption of normality has always been considered suspect in univariate setting. The assumption of multivariate normality is more dubious since it requires joint normality in addition to marginal normality of the components. It has also been established by several recent studies, e.g. see Chase and Bulgren (1971), Mardia (1975), that if the assumption of multivariate normality is inappropriate then Hotelling's T^2 is likely to be either too conservative or invalid. The same may be expected to be true of other multivariate procedures. The purpose of this paper is to examine multivariate normality of some well known data sets which are commonly used to illustrate normal theory methods.

There exist many exploratory and confirmatory methods for assessing normality of univariate data. These include graphical methods such as q-q plots, p-p plots and a variety of formal tests of the simple and composite hypothesis of normality; see D'Agostino and Stephens (1986). Many of the formal tests of normality including the Shapiro-Wilk test are now part of commonly used statistical software packages such as SAS, S-plus, etc. Many analogous procedures for assessing multivariate normality which are part of the literature require some familiarity with multivariate theory and few of them are incorporated into standard software packages. For an account of implementing some of the procedures using FORTRAN programs and SAS procedures see Looney (1995). Romeu and Ozturk (1993) have compared ten tests for assessing the assumption of multivariate normality and have provided a ranking of the procedures under various types of alternatives. They also highlight the fact that the convergence to the asymptotics of several tests of multivariate normality is very slow and for drawing valid conclusions often requires sample sizes to be as large as 200.

In Section 20.2, we first summarize and present a relatively up to date account of the proposals in the literature for testing the assumption of multivariate normality. A list of some of the famous data sets along with the results of applying some of the tests of Section 20.2 are presented in Section

20.3. Finally in Section 20.4 we present the conclusions and miscellaneous remarks.

20.2 TESTS OF MULTIVARIATE NORMALITY

The tests of multivariate normality can be broadly classified into four classes (i) tests based on Mahalanobis distances (ii) tests based on multivariate measures of skewness and kurtosis (iii) tests based on generalizing the univariate test procedures for multivariate setting, and (iv) tests based on transformations and goodness-of-fit.

Tests based on Mahalanobis distance

The basic role of Mahalanobis' D^2 in testing multivariate normality is to make it a univariate problem. If $X_1, X_2, ..., X_n$ is a random sample from a p-variate population then

$$D_i^2 = (X_i - \overline{X})'S^{-1}(X_i - \overline{X}), \qquad \text{for} \qquad i = 1, 2, ..., n$$

where $\overline{X}$ and S represent the sample mean vector and sample covariance matrix, respectively, are the Mahalanobis' distances. The distances D_i^2's are well known to be asymptotically independent χ_p^2 variates, e.g. see Mardia (1977). Healy (1968) proposes plotting, a version of q-q plot, $(D_i^{2/3})$ against their expected order statistics to assess multivariate normality of the data X. Further, Small (1978) uses the fact that $u_i = nD_i^2/(n-1)^2$ is a beta variate, B[a,b], where $a = d/2$ and $b = (n - d - 1)/2$, and proposes plotting the order statistics $u_{(i)}$ against u_i^*, where u_i^* is the solution of

$$\int_0^{u_i} \frac{\Gamma(\alpha)\Gamma(\beta)}{\Gamma(\alpha + \beta)} \ y^{\alpha-1}(1 - y)^{\beta-1}dy \quad = \quad \frac{i - \alpha}{n - \alpha - \beta + 1} \ .$$

Alternatively, Mudholkar, McDermott and Srivastava (1992) use the approximate normality of a refinement $(D_i^2)^h$, where $h = 1/3 + .11/p$, to develop the multivariate version Z_p of the Lin and Mudholkar's (1980) z-test based upon the characteristic independence of the sample mean and variance of a normal population. Specifically, they propose the test statistic $T_{n,p} = (Z_{n,p} - \mu_{n,p})/\sigma_{n,p} \sim N(0,1)$, where $\mu_{n,p} = A_1(p)/n - A_2(p)/n^2$, $A_1(p) = -1.0/p - .52p$ and $A_2(p) = .8p^2$ and $\sigma_{n,p} = B_1(p)/n - B_2(p)/n^2$, $B_1(p) = (3 - 1.67/p + .51/p^2)$, $B_2(p) = 1.8p - 9.75/p^2$, and reject the hypothesis of p-variate normality for large values of $| T_{n,p} |$. They further examine the finite sample behavior of the statistic empirically through an extensive simulation study.

Tests based on skewness and kurtosis

The coefficients skewness and kurtosis of multivariate distributions have been defined in several different ways but $\beta_{1,p}$ and $\beta_{2,p}$ measures defined by Mardia (1970, 1974, 1975, 1978) are probably most widely used and accepted. Mardia (1974) shows that the sample analogs $b_{1,p}$ and $b_{2,p}$ are distributed approximately as,

$$\frac{1}{6}n\,b_{1,p} \sim \chi_f^2 \quad \text{and} \quad \{b_{2,p} - p(p+2)\}/\{8p(p+2)/n\}^{1/2} \sim N(0,1)$$

where $f = \frac{1}{6}p(p+1)(p+2)$, and uses them to develop test of multivariate normality. Furthermore, Mardia and Foster (1983) combine the measures of skewness and kurtosis in various ways to construct omnibus tests of multivariate normality. After examining various alternatives using simulation methods they recommend using their S_W^2 or C_W^2 based on the Wilson-Hilferty transformation of a χ^2 variate, while showing a preference for C_W^2. Horswell and Looney (1992) note a necessity of using empirical rather than asymptotic critical values of the test statistics in order to obtain adequate type I error control and better power properties. Horswell and Looney (1993) also provide an account of the limitations of skewness coefficients in assessing the departures from univariate and multivariate normality.

Small (1980) considers the vector of sample values of marginal coefficients of skewness ($\mathbf{b_1}$) and kurtosis ($\mathbf{b_2}$), with covariance matrices Ω_1 and Ω_2, respectively, and uses them directly to test multivariate normality. Bowman and Shenton (1975), however, propose a normalizing transformation,

$$W_1 = \delta_1 \sinh^{-1}(\frac{\mathbf{b_1}}{\lambda_1}), \quad W_2 = \gamma_2 \mathbf{1_p} + \delta_2 \sinh^{-1}(\frac{[\mathbf{b_2} - \zeta \mathbf{1_p}]}{\lambda_2})$$

and the test statistics

$$Q_i = W_i' \Lambda_i^{-1} W_i \qquad \text{for} \qquad i = 1, 2,$$

where the diagonal elements of Λ_1 and Λ_2 are unity and the (j,k)th off-diagonal elements are the third and fourth powers of r_{jk}, the sample correlation coefficient between the j-th and k-th components. They approximate the null distribution of Q_i's by a χ_d^2 variate for samples of size 29 or more and $2 \le p \le 8$.

Malkovich and Afifi (1973) use the fact that $\mathbf{Y}_{(p \times 1)}$ is distributed as $N(\mu, \Sigma)$ if and only if $\mathbf{C'Y}$ is distributed as $N(\mathbf{C'}\mu, \mathbf{C'\Sigma C})$ for all constant vectors $\mathbf{C}$. Then they define the multivariate measures of skewness and kurtosis by maximizing the coefficients of skewness and kurtosis for $\mathbf{C'Y}$ for all values of $\mathbf{C}$.

Tests based on adaptation of the univariate procedures

Royston (1983) extends Shapiro-Wilk W test for testing univariate normality to the multivariate setting and proposes the test statistic $G = \Sigma_1^p k_i /p$, where $k_i = \{\Phi^{-1} [\frac{1}{2}\Phi(-z_i)]\}^2, i = 1, 2, ..., p$, $z = ((1 - W)^\lambda - \mu)/\sigma$, and Φ is a standard normal c.d.f. Its null distribution is adequately approximated by a scaled χ^2 variate, although, a bit conservative for the bivariate and trivariate populations.

Srivastava (1984) applies the principal component analysis for transforming the p-dimensional normal data into p independent normal samples and suggests using the normal probability plots or Shapiro-Wilk and Vasicek tests for assessing p variate normality. Srivastava and Hui (1987) propose two statistics M_1 and M_2 as generalizations of Shapiro-Wilk statistic, based respectively on Fisher and Tippett combination of independent p-values, for assessing multivariate normality. They find the type I error control of the tests to be slightly inflated even for sample of size 50 and indicate a preference for M_1 over M_2.

Stepwise procedure: Mudholkar, Srivastava and Lin (1995) use an orthogonal Helmert's matrix to decompose the problem of testing the hypothesis of multivariate normality into p independent problems of testing univariate normality. Then, they combine the p-values of the Shapiro-Wilk test applied to the component problems using one of the combination methods such as Fisher, Logit, Liptak and Tippett and propose the statistics W_L^*, W_L^*, W_N^*, and W_T^* , respectively. They also suggest a simplified recursive approach based on stepwise residuals and propose the statistics W_F, W_L, W_N and W_T , corresponding to Fisher, Logit, Liptak and Tippett method of combining p-values, respectively.. They note through an extensive simulation study that the type I error control for the exact tests W^*'s and their simplified versions W's are essentially equivalent and satisfactory, even for samples of size 20, and recommend using the statistics W's for testing the hypothesis of multinormality. However, it may be noted that the combination statistics W's require an ordering among the response variables which, in practice, may be available a priori. In the absence of availability of such an ordering the tests may be used in a data analytic fashion. They further conclude that among the combination statistics the one based upon Fisher's method of combination is generally superior in terms of power, Logit method being a close second. The two combinations based on Liptak and Tippett methods often have considerably lower power. They recommend the test based on Fisher combination statistic.

Tests based on transformations and goodness-of-fit

The idea is basically to transform the multivariate problem into a single univariate problem and exploit the properties of the test procedures well established in the univariate setting. Rincon-Gallardo, Quesenberry and O'Reilly (1979) use the probability integral transform and obtain a set of i.i.d. uniform random variables. Then they test the hypothesis of multivariate normality by testing the simple hypothesis that the transformed variables are from U(0,1). Hasofer and Stein (1990) apply the Gram-Schmidt orthogonalization procedure to the data from a p-variate normal population and obtain p mutually uncorrelated vectors with zero means and unit variances. Then to test the p-variate normality they test the univariate normality of all the transformed vectors and their independence.

20.3 DUBIOUS NORMALITY OF SOME WELL KNOWN DATA

We examined a large number of multivariate data sets in the literature using the battery of tests for multivariate normality described in the previous section. Not surprisingly, for every data set different tests showed varying, sometimes very different, degrees of agreement with the assumption of multivariate normality. The following discussion involving a selection of famous data represents a general picture. A summary of the results appears in Table 20.1.

TABLE 20.1   Validity of multivariate normal model with respect to the above tests

Data Set	Range of p-values	
	Minimum	Maximum
Cork Data	.006	.8439
Iris Setosa	.0003	.1142
Iris Versicolor	.0042	.6887
Iris Virginica	.0310	.4106
Hematology Data	.00003	.08
Peruivan Indian Data	<.00001	.0933

Cork data set [Rao (1948)]

This well known data set, appearing in Rao (1948), was used for illustrating Hotelling's T^2 test, also see Mardia (1994), Seber (1984). The weights of bark deposits of 28 trees in the four directions north (N), east (E), south (S) and west (W) were measured and the interest was in testing the hypothesis $H_0 : \mu_1 = \mu_2 = \mu_3 = \mu_4$, that is, the average weight of bark deposits were same in all the directions. Rao (1973) was particularly interested in testing the significance of the contrasts

$$U_1 = (N + S) - (E + W), \quad U_2 = N - S, \quad U_3 = E - W$$

using Hotelling's T^2 test. Hotelling's T^2 test would be justified under the assumption of four variate normality.

Pearson (1956) notes, and Seber (1984) concurs, that the weights have asymmetrical distributions. However, Mardia (1975) finds the values of $b_{1,4} = 4.476$ and $b_{2,4} = 22.957$ not significant and does not contain enough evidence against the assumption of four-variate normality. The value of his omnibus test $S_W^2 = .3393$ with a p-value of .8439 also did not caste any doubt on the assumption of four variate normality. Srivastava and Hui (1987) further examined this data and for their statistics obtain $M_1 = 19.974$ and $M_2 = .8947$ which gives p-values of .01 and .037 suggesting significant nonnormality. The value of statistic $T_{n,p} = 1.8774$, proposed by Mudholkar, McDermott and Srivastava (1992), gives a p-value of .0302 and the application of combination statistics, due to Mudholkar, Srivastava, and Lin (1995), yields p-values as small as .006 again casting doubt on the assumption of four variate normality.

Iris data (Edgar Anderson's data [Fisher (1936)])

These are the best known classical data sets due to Edgar Anderson. They have been studied by many workers including Fisher (1936), Small (1980), Royston (1993), Rao (1973), and Anderson (1984) for illustrating various multivariate normal methods. The data consist of four measurements (Sepal Length, Sepal Width, Petal Length and Petal Width) on 50 plants for each of the 3 varieties of Iris (Setosa, Versicolor and Virginica). It would be natural to check if the assumption of multivariate normality is appropriate for these data. The data are reproduced in Anderson (1984) and are also available in S-PLUS statistical software package.

Iris (Setosa): Small (1980) assessed the assumption of multivariate normality using his statistics Q_1, Q_2, and $Q_1 + Q_2$ and concluded that the data indicated considerable skewness, particularly in the fourth component, and suggested a departure from the four-variate normality. Royston (1983) obtains a p-value of p<<<.001 corresponding to the statistic G and providing a strong evidence against normality. However, Mardia and Foster (1983) obtain $S_w^2 = 4.3404$, which with a p-value of .1142 provides no evidence against the four variate normality. Similarly, the $T_{n,p}$ statistic, due to Mudholkar, McDermott and Srivastava (1992), gives a p-value of .1464 providing insufficient evidence against multinormality. On the other hand, the combination statistic, due to Mudholkar, Srivastava and Lin (1995), yields highly significant p-values as small as .0003. This once again provides strong evidence against the assumption of multivariate normality. The fact that most of the nonnormality manifests in the fourth component, as observed by several other authors, is confirmed by looking at the p-values of the combination orders where X_4 variable was taken into consideration first.

Iris (Versicolor): The marginal plots of these data are reasonably normal and hence have not been used to illustrate the tests of multivariate normality. However, since they have been used in illustrating the multivariate methods it seemed interesting to see the results of several tests of multivariate normality applied to this data set. The value of $S_w^2 = .7459$ gives a p-value of .6887 providing strong evidence in favor of the four variate normality. The value of the $Z_p = -.3449$ gives a p-value of .6349 which gives strong indication of data being consistent with the assumption of multivariate normality. However, when we apply the combination statistics for the orders (X_4, X_1, X_2, X_3) and (X_4, X_3, X_2, X_1) we get p-values of .0042 and .0044, respectively. This once again indicates that the assumption or normality may not be appropriate and that most of the nonnormality may manifest in the fourth component only.

Iris (Virginica): After seeing that even Versicolor variety of Iris showed some signs of deviations from normality, it was of interest to check the assumption of multivariate normality for the variety Virginica. The p-value of .4106 corresponding to Mardia and Foster's (1980) S_w^2 test suggests that the assumption of four dimensional normality may be appropriate. But the value of $T_{50,4} = 1.8656$ gives a p-value of .0310 which indicates a moderate departure from the normality assumption. The p-value corresponding to the Fisher combination statistics for combination order (X_4, X_3, X_2, X_1) is .0478, providing marginal evidence against nonnormality.

We can, therefore, conclude that the assumption of multivariate normality may not be appropriate for any of the three varieties of Iris and it seems that most of the nonnormality may manifest in the third or fourth component, i.e. petal length or petal width, respectively. However, the violation of the assumption may be more prominent in varieties Setosa and Versicolor, whereas the variety Virginica shows only a mild deviation from normality.

Hematology data [Royston (1983)]

These data are taken from Chatterjee, Royston and Smith (1982), and are considered by Royston (1983). They consist of six variables, HAEMO (X_1) = haemoglobin concentration, PCV (X_2) = packed cell volume, WBC (X_3) = white blood count, LYMPHO (X_4) = lymphocyte count, NEUTRO (X_5) = neutrophil count, and LEAD (X_6) = serum lead concentration on 103 West Indian or African workers in a car assembly plant and are reproduced in Table 2 of Royston (1983). These data are used by Smith (1982) and Looney (1995) for illustrating test of multivariate normality. As in Royston, we transformed the variables WBC, LYMPHO, NEUTRO and LEAD to a logarithmic scale. The value of $T_{103,6} = 3.2686$ with a p-value of .0005 strongly suggests a deviation from six normality assumption. Further, examination using Fisher combination method for several combination orders gives p-values ranged from .00003, to .00051. The above p-values indicate a clear departure from the assumption of multivariate normality. Royston (1983), on the other hand, using G statistic gets a p-value of .08 and does not provide significant evidence against departure from normality. Looney (1995) also considers this data but he considers Box-Cox transformations of all the variables except PCV and applies the test statistics H due to Royston (1983), Q_1, Q_2, and $Q_3 = Q_1 + Q_2$ due to Small (1980), b_{1p} and b_{2p} due to Srivastava (1984) and M_1 and M_2 due to Srivastava and Hui (1987) and concludes that the original variables indicate strong departure from normality. For the transformed variables the statistics proposed by Srivastava and Srivastava and Hui indicate departures from multinormality.

Peruvian Indian data

These data, regarding the heights (in mm) and weights (in kg) of 39 Peruvian Indians, are available in Seber (1984, p. 64). The interest is in testing $H_0 : \mu = (\mathbf{63.64}, \mathbf{1615.38})' = \mu_0$. Under the assumption of bivariate normality the desired hypothesis can be tested by a straight forward application of Hotelling's T^2. Seber (1984) remarks that the probability plot of weights is somewhat S shaped and also there is an outlying observation. However, in the context of using Hotelling's T^2, it would be appropriate to check the assumption of bivariate normality. The p-value corresponding to $S_w^2 = 4.7442$ is .0933, which provides some evidence in favor of mild departure from bivariate normality. By applying Z_p statistic we get the value of $T_{39,2} = 1.7521$ which gives a p-values of .0399 indicating that the assumption of normality may not be appropriate. By applying the Fisher's combination statistic for the combination orders (X_1, X_2) and (X_2, X_1) we get the $W_F = 8.6704$ and 27.1450 with the corresponding p-values of .0699 and $<.0001$ providing strong evidence against the assumption of bivariate normality.

20.4　CONCLUSIONS

It is clear from the above illustrations that in practice multivariate normality may be too strong an assumption. Even if not rejected by one goodness of fit test it may be rejected by another, and should be accepted with some reservation. It is, therefore, important to understand the effects of nonnormality on the normal theory inference methods and to develop methods which would be robust with moderate size samples. As noted by Box (1953) in univariate case, and confirmed by Layard (1972) in the multivariate case, the normal theory methods for the variances and covariance matrices are even asymptotically nonrobust. The univariate studies show that the normal theory methods for testing the means are asymptotically robust and in the case of moderate size samples have reasonable validity robustness, i.e. type I error control, but they lack efficiency robustness, i.e. have substantially lower power in relation to robust alternatives; e.g. see Mudholkar, Mudholkar and Srivastava (1991) and Srivastava, Mudholkar and Mudholkar (1992). An examination of a multivariate case suggests that these findings as well hold in multivariate setting, that of a co-ordinatewise trimmed means alternative to Hotelling's T^2 considered by Mudholkar and Srivastava (1995), and Mudholkar and Srivastava (1996). Obviously much developmental work in the multivariate case lies ahead. In summary it is safe to conclude that in practical terms the multivariate normality is, in general, elusive and the assumption illusory.

Acknowledgements The research work of Deo Kumar Srivastava was in part supported by the Grant CA 21765 and the American Lebanese Syrian Associated Charities.

REFERENCES

Anderson, T. W. (1984). *An Introduction to Multivariate Statistical Analysis*, John Wiley & Sons, Inc., New York.

Bowman, K. O. and Shenton, L. R. (1975). Omnibus test contours for departures from normality based on $\sqrt{b_1}$ and b_2, *Biometrika*, **62**, 243–249.

Box, G. E. P. (1953). Non-normality and tests on variances, *Biometrika*, **40**, 318–335.

Chase, G. R. and Bulgren, G. (1971). A Monte Carlo investigation of the robustness of T^2, *Journal of the American Statistical Association*, **66**, 499–502.

Chatterjee, D. S., Royston, J. P. and Smith, H. (1982). A Healthy survey of paint sprayers in a car assembly plant, *British Journal of Internal Medicine*.

D'Agostino, R. B. and Stephens, M. A. (Eds.) (1986). Goodness-of-fit Techniques, *Marcel Dekker*, New York.

Fisher, R. A. (1936). The use of multiple measurements in taxonomic problems, *Annals of Eugenics*, **7**, 179–188.

Geary, R. C. (1947). Testing for normality, *Biometrika*, **34**, 209–242.

Hasofer, A. M. and Stein, G. Z. (1990). Testing for multivariate normality after coordinate transformation, *Communications in Statistics—Theory and Methods*, **19**, 1403–1418.

Healy, M. J. R. (1968). Multivariate normal plotting, *Applied Statistics*, **17**, 157–161.

Horswell, R. L. and Looney, S. W. (1992). A comparison of tests for multivariate normality that are based on measures of multivariate skewness and kurtosis, *Journal of Statistical Computation and Simulation*, **42**, 21–38.

Horswell, R. L. and Looney, S. W. (1993). Diagnostic limitations of skewness coefficients in assessing departures from univariate and multivariate normality, *Communications in Statistics—Theory and Methods*, **22**, 437–459.

Kendall, M. G. (1961). *A Course in Multivariate Analysis*, Hafner Publishing Company, New York.

Layard, M. W. J. (1972). Large sample tests for the equality of two covariance matrices, *Annals of Mathematical Statistics*, **43**, 123–141.

Lin, C. C. and Mudholkar, G. S. (1980). A simple test for normality against asymmetric alternatives, *Biometrika*, **67**, 455–461.

Looney, S. W. (1995). How to use test for univariate normality to assess nultivariate normality, *The American Statistician*, **49**, 64–70.

Malkovich, J. F. and Afifi, A. A. (1973). On tests for multivariate normality, *Journal of the American Statistical Association*, **68**, 176–179.

Mardia, K. V. (1970). Measures of multivariate skewness and kurtosis with applications, *Biometrika*, **57**, 519–530.

Mardia, K. V. (1974). Applications of some measures of multivariate skewness and kurtosis in testing normality and robustness studies, *Sankhyā, Series B*, **36**, 115–128.

Mardia, K. V. (1975). Assessment of multinormality and the robustness of Hotelling's T^2, *Applied Statistics*, **24**, 163–171.

Mardia, K. V. (1977). Mahalanobis distance and angles. In, *Multivariate Analysis-IV* (Ed., P. R. Krishnaiah), North Holland, Amsterdam.

Mardia, K. V. (1978). Some properties of classical multidimensional scaling, *Communications in Statistics—Theory and Methods*, **7**, 1233–1241.

Mardia, K. V. and Foster, K. (1983). Omnibus tests of multinormality based on skewnes and kurtosis, *Communications in Statistics—Theory and Methods* , **12**, 207–221.

Mudholkar, G. S., McDermott, D. and Srivastava, D. K. (1992). A test of p-variate normality, *Biometrika*, **79**, 850–854.

Mudholkar, A, Mudholkar, G. S. and Srivastava, D. K. (1991). A construction and appraisal of pooled trimmed-t statistics, *Communications in Statistics—Theory and Methods*, **20**, 1345–1359.

Mudholkar, G. S., Srivastava, D. K. and Lin, C. T. (1995). Some p-variate adaptations of the Shapiro-Wilk test of normality, *Communications in Statistics—Theory and Methods*, **24**, 953–985.

Mudholkar, G. S. and Srivasatva, D. K. (1995). Trimmed $\widetilde{T}^2$: A robust analog of Hotelling's T^2, *Technical Report* 95/05, Department of Biostatistics, University of Rochester, NY.

Mudholkar, G. S. and Srivastava, D. K. (1996). Robust analogs of Hotelling's two-sample T^2 *Submitted for publication*.

Pearson, E. S. (1956). Some aspects of the geometry of statistics, *Journal of the Royal Statistical Society*, **119**, 125–146.

Rao, C. R. (1948). Tests of significance in multivariate analysis, *Biometrika*, **35**, 58–79.

Rao, C. R. (1962). *Advanced Statistical Methods in Biometric Research*, John Wiley & Sons, New York.

Rao, C. R. (1973). *Linear Statistical Inference and its Applications*, John Wiley & Sons, New York.

Rincon-Gallardo, S., Quesenberry, C. P. and O'Reilly, F. J. (1979). Conditional probability integral transformations and goodness-of-fit tests for multivariate normal distributions, *The Annals of Statistics*, **5**, 1052–1057.

Romeu, J. L. and Ozturk A. (1993). A comparative study of goodness-of-fit tests for multivariate normality, *Journal of Multivariate Analysis*, **46**, 309–334.

Royston, J. P. (1983). Some techniques for assessing multivariate normality based on the Shapiro-Wilk W, *Applied Statistics*, **32**, 121–133.

Seber, G. A. F. (1984). *Multivariate Observations*, John Wiley & Sons, New York.

Small, N. J. H. (1978). Plotting squared radii, *Biometrika*, **65**, 657–658.

Small, N. J. H. (1980). Marginal skewness and kurtosis in testing multivariate normality, *Applied Statistics*, **29**, 85–87.

S-PLUS. (1995). *Advanced Data Analysis Software*, MathSoft, Seattle, WA.

Srivastava, D. K., Mudholkar, A. and Mudholkar, G. S. (1992). Assessing the significance of difference between two quick estimates of location, *Journal of Applied Statistics*, **19**, 405–416.

Srivastava, M. S. (1984). A measure of skewness and kurtosis and a graphical method for testing multivariate normality, *Statistics and Probability Letters*, **2**, 263–267.

Srivastava, M. S. and Hui, T. K. (1987). On assessing multivariate normality based on Shapiro-Wilk W statistic, *Statistics and Probability Letters*, **5**, 15–18.

Part IV
Bayesian Inference

CHAPTER 21

CHARACTERIZATIONS OF TAILFREE AND NEUTRAL TO THE RIGHT PRIORS

R. V. RAMAMOORTHI   L. DRAGHICI   J. DEY

Michgan State University, East Lansing, MI

Abstract: Let $X_1, X_2, \ldots$ be exchangeable random variables. By de Finetti's Theorem, there exists a prior Π such that $P \sim \Pi$ and given P, $X_1, X_2, \ldots$ are i.i.d. P. We give necessary and sufficient conditions for Π to be (a) tailfree and (b) neutral to the right. En route to such characterization we also obtain characterizations in terms of the posterior.

Keywords and phrases: Neutral to the right prior, Beta-Stacy process, de Finetti's theorem, Polya tree process

21.1 INTRODUCTION

A random distribution function F is said to be neutral to the right (NR) if for any $t_1 < t_2 < \cdots < t_k$, $F(t_1), \frac{F(t_2)}{1-F(t_1)}, \ldots, \frac{F(t_k)}{1-F(t_{k-1})}$ are independent. Doksum (1974) introduced these priors and showed that if the process is NR then so is the posterior given n observations $X_1, \ldots, X_n$. The Beta and Beta-Stacy processes developed respectively by Hjort (1990) and Walker and Muliere (1997) are interesting examples of NR priors. These examples demonstrate the use of NR priors as a useful concept in Bayesian Nonparametrics, especially in the context of right censored data.

Tailfree priors [Freedman (1963), Fabius (1964) and Doksum (1974)] are similar to NR processes in the sense that they too require independence of certain conditional probabilities. We will give a formal definition later, but just note here that the Polya tree processes studied by Mauldin, Sudderth and Williams (1992) and Lavine (1992, 1994) show that the property of "tailfreedom" can be used to construct interesting classes of tractable priors.

In this note we provide characterization of NR and tailfree priors. In addition, we characterize situations when the prior given by de Finetti's Theorem would be NR or tailfree.

A comprehensive account of NR and tailfree priors, and de Finetti's Theorem can be found in Schervish (1995).

21.2   TAILFREE PRIORS

We begin by considering the multinomial case. Let $\mathcal{X}$ be a finite set. Construct a nested sequence of partitions $\mathbf{T}_i, i = 1, \ldots, k$ where

$$\mathbf{T}_1: \quad B_0, \quad B_1. \qquad B_0 \cap B_1 = \phi, \quad B_0 \cup B_1 = \mathcal{X}$$
$$\mathbf{T}_2: \quad B_{00}, \quad B_{01}, \quad B_{10}, \quad B_{11}.$$

where B_{00}, B_{01} is a partition of B_0 and B_{10}, B_{11} is a partition of B_1. Similarly $\mathbf{T}_3$ consists of partitions of elements of $\mathbf{T}_2$ and so on. For each i, let $E_i = \{0, 1\}^i$. We can then conveniently write the partition $\mathbf{T}_i$ as $\{B_\epsilon : \epsilon \in E_i\}$. Assume without loss of generality that $\mathbf{T}_k$ consists of singletons.

Let $\mathbf{M}(\mathcal{X})$ be the set of probability measures on $\mathcal{X}$ and let Π be a prior on $\mathbf{M}(\mathcal{X})$. Let $X_1, X_2, \ldots$ be a sequence of observations in $\mathcal{X}$ which are, given $P \in \mathbf{M}(\mathcal{X})$, independent with common distribution P.

For $\epsilon \in E_i$, let $N_{i,\epsilon}^n$ be the number of observations out of $X_1, \ldots, X_n$ which fall in B_ϵ. Formally $N_{i,\epsilon}^n = \sum_{i=1}^n I_{B_\epsilon}(X_i)$. Denote by N_i^n the vector $(N_{i,\epsilon}^n : \epsilon \in E_i)$.

For the prior Π, $\Pi_{X_1,\ldots,X_n}$ will stand for the posterior given $X_1, \ldots, X_n$ and $\Pi_{N_i^n}$ for the posterior given N_i^n. For a function g on $\mathbf{M}(\mathcal{X})$, we will write $\mathcal{L}(g(P) \mid \Pi_{X_1,\ldots,X_n})$ to denote the 'law' or distribution of $g(P)$ under $\Pi_{X_1,\ldots,X_n}$. Similar notation will be followed with measures $\Pi_{N_i^n}$.

Definition 21.2.1 A prior Π on $\mathbf{M}(\mathcal{X})$ is said to be tailfree if, under Π, the sets

$$\{P(B_0)\}, \{P(B_{00} \mid B_0), P(B_{10} \mid B_1)\}, \ldots \{P(B_{\epsilon 0} \mid B_\epsilon) : \epsilon \in E_i\}$$

are independent for all i.

Theorem 21.2.1 *Suppose* $\Pi\{0 < P(B_\epsilon) < 1\} = 1$ *for all* $\epsilon \in E_i, i \geq 1$. *Then the following are equivalent.*

1. Π is tailfree;

2. For all n and all $i = 1, \dots, k$,

$$\mathcal{L}\left(\{P(B_\epsilon) : \epsilon \in E_i\} \mid \Pi_{X_1,\dots,X_n}\right) = \mathcal{L}\left(\{P(B_\epsilon) : \epsilon \in E_i\} \mid \Pi_{N_i^n}\right).$$

PROOF Under the assumption, note that $\Pi_{X_1,\dots,X_n}$ is uniquely determined and hence the equality in (2) is pointwise and not just almost everywhere.

(1) $\Rightarrow$ (2) is well known and follows from noting that under $\Pi_{X_1,\dots,X_n}$, $\{P(B_\epsilon) : \epsilon \in E_i\}$ has the density

$$\frac{\prod_{\epsilon \in E_i} P(B_\epsilon)^{N_{i,\epsilon}^n}}{\int_{\mathbf{M}(\mathcal{X})} \prod_{\epsilon \in E_i} P(B_\epsilon)^{N_{i,\epsilon}^n} d\Pi(P)}.$$

For (2) $\Rightarrow$ (1) we first prove a lemma.

Lemma 21.2.1 *For any i, under (2),*

(a) $\{P(B_\epsilon) : \epsilon \in E_i\}$ and $\{P(B_{\epsilon 0} \mid B_\epsilon) : \epsilon \in E_i\}$ are independent.

(b) $\{P(B_{\epsilon 0} \mid B_\epsilon) : \epsilon \in \cup_{j=0}^{i-1} E_j\}$ and $\{P(B_{\epsilon 0} \mid B_\epsilon) : \epsilon \in E_i\}$ are independent.

Here E_0 is the empty set and $P(B_{\epsilon 0} \mid B_\epsilon), \epsilon \in E_0$ stands for $P(B_0)$.

PROOF OF LEMMA Since $\{P(B_\epsilon) : \epsilon \in E_i\}$ determines $\{P(B_\epsilon) : \epsilon \in E_j\}$ for any $j \leq i$, quantities like $P(B_{\epsilon 0} \mid B_\epsilon)$ for $\epsilon \in E_j, j < i$ are functions of $\{P(B_\epsilon) : \epsilon \in E_i\}$. Hence (b) is an immediate consequence of (a).

To prove (a) first note that (2) gives the conditional independence of $\{P(B_\epsilon) : \epsilon \in E_i\}$ and $(X_1, \dots, X_n)$ given N_i^n, which we write as

$$\{P(B_\epsilon) : \epsilon \in E_i\} \underset{N_i^n}{\perp} (X_1, \dots, X_n)$$

from where it follows that

$$\{P(B_\epsilon) : \epsilon \in E_i\} \underset{N_i^n}{\perp} N_{i+1}^n$$

and hence that

$$\mathcal{L}\left(\{P(B_\epsilon) : \epsilon \in E_i\} \mid \Pi_{N_i^n}\right) = \mathcal{L}\left(\{P(B_\epsilon) : \epsilon \in E_i\} \mid \Pi_{N_{i+1}^n}\right). \quad (21.2.1)$$

To establish (a) it is enough to show that for any collection $\{n_\epsilon : \epsilon \in E_i\}$ of integers

$$\mathbf{E}_\Pi \left\{ \prod_{\epsilon \in E_i} [P(B_{\epsilon 0} \mid B_\epsilon)]^{n_\epsilon} \mid \{P(B_\epsilon) : \epsilon \in E_i\} \right\} = constant \qquad a.e. \quad \Pi.$$

Let $n = \sum_{\epsilon \in E_i} n_\epsilon$. Consider the posterior density of $\{P(B_\epsilon) : \epsilon \in E_i\}$ given $N_{i,\epsilon}^n = n_\epsilon$ and its posterior density given $N_{i+1,\epsilon 0}^n = n_\epsilon$, $N_{i+1,\epsilon 1}^n = 0$. Since by (21.2.1) these two are equal, we have

$$\mathbf{E}_\Pi \left\{ \frac{\prod_{\epsilon \in E_i} [P(B_\epsilon)]^{n_\epsilon} [P(B_{\epsilon 0} \mid B_\epsilon)]^{n_\epsilon}}{\int_{\mathbf{M}(\mathcal{X})} \prod_{\epsilon \in E_i} [P(B_\epsilon)]^{n_\epsilon} [P(B_{\epsilon 0} \mid B_\epsilon)]^{n_\epsilon} d\Pi(P)} \mid \{P(B_\epsilon) : \epsilon \in E_i\} \right\}$$

$$= \frac{\prod_{\epsilon \in E_i} [P(B_\epsilon)]^{n_\epsilon}}{\int_{\mathbf{M}(\mathcal{X})} \prod_{\epsilon \in E_i} [P(B_\epsilon)]^{n_\epsilon} d\Pi(P)}$$

which yields

$$\mathbf{E}_\Pi \left\{ \prod_{\epsilon \in E_i} [P(B_{\epsilon 0} \mid B_\epsilon)]^{n_\epsilon} \mid \{P(B_\epsilon) : \epsilon \in E_i\} \right\}$$

$$= \frac{\int_{\mathbf{M}(\mathcal{X})} \prod_{\epsilon \in E_i} [P(B_{\epsilon 0})]^{n_\epsilon} d\Pi(P)}{\int_{\mathbf{M}(\mathcal{X})} \prod_{\epsilon \in E_i} [P(B_\epsilon)]^{n_\epsilon} d\Pi(P)}.$$

$\square$

Returning to the proof of the Theorem, (2) $\Rightarrow$ (1) now follows by applying the lemma successively for $i = k-1, k-2, \ldots, 1$. $\square$

We now turn our attention to de Finetti's Theorem, which states that, if $X_1, X_2, \ldots$ is a sequence of exchangeable random variables under a measure μ, then there is a unique prior Π on $\mathbf{M}(\mathcal{X})$ such that, for any n,

$$\mu\{X_1 \in B_1, \ldots, X_n \in B_n\} = \int_{\mathbf{M}(\mathcal{X})} \prod_{i=1}^n P(B_i) d\Pi(P).$$

The question we address is: what additional condition on μ would ensure that Π is tailfree. The last theorem can be used to provide an answer to this.

For each i, let $T_i(X)$ be the vector $(I_{B_\epsilon}(X) : \epsilon \in E_i)$. Let $\mu_{X_1 \ldots X_n}$ be the predictive distribution of $X_{n+1}, X_{n+2}, \ldots$ given $X_1, \ldots, X_n$.

Theorem 21.2.2 *Let $X_1, X_2, \ldots$ be an exchangeable sequence under μ, and let Π be the corresponding prior obtained from de Finetti's Theorem. If, for every $B_\epsilon, \epsilon \in \cup E_i$, $\lim_{n \to \infty} \mu\{X_1 \in B_\epsilon^c, \ldots, X_n \in B_\epsilon^c\} = 0$, then the following are equivalent*

1. *Π is tailfree;*

2. *For all n and all $i \geq 1$,*

$$\mathcal{L}(T_i(X_{n+1}) \mid \mu_{X_1 \ldots X_n}) = \mathcal{L}(T_i(X_{n+1}) \mid \mu_{T_i(X_1) \ldots T_i(X_n)}).$$

PROOF Condition (i) of the theorem ensures that $\Pi\{0 < P(B_\epsilon) < 1\} = 1$ for all $\epsilon \in E_i, i \geq 1$, and therefore, Theorem 1 can be applied. (1) $\Rightarrow$ (2) is easy to show since, for any $B_\epsilon \in \mathbf{T}_i$,

$$\mu_{X_1 \ldots X_n}(X_{n+1} \in B_\epsilon) = \int_{\mathbf{M}(\mathcal{X})} P(B_\epsilon) d\Pi_{X_1, \ldots, X_n}(P).$$

If Π is tailfree, then from Theorem 21.2.1, we have

$$\mathcal{L}(\{P(B_\epsilon) : \epsilon \in E_i\} \mid \Pi_{X_1, \ldots, X_n}) = \mathcal{L}(\{P(B_\epsilon) : \epsilon \in E_i\} \mid \Pi_{T_i(X_1) \ldots T_i(X_n)}),$$

from which (2) follows easily.

To show (2) $\Rightarrow$ (1), by Theorem 21.2.1, it is enough to show that

$$\mathcal{L}(\{P(B_\epsilon) : \epsilon \in E_i\} \mid \Pi_{X_1, \ldots, X_n}) = \mathcal{L}(\{P(B_\epsilon) : \epsilon \in E_i\} \mid \Pi_{T_i(X_1) \ldots T_i(X_n)}),$$

or equivalently, for any collection $\{n_\epsilon : \epsilon \in E_i\}$ of positive intergers

$$\int_{\mathbf{M}(\mathcal{X})} \prod_{\epsilon \in E_i} [P(B_\epsilon)]^{n_\epsilon} d\Pi_{X_1, \ldots, X_n}(P)$$

$$= \int_{\mathbf{M}(\mathcal{X})} \prod_{\epsilon \in E_i} [P(B_\epsilon)]^{n_\epsilon} d\Pi_{T_i(X_1) \ldots T_i(X_n)}(P).$$

Since for every n, by (2) of the theorem, for fixed i,

$$X_1, \ldots, X_n \underset{T_i(X_1), \ldots, T_i(X_n)}{\perp} T_i(X_{n+1}),$$

it is easy to see that, for every m,

$$X_1, \ldots, X_n \underset{T_i(X_1), \ldots, T_i(X_n)}{\perp} T_i(X_{n+1}), \ldots, T_i(X_{n+m}). \qquad (21.2.2)$$

Now let $m = \sum_{\epsilon \in E_i} n_\epsilon$ and, given $X_1, \ldots, X_n$, consider the conditional probability that out of the next m observations n_ϵ fall in B_ϵ for $\epsilon \in E_i$. This is given by

$$\int_{\mathbf{M}(\mathcal{X})} \prod_{\epsilon \in E_i} [P(B_\epsilon)]^{n_\epsilon} d\Pi_{X_1, \ldots, X_n}(P)$$

and, by equation (21.2.2) above, is equal to

$$\int_{\mathbf{M}(\mathcal{X})} \prod_{\epsilon \in E_i} [P(B_\epsilon)]^{n_\epsilon} d\Pi_{T_i(X_1) \ldots T_i(X_n)}(P).$$

$\square$

Remark The last two theorems can be extended immediately to tailfree priors on $\mathbf{M}(\mathbb{R})$, the set of probability measures on $\mathbb{R}$. Towards this let $\{\mathbf{T}_i : i \geq 1\}$ be a nested sequence of partitions of $\mathbb{R}$ by intervals such that $\cup_i \mathbf{T}_i$ generates the Borel σ-algebra on $\mathbb{R}$. A prior Π on $\mathbf{M}(\mathbb{R})$ is said to be tailfree if the families

$$\{P(B_0)\}, \{P(B_{00} \mid B_0), P(B_{10} \mid B_1)\}, \ldots, \{P(B_{\epsilon 0} \mid B_\epsilon) : \epsilon \in E_i\}$$

are independent for all i. It is easy to see that Theorems 21.2.1 and 21.2.2 hold if $\mathcal{X}$ is replaced by $\mathbb{R}$.

21.3   NEUTRAL TO RIGHT PRIORS

In this section we provide a characterization of neutral to the right priors in the same spirit as done for tailfree priors in the last section.

Let $\mathcal{F}$ be the set of distribution functions on $\mathbb{R}$ endowed with the σ-algebra that makes the functions $F \mapsto F(t)$ measurable for all $t \in \mathbb{R}$. A prior Π on $\mathcal{F}$ is said to be NR if, for all $t_1 < \cdots < t_k$, under Π,

$$\bar{F}(t_1), \frac{\bar{F}(t_2)}{\bar{F}(t_1)}, \ldots, \frac{\bar{F}(t_k)}{\bar{F}(t_{k-1})}$$

are independent, where $\bar{F}(t) = 1 - F(t)$.

As before, let $X_1, X_2, \ldots$ be a sequence of random variables which are, given $F \in \mathcal{F}$, i.i.d. F.

For each n, define the process N_n by

$$N_n(t) = \sum_{i=1}^{n} I_{\{X_i > t\}}.$$

For each T in $\mathbb{R}$, let $N_n^T = \{N_n(t) : t \leq T\}$. As before, $\Pi_{X_1,\ldots,X_n}$ will stand for the posterior given $X_1,\ldots,X_n$, and $\Pi_{N_n^T}$ for the posterior given $\{N_n(t) : t \leq T\}$.

Theorem 21.3.1 *Suppose Π is a prior such that, for all t, $\Pi\{F : 0 < F(t) < 1 \, \forall \, t\} = 1$. Then the following are equivalent*

1. *Π is NR;*

2. *For all $T \in \mathbb{R}$,*

$$\mathcal{L}(\{F(t) : t \leq T\} \mid \Pi_{X_1,\ldots,X_n}) = \mathcal{L}(\{F(t) : t \leq T\} \mid \Pi_{N_n^T});$$

3. *For all $t_1 < \cdots < t_k < t_{k+1}$,*

$$\mathcal{L}\left(F(t_1),\ldots,F(t_k) \mid \Pi_{N_n(t_1),\ldots,N_n(t_{k+1})}\right)$$
$$= \mathcal{L}\left(F(t_1),\ldots,F(t_k) \mid \Pi_{N_n(t_1),\ldots,N_n(t_k)}\right)$$

Remark Note that (2) is a statement that holds almost everywhere with respect to the marginal distribution of $X_1,\ldots,X_n$. However, under our assumption, (3) holds everywhere. Interpret (2) as, there exists a version of the L.H.S. equal to the R.H.S. everywhere.

PROOF (1) $\Rightarrow$ (2) is well known. See for instance Doksum (1974).

To see (3) $\Rightarrow$ (1), we will argue as in Theorem 21.2.1 that for fixed $t_1 < \cdots < t_{k+1}$,

$$\{\bar{F}(t_1),\ldots,\bar{F}(t_k)\} \quad \text{is independent of} \quad \frac{\bar{F}(t_{k+1})}{\bar{F}(t_k)}.$$

This would then show that

$$\left\{\bar{F}(t_1), \frac{\bar{F}(t_2)}{\bar{F}(t_1)}, \ldots, \frac{\bar{F}(t_k)}{\bar{F}(t_{k-1})}\right\} \quad \text{is independent of} \quad \frac{\bar{F}(t_{k+1})}{\bar{F}(t_k)}.$$

Let n be any integer. Since the posterior density of $(F(t_1),\ldots,F(t_k))$ given $\{N_n(t_1) = n, \ldots, N_n(t_k) = n\}$ and the posterior density given $\{N_n(t_1) = n, \ldots, N_n(t_k) = n, N_n(t_{k+1}) = n\}$ are equal, then so are the densities of $\left(\bar{F}(t_1), \frac{\bar{F}(t_2)}{\bar{F}(t_1)}, \ldots, \frac{\bar{F}(t_k)}{\bar{F}(t_{k-1})}\right)$. This gives, with $t_0 = -\infty$,

$$\frac{\prod_{i=1}^k [\frac{\bar{F}(t_i)}{\bar{F}(t_{i-1})}]^n}{\int_{\mathcal{F}} \prod_{i=1}^k [\frac{\bar{F}(t_i)}{\bar{F}(t_{i-1})}]^n d\Pi(F)} = \frac{\mathbb{E}_\Pi\left[\prod_{i=1}^{k+1} [\frac{\bar{F}(t_i)}{\bar{F}(t_{i-1})}]^n \mid \bar{F}(t_1),\ldots,\bar{F}(t_k)\right]}{\int_{\mathcal{F}} \prod_{i=1}^{k+1} [\frac{\bar{F}(t_i)}{\bar{F}(t_{i-1})}]^n d\Pi(F)}$$

and hence that

$$\mathbf{E}_\Pi \left[[\frac{\bar{F}(t_{k+1})}{\bar{F}(t_k)}]^n \mid \bar{F}(t_1), \ldots, \bar{F}(t_k) \right] = constant \qquad a.e. \quad \Pi.$$

Since this holds for all n, $\frac{\bar{F}(t_{k+1})}{\bar{F}(t_k)}$ is independent of $\{\bar{F}(t_1), \ldots, \bar{F}(t_k)\}$. Repeating the argument with k replaced by $j = k-1, k-2, \ldots, 1$ the result follows.

We now prove (2) $\Rightarrow$ (3). For this, note that since, by (2),

$$\{F(t_1), \ldots F(t_k)\} \underset{N_n(t):t\leq t_k}{\perp} \{N_n(t) : t \geq 0\}$$

it follows that

$$\{F(t_1), \ldots F(t_k)\} \underset{N_n(t):t\leq t_k}{\perp} \{N_n(t_1), \ldots N_n(t_{k+1})\}$$

Now, for any measurable function g of $(F(t_1), \ldots F(t_k))$, $\mathbf{E}[g \mid N_n(t_1), \ldots, N_n(t_{k+1})]$ is measurable both with respect to the σ-algebra generated by $\{N_n(t_1), \ldots, N_n(t_{k+1})\}$ and $\sigma\{N_n(t) : t \leq t_k\}$ and is hence measurable with respect to their intersection. But this intersection is precisely $\sigma\{N_n(t_1), \ldots, N_n(t_k)\}$ which concludes the proof. $\qquad \square$

To reinterpret the above theorem in the context of de Finetti's Theorem, let $X_1, X_2, \ldots$ be exchangeable under μ. The following result gives the condition on μ such that the corresponding de Finetti prior Π is NR.

Theorem 21.3.2 *Assume that, for every t, as $n \to \infty$*

$$\mu\{X_1 < t, \ldots, X_n < t\} \to 0 \quad and \quad \mu\{X_1 > t, \ldots, X_n > t\} \to 0.$$

Then the following are equivalent

1. Π is NR;

2. For all $t \in \mathbb{R}$ and $n \geq 1$,

$$\mu_{X_1, \ldots, X_n}\{X_{n+1} > t\} = \mu_{N_n^t}\{X_{n+1} > t\}.$$

PROOF   Since $\mu_{X_1 \ldots X_n}\{X_{n+1} > t\} = \int_{\mathcal{F}} \bar{F}(t) d\Pi_{X_1, \ldots, X_n}(F)$, in view of Theorem 21.3.1, (2) immediately follows from (1).

To see (2) $\Rightarrow$ (1), fix $s \in \mathbb{R}$ and define $T(x) = xI_{(-\infty,s]}(x) + (s+1)I_{(s,\infty)}(x)$. Then (2) implies that

$$X_1, \ldots, X_n \underset{N_n^s}{\perp} T(X_{n+1}), \ldots, T(X_{n+m}).$$

To prove this claim, consider $t_1 < \cdots < t_k < t_{k+1} = s$. Since $\sigma\{N_n^s\} \supseteq \sigma\{N_n^{t_{k+1}}\} \supseteq \cdots \supseteq \sigma\{N_n^{t_1}\}$, each of the events $\{X_{n+1} > t_i\}$, $i = 1, \ldots, k+1$, is conditionally independent of $X_1, \ldots, X_n$ given N_n^s and hence the function $T_k(X_{n+1}) = \sum_{i=1}^{k+1} t_i I_{(t_{i-1}, t_i]}(X_{n+1}) + (s+1)I_{(s,\infty)}(X_{n+1})$ is conditionally independent of $X_1, \ldots, X_n$ given N_n^s. Letting $t_1, \ldots, t_k$ run through a countable dense set in $(-\infty, s]$, we get that

$$X_1, \ldots, X_n \underset{N_n^s}{\perp} T(X_{n+1}).$$

A simple induction argument then yields the claim.

Now, fix $t_1 < \cdots < t_k$. Given integers $n_1, \ldots, n_k$, set $m = n_1 + \cdots + n_k$ and, given $X_1, \ldots, X_n$, consider the predictive probability of the event that of the next m observations, n_i observations fall in $(t_{i-1}, t_i]$ for $i = 1, \ldots, k$.

Since the event $\{X_{n+j} \in (t_{i-1}, t_i]\}$ is same as the event $\{T(X_{n+j}) \in (t_{i-1}, t_i]\}$ for all $i = 1, \ldots, k$ and $s \geq t_k$, we have

$$\int_{\mathcal{F}} [1 - \bar{F}(t_1)]^{n_1} [\bar{F}(t_1) - \bar{F}(t_2)]^{n_2} \ldots [\bar{F}(t_{k-1}) - \bar{F}(t_k)]^{n_k} d\Pi_{X_1, \ldots, X_n}(F)$$

$$= \int_{\mathcal{F}} [1 - \bar{F}(t_1)]^{n_1} [\bar{F}(t_1) - \bar{F}(t_2)]^{n_2} \ldots [\bar{F}(t_{k-1}) - \bar{F}(t_k)]^{n_k} d\Pi_{N_n^s}(F).$$

This shows that the distribution of $\bar{F}(t_1), \bar{F}(t_1) - \bar{F}(t_2), \ldots, \bar{F}(t_{k-1}) - \bar{F}(t_k)$ given $X_1, \ldots, X_n$ is same as that given N_n^s for all $s \geq t_k$. Since $\bar{F}(t_1), \ldots, \bar{F}(t_k)$ is a function of these quantities, the same is true for $\bar{F}(t_1), \ldots, \bar{F}(t_k)$ and hence (1) follows from Theorem 21.3.1. □

In a recent work, yet to appear, Walker and Muliere (1999) also obtained the same result. Their condition on μ is expressed in terms of the expected instantaneous hazard rate under the de Finetti prior and corresponding posteriors. When the set of values for the observations, $\mathcal{X}$, is finite they explicitly provide the condition on the predictive distributions.

21.4 NR PRIORS FROM CENSORED OBSERVATIONS

In this section we provide a characterization of neutral to the right priors with respect to the posterior in the presence of right censored observations.

Suppose as before, that $F \sim \Pi$ and given F, $X_1, X_2, \ldots$ are independent and identically distributed as F. Here $X_1, X_2, \ldots$ are thought of as survival times. Let $c_1, c_2, \ldots$ be constants. These are our censoring times. For each i, we only get to observe:

$$Z_i = \min(X_i, c_i) \quad \text{and} \quad \Delta_i = I_{\{X_i \leq c_i\}}$$

Therefore, $\Delta_i = 0$ means Z_i is a right censored observation.

Define the observation processes:

$$N_i^{(n)}(t) = \sum_{j=1}^{n} I_{\{Z_j \le t, \Delta_j = i\}}, \qquad i = 0, 1$$

and let $N^{(n)}(t) = \left(N_1^{(n)}(t), N_0^{(n)}(t)\right)$ and also let $A_n^T = \{t \le T : N_0^{(n)}(t) - N_0^{(n)}(t-) > 0\}$.

Under this set-up, the following theorem characterizes NR priors.

Theorem 21.4.1 *Suppose* Π *is a prior such that, for all* t, $\Pi\{F : 0 < F(t) < 1 \,\forall\, t\} = 1$. *Then the following are equivalent*

1. Π *is NR;*

2. *For all* $T \in \mathbb{R}$ *and* $n \ge 1$,

$$\mathcal{L}(\{F(t) : t \le T\} \mid \Pi_{\{N^{(n)}(t) : t \in \mathbb{R}\}})$$
$$= \mathcal{L}(\{F(t) : t \le T\} \mid \Pi_{\{N^{(n)}(t) : t \le T\}});$$

3. *For all* $n \ge 1$, $k \ge 1$ *and* $t_1 < \cdots < t_k < t_{k+1}$, *the distribution of* $\bar{F}(t_1), \ldots, \bar{F}(t_k)$ *under the posterior given* $\{N_1^{(n)}(t_1), \ldots, N_1^{(n)}(t_{k+1})\}$, $\{N^{(n)}(t) : t \le t_k\}$ *is the same as that under the posterior given* $\{N_1^{(n)}(t_1), \ldots, N_1^{(n)}(t_k)\}$, $\{N^{(n)}(t) : t \le t_k\}$.

PROOF (1) $\Rightarrow$ (2) is due to Ferguson and Phadia (1979). We omit the proof here.

To prove (2) $\Rightarrow$ (3) note that, by (2), $\{F(t_1), \ldots F(t_k)\}$ is conditionally independent of $\{N^{(n)}(t) : t \ge 0\}$ given $\{N^{(n)}(t) : t \le t_k\}$. Therefore $\{F(t_1), \ldots F(t_k)\}$ is conditionally independent of $\{N_1^{(n)}(t_1), \ldots, N_1^{(n)}(t_{k+1})\}$ and $\{N_0^{(n)}(t) : t \le t_k\}$ given $\{N^{(n)}(t) : t \le t_k\}$.

Then, for any measurable function g of $(F(t_1), \ldots F(t_k))$,

$$\mathbf{E}[g \mid N_1^{(n)}(t_1), \ldots, N_1^{(n)}(t_{k+1})]$$

is measurable both with respect to $\sigma\{N_1^{(n)}(t_1), \ldots, N_1^{(n)}(t_{k+1})\} \vee \sigma\{N_0^{(n)}(t) : t \le t_k\}$ and $\sigma\{N_n(t) : t \le t_k\}$. Hence it is also measurable with respect to their intersection, which is $\sigma\{N_1^{(n)}(t_1), \ldots, N_1^{(n)}(t_k)\} \vee \sigma\{N_0^{(n)}(t) : t \le t_k\}$.

To see (3) $\Rightarrow$ (1), we will show that for arbitrary $t_1 < \cdots < t_{k+1}$,

$$\{\bar{F}(t_1), \ldots, \bar{F}(t_k)\} \quad \text{is independent of} \quad \frac{\bar{F}(t_{k+1})}{\bar{F}(t_k)}.$$

Fix $y_1 < \cdots < y_r \le t_k$. Let $A_n^{t_k} = \{y_1, \ldots, y_r\}$ and let $s_1 < \cdots < s_{k+r+1}$ be points such that

$$A_n^{t_k} \cup \{t_1, \ldots, t_{k+1}\} = \{s_1, \ldots, s_{k+r+1}\}$$

Note that $s_{k+r} = t_k$ and $s_{k+r+1} = t_{k+1}$. Now, for integers $m_1 \le \cdots \le m_r$, consider the set

$$B = \{N_0^{(n)}(t) = \sum_{j=1}^r (m_j - m_{j-1}) I_{[y_j, \infty)}(t), t \le t_k\}$$

Thus, on B, for each $j = 1, \ldots, r$, $(m_j - m_{j-1})$ observations are censored at the point y_j. Let $C = \{N_1^{(n)}(s_j) = 0, j = 1, \ldots, k + r\}$ and let $C' = \{N_1^{(n)}(s_j) = 0, j = 1, \ldots, k + r + 1\}$.

Since the posterior density of $(F(t_1), \ldots, F(t_k))$ given $C \cap B$ and the posterior density given $C' \cap B$ are equal, then so are the densities of $\left(\bar{F}(t_1), \frac{\bar{F}(t_2)}{\bar{F}(t_1)}, \ldots, \frac{\bar{F}(t_k)}{\bar{F}(t_{k-1})}\right)$. This gives,

$$\frac{\prod_{i=1}^{k+r} [\frac{\bar{F}(s_i)}{\bar{F}(s_{i-1})}]^{n-m_{i-1}}}{\int_{\mathcal{F}} \prod_{i=1}^{k+r} [\frac{\bar{F}(s_i)}{\bar{F}(s_{i-1})}]^{n-m_{i-1}} d\Pi(F)}$$

$$= \frac{\mathbf{E}_\Pi \left[\prod_{i=1}^{k+r+1} [\frac{\bar{F}(s_i)}{\bar{F}(s_{i-1})}]^{n-m_{i-1}} \mid \bar{F}(s_1), \ldots, \bar{F}(s_{k+r})\right]}{\int_{\mathcal{F}} \prod_{i=1}^{k+r+1} [\frac{\bar{F}(s_i)}{\bar{F}(s_{i-1})}]^{n-m_{i-1}} d\Pi(F)}$$

and hence that

$$\mathbf{E}_\Pi \left[[\frac{\bar{F}(s_{k+r+1})}{\bar{F}(s_{k+r})}]^{n-m_k} \mid \bar{F}(s_1), \ldots, \bar{F}(s_{k+r})\right] = constant \qquad a.e. \ \ \Pi.$$

Since this holds for all positive integral values of $n - m_k$, $\frac{\bar{F}(t_{k+1})}{\bar{F}(t_k)}$ is independent of $\{\bar{F}(t_1), \ldots, \bar{F}(t_k)\}$. Repeating the argument with k replaced by $k - 1, k - 2, \ldots, 1$ the result follows.   □

REFERENCES

Doksum, K. (1974). Tailfree and neutral random probabilities and their posterior distributions, *Annals of Probability*, **2**, 183–201.

Fabius, J. (1964). Asymptotic behavior of Bayes' estimates, *Annals of Mathematical Statistics*, **35**, 846–856.

Ferguson, T. S. (1974). Prior distributions on spaces of probability measures, *Annals of Statistics*, **2**, 615–629.

Ferguson, T. S. and Phadia, E. G. (1979). Bayesian nonparametric estimation based on censored data, *Annals of Statistics*, **7**, 163–186.

Freedman, D. A. (1963). On the asymptotic behaviour of Bayes' estimates in the discrete case, *Annals of Mathematical Statistics*, **34**, 1386–1403.

Hjort, N. L. (1990). Nonparametric Bayes estimators based on beta processes in models for life history data, *Annals of Statistics*, **18**, 1259–1294.

Lavine, M. (1992). Some aspects of Polya tree distributions for statistical modeling, *Annals of Statistics*, **20**, 1222–1235.

Lavine, M. (1994). More aspects of Polya tree distributions for statistical modelling, *Annals of Statistics*, **22**, 1161–1176.

Mauldin, R. D., Sudderth, W. D. and Williams, S. C. (1992). Polya trees and random distributions, *Annals of Statistics*, **20**, 1203–1221.

Schervish, M. J. (1995). *Theory of Statistics*, Springer Series in Statistics, Springer-Verlag, New-York.

Walker, S. and Muliere, P. (1997). Beta-Stacy processes and a generalization of the Polya-urn scheme, *Annals of Statistics*, **25**, 1762–1780.

Walker, S. and Muliere, P. (1999). A characterisation of a neutral to the right prior via an extension of Johnson's sufficientness postulate, *Annals of Statistics*, **27**.

CHAPTER 22

EMPIRICAL BAYES ESTIMATION AND TESTING FOR A LOCATION PARAMETER FAMILY OF GAMMA DISTRIBUTIONS

N. BALAKRISHNAN   YIMIN MA

McMaster University, Hamilton, Ontario, Canada
University of Regina, Regina, Saskatchewan, Canada

Abstract: In this chapter, we consider a location parameter family of gamma distributions with θ as the location parameter and $\alpha \geq 2$ as the fixed shape parameter. The empirical Bayes estimator for θ and the empirical Bayes testing rule for the two-action problem $H_0 : \theta \leq \theta_0$ vs. $H_1 : \theta > \theta_0$ are studied. Under some moment conditions on the prior distribution G, the convergence rates for the proposed empirical Bayes estimator and the empirical Bayes testing rule are established.

Keywords and phrases: Convergence rate, empirical Bayes, linear loss, location parameter, squared-error loss, two-action problem

22.1 INTRODUCTION

The empirical Bayes approach, formulated originally by Robbins (1955), has been used rather extensively for various statistical problems by many authors including Robbins (1963, 1964), Johns and Van Ryzin (1971, 1972), Lin (1975), Singh (1979), and Singh and Wei (1992). For certain nonexponential families of distributions, the empirical Bayes method has been applied by Fox (1978), Van Houwelingen (1987), Nogami (1988), Datta (1991), Huang (1995), and Singh (1995), among others.

The gamma distribution with location parameter is useful in many areas of application including survival analysis, life-testing, and reliability theory (in these cases, the location parameter is often referred to as the 'threshold parameter'); for example, see Balakrishnan and Cohen (1991), and Johnson, Kotz, and Balakrishnan (1994). For this distribution model, Fox (1978) studied empirical Bayes estimation of the location parameter under squared-error loss; however, Fox did not examine the convergence rate of that empirical Bayes estimator.

In this chapter, we consider the empirical Bayes estimation of the location parameter as well as the empirical Bayes two-action testing problem. In Section 22.2 we formulate the two problems and derive the Bayes estimator and the Bayes testing rule. In Section 22.3, we derive the empirical Bayes estimator of the location parameter and the empirical Bayes testing rule for the two-action problem. In Sections 22.4 and 22.5, we examine the asymptotic optimality properties of the proposed empirical Bayes estimator and the empirical Bayes testing rule, respectively.

22.2   BAYES ESTIMATOR AND BAYES TESTING RULE

Consider the family of gamma distributions (with location parameter θ and shape parameter α) with conditional density

$$f(x \mid \theta) = \frac{1}{\Gamma(\alpha)} \, e^{-(x-\theta)} \, (x - \theta)^{\alpha-1}, \qquad x \geq \theta, \ \theta > 0, \qquad (22.2.1)$$

where $\Gamma(\cdot)$ is the complete gamma function and $\alpha \geq 2$ is fixed. In life-testing situations, θ in (22.2.1) is interpreted as 'minimum guaranteed life-time'.

22.2.1   Bayes Estimation

Under the square-error loss, it is known that the Bayes estimator relative to the prior $G(\theta)$ is

$$\phi_G(x) = \frac{\int \theta \, f(x \mid \theta) \, dG(\theta)}{\int f(x \mid \theta) \, dG(\theta)} \ . \qquad (22.2.2)$$

For the gamma model in (22.2.1), Fox (1978) obtained the Bayes estimator of θ as

$$\phi_G(x) = E(\theta \mid x) = x - \psi(x) \ , \qquad (22.2.3)$$

where

$$\psi(x) = \frac{\alpha \int_0^x e^{-(x-t)} dF(t)}{f(x)} \equiv \frac{d}{dx} \frac{w(x)}{f(x)} \qquad (22.2.4)$$

with

$$f(x) = \int_0^x f(x \mid \theta) \, dG(\theta)$$

being the density function of X and $F(x)$ the corresponding cumulative distribution function.

22.2.2   Bayes Testing

Consider the problem of testing the hypothesis $H_0 : \theta \le \theta_0$ vs. $H_1 : \theta > \theta_0$ with the linear loss function

$$\begin{aligned} L(a_i, \theta) &= b \max(0, \theta - \theta_0) & \text{if } i = 0 \\ &= b \max(0, \theta_0 - \theta) & \text{if } i = 1 , \end{aligned} \qquad (22.2.5)$$

where θ_0 is a given positive constant and a_i is the action in favour of H_i, $i = 0, 1$. $L(a_i, \theta)$ denotes the loss when action a_i is taken $(i = 0, 1)$, and b is a positive constant. Let

$$d(x) = \Pr\{\text{accepting } H_0 \mid x\} \qquad (22.2.6)$$

be the decision rule for the two-action problem considered. Then the Bayes risk associated with $d(x)$ under prior $G(\theta)$ is given by [Johns and Van Ryzin (1971, 1972)]

$$r(G, d) = b \int \alpha(x) d(x) \, dx + C_G , \qquad (22.2.7)$$

where

$$\begin{aligned} \alpha(x) &= \int \theta f(x \mid \theta) \, dG(\theta) - \theta_0 f(x) \\ &= (x - \theta_0) f(x) - \alpha \int_0^x e^{-(x-t)} dF(t) \\ &= (x - \theta_0) f(x) - w(x) \end{aligned} \qquad (22.2.8)$$

and

$$C_G = \int L(a_1, \theta) \, dG(\theta) . \qquad (22.2.9)$$

From (22.2.7), a Bayes testing rule $d_G(x)$ is then given by

$$\begin{aligned} d_G(x) &= 1 & \text{if } \alpha(x) \le 0 \\ &= 0 & \text{if } \alpha(x) > 0 . \end{aligned} \qquad (22.2.10)$$

22.3 EMPIRICAL BAYES ESTIMATOR AND EMPIRICAL BAYES TESTING RULE

Since the Bayes estimator and the Bayes testing rule presented in the last section are both dependent on the prior distribution $G(\theta)$ which may not be known, we adopt the empirical Bayes approach in this section.

Let x_i and θ_i $(i = 1, 2, \ldots, n)$ denote the observation and the location parameter at stage i, and that (conditional on θ_i) x_i follows a gamma distribution (with location parameter θ_i and shape parameter α) with density

$$f(x_i \mid \theta_i) = \frac{1}{\Gamma(\alpha)} \, e^{-(x_i - \theta_i)}(x_i - \theta_i)^{\alpha-1}, \qquad x_i \geq \theta_i, \; \theta_i > 0. \quad (22.3.11)$$

We assume that $\theta_1, \theta_2, \ldots, \theta_n$ are i.i.d. with the unknown prior distribution $G(\theta)$ and denote $x_{n+1} = x$ for the observation at the present stage.

Based on the past data, *viz.*, $x_1, x_2, \ldots, x_n$, we define the estimator for the function $w(x)$ in (2.4) as

$$w_n(x) = \frac{\alpha}{n} \sum_{i=1}^{n} e^{-(x-x_i)} I_{(0,x)}(x_i). \quad (22.3.12)$$

Further, let k_r be the class of all Borel measurable real-valued bounded functions vanishing off $(0, 1)$ such that

$$\int k(y)dy = 1, \quad \int y^\ell k(y)dy = 0 \text{ for } \ell = 1, \ldots, r - 1, \quad (22.3.13)$$

where r is an arbitrary, but fixed, positive integer. Then, define the kernel estimator for the density function $f(x)$ as

$$f_n(x) = \frac{1}{nh_n} \sum_{i=1}^{n} k\left(\frac{x_i - x}{h_n}\right) \quad (22.3.14)$$

where h_n is a positive function of n such that $h_n \to 0$ and $nh_n \to \infty$ as $n \to \infty$. These kernel estimators have been used by Johns and Van Ryzin (1972) and Singh (1977, 1979).

22.3.1 Empirical Bayes Estimator

Note that under the statistical model (22.2.1), $0 \leq \phi_G(x) = E(\theta|x) \leq x$, then, from (22.2.3), $0 \leq \psi(x) \leq x$. Utilizing $w_n(x)$ and $f_n(x)$ defined in (22.3.12) and (22.3.14), respectively, we propose the empirical Bayes estimator for the location parameter θ as [see Eqs. (22.2.3) and (22.2.4)]

$$\phi_n(x) = x - 0 \vee \left(\frac{w_n(x)}{f_n(x)}\right) \wedge x$$

$$\stackrel{d}{=} x - \psi_n(x), \tag{22.3.15}$$

where $a \vee b = \max(a, b)$ and $a \wedge b = \min(a, b)$.

22.3.2   Empirical Bayes Testing Rule

Next, we propose an empirical Bayes testing rule [from Eqs. (22.2.8) and (22.2.10)] as follows.

Let

$$\alpha_n(x) = (x - \theta_0)f_n(x) - w_n(x) . \tag{22.3.16}$$

Then, the empirical Bayes testing rule is given by

$$\begin{aligned} d_n(x) &= 1 \quad \text{if } \alpha_n(x) \leq 0 \\ &= 0 \quad \text{if } \alpha_n(x) > 0 . \end{aligned} \tag{22.3.17}$$

22.4   ASYMPTOTIC OPTIMALITY OF THE EMPIRICAL BAYES ESTIMATOR

Under the squared-error loss function, the Bayes risk of the empirical Bayes estimator $\phi_n(x)$ in (22.3.15) and the Bayes estimator $\phi_G(x)$ in (22.2.3) are

$$R(G, \phi_n(x)) = E(\theta - \phi_n(x))^2 \tag{22.4.18}$$

and

$$R(G) = R(G, \phi_G(x)) = E(\theta - \phi_G(x))^2 . \tag{22.4.19}$$

Since $\phi_G(x)$ is the Bayes estimator of θ, we have

$$E_n\{R(G, \phi_n(x))\} - R(G) \geq 0 , \tag{22.4.20}$$

where E_n denotes the expectation with respect to $(x_1, \ldots, x_n)$. It is known that

$$\begin{aligned} E_n\{R(G, \phi_n(x))\} - R(G) &= \int f(x)E_n(\phi_n(x) - \phi_G(x))^2 dx \\ &= E_n^*(\phi_n(x) - \phi_G(x))^2 , \end{aligned} \tag{22.4.21}$$

where E_n^* denotes the expectation with respect to $(x, x_1, \ldots, x_n)$.

Definition 22.4.1 A sequence of empirical Bayes estimators $\{\psi_n\}$ is said to be asymptotically optimal at least of order α_n relative to the prior G if

$$E_n\{R(G, \psi_n(x))\} - R(G) \leq O(\alpha_n) \qquad \text{as } n \to \infty , \qquad (22.4.22)$$

where $\{\alpha_n\}$ is a sequence of positive numbers such that $\lim_{n\to\infty} \alpha_n = 0$.

In order to examine the asymptotic optimality of the empirical Bayes estimator $\phi_n(x)$ proposed in (22.3.15), we need the following lemmas.

Lemma 22.4.1 *Let $y, z \neq 0$ and $L > 0$ be real numbers, and Y and Z be two real-valued random variables. Then, for any $0 < \tau \leq 2$*

$$E\left\{\left[\left(\frac{y}{z} - \frac{Y}{Z}\right) \wedge L\right]^\tau\right\}$$
$$\leq 2|z|^{-\tau}\left[E\left(|y - Y|^\tau\right) + \left(\left|\frac{y}{z}\right| + L\right)^\tau E\left(|z - Z|^\tau\right)\right]. \quad (22.4.23)$$

PROOF   This is Lemma 3.1 in Singh and Wei (1992).

Lemma 22.4.2 *(a) Let $f_n(x)$ be defined by (22.3.14) with kernel function $k \in k_{[\alpha-1]}$, α is the shape parameter, if $h_n = n^{-1/(2[\alpha-1]+1)}$, then for any $0 < \delta \leq 2$,*

$$E(|f_n(x) - f(x)|^\delta) \leq O(n^{-\delta[\alpha-1]/(2[\alpha-1]+1)}). \qquad (22.4.24)$$

(b) Let $w_n(x)$ be defined by (22.3.12), then for any $0 < \delta \leq 2$,

$$E(|w_n(x) - w(x)|^\delta) \leq O(n^{-\delta/2}). \qquad (22.4.25)$$

PROOF   (a) Since

$$f(x) = \int_0^x \frac{1}{\Gamma(\alpha)}(x - \theta)^{\alpha-1}e^{-(x-\theta)}dG(\theta)$$

and

$$f^{(r)}(x) = \sum_{i=0}^r (-1)^i \binom{r}{i} \int_0^x \frac{1}{\Gamma(\alpha)}(x - \theta)^{\alpha-1+i-r}e^{-(x-\theta)}dG(\theta),$$

it is easy to show that both $f(x)$ and $f^{([\alpha-1])}(x)$ are finite because the function $v(x) = x^a e^{-x}$, $x > 0$, $a > 0$, attains maximum value at $x = a$. Then with $h_n = n^{-1/(2[\alpha-1]+1)}$, by Corollary 3.3.4 of Singh (1977), we obtain

$$E(|f_n(x) - f(x)|^2) \leq O(n^{-2[\alpha-1]/(2[\alpha-1]+1)}).$$

Next, for any $0 < \delta \leq 2$, by Hölder's inequality,

$$E(|f_n(x) - f(x)|^\delta) \leq [E(|f_n(x) - f(x)|^2)]^{\delta/2} \leq O(n^{-\delta[\alpha-1]/(2[\alpha-1]+1)}).$$

(b) Since $w_n(x)$ is an unbiased estimator of $w_n(x)$, then

$$E(|w_n(x) - w(x)|^2) = \text{Var}(w_n(x)) \leq \frac{\alpha^2}{n}$$

and

$$E(|w_n(x) - w(x)|^\delta) \leq O(n^{-\delta/2}).$$

The following theorem presents the convergence rate of the empirical Bayes estimator $\phi_n(x)$ proposed in (22.3.15). In the rest of the chapter, we shall use c_1, c_2 and c to denote some positive constants which may be different with the same notation.

Theorem 22.4.1 *Let $\{\phi_n(x)\}$ be the sequence of empirical Bayes estimator of θ proposed in (22.3.15); if for $0 < \delta < 1$,*

$$E\left(\theta^{3/(1-\delta)}\right) < \infty, \tag{22.4.26}$$

then with the choice of $h_n = n^{-1/(2[\alpha-1]+1)}$, we have

$$E_n\{R(G, \phi_n(x)))\} - R(G) \leq O(n^{-\delta[\alpha-1]/(2[\alpha-1]+1)}). \tag{22.4.27}$$

PROOF From the definition of $\psi_n(x)$, we have by Lemma 22.4.1 and Lemma 22.4.2,

$$
\begin{aligned}
E_n^*(|\psi_n(x) - \psi(x)|^2) &\leq E_n^*\left[\left(\left|\frac{w_n(x)}{f_n(x)} - \frac{w(x)}{f(x)}\right| \wedge x\right)^2\right] \\
&\leq E_n^*\left[x^{2-\delta}\left(\left|\frac{w_n(x)}{f_n(x)} - \frac{w(x)}{f(x)}\right| \wedge x\right)^\delta\right] \\
&\leq E\{2x^{2-\delta}f^{-\delta}[E(|w_n(x) - w(x)|^\delta) \\
&\qquad + \left(\frac{w(x)}{f(x)} + x\right)^\delta E(|f_n(x) - f(x)|^\delta)]\} \\
&\leq c_1 A_1(n^{-\delta/2}) + c_2 A_2(n^{-\delta[\alpha-1]/(2[\alpha-1]+1)}),
\end{aligned}
$$

where

$$A_1 = E(x^{2-\delta}f^{-\delta}) = \int x^{2-\delta}\left(\int f(x|\theta)dG(\theta)\right)^{1-\delta}dx$$

$$A_2 = E(x^2 f^{-\delta}) = \int x^2\left(\int f(x|\theta)dG(\theta)\right)^{1-\delta}dx.$$

To show that A_2 is finite, we use Hölder's inequality to observe that

$$\int_0^1 (x^2) \left(\int_0^x f(x|\theta)dG(\theta) \right)^{1-\delta} dx$$

$$\leq \left[\int_0^1 \left(\int_0^x f(x|\theta)dG(\theta) \right) dx \right]^{1-\delta} \left[\int_0^1 (x^{2/\delta})dx \right]^{\delta}$$

$$\leq c \left[\int_0^1 \left(\int_\theta^1 f(x|\theta)dx \right) dG(\theta) \right]^{1-\delta} < \infty$$

and

$$\int_1^\infty (x^2) \left(\int_0^x f(x|\theta)dG(\theta) \right)^{1-\delta} dx$$

$$= \int_1^\infty (x^{-1}) \left[x^3 \left(\int_0^x f(x|\theta)dG(\theta) \right)^{1-\delta} \right] dx$$

$$\leq \left[\int_1^\infty (x^{-1})^{1/\delta}dx \right]^{\delta} \left[\int_1^\infty x^{3/(1-\delta)} \left(\int_0^x f(x|\theta)dG(\theta) \right) dx \right]^{1-\delta}$$

$$\leq c \left(\int_0^\infty \int_\theta^\infty x^{3/(1-\delta)} f(x|\theta)dx \, dG(\theta) \right)^{1-\delta}$$

$$= c \left(\int_0^\infty \int_\theta^\infty x^{3/(1-\delta)} \frac{1}{\Gamma(\alpha)} (x-\theta)^{\alpha-1} e^{-(x-\theta)}dx \, dG(\theta) \right)^{1-\delta}$$

$$\leq c \left\{ \int_0^\infty \int_\theta^\infty \left[c_1 (x-\theta)^{3/(1-\delta)} + c_1 \theta^{3/(1-\delta)} \right] \right.$$

$$\left. \times \frac{1}{\Gamma(\alpha)} (x-\theta)^{\alpha-1} e^{-(x-\theta)}dx \, dG(\theta) \right\}^{1-\delta}$$

$$\leq \left[\int_0^\infty \left(c_1 + c_2 \theta^{3/(1-\delta)} \right) dG(\theta) \right]^{1-\delta} < \infty$$

by the assumption. Similarly, we can show that A_1 is also finite under the assumption. Then, with the choice of $h_n = n^{-1/(2[\alpha-1]+1)}$, we have

$$E_n^*(|\psi_n(x) - \psi(x)|^2) \leq cn^{-\delta[\alpha-1]/(2[\alpha-1]+1)}.$$

Therefore, we obtain from the definition of $\phi_n(x)$,

$$E_n\{R(G, \phi_n)\} - R(G) = E_n^*(|\psi_n(x) - \psi(x)|^2)$$
$$\leq O(n^{-\delta[\alpha-1]/(2[\alpha-1]+1)}).$$

Hence, the theorem.

22.5 ASYMPTOTIC OPTIMALITY OF THE EMPIRICAL BAYES TESTING RULE

In this section, we examine the convergence rate of the empirical Bayes testing rule $d_n(x)$ proposed in (22.3.16). From (22.2.7), we have the Bayes risk associated with the empirical Bayes testing rule $d_n(x)$ and the Bayes testing rule $d_G(x)$ as

$$r(G, d_n) = b \int \alpha(x) \, d_n(x) \, dx + C_G \qquad (22.5.28)$$

and

$$r(G) = r(G, d_G) = b \int \alpha(x) \, d_G(x) \, dx + C_G , \qquad (22.5.29)$$

respectively. Obviously,

$$E_n\{r(G, d_n)\} - r(G) \geq 0 \qquad (22.5.30)$$

since the Bayes testing rule d_G achieves the minimum Bayes risk $r(G)$; the expectation E_n is taken with respect to $(x_1, x_2, \ldots, x_n)$.

From Lemma 1 of Johns and Van Ryzin (1972), we have

$$E_n\{r(G, d_n)\} - r(G)$$

$$\leq b \int |\alpha(x)| \Pr\{|\alpha_n(x) - \alpha(x)| \geq |\alpha(x)|\} dx$$

$$\leq b \int |\alpha(x)|^{1-\gamma} \, E_n(|\alpha_n(x) - \alpha(x)|^\gamma) dx \qquad (22.5.31)$$

for $0 < \gamma < 1$.

Definition 22.5.1 A sequence of empirical Bayes testing rule $\{\delta_n\}$ is said to be asymptotically optimal at least of order β_n relative to the unknown prior G if

$$E_n\{r(G, \delta_n)\} - r(G) \leq O(\beta_n) \qquad \text{as } n \to \infty , \qquad (22.5.32)$$

where β_n is a sequence of positive numbers such that $\lim_{n \to \infty} \beta_n = 0$.

Now we present the convergence rate of the empirical Bayes testing rule $d_n(x)$ in (22.3.16) in the following theorem.

Theorem 22.5.1 Let $\{d_n(x)\}$ be the sequence of empirical Bayes testing rule proposed in (3.7); if for $0 < \delta < 1$,

$$E\left(\theta^{2/(1-\delta)}\right) < \infty, \qquad (22.5.33)$$

then with the choice of $h_n = n^{-1/(2[\alpha-1]+1)}$, *we have*

$$E_n\{r(G, d_n)\} - r(G) \le O(n^{-\delta[\alpha-1]/(2[\alpha-1]+1)}). \qquad (22.5.34)$$

PROOF With

$$E_n\{r(G, d_n)\} - r(G) \le b \int |\alpha(x)|^{1-\delta} E_n(|\alpha_n(x) - \alpha(x)|^{\delta}) dx$$

and since

$$\begin{aligned} \alpha(x) &= \int \theta f(x|\theta) dG(\theta) - \theta_0 f(x) \\ &= (x - \theta_0) f(x) - w(x) \end{aligned}$$

and

$$\alpha_n(x) = (x - \theta_0) f_n(x) - w_n(x),$$

we have, by Lemma 22.4.2,

$$\begin{aligned} &E_n\{r(G, d_n)\} - r(G) \\ \le\ & b \int \left| \int \theta f(x|\theta) dG(\theta) - \theta_0 f(x) \right|^{1-\delta} \\ & \times E_n(|x(f_n(x) - f(x)) - \theta_0(f_n(x) - f(x)) - (w_n(x) - w(x))|^{\delta}) dx \\ \le\ & O(n^{-\delta[\alpha-1]/(2[\alpha-1]+1)})(A_1 + A_2 + A_3 + A_4), \end{aligned}$$

where

$$\begin{aligned} A_1 &= \int x^{\delta} \left(\int \theta f(x|\theta) dG(\theta) \right)^{1-\delta} dx \\ A_2 &= \int \left(\int \theta f(x|\theta) dG(\theta) \right)^{1-\delta} dx \\ A_3 &= \int x^{\delta} \left(\int f(x|\theta) dG(\theta) \right)^{1-\delta} dx \\ A_4 &= \int \left(\int f(x|\theta) dG(\theta) \right)^{1-\delta} dx. \end{aligned}$$

To show that A_1 is finite, we use Hölder's inequality to observe

$$\int_0^1 (x^{\delta}) \left[\left(\int_0^x \theta f(x|\theta) dG(\theta) \right)^{1-\delta} \right] dx$$

$$\leq \left[\int_0^1 x\,dx\right]^\delta \left[\int_0^1 \left(\int_0^x \theta f(x|\theta)dG(\theta)\right)dx\right]^{1-\delta}$$

$$\leq c\left[\int_0^1 \left(\int_\theta^1 \theta f(x|\theta)dx\right)dG(\theta)\right]^{1-\delta}$$

$$\leq c\left(\int_0^\infty \theta dG(\theta)\right)^{1-\delta} < \infty$$

and

$$\int_0^\infty (x^\delta)\left[\left(\int_0^x \theta f(x|\theta)dG(\theta)\right)^{1-\delta}\right]dx$$

$$= \int_1^\infty (x^{-1})\left[x^{1+\delta}\left(\int_0^x \theta f(x|\theta)dG(\theta)\right)^{1-\delta}\right]dx$$

$$\leq \left[\int_1^\infty (x^{-1})^{1/\delta}dx\right]^\delta \left[\int_1^\infty \left(x^{1+\delta/(1-\delta)}\right)\left(\int_0^x \theta f(x|\theta)dG(\theta)\right)^{1-\delta}dx\right]^{1-\delta}$$

$$\leq c\left[\int_0^\infty \int_\theta^\infty \left(x^{1+\delta/(1-\delta)}\right)(\theta f(x|\theta))dx\,dG(\theta)\right]^{1-\delta}$$

$$= c\left[\int_0^\infty \int_\theta^\infty \theta x^{1+\delta/(1-\delta)}\frac{1}{\Gamma(\alpha)}(x-\theta)^{\alpha-1}e^{-(x-\theta)}dx\,dG(\theta)\right]^{1-\delta}$$

$$\leq c\left\{\int_0^\infty \int_\theta^\infty \theta\left[c_1(x-\theta)^{1+\delta/(1-\delta)} + c_1\theta^{1+\delta/(1-\delta)}\right]\right.$$

$$\left.\times \frac{1}{\Gamma(\alpha)}(x-\theta)^{\alpha-1}e^{-(x-\theta)}dx\,dG(\theta)\right\}^{1-\delta}$$

$$\leq \left[\int_0^\infty \left(c_1\theta + c_2\theta^{2/(1-\delta)}\right)dG(\theta)\right]^{1-\delta} < \infty$$

by the assumption. We can similarly show that A_2, A_3 and A_4 are also finite. Then, we finally obtain that

$$E_n\{r(G, d_n)\} - r(G) \leq O\left(n^{-\delta[\alpha-1]/(2[\alpha-1]+1)}\right).$$

Hence, the theorem.

REMARKS

1. The location parameter family of gamma distributions in (22.2.1) is not included in the typical truncation parameter density family

as Datta (1991) and Huang (1995) considered in their papers. The importance of (22.2.3) and (22.2.4) by Fox (1978) is that it gives us an explicit expression for the Bayes estimator $\phi_G(x)$ in terms of the marginal distribution of x, which enables us to estimate $\phi_g(x)$ from the past observations $x_1, x_2, \ldots, x_n$.

2. The convergence rates in Theorem 22.4.1 and Theorem 22.5.1 are dependent on δ, $0 < \delta < 1$, and the shape parameter α. If α is bigger, then the convergence rates re faster. If conditions of Theorem 22.4.1 and Theorem 22.5.1 are satisfied for δ arbitrarily close to 1, then the convergence rates can be arbitrarily close to $O(n^{-[\alpha-1]/(2[\alpha-1]+1)})$.

3. For the more general location parameter family of gamma distributions when location parameter $\theta \in (-\infty, \infty)$ instead of $\theta \in (0, \infty)$ as in (22.2.1), the relation $0 \leq \phi_G(x) = E(\theta|x) \leq x$ is no longer true, we can just get $\phi_G(x) \leq x$. The we can similarly propose empirical Bayes estimator for both θ and empirical Bayes testing rule for the two-action problem and obtain convergence rates under some moment conditions on the prior G. But, since the distribution family (22.2.1) is more useful in application, we just present the asymptotic optimality results for the distribution family (22.2.1) in Theorem 22.4.1 and Theorem 22.5.1 in this chapter.

Acknowledgements The first author thanks the Natural Sciences and Engineering Research Council of Canada and the second author thanks the Canadian International Development Agency for supporting this research. The authors also thank Mrs. Debbie Iscoe for the excellent typing of the manuscript.

REFERENCES

Balakrishnan, N. and Cohen, A. C. (1991). *Order Statistics and Inference: Estimation Methods*, Academic Press, San Diego.

Datta, S. (1991). nonparametric empirical Bayes estimation with $O(n^{-1/2})$ rate of truncation parameter, *Statistics & Decisions*, **9**, 45–61.

Fox, R. (1978). Solutions to empirical Bayes squared error loss estimation problems, *Annals of Statistics*, **6**, 846–853.

Huang, S. Y. (1995). Empirical Bayes testing procedures in some nonexponential families using asymmetric linear loss function, *Journal of Statistical Planning and Inference*, **46**, 293–309.

Johns, M. V. and Van Ryzin, J. (1971). Convergence rates for empirical Bayes two-action problem I. Discrete case, *Annals of Mathematical Statistics*, **42**, 1521–1539.

Johns, M. V. and Van Ryzin, J. (1972). Convergence rates for empirical Bayes two-action problem II. Continuous case, *Annals of Mathematical Statistics*, **43**, 934–947.

Johnson, N.L., Kotz, S. and Balakrishnan, N. (1994). *Continuous Univariate Distributions - Vol. 1*, Second edition, John Wiley & Sons, New York.

Lin, P. E. (1975). Rates of convergence in empirical Bayes estimation problems: continuous case, *Annals of Statistics*, **3**, 155–164.

Nogami, Y. (1988). Convergence rates for empirical Bayes estimation in the uniform $U(0, \theta)$ distribution, *Annals of Statistics*, **16**, 1335–1341.

Robbins, H. (1955). An empirical Bayes approach to Statistics, *Proceedings of the Third Berkeley Symposium on Mathematical Statistics and Probability*, Vol. 1, pp. 157-163, University of California Press, Berkeley, CA.

Robbins, H. (1963). The empirical Bayes approach to testing statistical hypothesis, *Review of the International Statistical Institute*, **31**, 195–208.

Robbins, H. (1964). The empirical Bayes approach to statistical decision problems, *Annals of Mathematical Statistics*, **35**, 1–10.

Singh, R. S. (1977). Improvement on some known nonparametric uniformly consistent estimators of derivatives of a density, *Annals of Statistics*, **5**, 394–399.

Singh, R. S. (1979). Empirical Bayes estimation in Lebesgue exponential families with rates near the best possible rate, *Annals of Statistics*, **7**, 890–902.

Singh, R. S. (1995). Empirical Bayes linear loss hypothesis testing in a non-regular exponential family, *Journal of Statistical Planning and Inference*, **43**, 107–120.

Singh, R. S. and Wei, L. (1992). Empirical Bayes with rates and best rates of convergence in $u(x) \, c(\theta) \, \exp(-x/\theta)$-family: estimation case, *Annals of the Institute of Statistical Mathematics*, **44**, 435–449.

Van Houwelingen, J. C. (1987). Monotone empirical Bayes test for uniform distribution using the maximum likelihood estimator of a decreasing density, *Annals of Statistics*, **15**, 875–879.

CHAPTER 23

RATE OF CONVERGENCE FOR EMPIRICAL BAYES ESTIMATION OF A DISTRIBUTION FUNCTION

TACHEN LIANG

Wayne State University, Detroit, MI

Abstract: The paper considers nonparametric empirical Bayes estimation of a distribution function having a Dirichlet process prior $\mathcal{D}(\alpha)$. For the case when the size $\alpha(R)$ is unknown, it is shown that the proposed empirical Bayes estimators are asymptotically optimal with order $O(n^{-1})$, where n is the number of data at hand for the present estimation problem. Therefore, the results of Korwar and Hollander (1976) and Ghosh, Lahiri and Tiwari (1989), in which $\alpha(R)$ is assumed to be known and a rate of order $O(n^{-1})$ is achieved, are extended to $\alpha(R)$ unknown case.

Keywords and phrases: Asymptotically optimal, Dirichlet process, empirical Bayes, regret Bayes risk, rate of convergence

23.1 INTRODUCTION

Consider a pair of random elements $(F, \mathbf{X})$, where F is a random distribution, distributed according to the Dirichlet process with parameter α, denoted by $\mathcal{D}(\alpha)$, and $\mathbf{X} = (X_1, \ldots, X_m)$ is a sample of size m arising from the distribution F. Here, α is a finite, nonnull measure defined on the Borel σ-field $\mathcal{B}$ of the real line R. We are interested in the estimation of the distribution function F. Let both the parameter space and the action space be the sets of all distribution functions on R. For an estimator G of

331

the distribution function F, we consider the integrated squared error loss function

$$L(F, G) = \int_R [G(t) - F(t)]^2 dW(t), \qquad (23.1.1)$$

where $W(t)$ is a given, finite measure over the measurable space $(R, \mathcal{B})$.

Under the precedingly described statistical model, Ferguson's (1973) Bayes estimator of F based on $\mathbf{X}$ is given by

$$G_\alpha(t) = (1 - p)F_0(t) + p\widehat{F}(t), \qquad (23.1.2)$$

where $p = m/[m + \alpha(R)]$, $F_0(t) = \alpha((-\infty, t])/\alpha(R)$, and $\widehat{F}(t)$ is the empirical distribution function of $\mathbf{X}$. The Bayes risk of the Bayes estimator G_α is:

$$r(G_\alpha, \alpha) = E_{\mathbf{X}}\{ \int_R E_{\alpha|\mathbf{X}} [G_\alpha(t) - F(t)]^2 dW(t) \}, \qquad (23.1.3)$$

where $E_{\alpha|\mathbf{X}}$ denotes the expectation taken with respect to the posterior distribution of the Dirichlet process $\mathcal{D}(\alpha)$ given $\mathbf{X}$ and $E_{\mathbf{X}}$ is the expectation taken with respect to the marginal probability measure generated by $\mathbf{X}$. Note that the posterior distribution of the Dirichlet process $\mathcal{D}(\alpha)$ given $\mathbf{X}$ is a Dirichlet process $\mathcal{D}(\alpha(\mathbf{X}))$ with parameter $\alpha(\mathbf{X}) = \alpha + \sum_{i=1}^m \delta_{X_i}$, where for each $x \in R$, and $A \in \mathcal{B}$, $\delta_x(A) = 1$ if $x \in A$, $= 0$ otherwise.

When the parameter $\alpha(\cdot)$ is unknown, it is not possible to apply the Bayes estimator G_α for the estimation problem. In such a situation, Korwar and Hollander (1976), Zehnwirth (1981) and Ghosh, Lahiri and Tiwari (1989) have studied this estimation problem via the empirical Bayes approach, respectively. When the size $\alpha(R)$ is known, Korwar and Hollander (1976) and Ghosh, Lahiri and Tiwari (1989) proved that their proposed empirical Bayes estimators are asymptotically optimal of order $O(n^{-1})$, respectively, where n is the number of data at hand for the current estimation problem. When the size $\alpha(R)$ is unknown, Zehnwirth (1981) and Ghosh, Lahiri and Tiwari (1989) respectively, have studied certain empirical Bayes estimators. However, neither of the two papers discussed the rate of convergence associated with the proposed empirical Bayes estimators.

In this paper, we investigate the asymptotic optimality of two empirical Bayes estimators. It is shown that under the regularity condition that $\int x^4 d\alpha(-\infty, x] < \infty$, the underlying empirical Bayes estimators are asymptotically optimal of order $O(n^{-1})$. Hence, the results of Korwar and Hollander (1976) and Ghosh, Lahiri and Tiwari (1989) are extended to $\alpha(R)$ unknown case.

23.2   THE EMPIRICAL BAYES ESTIMATORS

In the empirical Bayes framework, let $(F_i, \mathbf{X}_i), 1 = 1, 2, \ldots$, be a sequence of independent pairs of random elements, where for each $i = 1, 2, \ldots, \mathbf{X}_i = (X_{i1}, \ldots, X_{im_i})$ is a sample of size m_i taken, at stage i, from a distribution F_i, and $F_1, F_2, \ldots$ are iid random distributions, having the common Dirichlet process prior $\mathcal{D}(\alpha)$. At stage n, let $\mathbf{X}(n-1) = (\mathbf{X}_1, \ldots, \mathbf{X}_{n-1})$ denote the historical data, and $\mathbf{X}_n$ the present sample. We want to estimate the current distribution function F_n based on $\mathbf{X}_n$ and $\mathbf{X}(n-1)$. Such an estimator is called an empirical Bayes estimator, and is denoted by G_n. Note that the data $\mathbf{X}(n)$ is implicitly contained in the subscript n. The conditional Bayes risk of the empirical Bayes estimator G_n given $\mathbf{X}(n-1)$ is:

$$r(G_n, \alpha | \mathbf{X}(n-1)) = E_{\mathbf{X}_n}\{ \int_R E_{\alpha | \mathbf{X}_n}[G_n(t) - F_n(t)]^2 dW(t)\},$$

and the unconditional Bayes risk of G_n is:

$$
\begin{aligned}
r(G_n, \alpha) &= E_{\mathbf{X}(n-1)} r(G_n, \alpha | \mathbf{X}(n-1)) \\
&= E_{\mathbf{X}_n}\{ \int_R E_{\mathbf{X}(n-1)} E_{\alpha | \mathbf{X}_n}[G_n(t) - F_n(t)]^2 dW(t)\}
\end{aligned}
$$

$$(23.2.4)$$

where the expectation $E_{\mathbf{X}(n-1)}$ is taken with respect to the marginal distribution of $\mathbf{X}(n-1)$.

Let $G_{n,\alpha}$ denote the Bayes estimator of F_n given $\mathbf{X}_n$. That is,

$$G_{n,\alpha}(t) = (1 - p_n)F_0(t) + p_n \hat{F}_n(t), \qquad (23.2.5)$$

where $p_i = m_i / [m_i + \alpha(R)]$, $\hat{F}_i(t)$ is the empirical distribution function of $\mathbf{X}_i, i = 1, \ldots, n$. Then, $r(G_n, \alpha | \mathbf{X}(n-1)) \geq r(G_{n,\alpha}, \alpha)$ for all $\mathbf{X}(n-1)$ and for all n. Hence $r(G_n, \alpha) - r(G_{n\alpha}, \alpha) \geq 0$ for all n. This nonnegative regret Bayes risk $r(G_n, \alpha) - r(G_{n,\alpha}, \alpha)$ is used as a measure of performance of the empirical Bayes estimator G_n.

A sequence of empirical Bayes estimators $\{G_n\}$ is said to be asymptotically optimal relative to the Dirichlet process $\mathcal{D}(\alpha)$ if $r(G_n, \alpha) - r(G_{n,\alpha}, \alpha) = o(1)$. $\{G_n\}$ is said to be asymptotically optimal of order $\{\beta_n\}$ relative to the Dirichlet process $\mathcal{D}(\alpha)$ if $r(G_n, \alpha) - r(G_{n,\alpha}, \alpha) = O(\beta_n)$ where $\{\beta_n\}$ is a sequence of positive numbers such that $\lim_{n \to \infty} \beta_n = 0$.

For $m_1 = \cdots = m_n = m$ and $\alpha(R)$ be known case, Korwar and Hollander (1976) investigated the asymptotic optimality of an empirical Bayes estimator, say G_n^{KH}, and established that $\{G_n^{KH}\}$ is asymptotically optimal of order $O(n^{-1})$. Note that in the case where $m_1 = \cdots = m_n = m, G_{n,\alpha} = G_\alpha$ for all $n = 1, 2, \ldots$. For $m_1 = \cdots = m_n = m$ but $\alpha(R)$ unknown case, Zehnwirth (1981) proposed an empirical Bayes estimator G_n^Z and proved that G_n^Z possesses the asymptotic optimality. However, the rate of convergence of $\{G_n^Z\}$ was not discussed. For general sample sizes case where $m_1, \cdots, m_n$ may not be equal, Ghosh, Lahiri and Tiwari (1989) proposed empirical Bayes estimators for F_n for each of the two cases: $\alpha(R)$ being known or unknown. For $\alpha(R)$ known case, their proposed empirical Bayes estimator, say G_n^{GLT1}, is asymptotically optimal of order $O(n^{-1})$. However for the $\alpha(R)$ unknown case, they only established the asymptotic optimality of their proposed empirical Bayes estimator, say G_n^{GLT2}, and no rate of convergence regarding $\{G_n^{GLT2}\}$ is discussed.

In this chapter, we study empirical Bayes estimators for F_n for $\alpha(R)$ unknown case. First, we introduce certain notations as follows. Let

$$
\begin{cases}
\bar{X}_{i\cdot} & = \frac{1}{m_i}\sum_{\ell=1}^{m_i} X_{i\ell}, \bar{X}_{\cdot\cdot} = \sum_{i=1}^n m_i \bar{X}_{i\cdot}/(\sum_{i=1}^n m_i); \\
MS_W & = \sum_{i=1}^n \sum_{\ell=1}^{m_i} (X_{i\ell} - \bar{X}_{i\cdot})^2/\sum_{i=1}^n (m_i - 1); \qquad (23.2.6) \\
MS_B & = \sum_{i=1}^n m_i (\bar{X}_{i\cdot} - \bar{X}_{\cdot\cdot})^2/(n-1).
\end{cases}
$$

A straightforward computation yields that

$$
\begin{cases}
E_{\mathbf{X}(n)}[MS_W] & = E_\alpha[Var_F(X)], \\
E_{\mathbf{X}(n)}[MS_B] & = E_\alpha[Var_F(X)] + \frac{g(m_1,\ldots,m_n)}{(n-1)} Var_\alpha[E_F(X)],
\end{cases}
$$
$$(23.2.7)$$

where $g(m_1, \ldots, m_n) = \sum_{i=1}^n m_i - \frac{\sum_{i=1}^n m_i^2}{\sum_{i=1}^n m_i}$. Therefore,

$$
E_{\mathbf{X}(n)}[\frac{(n-1)}{g(m_1,\ldots,m_n)}(MS_B - MS_W)] = Var_\alpha[E_F(X)]. \qquad (23.2.8)
$$

By the identity $\alpha(R) = E_\alpha[Var_F(X)]/Var_\alpha[E_F(X)]$ [see Zehnwirth (1977)] and by (23.2.7) and (23.2.8),

$$
\begin{aligned}
p_i & = \frac{m_i}{m_i + \alpha(R)} \\
& = \frac{m_i Var_\alpha[E_F(X)]}{m_i Var_\alpha[E_F(X)] + E_\alpha[Var_F(X)]} \\
& \equiv p_{in}. \qquad (23.2.9)
\end{aligned}
$$

We use MS_W to estimate $E_\alpha[Var_F(X)]$. Since $Var_\alpha[E_F(X)] > 0$ while $MS_B - MS_W$ may be negative, we therefore use $\frac{(n-1)}{g(m_1,\ldots,m_n)}(MS_B - MS_W)^+$

to estimate $Var_\alpha[E_F(X)]$, where $y^+ = \max(0, y)$. For each $i = 1, \ldots, n$, we estimate $p_i = p_{in}$ by

$$\hat{p}_{in} = \frac{m_i(n-1)g^{-1}(m_1, \ldots, m_n)(MS_B - MS_W)^+}{m_i(n-1)g^{-1}(m_1, \ldots, m_n)(MS_B - MS_W)^+ + MS_W}. \qquad (23.2.10)$$

We consider two empirical estimators G_n^* and $\widehat{G}_n$ defined as follows.

Empirical Bayes estimator G_n^*

Let $F_{on}^*(t) = \frac{1}{n}\sum_{i=1}^n \widehat{F}_i(t)$. Then, define

$$G_n^*(t) = (1 - \hat{p}_{nn})F_{on}^*(t) + \hat{p}_{nn}\widehat{F}_n(t). \qquad (23.2.11)$$

Empirical Bayes estimator $\widehat{G}_n$

Let

$$\widehat{F}_{on}(t) = \begin{cases} \frac{\sum_{i=1}^n \hat{p}_{in}\widehat{F}_i(t)}{\sum_{i=1}^n \hat{p}_{in}} & \text{if } MS_B - MS_W > 0, \\ F_{on}^*(t) & \text{if } MS_B - MS_W \le 0. \end{cases}$$

Then, define

$$\widehat{G}_n(t) = (1 - \hat{p}_{nn})\widehat{F}_{on}(t) + \hat{p}_{nn}\widehat{F}_n(t). \qquad (23.2.12)$$

23.3   ASYMPTOTIC OPTIMALITY

From (23.1.3), (23.2.4) and by an argument similar to that in Theorem 2.4 of Korwar and Hollander (1976) or in Theorem 2 of Ghosh, Lahiri and Tiwari (1989), it follows that

$$\begin{aligned} 0 &\le r(G_n^*, \alpha) - r(G_{n,\alpha}, \alpha) \\ &= \int_R E_{\mathbf{X}(n)}[G_n^*(t) - G_{n,\alpha}(t)]^2 dW(t). \qquad (23.3.13) \end{aligned}$$

From (23.2.5) and (23.2.11)

$$G_n^*(t) - G_{\alpha,n}(t) = (\hat{p}_{nn} - p_{nn})[\widehat{F}_n(t) - F_{on}^*(t)] + (1 - p_{nn})[F_{on}^*(t) - F_0(t)].$$

Therefore,

$$\begin{aligned} [G_n^*(t) &- G_{n,\alpha}(t)]^2 \\ &\le 2(\hat{p}_{nn} - p_{nn})^2[\widehat{F}_n(t) - F_{on}^*(t)]^2 + 2(1 - p_{nn})^2[F_{on}^*(t) - F_0(t)]^2 \\ &\le 2(\hat{p}_{nn} - p_{nn})^2 + 2[F_{on}^*(t) - F_0(t)]^2 \qquad (23.3.14) \end{aligned}$$

since $0 \leq \widehat{F}_n(t), F_{on}^*(t), p_{nn} \leq 1$. Combining (23.3.13) and (23.3.14) yields

$$r(G_n^*, \alpha) - r(G_{n,\alpha}, \alpha)$$
$$\leq \int_R 2E_{\mathbf{X}(n)}[\hat{p}_{nn} - p_{nn}]^2 dW(t) + \int_R 2E_{\mathbf{X}(n)}[F_{on}^*(t) - F_0(t)]^2 dW(t).$$
$$(23.3.15)$$

To establish the asymptotic optimality, we need the following helpful lemmas.

Lemma 23.3.1

(a) $E_{\mathbf{X}(n)}\widehat{F}_i(t) = F_0(t)$, $i = 1, 2, \ldots, n$, $E_{\mathbf{X}(n)}F_{on}^*(t) = F_0(t)$.

(b) $Var_{\mathbf{X}(n)}[\widehat{F}_i(t)] = \frac{F_0(t)}{\alpha(R)+1}[\frac{\alpha([t,\infty))}{m_i} + \alpha((-\infty, t]) + 1] - [F_0(t)]^2$,
$\leq F_0(t)(1 - F_0(t))$, $i = 1, \ldots, n$.

and

$$Var(F_{on}^*(t)) = \frac{1}{n^2}\sum_{i=1}^{n} Var_{\mathbf{X}(n)}(\hat{F}_i(t))$$
$$= \frac{F_0(t)}{n^2[\alpha(R)+1]}\sum_{i=1}^{n}[\frac{\alpha((t,\infty])}{m_i} + \alpha((-\infty, t]) + 1] - \frac{[F_0(t)]^2}{n}.$$

The result of Lemma 23.3.1 can be obtained via a straightforward computation. The details are omitted here.

Lemma 23.3.2 *Let* $V = Var_\alpha[E_F(X)]$ *and* $E = E_\alpha[Var_F(X)]$. *Then,*

(a) $E_{\mathbf{X}(n)}[MS_W - E]^2 \leq \frac{4}{\sum_{i=1}^n(m_i-1)}E_{\mathbf{X}(n)}[X_{11}^4]$;

(b) $E_{\mathbf{X}(n)}[MS_B - E - \frac{g(m_1,\ldots,m_n)}{(n-1)}V]^2 \leq \frac{2\sum_{i=1}^n m_i^2}{(n-1)^2}E_{\mathbf{X}(n)}[X_{11}^4]$;

(c) $E_{\mathbf{X}(n)}[\frac{(n-1)}{g(m_1,\ldots,m_n)}(MS_B-MS_W)^+ - V]^2 \leq \frac{2(n-1)^2}{g^2(m_1,\ldots,m_n)}\{E_{\mathbf{X}(n)}[MS_B - E - \frac{g(m_1,\ldots,m_n)}{(n-1)}V]^2 + E_{\mathbf{X}(n)}[MS_W - E]^2\}$.

PROOF (a) First, for each $i = 1, \ldots, n$,

$$Var_{\mathbf{X}(n)}[\sum_{i=1}^{m_i}(X_{i\ell} - \overline{X}_{i\cdot})^2] \leq E_{\mathbf{X}(n)}[\sum_{i=1}^{m_i}(X_{i\ell} - \overline{X}_{i\cdot})^2]^2$$

$$\leq \; E_{\mathbf{X}(n)}[\sum_{i=1}^{m_i} X_{i\ell}^2]^2$$

$$\leq \; E_{\mathbf{X}(n)}[2\sum_{\ell=1}^{m_i} X_{i\ell}^4]$$

$$= \; 2m_i E_{\mathbf{X}(n)}[X_{i1}^4]$$

$$= \; 2m_i E_{\mathbf{X}(n)}[X_{11}^4].$$

Therefore, by (23.2.6) and (23.2.7),

$$E_{\mathbf{X}(n)}[MS_W - E]^2 \; = \; Var_{\mathbf{X}(n)}[MS_W]$$

$$= \; \frac{1}{[\sum_{i=1}^{n}(m_i - 1)]^2}\sum_{i=1}^{n} Var_{\mathbf{X}(n)}[\sum_{i=1}^{m_i}(X_{i\ell} - \overline{X}_{i.})^2]$$

$$\leq \; \frac{2\sum_{i=1}^{n} m_i}{[\sum_{i=1}^{n}(m_i - 1)]^2}E_{\mathbf{X}(n)}[X_{11}^4].$$

$$\leq \; \frac{4}{\sum_{i=1}^{n}(m_i - 1)}E_{\mathbf{X}(n)}[X_{11}^4].$$

(b) By (23.2.6) and (23.2.7) again,

$$E_{\mathbf{X}(n)}[MS_B - E - \frac{g(m_1,\ldots,m_n)}{(n-1)}V]^2$$

$$= Var_{\mathbf{X}(n)}[MS_B]$$

$$\leq \frac{1}{(n-1)^2}E_{\mathbf{X}(n)}[\sum_{i=1}^{n} m_i(\overline{X}_{i.} - \overline{X}_{..})^2]^2$$

$$\leq \frac{1}{(n-1)^2}E_{\mathbf{X}(n)}[\sum_{i=1}^{n} m_i\overline{X}_{i.}^2]^2$$

$$\leq \frac{2}{(n-1)^2}\sum_{i=1}^{n} m_i^2 E_{\mathbf{X}(n)}[\overline{X}_{i.}^4]$$

$$\leq \frac{2}{(n-1)^2}\sum_{i=1}^{n} m_i^2 E_{\mathbf{X}(n)}[X_{i1}^4]$$

$$= \frac{2\sum_{i=1}^{n} m_i^2}{(n-1)^2}E_{\mathbf{X}(n)}[X_{11}^4].$$

(c) Since $V > 0$, we have

$$E_{\mathbf{X}(n)}[\frac{(n-1)}{g(m_1,\ldots,m_n)}(MS_B - MS_W)^+ - V]^2$$

$$\leq \quad E_{\mathbf{X}(n)}[\frac{(n-1)}{g(m_1,\ldots,m_n)}(MS_B - MS_W) - V]^2$$

$$= \quad \frac{(n-1)^2}{g^2(m_1,\ldots,m_n)}E_{\mathbf{X}(n)}[(MS_B - E - \frac{g(m_1,\ldots,m_n)}{(n-1)}V)$$

$$-(MS_W - E)]^2$$

$$\leq \quad \frac{2(n-1)^2}{g^2(m_1,\ldots,m_n)}\{E_{\mathbf{X}(n)}[MS_B - E - \frac{g(m_1,\ldots,m_n)}{(n-1)}V]^2$$

$$+E_{\mathbf{X}(n)}[MS_W - E]^2\}.$$

$\square$

Lemma 23.3.3 *For each* $i = 1,\ldots,n,$

$$E_{\mathbf{X}(n)}[\hat{p}_{in} - p_{in}]^2 \leq \frac{18m_i^2 M_4}{(m_i V + E)^2}\{\frac{4b_n^2 \sum_{i=1}^n m_i^2}{(n-1)^2} + \frac{4(2b_n^2 + 1)}{\sum_{i=1}^n (m_i - 1)}\}$$

where $b_n = \frac{n-1}{g(m_1,\ldots,m_n)}$ *and* $M_4 = E_{\mathbf{X}(n)}[X_{11}^4].$

PROOF From (23.2.9) and (23.2.10), $0 \leq \hat{p}_{in}, p_{in} \leq 1$. Then by Singh's inequality [see Singh (1977)], and Lemma 23.3.2,

$$E_{\mathbf{X}(n)}[\hat{p}_{in} - p_{in}]^2$$

$$\leq \quad \frac{2}{(m_i V + E)^2}\left\{\begin{array}{l} m_i^2 E_{\mathbf{X}(n)}[b_n(MS_B - MS_W)^+ - V]^2 \\ +4E_{\mathbf{X}(n)}[m_i b_n(MS_B - MS_W)^+ + MS_W \\ -m_i V - E]^2 \end{array}\right\}$$

$$\leq \quad \frac{2}{(m_i V + E)^2}\left\{\begin{array}{l} m_i^2 E_{\mathbf{X}(n)}[b_n(MS_B - MS_W)^+ - V]^2 \\ +8m_i^2 E_{\mathbf{X}(n)}[b_n(MS_B - MS_W)^+ - V]^2 \\ +8E_{\mathbf{X}(n)}[MS_W - E]^2 \end{array}\right\}$$

$$\leq \quad \frac{18m_i^2}{(m_i V + E)^2}\left\{\begin{array}{l} E_{\mathbf{X}(n)}[b_n(MS_B - MS_W)^+ - V]^2 \\ +E_{\mathbf{X}(n)}[MS_W - E]^2 \end{array}\right\}$$

$$\leq \quad \frac{18m_i^2}{(m_i V + E)^2}\left\{\begin{array}{l} 2b_n^2 E_{\mathbf{X}(n)}[MS_B - E - b_n^{-1}V]^2 \\ +(2b_n^2 + 1)E_{\mathbf{X}(n)}[MS_W - E]^2 \end{array}\right\}$$

$$\leq \quad \frac{18m_i^2}{(m_i V + E)^2}\left\{ 4b_n^2 \frac{\sum_{i=1}^n m_i^2}{(n-1)^2}M_4 + \frac{(2b_n^2 + 1)4M_4}{\sum_{i=1}^n (m_i - 1)} \right\}.$$

$\square$

Theorem 23.3.1 *Suppose that*

(a) The Dirichlet process $\mathcal{D}(\alpha)$ satisfies that $\int x^4 d\alpha(-\infty, x]) < \infty$;

(b) $2 \leq m_i \leq m$ for all $i = 1, 2, \ldots$, where the value of m is independent of $i = 1, 2, \ldots$.

Then, $r(G_n^*, \alpha) - r(G_{n,\alpha}, \alpha) = O(n^{-1})$.

PROOF   From Lemma 23.3.1, and since $W(t)$ is a finite measure on $(R, \mathcal{B})$, so,

$$\int_R E_{\mathbf{X}(n)}[F_{on}^*(t) - F_0(t)]^2 dW(t) = O(n^{-1}).$$

Next, note that $M_4 = E_{\mathbf{X}(n)}[X_{11}^4] = \int x^4 d\alpha((-\infty, x])/\alpha(R) < \infty$. This result and the assumption that $2 \leq m_i \leq m$ for all $i = 1, 2, \ldots$ together imply $E_{\mathbf{X}(n)}[\hat{p}_{in} - p_{in}]^2 = O(n^{-1})$ for all $i = 1 \ldots n$.

Now, the theorem follows from the above results and (23.3.15). $\square$

Next, we investigate the asymptotic optimality of the empirical Bayes estimator $\widehat{G}_n$. Similarly,

$$r(\widehat{G}_n, \alpha) - r(G_{n,\alpha}, \alpha)$$

$$\leq \int_R 2E_{\mathbf{X}(n)}[\hat{p}_{nn} - p_{nn}]^2 dW(t) + \int_R 2E_{\mathbf{X}(n)}[\hat{F}_{on} - F_0(t)]^2 dW(t)$$

$$= \int_R 2E_{\mathbf{X}(n)}[\hat{p}_{nn} - p_{nn}]^2 dW(t)$$

$$+ \int_R 2E_{\mathbf{X}(n)}\{[F_{on}^*(t) - F_0(t)]^2 I(MS_B - MS_W \leq 0)\}dW(t)$$

$$+ \int_R 2E_{\mathbf{X}(n)}\{[\frac{\sum \hat{p}_{in}\hat{F}_i(t)}{\sum \hat{p}_{in}} - F_0(t)]^2 I(MS_B - MS_W > 0)\}dW(t),$$

$$(23.3.16)$$

where $\int_R 2E_{\mathbf{X}(n)}[\hat{p}_{nn} - p_{nn}]^2 dW(t) = O(n^{-1})$ and

$$\int_R 2E_{\mathbf{X}(n)}\{[F_{on}^*(t) - F_0(t)]^2 I(MS_B - MS_W \leq 0)\}dW(t) = O(n^{-1})$$

which are obtained from the result of Theorem 23.3.1. Therefore, it suffices to investigate the asymptotic behavior of the third term of (23.3.16).

Note that

$$\left[\frac{\sum \hat{p}_{in} \hat{F}_i(t)}{\sum \hat{p}_{in}} - F_0(t)\right]^2$$

$$\leq 2\left[\frac{\sum \hat{p}_{in} \hat{F}_i(t)}{\sum \hat{p}_{in}} - \frac{\sum p_{in} \hat{F}_i(t)}{\sum p_{in}}\right]^2 + 2\left[\frac{\sum p_{in} \hat{F}_i(t)}{\sum p_{in}} - F_0(t)\right]^2 .$$

$$(23.3.17)$$

Hence,

$$E_{\mathbf{X}(n)}\left\{\left[\frac{\sum \hat{p}_{in} \hat{F}_i(t)}{\sum \hat{p}_{in}} - F_0(t)\right]^2 I(MS_B - MS_W > 0)\right\}$$

$$\leq 2E_{\mathbf{X}(n)}\left[\frac{\sum \hat{p}_{in} \hat{F}_i(t)}{\sum \hat{p}_{in}} - \frac{\sum p_{in} \hat{F}_i(t)}{\sum p_{in}}\right]^2$$

$$+2E_{\mathbf{X}(n)}\left[\frac{\sum p_{in} \hat{F}_i(t)}{\sum p_{in}} - F_0(t)\right]^2 , \qquad (23.3.18)$$

where

$$E_{\mathbf{X}(n)}\left[\frac{\sum p_{in} \hat{F}_i(t)}{\sum p_{in}} - F_0(t)\right]^2 = Var_{\mathbf{X}(n)}\left(\frac{\sum p_{in} \hat{F}_i(t)}{\sum p_{in}}\right)$$

$$= \frac{1}{(\sum p_{in})^2}\sum_{i=1}^{n} p_{in}^2 Var_{\mathbf{X}(n)}[\hat{F}_i(t)]$$

$$\leq \frac{(\sum p_{in}^2)}{(\sum p_{in})^2} F_0(t)(1 - F_0(t)), \qquad (23.3.19)$$

and by Singh's inequality and Lemma 23.3.3,

$$E_{\mathbf{X}(n)}\left[\frac{\sum \hat{p}_{in} \hat{F}_i(t)}{\sum \hat{p}_{in}} - \frac{\sum p_{in} \hat{F}_i(t)}{\sum p_{in}}\right]^2$$

$$\leq \frac{8}{(\sum p_{in})^2}\{E_{\mathbf{X}(n)}[\sum (\hat{p}_{in} - p_{in})\hat{F}_i(t)]^2 + E_{\mathbf{X}(n)}[\sum (\hat{p}_{in} - p_{in})]^2\}$$

$$\leq \frac{8}{(\sum p_{in})^2}\{2\sum_{i=1}^{n} E_{\mathbf{X}(n)}[(\hat{p}_{in}-p_{in})\hat{F}_i(t)]^2 + 2\sum_{i=1}^{n} E_{\mathbf{X}(n)}[\hat{p}_{in}-p_{in}]^2\}$$

$$\leq \frac{32}{(\sum p_{in})^2}\sum_{i=1}^{n} E_{\mathbf{X}(n)}[\hat{p}_{in}-p_{in}]^2$$

$$\leq \frac{32}{(\sum p_{in})^2}\sum_{i=1}^{n}\frac{72m_i^2 M_4}{(m_i V + E)^2}\left\{\frac{b_n^2 \sum m_i^2}{(n-1)^2} + \frac{2b_n^2+1}{\sum(m_i-1)}\right\} \qquad (23.3.20)$$

Under the assumption of Theorem 23.3.1 and from (23.3.19) and (23.3.20), we see that

$$E_{\mathbf{X}(n)}\left[\frac{\sum p_i \hat{F}_i(t)}{\sum p_{in}} - F_0(t)\right]^2 = O(n^{-1})$$

and

$$E_{\mathbf{X}(n)}\left[\frac{\sum \hat{p}_{in}\hat{F}_i(t)}{\sum \hat{p}_{in}} - \frac{\sum p_{in}\hat{F}_i(t)}{\sum p_{in}}\right]^2 = O(n^{-2}).$$

Therefore, we have the following theorem.

Theorem 23.3.2 *Under the assumptions of Theorem 23.3.1, $r(\hat{G}_n, \alpha) - r(G_{n,\alpha}, \alpha) = O(n^{-1})$.*

REFERENCES

Ferguson, T. S. (1973). A Bayesian analysis of some nonparametric problems, *Annals of Statistics*, 1, 209–230.

Ghosh, M., Lahiri, P. and Tiwari, R. C. (1989). Nonparametric empirical Bayes estimation of the distribution function and the mean, *Communications in Statistics—Theory and Methods*, 18, 121–146.

Korwar, R. M. and Hollander, M. (1976). Empirical Bayes estimation of a distribution function, *Annals of Statistics*, 4, 501–588.

Singh, R. S. (1977). Applications of estimators of a density and its derivatives to certain statistical problems, *Journal of the Royal Statistical Society, Series B*, 39, 357–363.

Zehnwirth, B. (1977). The mean credibility formula in a Bayes rule, *Scandinavian Actuarial Journal*, 4, 213–216.

Zehnwirth, B. (1981). A note on the asymptotic optimality of the empirical Bayes distribution function, *Annals of Statistics*, **9**, 221–224.

Part V
Selection Methods

CHAPTER 24

ON A SELECTION PROCEDURE FOR SELECTING THE BEST LOGISTIC POPULATION COMPARED WITH A CONTROL

SHANTI S. GUPTA

Purdue University, West Lafayette, IN

ZHENGYAN LIN

Hangzhou University, Hangzhou, China

XUN LIN

Purdue University, West Lafayette, IN

Abstract: In this paper we investigate the problem of selecting the best logistic population from $k(\geq 2)$ possible candidates. The selected population must also be better than a given control. We employ the empirical approach and develop a selection procedure. The performance (rate of convergence) of the proposed selection rule is also analyzed. We also carry out a simulation study to investigate the rate of convergence of the proposed empirical selection procedure. The results of the simulation study are provided in this chapter.

Keywords and phrases: Asymptotically optimal, empirical approach, selection procedure, logistic population, rate of convergence

345

24.1 INTRODUCTION

Logistic distributions have been widely used in studies that are related with growth processes. Berkson (1957) used the logistic distribution as a model to analyze quantal response. Plackett (1958) considered the use of the logistic distribution with life test data. The importance of the logistic distribution has resulted in numerous investigations involving the statistical aspects of the distribution. For example, Talacko (1956) showed that it could be a limiting distribution in various situations. Birnbaum and Dudman (1963), and Gupta and Shah (1965) studied its order statistics and their limiting properties. Gupta and Gnanadesikan (1966), and Gupta, Qureishi and Shah (1967) have considered the estimation of parameters of the logistic distribution, Gupta, Qureishi and Shah have constructed the best linear unbiased estimators of both location and scale parameters using order statistics.

It is now well recognized that the classical techniques for testing homogeneity hypotheses are inadequate to serve, in many practical situations, the experimenter's real purpose, which is to rank several competing populations or to select the best among them. Such realistic goals and formulations set the stage for the development of the ranking and selection theory. An important part of this development is the study of ranking and selection problems for specific parametric families of distributions including, of course, logistic distributions. Gupta and Han (1991) proposed an elimination type procedure based on the estimated sample means for selecting the best logistic population. In addition, Gupta and Han (1992) proposed another selection rule for selecting the best logistic population using the indifference zone approach. A very nice paper on ranking and selection procedures for the logistic populations is Panchapakesan (1992) which is published in "the Handbook of the Logistic Distribution", edited by Balakrishnan (1992). In this book one can find a good deal of recent developments related to the logistic distribution.

In this paper, we investigate the problem of selecting the best logistic population by using the observed sample medians. Assume that there are k independent logistic populations whose location parameters follow a prior normal distribution and the parameters of the prior normal distribution are unknown. Motivated by the empirical methodology, we propose an empirical selection procedure that is based on the past observed data.

24.2 FORMULATION OF THE SELECTION PROBLEM WITH THE SELECTION RULE

Let $\Pi_1, \ldots, \Pi_k$ be k independent logistic populations with unknown means $\theta_1, \ldots, \theta_k$. Let $\theta_{[1]} \leq \cdots \leq \theta_{[k]}$ denote the ordered values of the parameters $\theta_1, \ldots, \theta_k$. It is assumed that the exact pairing between the ordered and the unordered parameters is unknown. A population Π_i with $\theta_i = \theta_{[k]}$ is considered as the best population. For a given fixed control θ_0, population Π_i is defined to be good if the corresponding $\theta_i > \theta_0$, and bad otherwise. Our goal is to select the one which is the best among the k logistic populations and also good compared with the given standard θ_0. If there is no such population, we select none.

Let $\Omega = \{\underline{\theta} = (\theta_1, \ldots, \theta_k)\}$ be the parameter space and $\underline{a} = (a_0, \ldots, a_k)$ be an action, where $a_i = 0$, or 1, for $i = 0, 1, \ldots, k$, and $\sum_{i=0}^{k} a_i = 1$. For each $i = 1, \ldots, k$, $a_i = 1$ means population Π_i is selected as the best among the k candidates and also good compared with θ_0, while $a_i = 0$ means population Π_i is not selected either because it is not the best among the k candidates or because it is bad compared with the control. $a_0 = 1$ means that all the k populations are excluded as bad and none of these k logistic populations is selected. The following loss function will be considered:

$$L(\underline{\theta}, \underline{a}) = \max(\theta_{[k]}, \theta_0) - \sum_{i=0}^{k} a_i \theta_i.$$

For each $i = 1, \ldots, k$, let $X_{i1}, \ldots, X_{iM}$ be a sample of size M from the i-th logistic population $\Pi_i = L(\theta_i, \sigma_i^2)$ which has the following conditional density distribution given θ_i and σ_i^2

$$\frac{1}{\sigma_i} \frac{e^{-(x_i - \theta_i)/\sigma_i}}{(1 + e^{-(x_i - \theta_i)/\sigma_i})^2}, \qquad -\infty < x_i < \infty. \qquad (24.2.1)$$

For convenience, suppose (for now) M is an odd number, and we denote $M = 2s + 1$. Since logistic distribution is symmetric about its mean, the population mean and median are identical. We assume that for each $i = 1, \ldots, k$, the population median (and also the mean) θ_i is a realization of random variable Θ_i which follows a normal $N(\mu_i, \tau_i^2)$ prior distribution with parameters (μ_i, τ_i^2). The random variables $\Theta_1, \ldots, \Theta_k$ are mutually independent. σ_i^2, μ_i, τ_i^2 are unknown but fixed. In other words, σ_i^2, μ_i, τ_i^2 are fixed nuisance parameters. Let X_i be the median of $\{X_{i1}, \ldots, X_{iM}\}$, $i = 1, \ldots, k$, then the conditional distribution of X_i given (θ_i, σ_i^2) can be explicitly written out as follows:

$$f_i(x_i|\theta_i, \sigma_i^2) = \frac{(2s+1)!}{(s!)^2} \frac{1}{\sigma_i} \frac{(e^{-(x_i-\theta_i)/\sigma_i})^{s+1}}{(1 + e^{-(x_i-\theta_i)/\sigma_i})^{2s+2}}, \quad -\infty < x_i < \infty.$$

$$(24.2.2)$$

From (24.2.2) we see that the density function $f_i(x_i|\theta_i, \sigma_i^2)$ is symmetric about $\Theta_i = \theta_i$, therefore,

$$EX_i = E(E(X_i|\Theta_i = \theta_i)) = E\Theta_i = \mu_i. \qquad (24.2.3)$$

The posterior distribution density of Θ_i given $X_i = x_i$ is proportional to

$$\frac{(e^{-(x_i-\theta_i)/\sigma_i})^{s+1}}{(1 + e^{-(x_i-\theta_i)/\sigma_i})^{2s+2}} \cdot e^{-\frac{(\theta_i-\mu_i)^2}{2\tau_i^2}}, \qquad -\infty < \theta_i < \infty. \quad (24.2.4)$$

The selection procedure will be based on the sample medians X_i. An estimator of Θ_i given $X_i = x_i$ is the median of the posterior distribution of Θ_i. For $i = 1, \ldots, k$, denote $\varphi_i(x_i)$ to be the median of the posterior distribution of Θ_i given $X_i = x_i$.

Let $\underline{X} = (X_1, \ldots, X_k)$ and $\mathcal{X}$ be the sample space generated by $\underline{X}$. A selection procedure $\underline{d} = (d_0, \ldots, d_k)$ is a mapping defined on the sample space $\mathcal{X}$. For every $\underline{x} \in \mathcal{X}$, $d_i(\underline{x}), i = 1, \ldots, k$, is the probability of selecting population Π_i as the best among the k populations and also good compared with the given control θ_0, $d_0(\underline{x})$ is the probability of excluding all k populations as bad and selecting none. Also, $\sum_{i=0}^{k} d_i(\underline{x}) = 1$, for all $\underline{x} \in \mathcal{X}$.

We next derive a selection rule $d(\underline{x})$ based on the posterior median $\varphi_i(x_i), i = 1, \ldots, k$. For each $\underline{x} \in \mathcal{X}$, let $I(\underline{x}) = \{i|\varphi_i(x_i) = \max_{0 \le j \le k} \varphi_j(x_j), i = 0, 1, \ldots, k\}$, and $i^* = \min\{i|i \in I(\underline{x})\}$. Then based on $\varphi_i(x_i)$, a selection procedure $d(\underline{x}) = (d_0(\underline{x}), \ldots, d_k(\underline{x}))$ is constructed as follows:

$$\begin{cases} d_{i^*}(\underline{x}) = 1, \\ d_j(\underline{x}) = 0, \quad \text{for} \quad j \ne i^*. \end{cases} \qquad (24.2.5)$$

Under the preceding statistical model, the expected risk of the selection procedure $d(\underline{x})$ is denoted by $R(d(\underline{x}))$. Denote $h_i(\theta_i|\mu_i, \tau_i^2)$ to be the prior density function of Θ_i given (μ_i, τ_i^2), we have

$$R(d(\underline{x})) = -\int_{\mathcal{X}} [\sum_{i=0}^{k} d_i(\underline{x})\varphi_i(x_i)]f(\underline{x})d(\underline{x}) + C, \qquad (24.2.6)$$

where
$C = \int_\Omega \max(\theta_{[k]}, \theta_0) dH(\underline{\theta})$,
$H(\underline{\theta})$: the joint distribution of $\underline{\theta} = (\theta_1, \ldots, \theta_k)$,
$f_i(x_i) = \int_R f_i(x_i|\theta_i, \sigma_i^2) h_i(\theta_i|\mu_i, \tau_i^2) d\theta_i$,
$f(\underline{x}) = \Pi_{i=1}^k f_i(x_i)$,
$\varphi_0(x_0) = \theta_0$.
Note that sample median X_i is not a sufficient statistic for θ_i (the observation vector is a minimal sufficient statistic). So $d(\underline{x})$ may not be a Bayes rule. Also, the selection procedure $d(\underline{x})$ defined above depends on the unknown parameters (μ_i, τ_i^2), $i = 1, \ldots, k$ and the specific form of $\varphi_i(x_i)$. Since the parameters and the specific form of $\varphi_i(x_i)$ are both unknown, it is impossible to implement this selection procedure for the selection problem in practice.

To derive a practical selection rule, we assume there are past observations when the present selection is to be made. At time $l = 1, \ldots, n$, let X_{ijl} be the j-th observation from Π_i, that is, for each $i = 1, \ldots, k$, let

$$\Theta_{il} \sim N(\mu_i, \tau_i^2), \qquad l = 1, \ldots, n, \tag{24.2.7}$$

and

$$X_{ijl} \sim L(\theta_{il}, \sigma_i^2), \qquad j = 1, \ldots, M. \tag{24.2.8}$$

For $l = 1, \ldots, n$, denote $X_{i,l}$ to be the median of $(X_{i1l}, \ldots, X_{iMl})$, and

$$X_i(n) = \frac{1}{n} \sum_{l=1}^n X_{i,l}, \tag{24.2.9}$$

$$S_i^2(n) = \frac{1}{n-1} \sum_{l=1}^n (X_{i,l} - X_i(n))^2. \tag{24.2.10}$$

Then,

$$E(X_{i,l}) = E(E(X_{i,l}|\Theta_{il} = \theta_{il})) = E(\Theta_{il}) = \mu_i, \tag{24.2.11}$$

and

$$\begin{aligned}
Var(X_{i,l}) &= Var(E(X_{i,l}|\theta_{il})) + E(Var(X_{i,l}|\theta_{il})) \\
&= Var(\Theta_{il}) + E(Var(X_{i,l}|\theta_{il})) \\
&= \tau_i^2 + E(Var(X_{i,l}|\theta_{il})) \\
&< \infty.
\end{aligned} \tag{24.2.12}$$

Denote $\nu_i^2 = Var(X_{i,l})$. Since $(X_{i1}, \ldots, X_{in})$ are i.i.d., by the strong law of large numbers, we know that as $n \to \infty$,

$$\begin{cases} X_i(n) \longrightarrow \mu_i, & a.s. \\ S_i^2(n) \longrightarrow \nu_i^2, & a.s. \end{cases} \tag{24.2.13}$$

To derive an empirical selection procedure, we first consider the following lemmas.

Lemma 24.2.1 *Let $\{Y_i, 1 \leq i \leq m\}$ be m i.i.d. random observations from continuous distribution function F; also let $\hat{\xi}$ and ξ be the sample median of $\{Y_i, 1 \leq i \leq m\}$ and population median of F, respectively. Then, for any $\epsilon > 0$,*

$$P\{|\hat{\xi} - \xi| > \epsilon\} \leq 2e^{-2m\delta_\epsilon^2}, \tag{24.2.14}$$

where $\delta_\epsilon = \min\{F(\xi + \epsilon) - \frac{1}{2}, \frac{1}{2} - F(\xi - \epsilon)\}$.

This lemma is from Serfling (1980) and the proof can be found in it.

Back to our selection problem. Put

$$\sigma' = \min_{1 \leq i \leq k} \sigma_i, \quad \sigma^* = \max_{1 \leq i \leq k} \sigma_i.$$

$X_{i1}, \ldots, X_{iM}$ are i.i.d. from $L(\theta_i, \sigma_i^2)$, which has the following cumulative distribution function

$$F(t_i) = \frac{1}{1 + e^{-(t_i - \theta_i)/\sigma_i}} \qquad -\infty < t_i < \infty, \tag{24.2.15}$$

and for $0 < \epsilon \leq \sigma'$,

$$F(\theta_i + \epsilon) - \frac{1}{2} = \frac{1}{2} - F(\theta_i - \epsilon) = \frac{e^{\epsilon/\sigma_i} - 1}{2(e^{\epsilon/\sigma_i} + 1)} \geq \frac{\epsilon}{2(e+1)\sigma^*}. \tag{24.2.16}$$

Given $\Theta_i = \theta_i$, θ_i and X_i are the population median and sample median of $L(\theta_i, \sigma_i)$ respectively. We have, from Lemma 24.2.1,

$$P\{|X_i - \theta_i| > \epsilon\} \leq 2e^{\frac{-(2s+1)\epsilon^2}{2(e+1)^2\sigma^{*2}}}. \tag{24.2.17}$$

For any $0 < \epsilon \leq \sigma'$, denote $\mathcal{A}_i = \{\mathbf{x} \in \mathcal{X} : |x_i - \theta_i| \leq \epsilon\}$. We show that the conditional density of X_i given θ_i and σ_i^2 is approximately $N(\theta_i, \frac{2}{s+1}\sigma_i^2)$ as $s \to \infty$.

From (24.2.2), the conditional density of X_i given θ_i and σ_i^2 is

$$f_i(x_i|\theta_i, \sigma_i^2) = \frac{(2s+1)!}{(s!)^2} \frac{1}{\sigma_i} \frac{(e^{-(x_i-\theta_i)/\sigma_i})^{s+1}}{(1+e^{-(x_i-\theta_i)/\sigma_i})^{2s+2}}$$

$$= \frac{(2s+1)!}{(s!)^2} \frac{1}{\sigma_i} \frac{1}{(2+e^{-(x_i-\theta_i)/\sigma_i}+e^{(x_i-\theta_i)/\sigma_i})^{s+1}}.$$

$$(24.2.18)$$

By Stirling's formula, when s is large enough,

$$\frac{(2s+1)!}{(s!)^2} \approx \frac{2^{(2s+\frac{3}{2})}}{\sqrt{2\pi}}\sqrt{s+1}. \qquad (24.2.19)$$

Also choosing $\epsilon = \epsilon_s \downarrow 0$ to be a sequence of fixed numbers which tend to 0 as $s \to \infty$, by Taylor's polynomial expansion, we have

$$\log(2+e^{-(x_i-\theta_i)/\sigma_i}+e^{(x_i-\theta_i)/\sigma_i}) \approx \log 4 + \frac{1}{4}\frac{(x_i-\theta_i)^2}{\sigma_i^2} \qquad (24.2.20)$$

on $\mathcal{A}_i$. When $s \to \infty$, from (17),

$$P\{X \notin \mathcal{A}_i\} \le 2e^{\frac{-(2s+1)\epsilon^2}{2(\epsilon+1)^2\sigma^{*2}}} \longrightarrow 0. \qquad (24.2.21)$$

Therefore, we see that as $s \to \infty$,

$$f_i(x_i|\theta_i, \sigma_i^2) \approx \frac{1}{\sqrt{2\pi}\sqrt{2/s+1}}\frac{1}{\sigma_i}e^{-\frac{s+1}{4}\frac{(x_i-\theta_i)^2}{\sigma_i^2}}, \qquad (24.2.22)$$

that is, $f_i(x_i|\theta_i, \sigma_i^2)$ is approximately $N(\theta_i, \frac{2}{s+1}\sigma_i^2)$.

From above, we can see that for sufficiently large s, the conditional density of $X_{i,l}$ is approximately $N(\theta_i, \frac{2}{s+1}\sigma_i^2)$ given θ_i and σ_i. Since the prior distribution of θ_i is $N(\mu_i, \tau_i^2)$, the unconditional density of $X_{i,l}$ is approximately $N(\mu_i, \tau_i^2 + \frac{2}{s+1}\sigma_i^2)$.

For each population Π_i, let $W_i^2(n)$ be the measure of the overall sample variation for the past observations. That is,

$$\begin{cases} \bar{X}_{il} = \frac{1}{M}\sum_{j=1}^{M} X_{ijl}, \\ W_i^2(n) = \frac{1}{(M-1)n}\sum_{j=1}^{M}\sum_{l=1}^{n}(X_{ijl} - \bar{X}_{il})^2. \end{cases} \qquad (24.2.23)$$

Then we define, for $i = 1, \ldots, k$,

$$
\begin{cases}
\hat{\mu}_i = X_i(n), \\
\hat{\sigma}_i^2 = \frac{3}{\pi^2} W_i^2(n), \\
\hat{\nu}_i^2 = S_i^2(n), \\
\hat{\tau}_i^2 = \max(\hat{\nu}_i^2 - \frac{2}{s+1}\hat{\sigma}_i^2, 0).
\end{cases}
\tag{24.2.24}
$$

and

$$
\begin{cases}
\hat{\varphi}_i(x_i) = \begin{cases} (x_i\hat{\tau}_i^2 + \frac{2\hat{\sigma}_i^2}{s+1}\hat{\mu}_i)/\hat{\nu}_i^2, & \text{if } \hat{\nu}_i^2 - \frac{2}{s+1}\hat{\sigma}_i^2 > 0, \\ \hat{\mu}_i, & \text{if } \hat{\nu}_i^2 - \frac{2}{s+1}\hat{\sigma}_i^2 \le 0, \end{cases} \\
\hat{\varphi}_0(x_0) = \theta_0.
\end{cases}
\tag{24.2.25}
$$

Then for each $\underline{x} \in \mathcal{X}$, let $\hat{I}(\underline{x}) = \{i | \hat{\varphi}_i(x_i) = \max_{0 \le j \le k} \hat{\varphi}_j(x_j), i = 0, 1, \ldots, k\}$, and $\hat{i}^* = \min\{i | i \in \hat{I}(\underline{x})\}$. We propose the following selection procedure $d^{(n,s)}(\underline{x}) = (d_0^{(n,s)}(\underline{x}), \ldots, d_k^{(n,s)}(\underline{x}))$ as follows:

$$
\begin{cases}
d_{\hat{i}^*}^{(n,s)} = 1, \\
d_j^{(n,s)} = 0, \quad \text{for } j \ne \hat{i}^*.
\end{cases}
\tag{24.2.26}
$$

24.3   ASYMPTOTIC OPTIMALITY OF THE PROPOSED SELECTION PROCEDURE

Consider the selection procedure $d^{(n,s)}(\underline{x})$ constructed in (24.2.26). $d^{(n,s)}(\underline{x})$ is similar to selection rule $d(\underline{x})$ except that normal approximation is used to estimate the unknown prior parameters and the specific form of $\varphi_i(x_i)$ for $d^{(n,s)}(\underline{x})$. A natural question to ask is: How good is the selection rule $d^{(n,s)}(\underline{x})$ compared with $d(\underline{x})$? Let $R(d^{(n,s)}(\underline{x}))$ be the conditional expected risk given the past observations $\{X_{ijl}, i = 1, \ldots, k, j = 1, \ldots, M, \text{ and } l = 1, \ldots, n\}$, then

$$
R(d^{(n,s)}(\underline{x})) = - \int_{\mathcal{X}} [\sum_{i=0}^{k} d_i^{(n,s)}(\underline{x})\varphi_i(x_i)] f(\underline{x}) d(\underline{x}) + C.
\tag{24.3.27}
$$

Since $d^{(n,s)}(\underline{x})$ is mimicking $d(\underline{x})$, $R(d^{(n,s)}(\underline{x})) - R(d(\underline{x}))$ should be close to 0 if the empirical selection rule works well. Note that $R(d^{(n,s)}(\underline{x})) - R(d(\underline{x}))$ can be negative because $d(\underline{x})$ is not a Bayes rule. Therefore, we use the overall integrated risk $E|R(d^{(n,s)}(\underline{x})) - R(d(\underline{x}))| \geq 0$ as a measure of the performance of the selection procedure $d^{(n,s)}(\underline{x})$, where E is the expectation taken with respect to the past observations $\{X_{ijl}\}$.

We first state some facts about $\varphi_i(x_i)$, the posterior median of θ_i given $X_i = x_i$ and μ_i. From the definition of $\varphi_i(x_i)$, we can see that $\varphi_i(x_i)$ is between x_i and μ_i. Besides,

Lemma 24.3.1 *When s is large enough, for $1 \leq i \leq k$,*

$$|\varphi_i(x_i) - x_i| \leq 2\sigma_i\sqrt{\frac{\log s}{s}}. \tag{24.3.28}$$

PROOF We only prove $\varphi_i(x_i) \leq x_i + 2\sigma_i\sqrt{\frac{\log s}{s}}$ here. The proof of $\varphi_i(x_i) \geq x_i - 2\sigma_i\sqrt{\frac{\log s}{s}}$ is similar. To prove $\varphi_i(x_i) \leq x_i + 2\sigma_i\sqrt{\frac{\log s}{s}}$, it suffices to show that

$$
\begin{aligned}
&\int_{x_i+2\sigma_i\sqrt{\frac{\log s}{s}}}^{\infty} f_i(x_i|\theta_i, \sigma_i^2)h_i(\theta_i|\mu_i, \tau_i^2)d\theta_i \\
&= \int_{x_i+2\sigma_i\sqrt{\frac{\log s}{s}}}^{\infty} \frac{(2s+1)!}{(s!)^2}\frac{1}{\sigma_i}\frac{1}{\sqrt{2\pi}\tau_i}\frac{(e^{(\theta_i-x_i)/\sigma_i})^{s+1}}{(1+e^{(\theta_i-x_i)/\sigma_i})^{2s+2}}\cdot e^{-\frac{(\theta_i-\mu_i)^2}{2\tau_i^2}}d\theta_i \\
&= \int_{2\sqrt{\frac{\log s}{s}}}^{\infty} \frac{1}{\sqrt{2\pi}\tau_i}\frac{(2s+1)!}{(s!)^2}\left(\frac{e^{\theta}}{(1+e^{\theta})^2}\right)^{s+1}\cdot e^{-\frac{(\sigma_i\theta+x_i-\mu_i)^2}{2\tau_i^2}}d\theta \longrightarrow 0,
\end{aligned}
\tag{24.3.29}
$$

as $s \to \infty$. We first show

$$t(\theta, s) := \frac{(2s+1)!}{(s!)^2}\left(\frac{e^{\theta}}{(1+e^{\theta})^2}\right)^{s+1} \longrightarrow 0 \qquad \text{as } s \to \infty, \tag{24.3.30}$$

uniformly for $\theta \geq 2\sqrt{\frac{\log s}{s}}$. Obviously it is enough to consider the case of $\theta = 2\sqrt{\frac{\log s}{s}}$ since $t(\theta, s)$ is decreasing on $\theta > 0$. When $\theta = 2\sqrt{\frac{\log s}{s}}$ and s is large enough, by Taylor's formula,

$$\log(1 + e^{\theta}) = \log 2 + \frac{1}{2}\theta + \frac{1}{8}\theta^2 + o(\theta^2), \tag{24.3.31}$$

and by (24.2.19), when s is large enough,

$$\log\frac{(2s+1)!}{(s!)^2} \leq 2(s+1)\log 2 + \frac{1}{2}\log(s+1). \tag{24.3.32}$$

From (24.3.31) and (24.3.32), we obtain that

$$
\begin{aligned}
\log t(\theta, s) &= (s+1)[\theta - 2\log(1+e^\theta)] + \log\frac{(2s+1)!}{(s!)^2} \\
&\leq -2(s+1)\log 2 - \frac{s+1}{4}\theta^2 + 2(s+1)\log 2 + \frac{1}{2}\log s + o(s\theta^2) \\
&= -(\frac{s+1}{s} - \frac{1}{2})\log s + o(\log s) \longrightarrow -\infty, \qquad (24.3.33)
\end{aligned}
$$

as $s \to \infty$. Therefore, (24.3.30) is proved, from which we can immediately see that (24.3.29) holds true. It completes the proof of Lemma 24.3.1. $\square$

The next lemma is well known and can be found in Baum and Katz (1965).

Lemma 24.3.2 *Let $X_1, \ldots, X_n$ be i.i.d. random variables with mean 0. Suppose for $\alpha > 1$, $E|X_i|^\alpha < \infty$, for $i = 1, \ldots, n$, then for any $\epsilon > 0$,*

$$
P\{|\sum_{i=1}^n X_i/n| \geq \epsilon\} = o(n^{-(\alpha-1)}). \qquad (24.3.34)
$$

As a consequence of Lemma 24.3.2, we have

Lemma 24.3.3 *Let $X_1, \ldots, X_n$ be independent random variables, with mean $EX_i = \mu$ and variance $VarX_i = \sigma^2$, for $i = 1, \ldots, n$. Also let $\bar{X} = \frac{1}{n}\sum X_i$ and $S_n^2 = \frac{1}{n-1}\sum(X_i - \bar{X})^2$. Suppose for $i = 1, \ldots, n$ and a fixed number $\alpha > 2$, $E|X_i|^\alpha < \infty$, then for any $\epsilon > 0$,*

$$
P\{|S_n^2 - \sigma^2| \geq \epsilon\} = o(n^{-(\alpha/2-1)}). \qquad (24.3.35)
$$

Since $EX_{i.l}^4 < \infty$, for any $\epsilon > 0$, by Lemma 24.3.2,

$$
P\{|\hat{\mu}_i - \mu_i| \geq \epsilon\} = o(n^{-3}), \qquad (24.3.36)
$$

also by Lemma 24.3.3,

$$
P\{|\hat{\nu}_i^2 - \nu_i^2| \geq \epsilon\} = o(n^{-1}). \qquad (24.3.37)
$$

Similarly, we have for any $\epsilon > 0$,

$$
P\{|\hat{\sigma}_i^2 - \sigma_i^2| \geq \epsilon\} = o(n^{-1}). \qquad (24.3.38)
$$

When s is large enough, $\nu_i^2 - \frac{2}{s+1}\sigma_i^2 > 0$. Therefore, from (24.3.37) and (24.3.38), when s is sufficiently large,

$$
P\{\hat{\nu}_i^2 - \frac{2}{s+1}\hat{\sigma}_i^2 \leq 0\} = o(n^{-1}). \qquad (24.3.39)
$$

Besides, $\tau_i^2 = \nu_i^2 - E(Var(X_{i,l}|\theta_i))$ by (24.2.12) and

$$
\begin{aligned}
E(Var(X_{i,l}|\theta_i)) &= \int_{-\infty}^{\infty} (x_{il} - \theta_i)^2 \frac{(2s+1)!}{(s!)^2} \frac{1}{\sigma_i} \frac{(e^{-(x_{il}-\theta_i)/\sigma_i})^{s+1}}{(1+e^{-(x_{il}-\theta_i)/\sigma_i})^{2s+2}} dx_{il} \\
&= \sigma_i \int_{-\infty}^{\infty} x^2 \frac{(2s+1)!}{(s!)^2} \left(\frac{e^x}{(1+e^x)^2} \right)^{s+1} dx. \qquad (24.3.40)
\end{aligned}
$$

We have

Lemma 24.3.4

$$
\int_{-\infty}^{\infty} x^2 \frac{(2s+1)!}{(s!)^2} \left(\frac{e^x}{(1+e^x)^2} \right)^{s+1} dx = o\left(\sqrt{\frac{\log s}{s}} \right). \qquad (24.3.41)
$$

PROOF

$$
\begin{aligned}
&\int_{-\infty}^{\infty} x^2 \frac{(2s+1)!}{(s!)^2} \left(\frac{e^x}{(1+e^x)^2} \right)^{s+1} dx \\
&= 2\int_{0}^{\infty} x^2 \frac{(2s+1)!}{(s!)^2} \left(\frac{e^x}{(1+e^x)^2} \right)^{s+1} dx \\
&= 2\left(\int_{0}^{\sqrt{8\frac{\log s}{s}}} + \int_{\sqrt{8\frac{\log s}{s}}}^{3} + \int_{3}^{\infty} \right) x^2 \frac{(2s+1)!}{(s!)^2} \left(\frac{e^x}{(1+e^x)^2} \right)^{s+1} dx \\
&:= T_1 + T_2 + T_3. \qquad (24.3.42)
\end{aligned}
$$

By Stirling's formula, when s is large enough,

$$
\begin{aligned}
T_1 &= 2\frac{(2s+1)!}{(s!)^2} \int_{0}^{\sqrt{8\frac{\log s}{s}}} x^2 \left(\frac{e^x}{(1+e^x)^2} \right)^{s+1} dx \\
&\leq 2 \cdot 2^{2(s+1)} \sqrt{s+1} \cdot 2^{-2(s+1)} \int_{0}^{\sqrt{8\frac{\log s}{s}}} x^2 dx \\
&\leq \sqrt{s+1} \left(8\frac{\log s}{s} \right)^{3/2} \\
&= o\left(\sqrt{\frac{\log s}{s}} \right). \qquad (24.3.43)
\end{aligned}
$$

Using the same approach as in the proof of Lemma 24.3.1, we have

$$
T_2 = 2\frac{(2s+1)!}{(s!)^2} \int_{\sqrt{8\frac{\log s}{s}}}^{3} x^2 \left(\frac{e^x}{(1+e^x)^2} \right)^{s+1} dx
$$

$$\leq \ 2\frac{(2s+1)!}{(s!)^2}\left(\frac{e^{\sqrt{8\frac{\log s}{s}}}}{(1+e^{\sqrt{8\frac{\log s}{s}}})^2}\right)^{s+1}\int_{\sqrt{8\frac{\log s}{s}}}^{3}x^2 dx$$

$$= \ o\left(\sqrt{\frac{\log s}{s}}\right). \qquad\qquad (24.3.44)$$

Moreover,

$$T_3 \ = \ 2\frac{(2s+1)!}{(s!)^2}\int_3^\infty x^2\left(\frac{e^x}{(1+e^x)^2}\right)^{s+1}dx$$

$$\leq \ 2\frac{(2s+1)!}{(s!)^2}\int_3^\infty x^2 e^{-(s+1)x}dx$$

$$= \ o(\sqrt{\frac{\log s}{s}}). \qquad\qquad (24.3.45)$$

This completes the proof of Lemma 24.3.4. $\square$

From Lemma 24.3.4, we observe that when s is sufficiently large,

$$E(Var(X_{i,l}|\theta_i)) = o\left(\sqrt{\frac{\log s}{s}}\right), \qquad\qquad (24.3.46)$$

and therefore, by (24.3.37), (24.3.39) and the definition of $\hat{\tau}_i^2$, for $\epsilon \geq c\sqrt{\frac{\log s}{s}}$, where $c > 0$,

$$P\{|\hat{\tau}_i^2 - \tau_i^2| \geq \epsilon\} = o(n^{-1}), \qquad\qquad (24.3.47)$$

and furthermore,

$$P\{\hat{\nu}_i^2/\hat{\tau}_i^2 \leq \nu_i^2/(2\tau_i^2)\} = o(n^{-1}). \qquad\qquad (24.3.48)$$

Next we investigate the overall integrated risk $E|R(d^{(n,s)}(\underline{x})) - R(d(\underline{x}))|$. Let $P_{n,s}$ be the probability measure generated by the past observations $X_{ijl}, i = 1, \ldots k, j = 1, \ldots M$ and $l = 1, \ldots, n$.

$$E|R(d^{(n,s)}(\underline{x})) - R(d(\underline{x}))|$$

$$\leq \ \sum_{i=0}^{k}\sum_{j=0}^{k}\int_{\mathcal{X}} P_{n,s}\{i^* = i, \hat{i}^* = j\}|\varphi_i(x_i) - \varphi_j(x_j)|f(\underline{x})d\underline{x}$$

$$= \ \sum_{i=1}^{k}\int_{\mathcal{X}} P_{n,s}\{i^* = i, \hat{i}^* = 0\}|\varphi_i(x_i) - \theta_0|f(\underline{x})d\underline{x}$$

$$+\sum_{j=1}^{k}\int_{\mathcal{X}}P_{n,s}\{i^{*}=0,\hat{i}^{*}=j\}|\theta_{0}-\varphi_{j}(x_{j})|f(\mathbf{x})d\mathbf{x}$$

$$+\sum_{i=1}^{k}\sum_{j=1}^{k}\int_{\mathcal{X}}P_{n,s}\{i^{*}=i,\hat{i}^{*}=j\}|\varphi_{i}(x_{i})-\varphi_{j}(x_{j})|f(\mathbf{x})d\mathbf{x}$$

$$\leq 2\sum_{i=1}^{k}\int_{R}P_{n,s}\{|\hat{\varphi}_{i}(x_{i})-\varphi_{i}(x_{i})|>|\varphi_{i}(x_{i})-\theta_{0}|\}$$

$$\times|\varphi_{i}(x_{i})-\theta_{0}|f_{i}(x_{i})dx_{i}$$

$$+2\sum_{i=1}^{k}\sum_{j=1}^{k}\int_{R^{2}}P_{n,s}\{|\hat{\varphi}_{i}(x_{i})-\varphi_{i}(x_{i})|>\frac{|\varphi_{i}(x_{i})-\varphi_{j}(x_{j})|}{2}\}$$

$$\times|\varphi_{i}(x_{i})-\varphi_{j}(x_{j})|f_{i}(x_{i})f_{j}(x_{j})dx_{i}dx_{j}$$

$$:=\quad I_{1}+I_{2}. \tag{24.3.49}$$

For any $\epsilon > 0$, and $i, j = 1, \ldots, k$, let

$$\begin{cases} \mathcal{X}_{i}=\{x_{i}:|\varphi_{i}(x_{i})-\theta_{0}|\leq\epsilon\}, \\ \mathcal{X}_{ij}=\{(x_{i},x_{j}):|\varphi_{i}(x_{i})-\varphi_{j}(x_{j})|\leq\epsilon\}. \end{cases} \tag{24.3.50}$$

Then we have

$$I_{1}\quad=\quad 2\sum_{i=1}^{k}\int_{\mathcal{X}_{i}}P_{n,s}\{|\hat{\varphi}_{i}(x_{i})-\varphi_{i}(x_{i})|>|\varphi_{i}(x_{i})-\theta_{0}|\}$$

$$\times|\varphi_{i}(x_{i})-\theta_{0}|f_{i}(x_{i})dx_{i}$$

$$+2\sum_{i=1}^{k}\int_{R-\mathcal{X}_{i}}P_{n,s}\{|\hat{\varphi}_{i}(x_{i})-\varphi_{i}(x_{i})|>|\varphi_{i}(x_{i})-\theta_{0}|\}$$

$$\times|\varphi_{i}(x_{i})-\theta_{0}|f_{i}(x_{i})dx_{i}$$

$$\leq\quad 2\sum_{i=1}^{k}\int_{\mathcal{X}_{i}}\epsilon f_{i}(x_{i})dx_{i}$$

$$+2\sum_{i=1}^{k}\int_{R}P_{n,s}\{|\hat{\varphi}_{i}(x_{i})-\varphi_{i}(x_{i})|>\epsilon\}|\varphi_{i}(x_{i})-\theta_{0}|f_{i}(x_{i})dx_{i}.$$

$$\tag{24.3.51}$$

By Lemma 24.3.1, when s is large enough, $|\varphi_{i}(x_{i})-x_{i}|\leq 2\sigma_{i}\sqrt{\frac{\log s}{s}}$. From now on, we always set $\epsilon = 16\sigma^{*}\sqrt{\frac{\log s}{s}}$. Therefore, for sufficiently large s,

$$|x_i - \theta_0| \leq |\varphi_i(x_i) - x_i| + |\varphi_i(x_i) - \theta_0| \leq 2\epsilon \qquad (24.3.52)$$

on $\mathcal{X}_i$ and

$$
\begin{aligned}
\int_{\mathcal{X}_i} f_i(x_i)dx_i &\leq \int_{\{|x_i-\theta_0|\leq 2\epsilon\}} f_i(x_i)dx_i \\
&\leq \int_{\{|x_i-\theta_0|\leq 2\epsilon\}} \frac{1}{\sqrt{2\pi}\tau_i}dx_i \\
&= \frac{4\epsilon}{\sqrt{2\pi}\tau_i}.
\end{aligned}
\qquad (24.3.53)
$$

Thus,

$$
\begin{aligned}
I_1 \leq{} & \frac{8k}{\sqrt{2\pi}\tau_i}\epsilon^2 + 2\sum_{i=1}^{k}\int_R P_{n,s}\{|\hat{\varphi}_i(x_i) - \varphi_i(x_i)| > \epsilon\} \\
& \times [|\varphi_i(x_i) - \mu_i| + |\mu_i - \theta_0|]f_i(x_i)dx_i.
\end{aligned}
\qquad (24.3.54)
$$

Moreover,

$$
\begin{aligned}
I_2 ={} & 2\sum_{i=1}^{k}\sum_{j=1}^{k}\int_{\mathcal{X}_{ij}} P_{n,s}\{|\hat{\varphi}_i(x_i) - \varphi_i(x_i)| > \frac{|\varphi_i(x_i) - \varphi_j(x_j)|}{2}\} \\
& \times |\varphi_i(x_i) - \varphi_j(x_j)|f_i(x_i)f_j(x_j)dx_idx_j \\
& +2\sum_{i=1}^{k}\sum_{j=1}^{k}\int_{R^2-\mathcal{X}_{ij}} P_{n,s}\{|\hat{\varphi}_i(x_i) - \varphi_i(x_i)| > \frac{|\varphi_i(x_i) - \varphi_j(x_j)|}{2}\} \\
& \times |\varphi_i(x_i) - \varphi_j(x_j)|f_i(x_i)f_j(x_j)dx_idx_j \\
\leq{} & 2\epsilon\sum_{i=1}^{k}\sum_{j=1}^{k}\int_{\mathcal{X}_{ij}} f_i(x_i)f_j(x_j)dx_idx_j \\
& +2\sum_{i=1}^{k}\sum_{j=1}^{k}\int_{R^2} P_{n,s}\{|\hat{\varphi}_i(x_i) - \varphi_i(x_i)| > \frac{\epsilon}{2}\}|\varphi_i(x_i) - \varphi_j(x_j)| \\
& \times f_i(x_i)f_j(x_j)dx_idx_j.
\end{aligned}
\qquad (24.3.55)
$$

From (24.3.28), when s is large enough, $|\varphi_i(x_i) - x_i| \leq \epsilon$ and $|\varphi_j(x_j) - x_j| \leq \epsilon$. Therefore, when s is sufficiently large,

$$\{(x_i, x_j) : |\varphi_i(x_i) - \varphi_j(x_j)| \leq \epsilon\} \subset \{(x_i, x_j) : |x_i - x_j| \leq 3\epsilon\}. \quad (24.3.56)$$

Thus, similar to (24.3.53),

$$\int_{\mathcal{X}_{ij}} f_i(x_i)f_j(x_j)dx_i dx_j \leq \frac{6\epsilon}{\sqrt{2\pi}\min(\tau_i,\tau_j)}. \tag{24.3.57}$$

We observe that

$$I_2 \leq \sum_{i=1}^{k}\sum_{j=1}^{k} \frac{12\epsilon^2}{\sqrt{2\pi}\min(\tau_i,\tau_j)}$$

$$+2\sum_{i=1}^{k}\sum_{j=1}^{k}\int_{R^2} P_{n,s}\{|\hat{\varphi}_i(x_i) - \varphi_i(x_i)| > \frac{\epsilon}{2}\}$$

$$\times [|\varphi_i(x_i) - \mu_i| + |\varphi_j(x_j) - \mu_j| + |\mu_i - \mu_j|]f_i(x_i)f_j(x_j)dx_i dx_j. \tag{24.3.58}$$

From (24.3.54) and (24.3.58), it suffices to analyze the limiting behaviors of

$$\int_R P_{n,s}\{|\hat{\varphi}_i(x_i) - \varphi_i(x_i)| > \frac{\epsilon}{2}\}f_i(x_i)dx_i,$$
$$\int_R P_{n,s}\{|\hat{\varphi}_i(x_i) - \varphi_i(x_i)| > \frac{\epsilon}{2}\}|\varphi_i(x_i) - \mu_i|f_i(x_i)dx_i. \tag{24.3.59}$$

We first analyze $\int_R P_{n,s}\{|\hat{\varphi}_i(x_i) - \varphi_i(x_i)| > \frac{\epsilon}{2}\}f_i(x_i)dx_i$. Denote

$$\mathcal{Y}_i = \{x_i : |\varphi_i(x_i) - \theta_i| \leq \frac{\epsilon}{4}\},$$
$$\mathcal{Z}_i = \{x_i : |x_i - \theta_i| \leq \frac{\epsilon}{8}\}. \tag{24.3.60}$$

By Lemma 24.3.1, we know that when s is large enough, $|\varphi_i(x_i) - x_i| \leq \frac{\epsilon}{8}$. Therefore, for sufficiently large s, we have

$$R - \mathcal{Y}_i \subset R - \mathcal{Z}_i \tag{24.3.61}$$

and

$$\int_R P_{n,s}\{|\hat{\varphi}_i(x_i) - \varphi_i(x_i)| > \frac{\epsilon}{2}\}f_i(x_i)dx_i$$

$$\leq \int_R \left(\int_{R-\mathcal{Z}_i} P_{n,s}\{|\hat{\varphi}_i(x_i) - \varphi_i(x_i)| \right.$$

$$\left. > \frac{\epsilon}{2}\}f_i(x_i|\theta_i,\sigma_i^2)h_i(\theta_i|\mu_i,\tau_i^2)dx_i\right)d\theta_i$$

$$+ \int_R \left(\int_{\mathcal{Y}_i} P_{n,s}\{|\hat{\varphi}_i(x_i) - \varphi_i(x_i)| \right.$$

$$\left. > \frac{\epsilon}{2}\}f_i(x_i|\theta_i,\sigma_i^2)h_i(\theta_i|\mu_i,\tau_i^2)dx_i\right)d\theta_i$$

$$\leq \int_R \left(\int_{R-Z_i} f_i(x_i|\theta_i, \sigma_i^2) h_i(\theta_i|\mu_i, \tau_i^2) dx_i \right) d\theta_i$$

$$+ \int_R \left(\int_R P_{n,s}\{|\hat{\varphi}_i(x_i) - \theta_i| \geq \frac{\epsilon}{4}\} f_i(x_i|\theta_i, \sigma_i^2) h_i(\theta_i|\mu_i, \tau_i^2) dx_i \right) d\theta_i$$

$$\leq \int_R \left(\int_{|x_i - \theta_i| > \frac{\epsilon}{8}} f_i(x_i|\theta_i, \sigma_i^2) dx_i \right) h_i(\theta_i|\mu_i, \tau_i^2) d\theta_i$$

$$+ \int_R \left(\int_R P_{n,s}\{|\hat{\varphi}_i(x_i) - \theta_i| \geq \frac{\epsilon}{4}, \hat{\nu}_i^2 - 2\hat{\sigma}_i^2/(s+1) > 0\} \right.$$

$$\times f_i(x_i|\theta_i, \sigma_i^2) dx_i \Big) r h_i(\theta_i|\mu_i, \tau_i^2) d\theta_i$$

$$+ \int_R \left(\int_R P_{n,s}\{\hat{\nu}_i^2 - 2\hat{\sigma}_i^2/(s+1) \leq 0\} f_i(x_i|\theta_i, \sigma_i^2) dx_i \right)$$

$$\times h_i(\theta_i|\mu_i, \tau_i^2) d\theta_i$$

$$\leq 2 \int_R e^{-\frac{(2s+1)\epsilon^2}{128(e+1)^2\sigma^{*2}}} h_i(\theta_i|\mu_i, \tau_i^2) d\theta_i$$

$$+ \int_R \left(\int_R P_{n,s}\{|(x_i\hat{\tau}_i^2 + \frac{2\hat{\sigma}_i^2}{s+1}\hat{\mu}_i)/\hat{\nu}_i^2 - \theta_i| \geq \frac{\epsilon}{4}\} f_i(x_i|\theta_i, \sigma_i^2) dx_i \right)$$

$$\times h_i(\theta_i|\mu_i, \tau_i^2) d\theta_i + o(n^{-1})$$

$$\leq 2e^{-\frac{(2s+1)\epsilon^2}{128(e+1)^2\sigma^{*2}}}$$

$$+ \int_R \left(\int_R P_{n,s}\{|x_i - \theta_i| \geq \frac{\hat{\nu}_i^2}{\hat{\tau}_i^2}\frac{\epsilon}{8}\} f_i(x_i|\theta_i, \sigma_i^2) dx_i \right) h_i(\theta_i|\mu_i, \tau_i^2) d\theta_i$$

$$+ \int_R \left(\int_R P_{n,s}\{|\hat{\mu}_i - \theta_i| \geq \frac{(s+1)\hat{\nu}_i^2\epsilon}{16\hat{\sigma}_i^2}\} f_i(x_i|\theta_i, \sigma_i^2) dx_i \right)$$

$$\times h_i(\theta_i|\mu_i, \tau_i^2) d\theta_i + o(n^{-1})$$

$$\leq O(s^{-1})$$

$$+ \int_R \left(\int_R P_{n,s}\{|x_i - \theta_i| \geq \frac{\hat{\nu}_i^2}{\hat{\tau}_i^2}\frac{\epsilon}{8}, \frac{\hat{\nu}_i^2}{\hat{\tau}_i^2} \geq \nu_i^2/(2\tau_i^2)\} f_i(x_i|\theta_i, \sigma_i^2) dx_i \right)$$

$$\times h_i(\theta_i|\mu_i, \tau_i^2) d\theta_i$$

$$+ \int_R \left(\int_R P_{n,s}\{\frac{\hat{\nu}_i^2}{\hat{\tau}_i^2} \leq \nu_i^2/(2\tau_i^2)\} f_i(x_i|\theta_i, \sigma_i^2) dx_i \right) h_i(\theta_i|\mu_i, \tau_i^2) d\theta_i$$

$$+ \int_R \left(\int_R P_{n,s}\{|\hat{\mu}_i - \theta_i| \geq \frac{(s+1)\hat{\nu}_i^2\epsilon}{16\hat{\sigma}_i^2}, \hat{\nu}_i^2/\hat{\sigma}_i^2 \right.$$

$$\geq \nu_i^2/(2\sigma_i^2)\} f_i(x_i|\theta_i, \sigma_i^2) dx_i \Big) h_i(\theta_i|\mu_i, \tau_i^2) d\theta_i$$

$$+ \int_R \left(\int_R P_{n,s}\{\hat{\nu}_i^2/\hat{\sigma}_i^2 \leq \nu_i^2/(2\sigma_i^2)\} f_i(x_i|\theta_i, \sigma_i^2) dx_i \right) h_i(\theta_i|\mu_i, \tau_i^2) d\theta_i$$

$$+ o(n^{-1})$$

$$\leq O(s^{-1}) + \int_R \left(\int_{|x_i - \theta_i| \geq \frac{\nu_i^2}{2\tau_i^2} \frac{\epsilon}{16}} f_i(x_i|\theta_i, \sigma_i^2) dx_i \right) h_i(\theta_i|\mu_i, \tau_i^2) d\theta_i$$

$$+ o(n^{-1})$$

$$+ \int_R \left(\int_R P_{n,s}\{|\theta_i - \mu_i| \geq \frac{(s+1)\nu_i^2 \epsilon}{64\sigma_i^2}\} f_i(x_i|\theta_i, \sigma_i^2) dx_i \right)$$

$$\times h_i(\theta_i|\mu_i, \tau_i^2) d\theta_i$$

$$+ \int_R \left(\int_R P_{n,s}\{|\mu_i - \mu_i| \geq \frac{(s+1)\nu_i^2 \epsilon}{64\sigma_i^2}\} f_i(x_i|\theta_i, \sigma_i^2) dx_i \right)$$

$$\times h_i(\theta_i|\mu_i, \tau_i^2) d\theta_i$$

$$+ o(n^{-1})$$

$$\leq O(s^{-1}) + 2e^{-\frac{(2s+1)\nu_i^4 \epsilon^2}{2\times 32^2 (\epsilon+1)^2 \sigma^2 2\tau_i^4}} + \int_{|\theta_i - \mu_i| \geq \frac{(s+1)\nu_i^2 \epsilon}{64\sigma_i^2}} h_i(\theta_i|\mu_i, \tau_i^2) d\theta_i$$

$$+ o(n^{-3}) + o(n^{-1})$$

$$\leq O(s^{-1}) + e^{-\frac{(s+1)^2 \nu_i^4 \epsilon^2}{2\times 64^2 \sigma_i^4 \tau_i^2}} + o(n^{-1})$$

$$= O(s^{-1}) + o(n^{-1}). \qquad (24.3.62)$$

Using similar approach, we can obtain

$$\int_R P_{n,s}\{|\hat{\varphi}_i(x_i) - \varphi_i(x_i)| > \frac{\epsilon}{2}\}|\varphi_i(x_i) - \mu_i| f_i(x_i) dx_i = o(\frac{1}{n}) + O(\frac{1}{s}).$$

$$(24.3.63)$$

At the beginning of this paper, M is assumed to be an odd number. However, from the proof we can see that this condition can be dropped. In other words, no matter M is even or odd, the asymptotic property will hold true. Combining (24.3.49), (24.3.54), (24.3.58), (24.3.59), (24.3.62) and (24.3.63), we finally obtain the asymptotic property of the derived selection procedure.

Theorem 24.3.1 *The selection procedure $d^{(n,s)}(\underline{x})$ defined in (24.2.26) is asymptotically optimal with a convergence rate of order $o(\frac{1}{n}) + O(\frac{\log s}{s})$. That is,*

$$E|R(d^{(n,s)}(\underline{x})) - R(d(\underline{x}))| = o(\frac{1}{n}) + O(\frac{\log s}{s}). \qquad (24.3.64)$$

Theorem 24.3.1 establishes the rate of convergence of $E|R(d^{(n,s)}(\underline{x})) - R(d(\underline{x}))|$ as both n and s go to infinity in an additive form. This implies that $E|R(d^{(n,s)}(\underline{x})) - R(d(\underline{x}))|$ will converge to 0 when both n and s go to infinity.

24.4 SIMULATIONS

We carried out a simulation study to investigate the performance of the selection procedure $d^{(n,s)}(\underline{x})$. The overall integrated risk $E|R(d^{(n,s)}(\underline{x})) - R(d(\underline{x}))|$ is used as measure of the performance of the selection rule.

We consider the following case in which $k = 3$, that is, we have 3 logistic populations Π_1, Π_2 and Π_3 and we would like to use the proposed selection procedure to select the best population compared with a control.

The simulation scheme is described as follows:

(1) For each i, generate past observations as follows:

$$\left\{ \begin{array}{l} \text{for } l = 1, \ldots, n, \\[4pt] \text{(a) first generate } \Theta_{il} \text{ from normal distribution with} \\ \quad \text{density } N(\mu_i, \tau_i^2), \\[4pt] \text{(b) then generate } X_{ijl} \text{ from logistic distribution } L(\theta_{il}, \sigma_i). \end{array} \right. \quad (24.4.65)$$

(2) For each i, generate current observations Θ_i from $N(\mu_i, \tau_i^2)$ and $(X_{i1}, \ldots, X_{iM})$ i.i.d. from $L(\theta_i, \sigma_i)$.

(3) Based on the past observations X_{ijl}, and the present observations, we construct $d(\underline{x})$ and $d^{(n,s)}(\underline{x})$. Then compute the losses $L(d(\underline{x}))$ and $L(d^{(n,s)}(\underline{x}))$.

(4) Repeat Steps (2) and (3) 1000 times. Calculate the averages of the conditional losses $L(d(\underline{x}))$ and $L(d^{(n,s)}(\underline{x}))$, respectively. Denote the averages to be $\hat{R}(d(\underline{x}))$ and $\hat{R}(d^{(n,s)}(\underline{x}))$. Then compute the absolute difference

$$D = |\hat{R}(d^{(n,s)}(\underline{x})) - \hat{R}(d(\underline{x}))|. \quad (24.4.66)$$

(5) Repeat steps (1), (2), (3) and (4) 5000 times. The average of the Ds in (66), denoted by $D(n, s)$, is used as an estimator of the differences $E|R(d^{(n,s)}(\underline{x})) - R(d(\underline{x}))|$.

Tables 24.1, 24.2, and 24.3 give the results of simulation for the performance of the proposed empirical selection procedures. We choose $\theta_0 = 0.5$, $\mu_1 = 0.4$, $\mu_2 = 0.5$, $\mu_3 = 0.6$, $\tau_1 = \tau_2 = \tau_3 = 1$, and $\sigma_1 = \sigma_2 = \sigma_3 = 1$. The related figures are also attached.

Acknowledgements This research was supported in part by US Army Research Office, Grant DAAH04-95-1-0165 at Purdue University.

REFERENCES

Balakrishnan, N. (Ed.)(1992). *Handbook of the Logistic Distribution*, Marcel Dekker, New York.

Baum, L. E. and Katz, M. (1965). Convergence rates in the law of large numbers, *Transactions of the American Mathematics Society*, **120**, 108–123.

Berkson, J. (1957). Tables for the maximum likelihood estimate of the logistic function, *Biometrics*, **13**, 28–34.

Birnbaum, A. and Dudman, J. (1963). Logistic order statistics, *Annals of Mathematical Statistics*, **34**, 658–663.

Gupta, S. S. and Gnanadesikan, M. (1966). Estimation of the parameters of the logistic distribution, *Biometrika*, **53**, 565–570.

Gupta, S. S. and Han, S. (1991). An elimination type two-stage procedure for selecting the population with the largest mean from k logistic populations, *American Journal of Mathematical and Management Sciences*, **11**, 351–370.

Gupta, S. S. and Han, S. (1992). Selection and ranking procedures for logistic populations, In *Order Statistics and Nonparametrics: Theory and Applications* (Eds., P. K. Sen and I. A. Salama), pp. 377-404, Elsevier Science B. V., Amsterdam, The Netherlands.

Gupta, S. S. and Liang, T. (1999). On empirical Bayes simultaneous selection procedures for comparing normal populations with a standard, *Journal of Statistical Planning and Inference*, **77**, 73–88.

Gupta, S. S. and Panchapakesan, S. (1996). Design of experiments with selection and ranking goals, *Handbook of Statistics*, Vol. 13 (Eds., S. Ghosh and C. R. Rao), pp. 555–584, Elsevier Science B. V., Amsterdam, The Netherlands.

Gupta, S. S., Qureishi, A. S. and Shah, B. K. (1967). Best linear unbiased estimators of the parameters of the logistic distribution using order statistics, *Technometrics*, **9**, 43–56.

Gupta, S. S. and Shah, B. K. (1965). Exact moments and percentage points of the order statistics and the distribution of the range from the logistic distribution, *Annals of Mathematical Statistics*, **36**, 907–920.

Panchapakesan, S. (1992). Ranking and selection procedures, In *The Handbook of the Logistic Distribution* (Ed., N. Balakrishnan), pp. 145–167, Marcel Dekker, New York.

Plackett, R. L. (1959). The analysis of life test data, *Technometrics*, 1, 9–19.

Serfling, J. R. (1980). *Approximation Theorems of Mathematical Statistics*, John Wiley & Sons, New York.

Talacko, J. (1956). Perk's distributions and their role in the theory of Wiener's stochastic variates, *Trabajos de Estadistica*, 7, 159–174.

TABLE 24.1 Performance of the selection rule when $s = 5$

n	$D(n,s)$	$SE(D(n,s))$
5	0.08509841	0.027865659
10	0.04994583	0.017846225
15	0.02375784	0.009525936
20	0.01263961	0.006187404
30	0.00738123	0.003425718
40	0.00651246	0.003046143
50	0.00519491	0.002195022
60	0.00415648	0.001433615
70	0.00381372	0.001086059
80	0.00361584	0.000818471
90	0.00347570	0.000713619
100	0.00338605	0.000689365
125	0.00316974	0.000642861
150	0.00309849	0.000627380

TABLE 24.2 Performance of the selection rule when $s = 10$

n	$D(n,s)$	$SE(D(n,s))$
5	0.07304765	0.019237152
10	0.03923180	0.012678379
15	0.01681043	0.008253716
20	0.01064427	0.004316348
30	0.00517343	0.002579812
40	0.00325936	0.001308793
50	0.00216874	0.000845426
60	0.00157431	0.000393872
70	0.00128317	0.000251801
80	0.00107369	0.000137618
90	0.00098735	0.000113625
100	0.00089170	0.000093612
125	0.00084682	0.000079186
150	0.00081364	0.000072564

TABLE 24.3 Performance of the selection rule when $s = 50$

n	$D(n, s)$	$SE(D(n, s))$
5	0.05812975	0.010374924
10	0.01450924	0.007316987
15	0.00810957	0.003264550
20	0.00496331	0.001684319
30	0.00156783	0.000873342
40	0.00084367	0.000439320
50	0.00058916	0.000231186
60	0.00033976	0.000124923
70	0.00026253	0.000095978
80	0.00023609	0.000075646
90	0.00019546	0.000057012
100	0.00017925	0.000043645
125	0.00013687	0.000037202
150	0.00012795	0.000030269

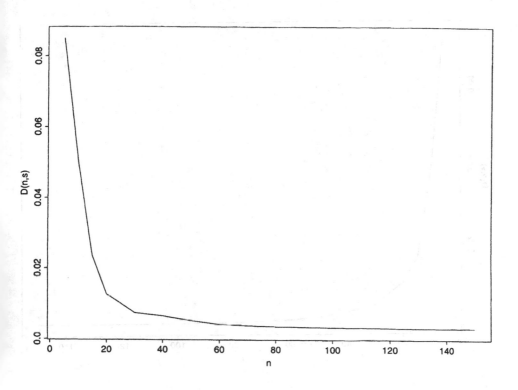

FIGURE 24.1 Graph for Table 24.1

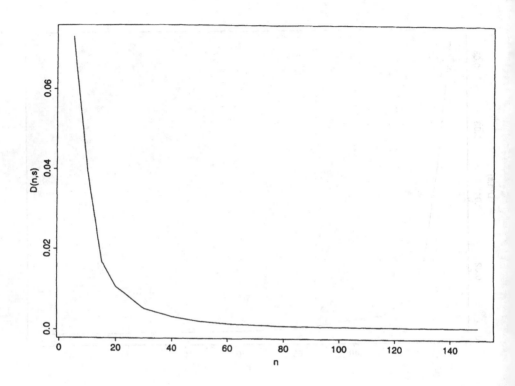

FIGURE 24.2 Graph for Table 24.2

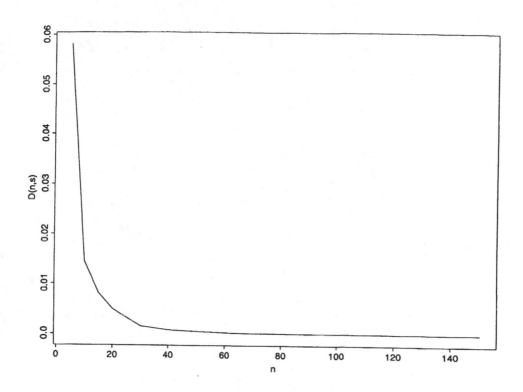

FIGURE 24.3 Graph for Table 24.3

CHAPTER 25

ON SELECTION FROM NORMAL POPULATIONS IN TERMS OF THE ABSOLUTE VALUES OF THEIR MEANS

KHALED HUSSEIN   S. PANCHAPAKESAN

Southern Illinois University, Carbondale, IL

Abstract: In this chapter, the problem of selecting the t normal populations (out of k) with the largest or the smallest absolute value of means. For this purpose, both indifference zone and subset selection approaches are considered.

Keywords and phrases: Signal-to-noise ratio, selection procedure, indifference zone approach, subset selection approach

25.1 INTRODUCTION

Let $\Pi_1, \ldots, \Pi_k$ ($k \geq 2$) be k independent normal populations with *unknown* means $\mu_1, \ldots, \mu_k$, respectively, and a common variance σ^2. Let $\theta_i = |\mu_i|$, $i = 1, \ldots, k$. The populations are ranked according to their θ-values. Let $\theta_{[1]} \leq \cdots \leq \theta_{[k]}$ denote the ordered θ_i. It is assumed that there is no prior knowledge about the correspondence between the ordered and the unordered θ_i. Our goal for most of this paper is to select the populations associated with the t largest or the t smallest θ-values, where $1 \leq t \leq k-1$. Consider selecting those with the t largest θ-values. These are referred to as the t *best* populations. Since the populations have the same variance σ^2, the problem is equivalent to selecting the populations associated with

371

t largest $\theta_i/\sigma = |\mu_i|/\sigma$, which is the signal-to-noise ratio well-known in communications theory. In terms of comparing k different electronic devices, our goal is to select the devices having the t largest signal-to-noise ratios. We also note that θ_i^2/σ^2 is the Mahalanobis distance between Π_i and $N(0, \sigma^2)$ population. In this context, our goal is to select the t farthest populations from $N(0, \sigma^2)$.

We now formulate our problem of selecting the t best using the two classical approaches, namely, the indifference zone (IZ) approach of Bechhofer (1954) and the subset selection (SS) approach of Gupta (1956, 1965). Let $\Omega = \{\boldsymbol{\theta} : \boldsymbol{\theta} = (\theta_1, \ldots, \theta_k),\ 0 \le \theta_i < \infty,\ i = 1, \ldots, k\}$ and $\Omega(\delta^*) = \{\boldsymbol{\theta} : \theta_{k-t+1]} - \theta_{k-t]} \ge \delta^*,\ \delta^* > 0,\ \boldsymbol{\theta} \in \Omega\}$ for any specified δ^*. In the IZ approach, one wants to select *exactly* t populations with a probability requirement that the probability of a correct selection (PCS) is at least P^* whenever $\boldsymbol{\theta} \in \Omega(\delta^*)$. Here a correct selection (CS) occurs when the t best populations are selected. Also, δ^* and P^* are specified in advance by the experimenter. For a meaningful problem, we take $\begin{pmatrix} k \\ t \end{pmatrix}^{-1} < P^* < 1$. For $P^* \le \begin{pmatrix} k \\ t \end{pmatrix}^{-1}$, the probability requirement can be satisfied by choosing t populations randomly. The region $\Omega(\delta^*)$ is called the *preference zone* (PZ). Its complement w.r.t. Ω is the so-called *indifference zone* (IZ). In the SS approach, we want to select a non-empty subset of the k populations so that the selected subset contains the t best populations (in which case, a correct selection is said to occur). The size of the selected subset is random subject to a minimum of t. Also, if there is more than one set as a contender for the t best, we assume that one of them is tagged as the set of t best populations. We again take $\begin{pmatrix} k \\ t \end{pmatrix}^{-1} < P^* < 1$. In either approach, we need to define a selection rule R which really has three parts: a sampling rule, a stopping rule for sampling, and a decision rule after stopping sampling. Denoting the PCS using the rule R by $P(CS|R)$, the rule R is valid if

$$P(CS|R) \ge P^* \quad \text{whenever } \boldsymbol{\theta} \in \Omega(\delta^*) \tag{25.1.1}$$

in the case of the IZ approach, and if

$$P(CS|R) \ge P^* \quad \text{for all } \boldsymbol{\theta} \in \Omega \tag{25.1.2}$$

in the case of the SS approach. For comparing valid rules under the IZ formulation, one naturally considers the sample sizes (or expected sample sizes) for the rules. For the SS approach, one can always select all the populations, yielding a valid rule! So we use operating characteristics such as $E(S)$, the expected size of the selected subset, $E(S) - PCS$, which is the expected number of non-best populations included in the selected subset, and $E(S)/PCS$.

In this paper, we briefly review (Sections 25.3 through 25.7) several significant results available in the literature under the IZ approach (Sections 25.3 and 25.5), the SS approach (Section 25.4) and an integrated formulation (Section 25.7). We present new results for simultaneously selecting the two extreme populations under the IZ (Section 25.8) as well as the SS approach (Section 25.9). We conclude (Section 25.10) with some directions for further investigations.

25.2 SOME PRELIMINARY RESULTS

Let $\overline{X}$ denote the mean of a random sample of size n from a normal population with mean μ and variance σ^2. Let $W = |\overline{X}|$ and $\theta = |\mu|$. Then it is easy to see that the cdf $H(w; \theta)$ and the density function $h(w, \theta)$ of W are given by

$$H(w; \theta) = \begin{cases} \Phi\left\{\frac{\sqrt{n}}{\sigma}(w - \theta)\right\} - \Phi\left\{\frac{\sqrt{n}}{\sigma}(-w - \theta)\right\}, & w \geq 0 \\ 0, & w < 0 \end{cases} \qquad (25.2.3)$$

and

$$h(w; \theta) = \begin{cases} \frac{\sqrt{n}}{\sigma}\left[\phi\left\{\frac{\sqrt{n}}{\sigma}(w - \theta)\right\} + \phi\left\{\frac{\sqrt{n}}{\sigma}(w + \theta)\right\}\right], & \\ & w \geq 0 \\ 0, & \text{otherwise} \end{cases} \qquad (25.2.4)$$

where Φ and ϕ are standard normal cdf and density function, respectively.

For $w \geq 0$ and $\theta > 0$, we have

$$\frac{\partial H(w; \theta)}{\partial \theta} = \frac{\sqrt{n}}{\sigma}\left[\phi\left\{\frac{\sqrt{n}}{\sigma}(w + \theta)\right\} - \phi\left\{\frac{\sqrt{n}}{\sigma}(w - \theta)\right\}\right]$$

which is < 0, since $\phi\left\{\frac{\sqrt{n}}{\sigma}(w - \theta)\right\} = \phi\left\{\frac{\sqrt{n}}{\sigma}|w - \theta|\right\} \geq \phi\left\{\frac{\sqrt{n}}{\sigma}(w + \theta)\right\}$. This shows that the family $\{H(w; \theta)\}$, $\theta \geq 0$, is stochastically increasing in θ.

We now continue with our selection problem under the two formulations mentioned.

25.3 INDIFFERENCE ZONE FORMULATION: KNOWN COMMON VARIANCE

For the goal of selecting the populations associated with the t largest θ_i's, Rizvi (1971) studied a fixed sample procedure based on samples of common size n from the k populations. Let $\overline{X}_i$ be the mean of the sample from Π_i and let $W_i = |\overline{X}_i|$, $i = 1, \ldots, k$. Rizvi (1971) proposed the rule

$$R_1 : \text{Select the populations that yielded} \atop \text{the } t \text{ largest sample means.} \qquad (25.3.5)$$

The design problem is to determine the minimum sample size n needed in order to satisfy the probability requirement (25.1.1) for given k, t, P^* and δ^*. Let $W_{(i)}$ denote the W_j from the population associated with $\theta_{[i]}$, $i = 1, \ldots, k$. Rizvi (1971) has shown that the infimum of $P(CS|R_1)$ over $\Omega(\delta^*)$ is attained for the configuration

$$\theta_{[1]} = \cdots = \theta_{[k-t]} = 0; \ \theta_{[k-t+1]} = \cdots = \theta_{[k]} = \delta^* \qquad (25.3.6)$$

which is called the least favorable configuration (LFC). At the LFC, the PCS given by

$$P_{LFC}(CS|R_1) = 2(k-t) \int_0^\infty [2\Phi(u) - 1]^{k-t-1}$$
$$\times [\Phi(-u + \lambda) + \Phi(-u - \lambda)]^t \phi(u) du, \quad (25.3.7)$$

where $\lambda = \sqrt{n}\,\delta^*/\sigma$. Let $\lambda = \lambda(k, t, P^*)$ be the solution of the equation obtained by equating the right side of (25.3.7) to P^*. Then the minimum sample size required to satisfy (25.1.1) is given by

$$n = \left\langle \left(\frac{\lambda\sigma}{\delta^*}\right)^2 \right\rangle, \qquad (25.3.8)$$

where $\langle s \rangle$ denotes the smallest integer $\geq s$.

Two notable special cases are: (A) $t = 1$ and (B) $t = k - 1$, corresponding to selecting the population with the largest and the smallest θ-value, respectively. For both cases, Rizvi (1971) has tabulated the PCS at the LFC for $k = 2(1)10$ and $\lambda = 0.0$ (0.5) 2.0 (0.2) 7.0, and the value of λ needed to determine n from (25.3.8) for $k = 2(1)10$ and $P^* = 0.5000, 0.7500, 0.9000, 0.9500, 0.9750, 0.9900, 0.9950, 0.9990, 0.9995, 0.9999$.

Rizvi (1963) has proved some optimality properties of the rule R_1. In particular, he has shown that R_1 is a most economical decision rule in the sense that there is no other single sample procedure with a smallest common sample size satisfying the probability requirement (25.1.1).

25.4   SUBSET SELECTION FORMULATION: KNOWN COMMON VARIANCE

As before, let $W_i = |\overline{X}_i|$, $i = 1, \ldots, k$, based on samples of a common size n. In this case, Rizvi (1971) considered only selecting the best ($t = 1$) population and selecting the worst (associated with the smallest θ_i) population. For selecting the best, Rizvi (1971) proposed the rule

$$R_2 : \text{Select } \Pi_i \text{ if, and only if,}$$
$$W_i \geq \max_{1 \leq j \leq k} W_j - d, \qquad (25.4.9)$$

where $d = d(n, k, P^*)$ is chosen as the smallest positive constant for which the probability requirement (25.1.2) is satisfied. The infimum of $P(CS|R_2)$ over Ω has been shown to occur when $\theta_{[1]} = \cdots = \theta_{[k]} = \theta \to \infty$. This infimum is given by

$$P_{LFC}(CS|R_2) = \int_{-\infty}^{\infty} \Phi^{k-1}(u + D)\phi(u)du, \qquad (25.4.10)$$

where $D = \sqrt{n}\,d/\sigma$. Thus, our constant d is obtained by solving for D the equation

$$\int_{-\infty}^{\infty} \Phi^{k-1}(u + D)\phi(u)du = P^*. \qquad (25.4.11)$$

The value of D satisfying (25.4.11) has been tabulated for different ranges of k and P^* by Bechhofer (1954), Gupta (1963), Milton (1963) and Gupta, Nagel and Panchapakesan (1973). Rizvi (1971) has shown that the supremum over Ω of the expected subset size, $E(S)$, is attained when $\theta_{[1]} = \cdots = \theta_{[k]} = 0$ and is given by

$$\sup_{\Omega} E(S) = 2k \int_0^{\infty} [2\Phi(u + \sqrt{n}\,d/\sigma) - 1]^{k-1}\phi(u)du. \qquad (25.4.12)$$

The above bound for $E(S)$ exceeds kP^* but is less than k.

In addition to satisfying the probability requirement (25.1.2), one can ask for the smallest common sample size n necessary to control $E(S)$ at some pre-assigned level. For a particular parametric configuration $\boldsymbol{\theta}_0$ in Ω. For example, $\boldsymbol{\theta}_0$ could be a slippage configuration [i.e. $\theta_{[k]} - \theta_{[i]} = \delta^*$, $i = 1, \ldots, k-1$, for some specified $\delta^* > 0$] or an equi-distance configuration [i.e. $\theta_{[i+1]} - \theta_{[i]} = \delta^*$ for some specified $\delta^* > 0$, $i = 1, \ldots, k-1$]. Then we want to find the smallest n such that $E(S|\boldsymbol{\theta}_0) \leq 1 + \eta$, for a specified $\eta \in (0, k-1)$. Alternatively, one can control the supremum of $E(S)$ over a subspace $\Omega(\xi)$ of Ω at a specified level. Rizvi (1971) considered $\Omega(\xi) =$

$\{\boldsymbol{\theta} : \theta_{[k]} - \theta_{[k-1]} \geq \xi > 0\}$. The problem then is to determine the smallest n needed so that $\sup\limits_{\Omega(\xi)} E(S) \leq 1 + \eta$ for some specified $\eta \in (0, k-1)$.

The problem of selecting a subset containing the population associated with the smallest θ_i is treated in an analogous manner. Rizvi (1971) studied the analogous rule

$$R_2' : \text{ Select } \Pi_i \text{ if, and only if,}$$
$$W_i \leq \min_{1 \leq j \leq k} W_j + d', \qquad (25.4.13)$$

where $d' = d'(n, k, P^*)$ is chosen as the smallest positive constant for which the probability requirement (25.1.2) is satisfied (with CS suitably modified). The infimum of $P(CS|R_2')$ over Ω has been shown by Rizvi (1971) to occur when $\theta_{[1]} = \cdots = \theta_{[k]} = \theta \to \infty$. This infimum is given by

$$P_{LFC}(CS|R_2') = \int_{-\infty}^{\infty} \Phi^{k-1}\left(u + \frac{\sqrt{n}\,d'}{\sigma}\right)\phi(u) \qquad (25.4.14)$$

which is same as the right side (25.4.10) if we let $D = \sqrt{n}\,d'/\sigma$. Thus the constant d' is same as d for R_2. Parallel results can be obtained regarding the supremum of $E(S)$.

Finally, both rules R_2 and R_2' possess an important monotonicity property. Let Q_i (Q_i') be the probability that the population associated with the parameter $\theta_{[i]}$ is included in the subset selected by R_2 (R_2'). Then, for $i < j$, $Q_i \leq Q_j$ and $Q_i' \geq Q_j'$.

25.5 INDIFFERENCE ZONE FORMULATION: UNKNOWN COMMON VARIANCE

In Section 25.3, we discussed the selection of the t best populations when the common variance σ^2 is known. When σ is unknown, it can be seen from (25.3.8) that we cannot determine without the knowledge of σ the sample size necessary for R_1 to meet the probability requirement (25.1.1). In this case, a single sample procedure that guarantees a minimum PCS does not exist. For the problem of selecting the best ($t = 1$), Jeyaratnam and Panchapakesan (1999) proposed a two-stage procedure analogous to that of Bechhofer, Dunnett and Sobel (1954) for selecting the normal population having the largest μ_i. Their procedure R_3 is described below.

Procedure R_3:

(1) First take a random sample of n_0 observations from each of the k populations. Let these be X_{ij}, $j = 1, \ldots, n_0$; $i = 1, \ldots, k$. Define

$$\overline{X}_i = \sum_{i=1}^{n_0} X_{ij}/n_0, \quad i = 1, \ldots, k,$$

$$S^2 = \sum_{i=1}^{k} \sum_{j=1}^{n_0} (X_{ij} - \overline{X}_i)^2 / k(n_0 - 1).$$

(2) Choose h such that

$$\int_0^\infty T(h\sqrt{u/\nu}) g_\nu(u) du = P^*,$$

where

$$T(\lambda) = 2(k-1) \int_0^\infty \{2\Phi(u) - 1\}^{k-2} \{\Phi(-u+\lambda) + \Phi(-u-\lambda)\} \phi(u) du,$$

$g_\nu(u)$ is the chi-square density with $\nu = k(n_0 - 1)$ degrees of freedom. Now, take a second sample of $(N - n_0)$ observations from each population, where

$$N = \max\{n_0, \langle (hS/\delta^*)^2 \rangle\}$$

and $\langle s \rangle$ denotes the smallest integer $\geq s$.

(3) Let $\overline{Y}_i$ denote the overall mean of the N observations from Π_i and let $W_i = |\overline{Y}_i|$, $i = 1, \ldots, k$. Select as the best the population that yields the largest W_i.

Jeyaratnam and Panchapakesan (1999) have tabulated the values of h for $P^* = 0.90, 0.95$, $k = 2(1)10$, and selected values of n_0 for each k. The expression for $E(N)$, the expected value of the total sample size N, is given (to within a quantity less than unity) by

$$E(N) = n_0 \Pr\left\{ \chi_\nu^2 < \nu n_0 \left(\frac{\delta^*}{h\sigma}\right)^2 \right\}$$

$$+ \left(\frac{h\sigma}{\delta^*}\right)^2 \Pr\left\{ \chi_{\nu+2}^2 > \nu n_0 \left(\frac{\delta^*}{h\sigma}\right)^2 \right\}$$

$$+ \tau \Pr\left\{ \chi_\nu^2 > \nu n_0 \left(\frac{\delta^*}{h\sigma}\right)^2 \right\},$$

where $0 \leq \tau < 1$ and χ_ν^2 is a chi-square variable with $\nu = k(n_0 - 1)$ degrees of freedom. Jeyaratnam and Panchapakesan (1999) have done some limited analysis of gain due to the knowledge of σ. The case of selecting the population associated with the smallest θ_i can be handled analogously.

25.6 SUBSET SELECTION FORMULATION: UNKNOWN COMMON VARIANCE

Under the subset selection formulation, when σ is unknown, one can define a single sample procedure. Based on random samples of common size n from the k populations, let $\overline{X}_i$, $i = 1, \ldots, k$, be the sample means and let S^2 denote the usual pooled unbiased estimator of σ^2 on $\nu = k(n-1)$ degrees of freedom. For selecting the population associated with $\theta_{[k]}$, Rizvi (1971) studied the rule

$$R_4 : \text{Select } \Pi_i \text{ if, and only if,}$$
$$W_i \geq \max_{1 \leq j \leq k} W_j - cS, \qquad (25.6.15)$$

where $c = c(n, k, P^*)$ is chosen as the smallest positive constant for which (25.1.2) is satisfied. The minimization of $P(CS|R_4)$ can be carried out using the steps involved in the case of R_2 in (25.4.9) by first conditioning on S. Equating the minimum $P(CS|R_4)$ to P^*, the constant c is given by

$$\int_0^\infty g_\nu(v) \left[\int_{-\infty}^\infty \Phi^{k-1}(u + Cv)\phi(u)du \right] dv = P^*, \qquad (25.6.16)$$

where $C = \sqrt{n}\, c/\sqrt{\nu}\, \sigma$ and $g_\nu(\cdot)$ is the chi-density with ν degrees of freedom. The constant C is a multiple of upper $100(1 - P^*)$ percentage point of the distribution of the studentized maximum of equally correlated ($\rho = 0.5$) normal variables. The need for these percentage points (with $0 < \rho < 1$) arises in several problems. So the value of C satisfying (25.6.16) can be obtained from the tables in Dunnett and Sobel (1954) [$k = 2$ only], Dunnett (1955), Gupta and Sobel (1957), Krishnaiah and Armitage (1966), and Gupta, Panchapakesan and Sohn (1985) for several selected values k, ν and P^*.

One can carry out an analysis of the behavior of $E(S)$ for R_4 parallel to that of R_2. Also, one can study the problem of selecting the population with the smallest θ_i by defining a rule R_4' be replacing d' by $c'S$ ($c' > 0$) in (25.4.13). However, these results seem to be not explicitly available in the literature.

25.7   AN INTEGRATED FORMULATION

In Section 25.1, we discussed the IZ and SS formulations. We will consider here the problem of selecting the best $(t = 1)$ population. We also assume that the common variance σ^2 is known. In the IZ approach, we control the PCS when $\boldsymbol{\theta} \in \Omega(\delta^*)$, but there is no probability requirement when $\boldsymbol{\theta} \in \Omega\backslash\Omega(\delta^*)$. On the other hand, when $\boldsymbol{\theta} \in \Omega(\delta^*)$ [meaning the best population stands sufficiently apart from the rest], the SS approach does not restrict the correct selection to selecting only the best population. An integrated formulation incorporating the features of both IZ and SS formulations has been studied by Chen and Sobel (1987a,b). A few other papers in this direction are Chen (1988), Chen and Panchapakesan (1994), and Jeyaratnam and Panchapakesan (1997). We discuss here the results of Jeyaratnam and Panchapakesan (1997) who considered the problem of selecting the normal population with the largest θ_i, assuming a common known variance σ^2.

When $\theta_{[k]} - \theta_{[k-1]} \geq \delta^*$ $(\boldsymbol{\theta} \in \Omega(\delta^*) = PZ)$, the population associated with $\theta_{[k]}$ is defined as the δ^*-*best*; otherwise when $\boldsymbol{\theta} \in IZ$, this population is defined as just *the best*. Our goal is to select the δ^*-best and none other when $\boldsymbol{\theta} \in PZ$, and select a non-empty subset containing the best population when $\boldsymbol{\theta} \in IZ$. Let CD_1 and CD_2 denote the correct decision when $\boldsymbol{\theta} \in PZ$ and $\boldsymbol{\theta} \in IZ$, respectively. It is required of any valid procedure that

$$P(CD_1) \geq P_1^* \text{ whenever } \boldsymbol{\theta} \in PZ \qquad (25.7.17)$$

and

$$P(CD_2) \geq P_2^* \text{ whenever } \boldsymbol{\theta} \in IZ, \qquad (25.7.18)$$

where $k^{-1} < P_i^* < 1$, $i = 1, 2$. Here δ^*, P_1^* and P_2^* are specified in advance.

As before, based on random samples of common size n, let $W_i = |\overline{X}_i|$, where $\overline{X}_i$ is the sample mean from Π_i, $i = 1, \ldots, k$. Let c and d be constants, with $0 < d < c$, to be determined (along with n) as functions of δ^*, P_1^*, P_2^* and k. Assume that c, d and n are already determined. Jeyaratnam and Panchapakesan (1997) proposed and studied the procedure

R_5 : If $W_{[k]} - W_{[k-1]} > c$, then select the population that yielded
the largest W_i as the δ^*-best population; otherwise, select
all populations Π_i for which $W_i \geq W_{[k]} - d$ and claim that
the best population is included in the selected subset.

Using a lower bound for the infimum of $P(CD_1|PZ)$, Jeyaratnam and Panchapakesan (1997) have shown that (25.7.17) is satisfied if we can choose

c such that

$$\int_0^\infty H^{k-1}(u;0)h(u+c;\delta^*)du \geq P_1^*, \qquad (25.7.19)$$

where $H(u;\theta)$ and $h(u;\theta)$ are defined in (25.2.3) and (25.2.4), respectively. They have also shown that $P(CD_2|IZ)$ attains its infimum when $\theta_{[1]} = \cdots = \theta_{[k]} = \theta \to \infty$. This shows that (25.7.18) is satisfied if

$$\int_{-\infty}^\infty \Phi^{k-1}\left(u + \frac{d\sqrt{n}}{\sigma}\right)\phi(u)du \geq P_2^*. \qquad (25.7.20)$$

In order to implement the rule R_5, we need to find n, c and d satisfying (25.7.19) and (25.7.20) for given k, P_1^* and P_2^* with the restriction that $d < c < \delta^*$. Jeyaratnam and Panchapakesan (1997) have discussed how, by a trial and error method, one can get a feasible solution, not necessarily unique. For comments regarding finding an optimal solution and some desirable properties of R_5, the reader is referred to their paper.

25.8   SIMULTANEOUS SELECTION OF THE EXTREME POPULATIONS: INDIFFERENCE ZONE FORMULATION AND KNOWN COMMON VARIANCE

We now consider a modified goal of selecting simultaneously the two populations associated with $\theta_{[k]}$ and $\theta_{[1]}$, individually identified as such. This is a special case of the general ranking problem of partitioning the set of k population into s groups, $I_1, \ldots, I_s$, such that the populations in I_i have smaller θ-values than those in the set I_{i+1}, $i = 1, \ldots, k - 1$. This general goal has been mentioned in Bechhofer (1954) but not investigated. Mishra (1986) has considered our special case ($s = 3$) for location parameters. In our present formulation, a correct selection occurs when the two target populations are selected and identified correctly. It is required of any valid rule R that

$$P(CS|R) \geq P^* \quad \text{whenever} \quad \boldsymbol{\theta} \in \Omega^*, \qquad (25.8.21)$$

where $\Omega^* \equiv \Omega(\delta_1^*, \delta_2^*) = \{\boldsymbol{\theta} : \theta_{[2]} - \theta_{[1]} \geq \delta_1^* > 0, \; \theta_{[k]} - \theta_{[k-1]} \geq \delta_2^* > 0\}$ is the preference zone, and the constants δ_1^* and δ_2^* are specified in advance along with the probability level P^* such that $[k(k-1)]^{-1} < P^* < 1$.

Based on random samples of common size n from the k populations, let $W_{[1]} \leq \cdots \leq W_{[k]}$ denote the ordered $W_i = |\overline{X}_i|$, $i = 1, \ldots, k$. We propose the rule

R_6 : Select the population that yields $W_{[1]}(W_{[k]})$ as
the population associated with $\theta_{[1]}(\theta_{[k]})$. $\qquad (25.8.22)$

Denoting by $W_{(i)}$ the W associated with the population having the parameter $\theta_{[i]}$, $i = 1, \ldots, k$, we see that

$$P(CS|R_6) = Pr\{W_{(1)} \leq W_{(i)} \leq W_{(k)}, \quad i = 2, \ldots, k-1\}$$

$$= \int_0^\infty \int_0^t \prod_{i=2}^{k-1} [H(t;\theta_{[i]}) - H(s;\theta_{[i]})] dH(s;\theta_{[1]})$$

$$\times \, dH(t;\theta_{[k]}) \tag{25.8.23}$$

The exact infimum over Ω^* of the double integral in (25.8.23) is not available for $k \geq 4$. For $k = 3$, we will obtain the LFC for $P(CS|R_6)$. In this case,

$$P(CS|R_6)$$
$$= Pr[W_{(1)} \leq W_{(2)} \leq W_{(3)}]$$
$$= \int_0^\infty \int_0^t [H(t;\theta_{[2]}) - H(s;\theta_{[2]})] h(s;\theta_{[1]})$$
$$\times h(t;\theta_{[3]}) ds\, dt$$
$$= \int_0^\infty \int_0^t H(t;\theta_{[2]}) h(s;\theta_{[1]}) h(t;\theta_{[3]}) ds\, dt$$
$$- \int_0^\infty \int_s^\infty H(s;\theta_{[2]}) h(s;\theta_{[1]}) h(t;\theta_{[3]}) dt\, ds$$
$$= \int_0^\infty H(t;\theta_{[2]}) H(t;\theta_{[1]}) h(t;\theta_{[3]}) dt$$
$$- \int_0^\infty H(s;\theta_{[2]}) [1 - H(s;\theta_{[3]})] h(s;\theta_{[1]}) ds. \tag{25.8.24}$$

Using (25.2.3) and (25.2.4) in (25.8.24) and letting $u = \sqrt{n}\, t/\sigma$ and $\alpha_i = \sqrt{n}\, \theta_{[i]}/\sigma$, $i = 1,2,3$, we obtain

$$P(CS|R_6)$$
$$= \int_0^\infty \{\Phi(u - \alpha_2) - \Phi(-u - \alpha_2)\}\{\Phi(u - \alpha_1) - \Phi(-u - \alpha_1)\}$$
$$\times \{\phi(u - \alpha_3) + \phi(u + \alpha_3)\} du$$
$$- \int_0^\infty \{\Phi(u - \alpha_2) - \Phi(-u - \alpha_2)\}\{\Phi(-u + \alpha_3) + \Phi(-u - \alpha_3)\}$$
$$\times \phi(u - \alpha_1) du$$
$$- \int_0^\infty \{\Phi(u - \alpha_2) - \Phi(-u - \alpha_2)\}\{\Phi(-u + \alpha_3) + \Phi(-u - \alpha_3)\}$$
$$\times \phi(u + \alpha_1) du$$
$$= I_1(\alpha_1) - I_2(\alpha_1) - I_3(\alpha_1), \text{ say.} \tag{25.8.25}$$

We want to establish that $P(CS|R_6)$ is non-increasing in α_1 when α_2 and α_3 are kept fixed. Towards this end, we find the derivative of the expression in (25.8.25) w.r.t. α_1. In doing so, we first change the variable in $I_2(\alpha_1)$ and $I_3(\alpha_1)$ by putting $y = u - \alpha_1$ and $y = u + \alpha_1$, respectively. After straightforward differentiation, we change back by the same transformations. This gives us

$$\frac{\partial}{\partial \alpha_1} P(CS|R_6)$$
$$= \int_0^\infty \{\Phi(-u + \alpha_3) + \Phi(-u - \alpha_3)\}$$
$$\times \{\phi(u - \alpha_2) - \phi(-u - \alpha_2)\}\{\phi(u + \alpha_1) - \phi(u - \alpha_1)\}du,$$

which is ≤ 0 since $\phi(u+\alpha_1) \leq \phi(u-\alpha_1)$. Thus $P(CS|R_6)$ is non-increasing in α_1 when α_2 and α_3 are kept fixed.

Similarly, we consider $P(CS|R_6)$ as a function of α_3 when α_1 and α_2 are kept fixed. It can be seen again by straightforward differentiation that

$$\frac{\partial}{\partial \alpha_3} P(CS|R_6)$$
$$= \int_0^\infty \{\Phi(u - \alpha_1) - \Phi(-u - \alpha_1)\}$$
$$\times \{\phi(u - \alpha_2) + \phi(-u - \alpha_2)\}\{\phi(u - \alpha_3) - \phi(u + \alpha_3)\}du,$$

which is ≥ 0 since $\phi(u-\alpha_3) \geq \phi(u+\alpha_3)$. Thus $P(CS|R_6)$ is non-decreasing in α_3 when α_1 and α_2 are kept fixed.

By combining the above two results, it is easy to see that the infimum of $P(CS|R_6)$ over Ω^* occurs at an LFC of the form:

$$\theta_{[1]} + \delta_1^* = \theta_{[2]} = \theta_{[3]} - \delta_2^* = \theta_0 \text{ (say).} \qquad (25.8.26)$$

Letting $\alpha = \sqrt{n}\,\theta_0/\sigma$ and $\delta_i = \sqrt{n}\,\delta_i^*/\sigma$, $i = 1, 2, 3$, $P(CS|R_6)$ at the LFC in (25.8.26) is given by

$$P_{LFC}(CS|R_6) = V_1(\alpha_0; \delta_1, \delta_2) + V_2(\alpha_0; \delta_1, \delta_2)$$
$$-V_3(\alpha_0; \delta_1, \delta_2) - V_4(\alpha_0; \delta_1, \delta_2), \qquad (25.8.27)$$

where

$$V_1(\alpha_0; \delta_1, \delta_2)$$
$$= \int_0^\infty \{\Phi(u - \alpha_0) - \Phi(-u - \alpha_0)\}$$

$$\times\{\Phi(u-\alpha_0+\delta_1)-\Phi(-u-\alpha_0+\delta_1)\}\phi(u-\alpha_0-\delta_2)du,$$

$$V_2(\alpha_0;\delta_1,\delta_2)$$

$$= \int_0^\infty \{\Phi(u-\alpha_0)-\Phi(-u-\alpha_0)\}$$

$$\times\{\Phi(u-\alpha_0+\delta_1)-\Phi(-u-\alpha_0+\delta_1)\}\phi(u+\alpha_0+\delta_2)du,$$

$$V_3(\alpha_0;\delta_1,\delta_2)$$

$$= \int_0^\infty \{\Phi(u-\alpha_0)-\Phi(-u-\alpha_0)\}$$

$$\times\{\Phi(-u+\alpha_0+\delta_2)+\Phi(-u-\alpha_0-\delta_2)\}\phi(u-\alpha_0+\delta_1)du,$$

$$V_4(\alpha_0;\delta_1,\delta_2)$$

$$= \int_0^\infty \{\Phi(u-\alpha_0)-\Phi(-u-\alpha_0)\}$$

$$\times\{\Phi(-u+\alpha_0+\delta_2)+\Phi(-u-\alpha_0-\delta_2)\}\phi(u+\alpha_0-\delta_1)du.$$

$$(25.8.28)$$

We now establish the monotonicity of $P(CS|R_6)$ at the *LFC* given in (25.8.26) as a function of θ_0 (or, equivalently α_0).

Lemma 25.8.1 *At the LFC given in (25.8.26), $P(CS|R_6)$ is non-decreasing in θ_0 for fixed δ_1^* and δ_2^*.*

PROOF First, put $y=u-\alpha_0-\delta_2$, $y=n+\alpha_0+\delta_2$, $y=u-\alpha_0+\delta_1$ and $y=u+\alpha_0-\delta_1$ in $V_i(\alpha_0;\delta_1,\delta_2)$, $i=1,\ldots,4$, respectively. Note that $V_2(\alpha_0;\delta_1,\delta_2)=-V_1(-\alpha_0;-\delta_1,-\delta_2)$ and $V_4(\alpha_0;\delta_1,\delta_2)=V_3(-\alpha_0;-\delta_1,-\delta_2)$. Let $Z_1(\alpha_0;\delta_1,\delta_2)=V_1'(\alpha_0;\delta_1,\delta_2)$ and $Z_2(\alpha_0;\delta_1,\delta_2)=V_3'(\alpha_0;\delta_1,\delta_2)$, where the prime (') denotes the derivative w.r.t. α_0. It can now be seen that $V_2'(\alpha_0;\delta_1,\delta_2)=Z_1(-\alpha_0;-\delta_1,-\delta_2)$ and $V_4'(\alpha_0;\delta_1,\delta_2)=Z_2(-\alpha_0;-\delta_1,-\delta_2)$. By carrying out routine differentiation and using the substitutions mentioned in the beginning of the proof, one can obtain

$$\frac{\partial}{\partial\alpha_0}P_{LFC}(CS|R_6)$$

$$= 2\int_0^\infty \{\Phi(u-\alpha_0+\delta_1)-\Phi(-u-\alpha_0+\delta_1)\}$$

$$\times\{\phi(-u-\alpha_0)\phi(u-\alpha_0-\delta_2)-\phi(-u+\alpha_0)\phi(u+\alpha_0+\delta_2)\}du$$

$$-2\int_0^\infty \{\Phi(-u+\alpha_0+\delta_2)+\Phi(-u-\alpha_0-\delta_2)\}$$

$$\times\{\phi(-u-\alpha_0)\phi(u-\alpha_0+\delta_1)-\phi(-u+\alpha_0)\phi(u+\alpha_0-\delta_1)\}du.$$

It is easy to check that

$$\phi(-u - \alpha_0)\phi(u - \alpha_0 - \delta_2) - \phi(-u + \alpha_0)\phi(u + \alpha_0 + \delta_2) \geq 0$$

and

$$\phi(-u - \alpha_0)\phi(u - \alpha_0 + \delta_1) - \phi(-u + \alpha_0)\phi(u + \alpha_0 - \delta_1) \leq 0.$$

Thus $P_{LFC}(CS|R_6)$ is non-decreasing in α_0. This proves the lemma.   □

Summarizing the preceding discussion regarding $P(CS|R_6)$ for $k = 3$ populations and using Lemma 25.8.1, we obtain the following theorem.

Theorem 25.8.2 *For the rule R_6 defined in (25.8.22) applied to $k = 3$ populations,*

$$\inf_{\theta \in \Omega^*} P(CS|R_6)$$

$$= \int_0^\infty \{\Phi(u - \delta_1) - \Phi(-u - \delta_1)\}\{\Phi(u) - \Phi(-u)\}$$

$$\times \{\phi(u - \delta_1 - \delta_2) + \phi(u + \delta_1 + \delta_2)\}du$$

$$-2 \int_0^\infty \{\Phi(u - \delta_1) - \Phi(-u - \delta_1)\}$$

$$\times \{\Phi(-u + \delta_1 + \delta_2)) + \Phi(-u - \delta_1 - \delta_2)\}\phi(u)du,$$

$$(25.8.29)$$

where $\delta_i = \sqrt{n}\,\delta_i^*/\sigma$, $i = 1, 2$.

For given P^*, δ_1^* and δ_2^*, we have to find the smallest n for which the right side of (7.9) is $\geq P^*$.

25.9   SIMULTANEOUS SELECTION OF THE EXTREME POPULATIONS: SUBSET SELECTION FORMULATION AND KNOWN COMMON VARIANCE

Our goal here is to select two non-empty subsets of the k given populations, namely, S_B which contains the population associated with $\theta_{[1]}$ (called the worst) and S_G which contains the population associated with $\theta_{[k]}$ (called the best). In the case of a tie for the best or the worst population, one of the contenders is assumed to be the best or the worst as the case may be.

Any selection of two subsets S_B and S_G consistent with our goal is called a correct selection. For any valid procedure R, we require that

$$P(CS|R) \geq P^* \text{ for all } \boldsymbol{\theta} \in \Omega, \tag{25.9.30}$$

where P^* is specified in advance such that $[k(k-1)]^{-1} < P^* < 1$. Continuing our notations of Section 25.8, we propose the rule

R_7 : Put Π_i in S_G if, and only if, $W_i \geq W_{[k]} - c_1$ and,
     Put Π_i in S_B if, and only if, $W_i \geq W_{[1]} - c_2$,

where c_1 and c_2 are non-negative constants to be chosen so that the requirement (25.9.30) is satisfied.

Now,

$$
\begin{aligned}
P(CS|R_7) &= \Pr\{W_{(k)} \geq W_{[k]} - c_1, W_{(1)} \leq W_{[1]} + c_2\} \\
&= \Pr\{W_{(1)} - c_2 \leq W_{(i)} \leq W_{(k)} + c_1, \quad i = 2, \ldots, k-1; \\
&\qquad\quad W_{(1)} \leq W_{(k)} + \min(c_1, c_2)\}.
\end{aligned}
$$

We are unable to obtain the exact infimum of $P(CS|R_7)$ over Ω. We get a conservative solution for (c_1, c_2), by using a lower bound for the PCS. By letting $A = \{W_{(i)} \leq W_{(k)} + c_1, \ i = 1, \ldots, k-1\}$ and $B = \{W_{(i)} \geq W_{(1)} - c_2, \ i = 2, \ldots, k\}$, we get

$$P(CS|R_7) = Pr(A \cap B) \geq P(A) + P(B) - 1.$$

We note that $P(A)$ and $P(B)$ are same as the PCS for the rules R_2 in (25.4.9) and R_2' in (25.4.13), respectively. Thus, from Rizvi (1971), it follows that $P(A)$ and $P(B)$ are minimized for a configuration with $\theta_{[1]} = \cdots = \theta_{[k]} = \theta$ (say), and that each is monotonically decreasing in θ. Consequently, the infimum of $P(A) + P(B) - 1$ is obtained by letting $\theta \to \infty$. This gives

$$
\begin{aligned}
\inf_{\theta \in \Omega} P(CS|R_7) \\
\geq \int_{\theta \in \Omega} \{P(A) + P(B) - 1\} \\
= \int_{-\infty}^{\infty} \Phi^{k-1}(y + d_1)\phi(y)dy + \int_{-\infty}^{\infty} \Phi^{k-1}(y + d_2)\phi(y)dy \\
= C(d_1, k) + C(d_2, k) - 1 \text{ (say)}, \tag{25.9.31}
\end{aligned}
$$

where $d_i = \sqrt{n}\, c_i/\sigma$, $i = 1, 2$. A conservative solution for (c_1, c_2) can now be obtained by solving

$$C(d_1, k) + C(d_2, k) - 1 = P^*. \tag{25.9.32}$$

The left side of (25.9.32) varies from $(2/k) - 1$ to 1 as d_1 and d_2 vary from 0 to ∞. So a solution (d_1, d_2) exists for any specified P^* between $[k(k-1)]^{-1}$ and 1. However, there is no unique solution. If we take $c_1 = c_2 = c$, then $d_1 = d_2 = \sqrt{n}\, c = d$ (say). Then d is given by

$$C(d, k) \equiv \int_{-\infty}^{\infty} \Phi^{k-1}(y + d)\phi(y)dy = \frac{1 + P^*}{2} . \qquad (25.9.33)$$

Since $C(d, k)$ increases in d for $k \geq 2$ from $1/k$ to 1, and since $(1 + P^*)/2 \geq (1/k)$, there is a unique solution for d. Equation (25.9.33) is the same as (25.4.11) with P^* replaced with $P_1^* = (1 + P^*)/2$. So the value of d satisfying (25.9.33) can be obtained different ranges of k and P_1^* from tables referred to following (25.4.11).

If G and B denote the sizes of the subsets S_G and S_B, respectively, then $E(G)$ and $E(B)$ are same as the expected subset sizes of the rules R_2 in (25.4.9) and R_2' in (25.4.13). Thus the suprema of $E(G)$ and $E(B)$ can be obtained from Rizvi (1971).

As in the case of the rule R_6 defined in (25.8.22), we can obtain the exact infimum over Ω of $P(CS|R_7)$ when $k = 3$ and $c_1 = c_2 = c$ (say). The method is similar to that for R_6. Hussein (1998) has shown that $P(CS|R_7)$ is minimized when $\theta_{[1]} = \theta_{[2]} = \theta_{[3]} = \theta$ (say) and that $P(CS|R_7)$ is decreasing in θ for fixed $d = \sqrt{n}\, c/\sigma$. By letting $\theta \to \infty$, we get

$$\inf_{\theta \in \Omega} P(CS|R_7) = 2 \int_{-\infty}^{\infty} \Phi^2(y + d)\phi(y)dy - \Phi\left[\frac{d}{\sqrt{2}}\right]. \qquad (25.9.34)$$

When $k = 3$ and $c_1 = c_2 = c$, the lower bound for $\inf P(CS|R_7)$ given in (25.9.31) becomes $2 \int_{-\infty}^{\infty} \Phi^2(y + d)\phi(y) - 1$ which is less than the exact infimum in (25.9.34) by the amount $1 - \Phi(d/\sqrt{2})$.

Finally, we note that the rule R_7 selects two subsets S_G and S_B which may overlap. It is desirable to ask for S_G and S_B which are mutually exclusive. This problem remains unsolved.

25.10   CONCLUDING REMARKS

As pointed out at the end of Section 25.9, the problem of simultaneously selecting two non-overlapping subsets S_G and S_B is important and it remains to be solved. When the common variance is known, several other problems arise naturally for further investigations. These are suggested by the literature available on the problem of selecting the normal population having the largest mean, assuming a common known variance. Under the indifference zone formulation, one can consider a two-stage procedure which uses a subset procedure at the first stage to screen poorer populations and goes to the second stage, if necessary, to select the best. This is analogous to the two-stage procedure of Tamhane and Bechhofer (1978) for selecting the normal population with the largest mean, assuming a common known variance. Other problems relate to several aspects such as unequal sample sizes, unequal and unknown variances, modified and generalized goals, and related problems of estimation of the PCS and estimation after selection. For discussions on these topics and for further references, see Gupta and Panchapakesan (1979, 1985), and Gupta (1991).

REFERENCES

Bechhofer, R. E. (1954). A single-sample multiple-decision procedure for ranking means of normal populations with known variances, *Annals of Mathematical Statistics*, **25**, 16–39.

Bechhofer, R. E., Dunnett, C. W. and Sobel, M. (1954). A two-sample multiple decision procedure for ranking means of normal populations with a common unknown variance, *Biometrika*, **41**, 170–176.

Chen, P. (1988). An integrated formulation for selecting the most probable multinomial cell, *Annals of the Institute of Statistical Mathematics*, **40** 615–625.

Chen, P. and Panchapakesan, S. (1994). An integrated formulation for selecting the best normal population and eliminating the bad ones, In *Proceedings in Computational Statistics: Short Communications* (Eds. R. Dutter and W. Grossmann), 11th Symposium on Computational Statistics, Vienna, Austria, 18–19, Vienna Institutes for Statistics *et al.*, Austria.

Chen, P. and Sobel, M. (1987a). An integrated formulation for selecting the t best of k normal populations, *Communications in Statistics—Theory and Methods*, **16**, 121–146.

Chen, P. and Sobel, M. (1987b). A new formulation for the multinomial selection problem, *Communications in Statistics—Theory and Methods*, **16**, 147–180.

Dunnett, C. W. (1955). A multiple comparison procedure for comparing several treatments with a control, *Journal of the American Statistical Association*, **50**, 1096–1121.

Dunnett, C. W. and Sobel, M. (1954). A bivariate generalization of Student's t-distribution with tables for certain special cases, *Biometrika*, **41**, 153–169.

Gupta, S. S. (1956). On a Decision Rule for a Problem in Ranking Means, *Ph.D. Dissertation* (Mimeo. Series No. 150), Institute of Statistics, University of North Carolina, Chapel Hill, NC.

Gupta, S. S. (1963). Probability integrals of the multivariate normal and multivariate t, *Annals of Mathematical Statistics*, **34**, 792–828.

Gupta, S. S. (1965). On some multiple decision (selection and ranking) rules, *Technometrics*, **7**, 225–245.

Gupta, S. S. (1991). Recent advances in statistical ranking and selection: Theory and methods, In *The Proceedings of the 1990 Taipei Symposium in Statistics* (Eds. M. T. Chao and P. E. Cheng), pp. 133–166, Institute of Statistical Science, Academia Sinica, Taipei.

Gupta, S. S., Nagel, K. and Panchapakesan, S. (1973). On the order statistics from equally correlated normal random variables, *Biometrika*, **60**, 403–413.

Gupta, S. S. and Panchapakesan, S. (1979). *Multiple Decision Procedures: Methodology of Selecting and Ranking Populations*, John Wiley & Sons, New York.

Gupta, S. S. and Panchapakesan, S. (1985). Subset selection procedures: Review and assessment, *American Journal of Mathematical and Management Sciences*, **5**, 235–311.

Gupta, S. S., Panchapakesan, S. and Sohn, J. K. (1985). On the distribution of the studentized maximum of equally correlated normal random variables, *Communications in Statistics—Simulation and Computation*, **14**, 103–135.

Gupta, S. S. and Sobel, M. (1957). On a statistic which arises in selection and ranking problems, *Annals of Mathematical Statistics*, **28**, 957–967.

Hussein, K. (1998). Simultaneous Selection and Ranking for Extreme Populations, *Ph.D. Dissertation*, Department of Mathematics, Southern Illinois University, Carbondale, IL.

Jeyaratnam, S. and Panchapakesan, S. (1997). An integrated formulation for selecting the best from several normal populations in terms of the absolute values of their means: Known variance case, In *Advances in Statistical Decision Theory and Methodology* (Eds. S. Panchapakesan and N. Balakrishnan), pp. 277–289, Birkhäuser, Boston.

Jeyaratnam, S. and Panchapakesan, S. (2000). Selecting from normal populations the one with the largest absolute mean: Common unknown variance case, In *Advances on Stochastic Simulation Methods* (Eds. N. Balakrishnan, V. B. Melas and S. Ermakov), pp. 283–292, Birkhäuser, Boston.

Krishnaiah, P. R. and Armitage, J. V. (1966). Tables for multivariate t distribution, *Sankhyā, Series B*, **28**, 31–56.

Milton, R. C. (1963). Tables of equally correlated multivariate normal probability integral, *Technical Report No. 27*, Department of Statistics, University of Minnesota, Minneapolis, MN.

Mishra, S. N. (1986). Simultaneous selection of extreme population means: Indifference zone formulation, *American Journal of Mathematical and Management Sciences*, **6**, 131–142.

Rizvi, M. H. (1963). Ranking and Selection Problems of Normal Populations Using the Absolute Values of Their Means: Fixed Sample Size Case, *Ph.D. Dissertation* (Technical Report No. 31), Department of Statistics, University of Minnesota, Minneapolis, MN.

Rizvi, M. H. (1971). Some selection problems involving folded normal population, *Technometrics*, **13**, 355–369.

Tamhane, A. C. and Bechhofer, R. E. (1977). A two-stage minimax procedure with screening for selecting the largest mean. *Communications in Statistics—Theory and Methods*, **6**, 1003–1033.

CHAPTER 26

A SELECTION PROCEDURE
PRIOR TO SIGNAL DETECTION

PINYUEN CHEN

Syracuse University, Syracuse, NY

Abstract: We study a procedure in the framework of ranking and selection theory to identify multivariate normal observations that have different covariance structures from a control covariance matrix. Simulation results are presented to illustrate that, for a sample of data contaminated with non-homogeneous observations, our selection procedure improves the performance of the hypothesis testing for a signal in terms of the probability of type II error while the level of significance is held at a constant.

Keywords and phrases: Complex multivariate normal, likelihood ratio principle, constant false alarm rate test, nonhomogeneous covariance structure

26.1 INTRODUCTION

In signal detection, one is interested in the problem of detection of a given radar signal s which is a complex vector in the presence of noise in transmission. The signal may be a set of voltages an electromagnetic wave from the selected search directions induces on a number of receiving elements. The actual observed data Y may be a pure noise vector n or the signal s plus a noise vector n. It is assumed that the noise follows a complex multivariate normal distribution with mean 0 and covariance matrix σ. Statistically, the model can be described as $Y = s + n$ where s is a specific signal and n is a noise random vector. The goal is to test the null hypothesis that $Y = n$ versus the alternative hypothesis that $Y = s + n$.

391

Reed, Mallett and Brennan (1974) discussed an adaptive procedure for the above detection problem in which two sets of input data are used, which are called the primary and secondary data. A radar receives primary data Y_0 which may or may not contain a signal, and secondary data $y_1, Y_2, \ldots, Y_n$ which are assumed to contain only noise, independent of and statistically identical to the noise components of the primary data. The goal is to test $H_0 : \mu = 0$ versus $H_1 : \mu = s$ where μ is the population mean of Y_0. Kelly (1986) used the likelihood ratio principle to derive a test statistic for the above hypothesis testing problem. The test possesses a desirable property that its false alarm rate (the probability of type I error) remains constant when the unknown covariance matrix of the secondary data varies. That is, it is a CFAR (constant false alarm rate) test. Khatri and Rao (1987) used a conditional argument to derive a test equivalent to Kelly's and they showed that the test is more powerful than the traditional Hotelling's T^2 test.

Numerical examples in Melvin, Wicks and Chen (1998) showed that a fundamental problem in radar that employs the above detection methods is that the environment is non- homogeneous. That is, the covariance matrix of the secondary data differs from that of the primary data. Data from a non-homogeneous covariance structure produces biased estimate of the covariance matrix, thereby leading to a severe degradation in detection performance. To resolve this problem, Chen and Wicks (1999) proposed a selection procedure which compares the covariance matrices of the secondary data with that of the primary data. It is used to identify and eliminate those observations that have different covariance structure from the secondary data. It retains homogeneous radar data for further investigation. As described in Chen and Wicks (1999), this procedure can be applied prior to the step of estimating the covariance matrix of the secondary data in Kelly (1986) and Khatri and Rao (1987). As a consequence, the CFAR property remains true in the detection process.

26.2 THE SELECTION PROCEDURE

We begin with some notations and definitions. If $A = a_{ij}$ is a matrix of complex numbers, the conjugate transpose of A is defined by $A* = (a_{ij}^*)$, where a_{ij}^* is the complex conjugate of a_{ij}. Let $z = x + iy$ be a complex random p-vector with mean θ and covariance matrix $Q = \sigma_1 + i\sigma_2$. Then z has a complex multivariate normal distribution with mean θ and covariance matrix Q, written as $z \sim CN_p(\theta, Q)$, if and only if

$$\begin{pmatrix} x \\ y \end{pmatrix} \sim N_{2p} \left(\begin{pmatrix} \theta_1 \\ \theta_2 \end{pmatrix}, \sigma \right)$$

where

$$\sigma = \frac{1}{2} \begin{pmatrix} \sigma_1 & -\sigma_2 \\ \sigma_2 & \sigma_1 \end{pmatrix}, \quad \theta = \theta_1 + i\theta_2.$$

[See, for example, Definition 2.9.2 in Srivastava and Khatri (1979)].

Let $Y_0 \sim CN_p(\mu, \sigma)$ denote the primary data which is received by a receiver and is to be tested for a specific signal s where s is a known vector. Let $Y_1, Y_2, \ldots, Y_n \sim CN_p(0, \sigma)$ be the secondary data which is to be used to estimate the unknown covariance matrix σ. The random vector Y_0 is independent of the secondary data. Let S denote n times the sample covariance matrix of the secondary data sample $Y_1, Y_2, \ldots, Y_n$. Therefore we have $S \sim CW_p(n, \sigma)$. As described earlier, the goal of a signal detection problem is to test

$$H_0 : \mu = 0 \quad \text{versus} \quad H_1 : \mu = s. \tag{26.2.1}$$

Kelly's likelihood ratio test statistic for (26.2.1) can be written as

$$\eta = \frac{|s^* S^{-1} Y_0|^2}{(s^* S^{-1} s)(1 + Y_0^* S^{-1} Y)} . \tag{26.2.2}$$

The null hypothesis is rejected for large observed η. A desirable property of the test statistic is that its distribution under $|H_0$ is independent of S. Therefore, the probability of type I error (false alarm rate) remains constant when the secondary data varies. It was shown in Khatri and Rao (1987) that under H_0, $\eta \sim \text{Beta}(1, n - p + 1)$, a Beta distribution with parameters 1 and $n - p + 1$. Following Reed, Mallett, and Brennan (1974)'s structure of radar data, Kelly's test also assumes an i.i.d. sample $Y_1, Y_2, \ldots, Y_n$ for the secondary data and an independently distributed primary data Y_0. However, the real airborne radar signal environment is not homogeneous in terms of their covariance structure, as has been determined at the United States Air Force's Rome Laboratory through extensive analysis of airborne radar data. In fact, the signal environment of an airborne platform can be as non-homogeneous as the visual environment appears to an airborne observer who might see, e.g., mountains, lakes, forests, roads, and vehicles, etc., from an aircraft window. [See, for example, Melvin, Wicks and Wicks (1998, p. 3)]. However, environment changes among clusters, but not within cluster. The data obtained from mountains may have different covariance structure from that of the data obtained from lakes. The goal in Chen and Wicks (1999) is to perform a data preprocessing procedure that will allow an experimenter to select among several cells which are potential target cells the ones that have the same covariance structure as that of a control secondary data. A selected cell, which should have the same covariance structure as the secondary data, later may either be used as a part of secondary data or its nearby cell may be tested for target.

Let $\pi_1, \pi_2, \ldots, \pi_k$ represent k p-variate complex normal populations, $i = 1, 2, \ldots, k$, and let π_0 be a control p-variate complex normal population. Those k populations are the resources of the k cells which may or may not have the same or similar covariance structures as the control population π_0 from which the secondary data are taken. Here "similarity" is defined in (26.2.3) and (26.2.4) and the paragraph after (26.2.4) later in this section. Thus, from each of the k experimental populations, only one observation is taken, and from the control population, n observations are taken. We assume that $\mu_i = 0$, $i = 0, 1, 2, \ldots, k$ since the k experimental populations are the cells which are assumed to have zero mean. In comparing the variances of two univariate normal populations with zero mean, a measure of similarity is the ratio function, $d(x, y) = x/y$, because the variance is a scale parameter. The covariance matrix of a multivariate normal random vector has similar properties as the variance of univariate normal random variable, especially in distribution theory. Here, we also use the ratio, $\sigma_i \sigma_0^{-1}$, of two covariance matrices as a distance measure in our study. Let $\lambda_{i,1} \geq \lambda_{i,2} \geq \cdots \geq \lambda_{i,p} > 0$ denote the ordered eigenvalues of $\sigma_i \sigma_0^{-1}$. We define the two disjoint and exhaustive subsets, Ω_G and Ω_B, of the set $\Omega = \{\pi_1, \pi_2, \ldots, \pi_k\}$, by using a pair of distance functions d_1 and d_2 defined as follows:

$$d_1(\sigma_i, \sigma_0) = \lambda_{i,1}; \quad d_2(\sigma_i, \sigma_0) = \lambda_{i,p}; \tag{26.2.3}$$

and

$$\begin{aligned} \Omega_B &= \{\pi_i \mid \delta_2^* \leq d_2(\sigma_i, \sigma_0) \text{ or } d_1(\sigma_i, \sigma_0) \leq \delta_1^*\} \\ \Omega_G &= \Omega - \Omega_B, \end{aligned} \tag{26.2.4}$$

where $\delta_1^* < \delta_2^*$ are pre-assigned positive real numbers which are used to define similar and dissimilar populations. Theoretically, the values of δ_1^* should be less than 1 and the value of δ_2^* should be greater than 1 since $\delta_1^* = \delta_2^* = 1$ is equivalent to the perfect case when the control population π_0 has exactly the same covariance matrix as that of the experimental populations π_i. A population is considered similar to a control population when the distance measures are close to unity. Our goal is to separate the populations obtained from the guard cells into two disjoint subsets, S_G and S_B. The separation is correct if $S_G \subset \Omega_G$, meaning that all populations included in selected subset S_G have similar covariance structure as the control population. It also means that all populations with significantly different covariance structures are eliminated. We require a procedure R that will satisfy the probability requirement that Pr(the separation is correct$| R$) = $\Pr(CS|R) \geq P^*$, where P^* is chosen in advance of the experiment such that $2^{-k} < p^* < 1$.

The procedure R defined in Chen and Wicks (1999) is as follows.

Procedure R For each population π_i $(i = 1, 2, \ldots, k)$, we first compute $T_i = (x_i^H S^{-1} x_i)/n$ where x_i's are the data vectors from experimental cells, x_i^H is the conjugate transpose of x_i, and S is the sample covariance matrix associated with population π_0. Then we partition the set of populations $\Omega = \{\pi_1, \pi_2, \ldots, \pi_k\}$ into two subsets S_G and S_B. The subset S_G consists of those populations π_i with $c \leq T_i \leq d$ where c and d are chosen such that the probability requirement $P(CS) \geq P^*$ is satisfied and $S_B = \Omega - S_G$. We claim that S_G consists of the populations that have covariance matrices that are the same as the covariance matrix of π_0.

The distribution of was derived in Chen, Melvin and Wicks (1999). Chen and Wicks (1999) showed that the so-called Least Favorable Configuration (LFC) under procedure R is given by

$$(\lambda_{1,1}, \lambda_{1,2}, \ldots, \lambda_{1.p}; \lambda_{2,1}, \lambda_{2,2}, \ldots, \ldots; \lambda_{k,1}, \ldots, \lambda_{k,p}) \qquad (26.2.5)$$

where

$$\lambda_{i,1} = \lambda_{i,2} = \cdots = \lambda_{i,p} = \delta_1^*, \quad i = 1, 2 \ldots, m;$$
$$\lambda_{j,1} = \lambda_{j,2} = \cdots = \lambda_{j,p} = \delta_2^*, \quad j = m + 1, 2, \ldots, k.$$

Under the LFC, the distributions of

$$\left(\frac{n - p + 1}{p}\right) \frac{T_i}{\delta_1^*} \quad i = 1, \ldots, m$$

and

$$\left(\frac{n - p + 1}{p}\right) \frac{T_j}{\delta_2^*} \quad j = m + 1, \ldots, k$$

become an F distribution with $2p$ and $2(n - p + 1)$ degrees of freedom.

A typical measure of the performance of the above procedure is the probability of a correct separation $P(CS)$. One would always require a large probability of a correct separation when the procedure is employed. To implement the procedure with a pre-determined probability requirement P^*, Chen and Wicks (1999) have shown that the procedure constants c and d have to satisfy the following integral equation:

$$
\begin{aligned}
\min_{\Omega}\{P(CS) \\
= & \min_{0 \leq m \leq k} P(T_i < c, \ i = 1, 2, \ldots, m; \ T_j > d, \ j = m+1, \ldots, k) \\
\geq & \min_{0 \leq m \leq k} \{1 - mP(T_i > c) - (k-m)P(T_i < d)\} \\
= & \min_{0 \leq m \leq k} \left\{1 - m\left(1 - F_{2p,2(n-p+1)}\left(\frac{(n-p+1)}{p}\frac{c}{\delta_1^*}\right)\right)\right. \\
& \left. -(k-m)F_{2p,2(n-p+1)}\left(\frac{(n-p+1)}{p}\frac{d}{\delta_2^*}\right)\right\}
\end{aligned}
\tag{26.2.6}
$$

where $F_{2p,2(n-p+1)}$ is the distribution function of an F distribution with $2p$ and $2(n-p+1)$ degrees of freedom. We also define $T_0 \equiv 0$ and $T_{k+1} \equiv 0$.

26.3 TABLE, SIMULATION STUDY AND AN EXAMPLE

We use the formula given in (26.2.6) to approximate the minimum sample sizes needed to achieve the P^*-requirement for the cases p (the number of components in a signal) $= 10$, and 20; $k = 3, 4, 5$; $\delta_1^*/c = 1/2, 1/3, 1/4$, and $1/5$; $\delta_2^*/d = 2, 3, 4$, and 5; $P^* = .90$, and $.95$. The results are given in Table 26.1 at the end of the paper. The choices of c and d of Procedure R are arbitrary. However, the values of c and d arrived at by taking a combination of δ_1^*/c and δ_2^*/d in Table 26.1 may not be admissible. For example, take $\delta_1^* = 1/3$, $\delta_2^* = 1.2$, $\delta_1^*/c = 1/2$ and $\delta_2^*/d = 2$. In this case, Table 26.1 gives a solution for n for $k = 5$, $p = 20$. However, this gives $c = 2/3$, $d = 0.6$, which is inadmissible because $c > d$. When we search for solution of n from Table 26.1, we need to make sure that our procedure constants c and d satisfy the condition $c < d$. Now we consider a simulation example.

Example Five test (or guard) cells are to be examined and to be compared with a sample of secondary cells. Each cell π_i is represented by a 20×1 random vector from a multivariate complex normal distribution with mean 0 and covariance matrix σ_i. The covariance matrix of the secondary cells is denoted by σ_i. The five test cells come from normal populations with covariance matrices σ_i such that

$\sigma_1 \sigma_0^{-1}$ = diag(2.8, 6.7, .06, .05, .08, .07, .06, .05, 1.68,

.09, 11.7, 9.6, .05, .08, .07, .06, .05, .08, .07, .06);

$\sigma_2 \sigma_0^{-1}$ = diag(.1, .1,..., .1);

$\sigma_3 \sigma_0^{-1}$ = diag(1.2, 2.5, 3.1, .8, 2.3, 5.4, 3, 2.9, 6.1, 3.3, 5.3, .5, .9,

7.3, 1.7, 5.5, 2.3, 3.1, 6.4, 5.5);

$\sigma_3 \sigma_0^{-1}$ = I;

$\sigma_5 \sigma_0^{-1}$ = diag(10, 10,..., 10).

Suppose we want to eliminate the test cell π_i if either the largest eigenvalue of $\sigma_i \sigma_0^{-1}$ is smaller than or equal to $\delta_1^* = .1$ or the smallest eigenvalue of $\sigma_i \sigma_0^{-1}$ is larger than or equal to $\delta_2^* = 10$. Then by choosing $c = .2$ and $d = 5$, we find from Table 26.1, for the case $k = 5$, $p = 20$, $\delta_1^*/c = 1/2$ and $\delta_2^*/d = 2$, that the required sample size is $n = 39$ for the secondary data to achieve $P^* = .90$. We simulated 100 trials of x_i ($i = 1,...,5$) and S from the multivariate complex normal populations with mean 0 and with respective covariance matrices satisfying the above conditions. Then for each trial, we calculate the test statistic $T_i = x_i^H S^{-1} x_i/n$. The results are plotted in Figure 26.1 at the end of the paper. From the definition of Procedure R given in Section 26.2, Cell π_i is retained if $.2 < T_i < 5$. It is clear from the figure that Cell 4 is always retained. Cell 2 and Cell 5 are always eliminated. Cell 1 and Cell 3 are retained most of the times. Notice that Cell 4 is a perfect cell while Cell 1 and Cell 3 are both considered good cells.

In our next simulation illustrations, we show, in Figures 26.2–26.5, the probability of the false alarm $(P(FA))$ and the probability of the detection $(P(D))$ when Kelly's adaptive detection algorithm is applied to three different data sets. The first data set is the perfect data set where all the observations in the secondary data are simulated from the same multivariate complex normal distribution as the primary data. The second data set is the contaminated data set where the secondary data includes some observations that were obtained from simulation of various multivariate complex normal distributions whose covariance matrices are significantly different from the covariance matrix of the primary data. The third data set is the screened data set which consists of those observations that were originally in the contaminated data set and were retained in the secondary data after our procedure R has been applied. We consider the following cases: n, the sample size of the secondary data, $= 25,...,50$; $p = 20$; and $s = (.5,...,.5)^*$ and $(1,...,1)^*$. The level of significance is set at .05 for all the cases considered. In Figures 26.2–26.5, the 'o"s are for the contaminated data set.

The 'x's are for the perfect data set, and the '+'s are for the screened data set. It is clear from the illustrations that Kelly's algorithm does not provide a constant false alarm rate (CFAR) for the contaminated data set and it always gives CFAR for the perfect data set and screened data set. The contribution of the non-homogeneous data screening proposed in our research is significant as we can see from the illustrations that the $P(D)$s for the screened data set always stay closely to those for the perfect data set while the $P(FA)$s are always below the level of significance.

REFERENCES

Chen, P., Melvin, W. L. and Wicks, M. C. (1999). Screening among multivariate normal data, *Journal of Multivariate Analysis*, **69**, 10–29.

Chen, P. and Wicks, M. C. (1999). Identifying non-homogenous multivariate normal observations, *Submitted for publication*.

Kelly, E. J. (1986). An adaptive detection algorithm, *IEEE Transactions on Aerospace & Electronic Systems*, **22**, 115–127.

Khatri, C. G. and Rao, C. R. (1987). Test For A specified signal when the noise covariance matrix is unknown, *Journal of Multivariate Analysis*, **22**, 177–188.

Melvin, W. L., Wicks, M. and Chen, P. (1998). Nonhomogeneity Detection Method and Apparatus for Improved Adaptive Signal Processing, USA Patent Number 5706013.

Reed, I. S., Mallett, J. D. and Brennan, L. E. (1974). Rapid convergence in adaptive arrays, *IEEE Transactions on Aerospace & Electronic Systems*, **10**, 853–863.

Srivastava, M. S. and Khatri, C. G. (1979). *An Introduction to Multivariate Statistics*, Elsevier Science Publishers, New York.

TABLE 26.1 Sample size n needed to achieve the P^* requirement

| $P^* = .90$ | | | | | $P^* = .95$ | | | |

$k = 3, p = 10$

δ_2^*/d δ_1^*/c	2	3	4	5	δ_2^*/d δ_1^*/c	2	3	4	5
1/2	*	21	21	21	1/2	*	*	22	22
1/3	*	18	18	18	1/3	*	19	19	19
1/4	16	16	16	16	1/4	*	17	17	17
1/5	15	15	15	15	1/5	16	16	16	16

$k = 3, p = 20$

δ_2^*/d δ_1^*/c	2	3	4	5	δ_2^*/d δ_1^*/c	2	3	4	5
1/2	38	38	38	38	1/2	39	39	39	39
1/3	33	33	33	33	1/3	34	34	34	34
1/4	30	30	30	30	1/4	31	31	31	31
1/5	28	28	28	28	1/5	29	29	29	29

$k = 4, p = 10$

δ_2^*/d δ_1^*/c	2	3	4	5	δ_2^*/d δ_1^*/c	2	3	4	5
1/2	*	21	21	21	1/2	*	*	22	22
1/3	*	18	18	18	1/3	*	19	19	19
1/4	*	17	17	17	1/4	*	17	17	17
1/5	16	16	16	16	1/5	16	16	16	16

$k = 4, p = 20$

δ_2^*/d δ_1^*/c	2	3	4	5	δ_2^*/d δ_1^*/c	2	3	4	5
1/2	38	38	38	38	1/2	*	40	40	40
1/3	33	33	33	33	1/3	34	34	34	34
1/4	31	31	31	31	1/4	32	32	32	32
1/5	29	29	29	29	1/5	30	30	30	30

TABLE 26.1 (cont'd)

$$k = 5, p = 10$$

δ_2^*/d δ_1^*/c	2	3	4	5	δ_2^*/d δ_1^*/c	2	3	4	5
1/2	*	*	22	22	1/2	*	*	23	23
1/3	*	18	18	18	1/3	*	19	19	19
1/4	*	17	17	17	1/4	*	18	18	18
1/5	16	16	16	16	1/5	*	17	17	17

$$k = 5, p = 20$$

δ_2^*/d δ_1^*/c	2	3	4	5	δ_2^*/d δ_1^*/c	2	3	4	5
1/2	39	39	39	39	1/2	*	40	40	40
1/3	34	34	34	34	1/3	35	35	35	35
1/4	31	31	31	31	1/4	32	32	32	32
1/5	29	29	29	29	1/5	30	30	30	30

A * sign shows that the probability requirement P^* is not satisfied by any sample size.

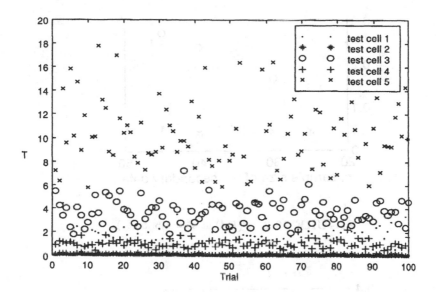

FIGURE 26.1 100 trials of T for 5 test cells x and a sample covariance S from $n = 39$ secondary cells

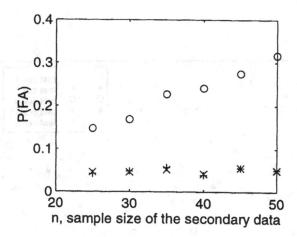

FIGURE 26.2 $P(FA)$ at $s^* = (.5 \ .5 \ \ldots)$

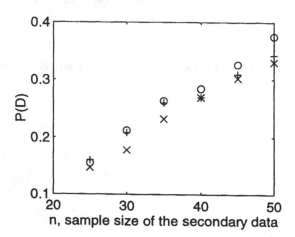

FIGURE 26.3 $P(D)$ at $s^* = (.5 \ .5 \ \ldots)$

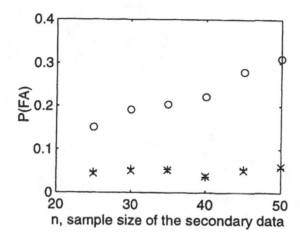

FIGURE 26.4 $P(FA)$ at $s^* = (1\ 1\ \ldots)$

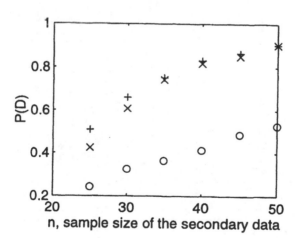

FIGURE 26.5 $P(D)$ at $s^* = (1\ 1\ \ldots)$

Part VI
Regression Methods

CHAPTER 27

TOLERANCE INTERVALS AND CALIBRATION IN LINEAR REGRESSION

YI-TZU LEE   THOMAS MATHEW

University of Maryland Baltimore County, Baltimore, MD

Abstract: In the linear regression model, we explore whether intervals obtained using the tolerance interval condition will satisfy a marginal property that arises in multiple use calibration. The marginal property of interest is the following. Suppose the same estimated regression line is used a large number of times in order to construct intervals for future observations, corresponding to possibly different values of the explanatory variable. At least a proportion α of such intervals are to contain the corresponding future observations with confidence γ. The problem is investigated numerically and the numerical results indicate that intervals derived using the tolerance interval condition will satisfy such a marginal property quite well. We have used this observation for the construction of multiple use confidence regions in the calibration problem. The results are illustrated using examples.

Keywords and phrases: Confidence level, conservative tolerance intervals, multiple use confidence region, simultaneous tolerance intervals

27.1 INTRODUCTION

This article addresses the problem of deriving tolerance intervals in the context of a normal linear regression model. Thus let $\mathbf{y}$ be an $n \times 1$ vector of observations having the distribution

$$\mathbf{y} \sim N(X\boldsymbol{\beta}, \sigma^2 I_n), \tag{27.1.1}$$

407

where X is a known $n \times m$ matrix, β, an $m \times 1$ vector, and $\sigma^2 > 0$, are both unknown parameters, and I_n denotes the $n \times n$ identity matrix. The rows of X are the values of an m–dimensional independent variable, assumed to be non-random. If the model contains an intercept term, then the first component of β can be taken as the intercept and, in this case, the first column of X will be a vector of ones.

Now let $y(\mathbf{x})$ denote a future observation corresponding to the value $\mathbf{x}'$ of the independent variable. Assume that

$$y(\mathbf{x}) \sim N(\mathbf{x}'\beta, \sigma^2), \tag{27.1.2}$$

where $y(\mathbf{x})$ is also assumed to be independent of $\mathbf{y}$ in (27.1.1). The problem that we shall address is the construction of tolerance intervals for a sequence of future observations $y(\mathbf{x})$, corresponding to possibly different values of $\mathbf{x}$. A tolerance interval (TI) for $y(\mathbf{x})$ is an interval such that with confidence γ, the interval is to contain at least a proportion α of the normal distribution of $y(\mathbf{x})$, for any fixed $\mathbf{x}$. On the other hand, simultaneous tolerance intervals (STI's) satisfy the following. Suppose the same estimated regression line is used a large number of times in order to construct intervals for future observations $y(\mathbf{x})$, corresponding to possibly different values of $\mathbf{x}$. With confidence γ, at least a proportion α of the $y(\mathbf{x})$ distribution is to be contained in the corresponding tolerance interval, simultaneously for every $\mathbf{x}$. The construction of a TI, or an STI, amounts to obtaining $k(\mathbf{x})$, a function of $\mathbf{x}$, so that the interval $\mathbf{x}'\hat{\beta} \pm k(\mathbf{x})s$ satisfies the condition required of a TI, or an STI. Here $\hat{\beta}$ denotes the least squares estimator $(X'X)^{-1}X'\mathbf{y}$ and $s^2 = \frac{1}{n-m}(\mathbf{y} - X\hat{\beta})'(\mathbf{y} - X\hat{\beta})$. Note that s^2 is the residual mean square based on the model (27.1.1). (In this article we shall assume without loss of generality that the $m \times m$ matrix $X'X$ is nonsingular).

The setup given above is the same as that in Mee, Eberhardt and Reeve (1991). A review of the literature on the construction of simultaneous tolerance intervals is given in this article and also in the recent article by Mee and Eberhardt (1996). Further references on the problem can be found in these articles. The usual approach for obtaining STI's is as follows: the proportion of the normal distribution of $y(\mathbf{x})$, contained in the interval $\mathbf{x}'\hat{\beta} \pm k(\mathbf{x})s$, is a function of $\mathbf{x}$. Compute $k(\mathbf{x})$ subject to the requirement that the minimum (with respect to $\mathbf{x}$) of this proportion is to be at least α, with confidence γ. This is the procedure followed by a number of authors, including Mee, Eberhardt and Reeve (1991).

The purpose of this article is to explore whether TI's satisfy the following marginal property. Suppose the same estimated regression line is used a large number of times in order to construct intervals for future observations

$y(\mathbf{x})$, corresponding to possibly different values of $\mathbf{x}$. At least a proportion α of such intervals are to contain the corresponding future observations with confidence γ. This is a weaker requirement than the STI condition. A theoretical investigation of this marginal property appears to be difficult. Consequently, we have studied the problem numerically. Our fairly extensive numerical results for the simple linear regression model indicate that TI's very nearly satisfy the above marginal property.

We would like to point out that the computation of TI's is quite straightforward. It turns out that the $k(\mathbf{x})$ required to compute the TI $\mathbf{x}'\hat{\beta} \pm k(\mathbf{x})s$ depends on $\mathbf{x}$ only through the scalar $d^2 = \mathbf{x}'(X'X)^{-1}\mathbf{x}$; see Mee, Eberhardt and Reeve (1991, Section 2). Hence we shall use the notation $k(d)$ instead of $k(\mathbf{x})$. For any value of d, $k(d)$ satisfying the TI condition can be evaluated using the program in Eberhardt, Mee and Reeve (1989). (For the simple linear regression model, we used a MATLAB program for the evaluation of $k(d)$). It will be convenient to obtain the functional form of $k(d)$, so that for any value of d that is of interest, the value of $k(d)$ is readily available. Our recommendation is to numerically evaluate $k(d)$ for a few selected values of d, and then fit a suitable function to the $k(d)$ values so obtained. This will provide the approximate functional form of $k(d)$. This is the approach used in Mathew and Zha (1996, 1997) for the construction of confidence regions in a calibration problem. We would like to emphasize that the numerical evaluation of a few $k(d)$ values and the fitting of a suitable function (usually a polynomial) to these values is computationally quite simple and fast. This will become clear from the later sections.

The rest of the article is organized as follows. In the next section, we shall give the mathematical expressions for the conditions to be satisfied by TI's and STI's, along with the marginal property mentioned above. In Section 27.3, we numerically investigate whether TI's satisfy the marginal property. The numerical study is carried out in the case of simple linear regression. In the numerical study, there are only two quantities that can vary: the value of n (the dimension of $\mathbf{y}$ in (1.1)) and the range for $d = \{\mathbf{x}'(X'X)^{-1}\mathbf{x}\}^{1/2}$. Assuming that $0 \le d \le \eta$ (a known upper bound) we have carried out the necessary simulations to conclude whether TI's satisfy the marginal property. By inverting intervals satisfying the marginal property, it is possible to obtain multiple use confidence regions in the calibration problem. This is briefly indicated in Section 27.4. Some concluding remarks are made in Section 27.5.

We conclude this section by recalling some of the observations in Mee and Eberhardt (1996). These authors report one Monte-Carlo study where it turns out that a TI satisfies the marginal property; see Table 4 and Section 4 in their article. In Section 5 of their article, Mee and Eberhardt

(1996) state that "....intervals having the shape of TI's seem to hold the most promise for achieving this goal" (i.e., achieving the requirements of multiple use confidence regions in calibration). The numerical study in this article suggests that this is indeed the case. The numerical results in Mathew, Sharma and Nordström (1998) also support this in the context of a multivariate calibration problem.

27.2 TOLERANCE INTERVALS, SIMULTANEOUS TOLERANCE INTERVALS AND A MARGINAL PROPERTY

As pointed out in the previous section, we want to construct intervals of the form $\mathbf{x}'\hat{\beta}\pm k(\mathbf{x})s$. Here $\hat{\beta} = (X'X)^{-1}X'\mathbf{y}$ and $s^2 = \frac{1}{n-m}(\mathbf{y}-X\hat{\beta})'(\mathbf{y}-X\hat{\beta})$. The quantity $k(\mathbf{x})$ is to be determined subject to the conditions that should be satisfied by the interval. Let

$$\mathbf{u} = \frac{\hat{\beta} - \beta}{\sigma}. \qquad (27.2.3)$$

Then

$$\mathbf{u} \sim N[0, (X'X)^{-1}], \ (n-m)v^2 = (n-m)s^2/\sigma^2 \sim \chi^2_{n-m}, \qquad (27.2.4)$$

where χ^2_{n-m} denotes the central chi square distribution with df $= n-m$. In order to state the conditions to be satisfied by a TI, let

$$
\begin{aligned}
C(\mathbf{x}) &= \mathbf{P}_{y(\mathbf{x})}\left[\mathbf{x}'\hat{\beta} - k(\mathbf{x})s \leq y(\mathbf{x}) \leq \mathbf{x}'\hat{\beta} + k(\mathbf{x})s \Big| \hat{\beta}, \ s\right] \\
&= \mathbf{P}_z\left[\mathbf{x}'\mathbf{u} - k(\mathbf{x})v \leq z \leq \mathbf{x}'\mathbf{u} + k(\mathbf{x})v \Big| \mathbf{u}, v\right], \qquad (27.2.5)
\end{aligned}
$$

where $z = [y(\mathbf{x}) - \mathbf{x}'\beta]/\sigma \sim N(0,1)$ and $\mathbf{u}$ and v are as defined in (27.2.3) and (27.2.4). From (27.2.5), it is clear that any probability statement regarding $C(\mathbf{x})$ depends on $\mathbf{u}$ and v. Hence we shall use the notation $C(\mathbf{x}; \mathbf{u}, v)$ instead of $C(\mathbf{x})$. Thus (27.2.5) can be written as

$$C(\mathbf{x}; \mathbf{u}, v) = \Phi(\mathbf{x}'\mathbf{u} + k(\mathbf{x})v) - \Phi(\mathbf{x}'\mathbf{u} - k(\mathbf{x})v), \qquad (27.2.6)$$

where $\Phi(.)$ denotes the standard normal cdf. The condition to be satisfied by a TI is

$$\mathbf{P}_{\mathbf{u},v}\left[C(\mathbf{x}; \mathbf{u}, v) \geq \alpha\right] = \gamma, \qquad (27.2.7)$$

for every **x**. The marginal property (marginal in **x**) that we are interested in is

$$P_{u,v}\left[\frac{1}{N}\sum_{i=1}^{N}C(\mathbf{x}_i; \mathbf{u}, v) \geq \alpha\right] = \gamma, \qquad (27.2.8)$$

for every sequence $\{\mathbf{x}_i\}$ and for every N. STI's, on the other hand, satisfy the following.

$$P_{u,v}\left[\min_{\mathbf{x}}C(\mathbf{x}; \mathbf{u}, v) \geq \alpha\right] = \gamma. \qquad (27.2.9)$$

We note that in practice, **x** in (27.2.7) and the $\mathbf{x}_i$'s in (27.2.8) won't be completely arbitrary. They will be bounded quantities with known bounds.

The interpretation of (27.2.7) is as follows. If a large number of future observations $y(\mathbf{x})$ are obtained, based on the normal distribution (27.1.2), corresponding to the same **x**, the interval $\mathbf{x}'\hat{\boldsymbol{\beta}} \pm k(\mathbf{x})s$ will contain at least a proportion α of these future observations with confidence γ. On the other hand, the condition (27.2.8) states that in a sequence of future observations $y(\mathbf{x}_i)$, corresponding to possibly different **x** values $\mathbf{x}_i$ ($i = 1, 2, 3, ...$), at least a proportion α of these observations will belong to the corresponding interval $\mathbf{x}_i'\hat{\boldsymbol{\beta}} \pm k(\mathbf{x}_i)s$, with confidence γ.

The formulas and arguments that we have presented so far in this section are quite standard and are available, for example, in Mee, Eberhardt and Reeve (1991) and Eberhardt and Mee (1996). The condition (27.2.8) is given in Mee and Eberhardt (1996). These authors have also discussed the significance of (27.2.8). In the next section, we propose to investigate the following. Suppose $k(\mathbf{x})$ is derived subject to (27.2.7), which is the condition for a TI. Does such a $k(\mathbf{x})$ satisfy (27.2.8)? What we have done is only a numerical study for the simple linear regression model. We shall first give the simplified versions of (27.2.7) and (27.2.8) for such models. For the simple linear regression model, the components of **y** in (27.1.1), say $y_1, y_2, ..., y_n$, are independently distributed with

$$y_i \sim N(\beta_0 + \beta_1 x_i, \sigma^2), \qquad (27.2.10)$$

where x_i's are the known values of the explanatory variable ($i = 1, 2, ..., n$). Also,

$$y(x) \sim N(\beta_0 + \beta_1 x, \sigma^2) \qquad (27.2.11)$$

denotes a future observation corresponding to the value x of the explanatory variable. Note that in the notation of (27.1.1) and (27.1.2), $\boldsymbol{\beta} = (\beta_0, \beta_1)'$ and $\mathbf{x} = (1, x)'$. Write

$$\bar{x} = \frac{1}{n}\sum_{i=1}^{n}x_i, \quad \bar{y} = \frac{1}{n}\sum_{i=1}^{n}y_i, \text{ and } S = \sum_{i=1}^{n}(x_i - \bar{x})^2. \qquad (27.2.12)$$

Let $\hat{\beta}_0$ and $\hat{\beta}_1$ denote the least squares estimators of β_0 and β_1, respectively, and $s^2 = \frac{1}{n-2} \sum_{i=1}^{n} (y_i - \hat{\beta}_0 - \hat{\beta}_1 x_i)^2$. Then

$$\mathbf{x}'\hat{\beta} = \hat{\beta}_0 + \hat{\beta}_1 x = \bar{y} + \hat{\beta}_1 (x - \bar{x}). \qquad (27.2.13)$$

Let

$$u_1 = \frac{\bar{y} - (\beta_0 + \beta_1 \bar{x})}{\sigma/\sqrt{n}} \sim N(0,1),$$

$$u_2 = \frac{\hat{\beta}_1 - \beta_1}{\sigma/\sqrt{S}} \sim N(0,1), \qquad (27.2.14)$$

where S is defined in (27.2.12). Also note that u_1 and u_2 are independently distributed. By using the definition of $C(\mathbf{x})$ in (27.2.5) [to be denoted by $C(x)$ for the model (27.2.11)], and by straightforward calculations, we get

$$
\begin{aligned}
C(x) &= P_{y(x)}\left[\hat{\beta}_0 + \hat{\beta}_1 x - k(x)s \le y(x) \le \hat{\beta}_0 + \hat{\beta}_1 x + k(x)s \Big| \hat{\beta}_0, \ \hat{\beta}_1, \ s\right] \\
&= P_{y(x)}\left[\bar{y} + \hat{\beta}_1 (x - \bar{x}) - k(x)s \le y(x)\right. \\
&\qquad\qquad \left. \le \bar{y} + \hat{\beta}_1 (x - \bar{x}) + k(x)s \Big| \bar{y}, \ \hat{\beta}_1, \ s\right] \\
&= P_z\left[\frac{u_1}{\sqrt{n}} + \frac{(x - \bar{x})}{\sqrt{S}} u_2 - k(x)v \le z\right. \\
&\qquad\qquad \left. \le \frac{u_1}{\sqrt{n}} + \frac{(x - \bar{x})}{\sqrt{S}} u_2 + k(x)v \Big| u_1, \ u_2, \ v\right] \\
&= \Phi\left(\frac{u_1}{\sqrt{n}} + cu_2 + k(x)v\right) - \Phi\left(\frac{u_1}{\sqrt{n}} + cu_2 - k(x)v\right), \qquad (27.2.15)
\end{aligned}
$$

where $z = [y(x) - \beta_0 - \beta_1 x]/\sigma \sim N(0,1)$, u_1 and u_2 are as defined in (27.2.14), $v = \frac{s^2}{\sigma^2}$, so that $(n-2)v^2 \sim \chi_{n-2}^2$ and

$$c = \frac{(x - \bar{x})}{\sqrt{S}}. \qquad (27.2.16)$$

In view of (27.2.15) and (27.2.16), we shall use the notation $k(c)$ instead of $k(x)$ and $C(c; u_1, u_2, v)$ instead of $C(x)$. Thus

$$C(c; u_1, u_2, v) = \Phi\left(\frac{u_1}{\sqrt{n}} + cu_2 + k(c)v\right) - \Phi\left(\frac{u_1}{\sqrt{n}} + cu_2 - k(c)v\right). \qquad (27.2.17)$$

Let δ denote an upper bound on $|c|$, i.e., $|c| \leq \delta$. Then the conditions (27.2.7) and (27.2.8) simplify to

$$P_{u_1,u_2,v}\left[C(c; u_1, u_2, v) \geq \alpha\right] = \gamma, \qquad (27.2.18)$$

for every c satisfying $|c| \leq \delta$, and

$$P_{u_1,u_2,v}\left[\frac{1}{N}\sum_{i=1}^{N} C(c_i; u_1, u_2, v) \geq \alpha\right] = \gamma, \qquad (27.2.19)$$

for every sequence $\{c_i\}$ satisfying $|c_i| \leq \delta$. Note that it is reasonable to assume a known upperbound for c. For example, if x in (27.2.11) is like the x_i's in (27.2.10), then $x_{\min} \leq x \leq x_{\max}$, where $x_{\min}$ and $x_{\max}$ denote the minimum and the maximum among the x_i's. From the definition of c in (27.2.16), it follows that

$$\frac{x_{\min} - \bar{x}}{\sqrt{S}} \leq c \leq \frac{x_{\max} - \bar{x}}{\sqrt{S}}.$$

Since $S = \sum_{i=1}^{n}(x_i - \bar{x})^2$, it is clear that $\frac{|x_{\min} - \bar{x}|}{\sqrt{S}}$ and $\frac{|x_{\max} - \bar{x}|}{\sqrt{S}}$ are both less than or equal to one. In other words $|c| \leq 1$ is a reasonable assumption. The actual upper bound for $|c|$ will depend on the x_i's and will perhaps be much less than one. In fact one should expect the upper bound on $|c|$ to get smaller as n becomes large, since S becomes large with n.

Now suppose $k(c)$ is derived subject to (27.2.18). In the next section we have numerically investigated whether such a $k(c)$ will satisfy (27.2.19). The problem essentially involves only two quantities that can vary, namely, n and δ (the upper bound on $|c|$). However, in order to carry out any numerical computation based on (27.2.19), one also has to decide upon a large enough value of N and, more importantly, one has to choose the sequence $\{c_i\}$, $i = 1, 2, ..., N$. Even for a fixed value of N, it is obviously quite time consuming to carry out simulations for a large number of different choices for the sequence $\{c_i\}$, $i = 1, 2, ..., N$. Note that if the c_i's are all equal, then (27.2.18) and (27.2.19) coincide. Thus, when the c_i's are nearly equal, (27.2.18) is likely to imply (27.2.19). Since we want to study if $k(c)$ satisfying (27.2.18) also satisfies (27.2.19), it appears that we should choose the c_i's to be as unequal as possible in the sequence $\{c_i\}$, $i = 1, 2, ..., N$. For our numerical results in the next section, we have chosen $N = 10,000$ and have made three choices of the sequence $\{c_i\}$, $i = 1, 2, ..., 10,000$. Our first choice consists of 10,000 equispaced c_i's, starting at $-\delta$, in the interval $[-\delta, \delta]$. Our second choice consists of 5,000 c_i's equal to zero and 5,000 c_i's equal to δ. The third choice consists of 5,000 c_i's equal to $-\delta$ and 5,000 c_i's equal to δ. Note that the third choice maximizes the variance among the c_i's $(i = 1, 2, ..., 10,000)$ subject to $|c_i| \leq \delta$. However, it is certainly not clear if such a choice is the least favorable one for the condition (27.2.19) to hold. In our computations, we have also made two choices for δ, namely, $\delta = 0.5$ and $\delta = 5$.

Since there is arbitrariness in the choice of the sequence $\{c_i\}$, an alternative is the following. Assume a probability distribution on the interval $[-\delta, \delta]$ and generate the sequence $\{c_i\}$, $i = 1, 2, ..., N$, according to this probability distribution. Suppose, for example, that the c_i's are generated independently according to a uniform distribution on $[-\delta, \delta]$. Then, by the strong law of large numbers, $\frac{1}{N} \sum_{i=1}^{N} C(c_i; u_1, u_2, v) \to \mathrm{E}_c\{C(c; u_1, u_2, v)\}$, with probability one, as $N \to \infty$. Here $\mathrm{E}_c(.)$ denotes expected value computed with respect to the uniform distribution for $c \in [-\delta, \delta]$. The above result is true for every fixed u_1, u_2 and v. Thus, assuming a uniform distribution for $c \in [-\delta, \delta]$, (27.2.19) becomes

$$\mathrm{P}_{u_1, u_2, v}\left[\mathrm{E}_c\{C(c; u_1, u_2, v)\} \geq \alpha\right] = \gamma. \qquad (27.2.20)$$

In our numerical results in the next section, we have also studied the condition (27.2.20) assuming a uniform distribution for $c \in [-\delta, \delta]$.

27.3 NUMERICAL RESULTS

We shall first give a small simplification of (27.2.18). Let

$$u = (\frac{1}{n} + c^2)^{-1/2}(\frac{u_1}{\sqrt{n}} + cu_2) \sim N(0, 1).$$

Hence $\frac{u_1}{\sqrt{n}} + cu_2 = (\frac{1}{n} + c^2)^{1/2}u$. In view of this observation and (27.2.17), (27.2.18) can be written as

$$\mathrm{P}_{u,v}\left[\Phi\left((\frac{1}{n} + c^2)^{1/2}|u| + k(c)v\right) - \Phi\left((\frac{1}{n} + c^2)^{1/2}|u| - k(c)v\right) \geq \alpha\right] = \gamma, \qquad (27.3.21)$$

where $u \sim N(0, 1)$ and $(n - 2)v^2 \sim \chi_{n-2}^2$. Note from (27.3.21) that $k(c) = k(-c)$. (It is easy to verify that (27.2.18) depends on u only through $|u|$). In other words, it is enough to obtain $k(c)$ satisfying (27.3.21) for $c \geq 0$. For our numerical calculations, we have assumed $|c| \leq \delta$ with two choices for δ: $\delta = 0.5$ and $\delta = 5$. As already pointed out, δ is expected to be much less than one in actual applications. However, we have included the choice $\delta = 5$ in our numerical results in order to see to what extent (27.2.18) will imply (27.2.19) when the c_i's vary over a rather wide interval.

The specific numerical calculations that we have carried out are the following. Starting with $c = 0$ and ending with $c = 5$, obtain the values of $k(c)$ satisfying (27.3.21) for 1001 equispaced values of c in the interval $[0, 5]$. A MATLAB program was used for this calculation. The program first computes $k(0)$ and uses $k(0)$ as a starting value for the computation of $k(c_1)$ for $c_1 > 0$. The value of $k(c_1)$ is used as a starting value for the computation of $k(c_2)$ for $c_2 > c_1$ etc. For the evaluation of $k(0)$, a starting value is required. Let $k_*(0)$ satisfy

$$ P_v\left[\Phi\left(k_*(0)v \right) - \Phi\left(-k_*(0)v \right) \geq \alpha \right] = \gamma. \qquad (27.3.22) $$

From (27.3.22), we immediately get

$$ k_*(0) = \frac{\sqrt{(n-2)} \times z[(1+\alpha)/2]}{\sqrt{\chi^2_{n-2}(1-\gamma)}}, \qquad (27.3.23) $$

where $z[(1+\alpha)/2]$ denotes the $100[(1+\alpha)/2]$th percentile of the standard normal distribution and $\chi^2_{n-2}(1-\gamma)$ denotes the $100(1-\gamma)$th percentile of the central chi square distribution with df $= n - 2$. Note that at $c = 0$, (27.3.21) is "approximately the same" as (27.3.22), especially when n is large. In any case, we have $k(0) > k_*(0)$ and $k_*(0)$ in (27.3.23) can be used as a starting value for computing $k(0)$. Now fit a function, perhaps a polynomial of appropriate degree, to the $k(c)$ values numerically obtained. The degree of the polynomial is decided so as to get a satisfactory fit. The values of $k(c)$ along with the fitted function can be plotted to verify that the fitted function is visually satisfactory. The fitted function is taken to be the function $k(c)$ that satisfies (27.2.18), or equivalently (27.3.21), for all values of c in the interval $[-5, 5]$. A few fitted functions are given in Appendix A corresponding to $\alpha = 0.90$, $\gamma = 0.95$ for $n - 2 = 5, 13, 38$ and 100 and corresponding to $\alpha = 0.95$, $\gamma = 0.99$ for $n - 2 = 38$. (Recall that $n - 2$ is the df of the chi square distribution of $(n-2)v^2$). Figures 27.1(a)–(d) give plots of the actual $k(c)$ values and these fitted functions. The $k(c)$ functions given in Appendix A contain powers of c up to and including $c^{6.5}$ and we have kept several decimal places in the coefficients of the various powers of c. This was done in order to get a visually satisfactory fit. As should be clear from Figures 27.1(a)–(d), the functions given in Appendix A provide very satisfactory fits to the $k(c)$ values.

27.3.1   The Simulation of (27.2.19) and (27.2.20)

Consider the following sequence of c_i's:

SEQUENCE 1(a) : $c_1 = -0.5, \ c_{i+1} = c_i + 0.0001, \ i = 1, 2, ..., 10,000.$

SEQUENCE 1(b) : $c_i = 0, i = 1, 2, ..., 5,000, \ c_i = 0.5,$

                             $i = 5001, ..., 10,000.$

SEQUENCE 1(c) : $c_i = -0.5, i = 1, 2, ..., 5,000, \ c_i = 0.5,$

                             $i = 5001, ..., 10,000.$

SEQUENCE 2(a) : $c_1 = -5, \ c_{i+1} = c_i + 0.001, \ i = 1, 2, ..., 10,000.$

SEQUENCE 2(b) : $c_i = 0, i = 1, 2, ..., 5,000, \ c_i = 5,$

                             $i = 5001, ..., 10,000.$

SEQUENCE 2(c) : $c_i = -5, i = 1, 2, ..., 5,000, \ c_i = 5,$

                             $i = 5001, ..., 10,000.$ (27.3.24)

Note that sequence 1(a) has 10001 equispaced values in the interval $[-0.5, 0.5]$, starting with -0.5 and ending with 0.5. Sequence 1(b) has 10,000 values, half of them being equal to 0 and the remaining half equal to 0.5. Sequence 1(c) has 10,000 values, half of them being equal to -0.5 and the remaining half equal to 0.5. Sequences 2(a), 2(b) and 2(c) are similarly defined for the interval $[-5, 5]$. For the sequences in (27.3.24), we have simulated

$$P_{u_1, u_2, v} \left[\frac{1}{N} \sum_{i=1}^{N} C(c_i; u_1, u_2, v) \geq \alpha \right], \qquad (27.3.25)$$

where $u_1 \sim N(0, 1)$, $u_2 \sim N(0, 1)$, $(n-2)v \sim \chi_{n-2}^2$ and these random variables are independent. $C(c_i; u_1, u_2, v)$ is given in (27.2.17). Here $N = 10,001$ for sequences 1(a) and 2(a) and $N = 10,000$ for the other sequences in (27.3.24). Thus we are essentially simulating the lhs of (27.2.19). For this purpose, we generated 10,000 values of the random variables u_1, u_2 and v. Tables 27.1a, 27.1b and 27.1c give the simulated values of (27.3.25). Table 27.1a gives the simulated values of (27.3.25) corresponding to the sequences 1(a) and 2(a) in (27.3.24), and Table 27.1b and Table 27.1c give the same for the sequences 1(b), 2(b) and 1(c), 2(c) in (3.4). If (27.2.18) implies (27.2.19), then the simulated values of (27.3.25) should all be close to γ. We see from Tables 27.1a, 27.1b and 27.1c that most of the simulated values are very close to γ or slightly more than γ. There are only very few values that are below γ, and none of them is unacceptably below the corresponding γ. This is especially true for $\delta = 0.5$. The conclusion is that if $k(c)$ satisfies the TI condition (27.2.18), then such a $k(c)$ meets the condition (27.2.19) quite well, at least for the sequences in (27.3.24).

For the simulation of (27.3.25) using the sequences 1(a) and 2(a), we used values of $k(c)$ obtained from the fitted functions. Note that in order to simulate (27.3.25), it is not necessary to obtain the functional form of $k(c)$; only its numerical values are required. However, with the sequences 1(a) and 2(a) in (27.3.24), this amounts to evaluating 10,001 $k(c)$ values numerically. To save time, we chose to evaluate only 1001 $k(c)$ values and then fit a function to the values so obtained. We note that the simulations reported in Mee and Eberhardt (1996) uses the individual $k(c)$ values instead of a fitted function. For the sequences 1(b), 1(c), 2(b) and 2(c) in (27.3.24), we used the actual $k(c)$ values in the simulation since only two $k(c)$ values are required, namely, $k(0)$ and $k(0.5)$ for the sequences 1(b) and and 1(c), and $k(0)$ and $k(5)$ for the sequences 2(b) and 2(c) (recall that $k(c) = k(-c)$).

Table 27.2 gives the simulated values of the lhs of (27.2.20), where, for given values of u_1, u_2 and v, $E_c[C(c; u_1, u_2, v)]$ was computed using 10,000 values of c generated based on the uniform distribution on the intervals $[-0.5, 0.5]$ and $[-5, 5]$. For each value of c, the value of $k(c)$ was obtained from the fitted function. For simulating the lhs of (27.2.20), 10,000 values of u_1, u_2, and v were generated. From Table 27.2, it is clear that the lhs of (27.2.20) is very close to γ and is slightly more than γ in most cases. Note that c being uniform on $[-0.5, 0.5]$ or $[-5, 5]$ implies that x has a uniform distribution on an appropriate interval; see (27.2.16) for the relationship between x and c. One can also simulate the lhs of (27.2.20) using some other distribution for x or c. However, we have carried out our simulation only using the uniform distribution.

The conclusion we draw from the numerical results is that it is enough to obtain $k(c)$ satisfying the TI condition in order to satisfy the requirement (27.2.19). Unfortunately, the numerical results do not reveal any pattern regarding the behavior of (27.3.25) when the df, α, γ and δ vary. One should expect that the simulated values of (27.3.25) would be less for the wider interval $[-5,5]$ than for the interval $[-0.5, 0.5]$. Most, though not all, of the simulated values support this.

TABLE 27.1 Simulated values of (27.3.25) for the sequences in (27.3.24) for $c_i \in [-\delta, \delta]$, with $\delta = 0.5, 5$, and df $= n - 2$

TABLE 27.1a Simulated values of (27.3.25) for the sequences 1(a) and 2(a) in (27.3.24)

	df	$\gamma = .90$		$\gamma = .95$		$\gamma = .99$	
		$\delta = .5$	$\delta = 5$	$\delta = .5$	$\delta = 5$	$\delta = .5$	$\delta = 5$
$\alpha = .75$	5	.902	.903	.950	.950	.992	.991
	13	.908	.906	.956	.954	.994	.990
	38	.914	.902	.961	.955	.994	.991
	100	.905	.900	.968	.957	.996	.990
$\alpha = .90$	5	.899	.900	.949	.948	.989	.990
	13	.904	.903	.954	.950	.991	.989
	38	.910	.900	.958	.951	.994	.989
	100	.915	.909	.961	.953	.994	.989
$\alpha = .95$	5	.896	.898	.949	.947	.992	.989
	13	.899	.900	.948	.951	.993	.992
	38	.909	.906	.956	.956	.992	.992
	100	.913	.910	.959	.952	.995	.989
$\alpha = .99$	5	.899	.885	.951	.943	.991	.988
	13	.892	.884	.953	.947	.991	.991
	38	.905	.895	.954	.949	.992	.991
	100	.905	.902	.957	.955	.995	.991

TABLE 27.1b Simulated values of (27.3.25) for the sequences 1(b) and 2(b) in (27.3.24)

	df	$\gamma = .90$		$\gamma = .95$		$\gamma = .99$	
		$\delta = .5$	$\delta = 5$	$\delta = .5$	$\delta = 5$	$\delta = .5$	$\delta = 5$
$\alpha = .75$	5	.900	.894	.950	.945	.991	.990
	13	.906	.898	.958	.952	.994	.992
	38	.916	.901	.960	.949	.995	.991
	100	.911	.901	.969	.953	.995	.988
$\alpha = .90$	5	.898	.885	.950	.938	.990	.986
	13	.901	.897	.953	.945	.991	.988
	38	.910	.898	.958	.949	.993	.989
	100	.923	.910	.965	.952	.994	.989
$\alpha = .95$	5	.897	.880	.946	.936	.991	.987
	13	.900	.886	.951	.944	.992	.991
	38	.902	.890	.955	.950	.992	.990
	100	.918	.907	.962	.951	.994	.990
$\alpha = .99$	5	.900	.874	.952	.936	.990	.987
	13	.899	.881	.949	.936	.991	.988
	38	.893	.886	.950	.941	.993	.991
	100	.908	.900	.957	.952	.994	.991

TABLE 27.1c Simulated values of (27.3.25) for the sequences 1(c) and 2(c) in (27.3.24)

	df	$\gamma = .90$		$\gamma = .95$		$\gamma = .99$	
	df	$\delta = .5$	$\delta = 5$	$\delta = .5$	$\delta = 5$	$\delta = .5$	$\delta = 5$
$\alpha = .75$	5	.901	.900	.954	.952	.992	.990
	13	.914	.906	.961	.953	.994	.991
	38	.910	.902	.959	.954	.993	.991
	100	.893	.897	.955	.953	.990	.988
$\alpha = .90$	5	.893	.896	.950	.950	.992	.992
	13	.903	.907	.956	.951	.992	.990
	38	.901	.894	.959	.954	.991	.989
	100	.907	.905	.957	.958	.991	.990
$\alpha = .95$	5	.894	.899	.946	.943	.989	.991
	13	.898	.904	.950	.952	.993	.991
	38	.902	.903	.953	.947	.991	.992
	100	.903	.905	.956	.954	.990	.988
$\alpha = .99$	5	.896	.898	.952	.949	.991	.991
	13	.892	.892	.951	.952	.992	.991
	38	.904	.901	.952	.949	.991	.991
	100	.901	.899	.951	.956	.991	.992

TABLE 27.2 Simulated values of the lhs of (27.2.20)

	df	$\gamma = .90$		$\gamma = .95$		$\gamma = .99$	
	df	$\delta = .5$	$\delta = 5$	$\delta = .5$	$\delta = 5$	$\delta = .5$	$\delta = 5$
$\alpha = .75$	5	.906	.897	.952	.948	.992	.993
	13	.910	.890	.955	.939	.992	.988
	38	.902	.883	.955	.937	.991	.985
	100	.900	.874	.954	.943	.990	.989
$\alpha = .90$	5	.897	.900	.948	.952	.989	.990
	13	.907	.897	.950	.949	.991	.989
	38	.900	.898	.952	.949	.990	.991
	100	.908	.899	.953	.951	.991	.990
$\alpha = .95$	5	.896	.904	.947	.950	.991	.991
	13	.900	.899	.952	.951	.992	.992
	38	.905	.894	.954	.947	.991	.990
	100	.908	.900	.953	.952	.989	.991
$\alpha = .99$	5	.888	.905	.947	.955	.991	.993
	13	.883	.894	.947	.948	.991	.991
	38	.897	.899	.949	.946	.991	.989
	100	.901	.902	.959	.952	.992	.991

Remark For the models (1.1) and (1.2), the computation of $k(\mathbf{x})$ satisfying (2.5) is quite similar to that of $k(c)$ satisfying (3.1). To see this, note that, in view of (2.2),

$$\mathbf{x'u} \sim N(0, d^2), \text{ where } d^2 = \mathbf{x'}(X'X)^{-1}\mathbf{x}. \tag{27.3.26}$$

Hence (2.4) can be written as

$$
\begin{aligned}
C(\mathbf{x}; \mathbf{u}, v) &= \Phi(\mathbf{x'u} + k(\mathbf{x})v) - \Phi(\mathbf{x'u} - k(\mathbf{x})v) \\
&= \Phi(d|u| + k(d)v) - \Phi(d|u| - k(d)v),
\end{aligned}
$$

where $u = \mathbf{x'u}/d \sim N(0, 1)$ and we use the notation $k(d)$ instead of $k(\mathbf{x})$. Thus, in (3.1), $d^2 = (\frac{1}{n} + c^2)$. (2.5) can now be written as

$$\mathbf{P}_{u,v}\left[\Phi(d|u| + k(d)v) - \Phi(d|u| - k(d)v) \geq \alpha\right] = \gamma. \tag{27.3.27}$$

Note that (3.1) and (3.7) are identical. In other words, the computation of $k(\mathbf{x})$ satisfying (2.5), or equivalently, the computation of $k(d)$ satisfying (3.7), does not present any difficulty. However, in order to verify whether $k(\mathbf{x})$ satisfying (2.5) also satisfies (2.6), we need to consider a sequence of vectors $\{\mathbf{x}_i\}$ ($i = 1, 2, 3, ...$) for computing the lhs of (2.6). The advantage in studying the condition (2.17), as opposed to (2.6), is that in (2.17), the sequence that we need to consider, namely, $\{c_i\}$ ($i = 1, 2, 3,....$), is a sequence of scalars. Perhaps one way to study (2.6) is to consider the $\mathbf{x}_i$'s belonging to a bounded region (which is usually the case) and then generate $\mathbf{x}_i$'s according to a probability distribution on this region. In other words, we can study a condition analogous to (2.18). In this article, we have not pursued this further.

27.3.2   An Example

Intervals satisfying (2.16) or (2.17) will be much shorter compared to the STI's that satisfy (2.7). In other words, $k(c)$ satisfying (2.16) or (2.17) will be smaller compared to $k(c)$ satisfying (2.7). Of course, the conditions (2.16), (2.17) and (2.7) express different requirements. Nevertheless, we shall now compare our $k(c)$ values with those obtained using the Mee, Eberhardt and Reeve (1991) procedure for the example in Lieberman and Miller (1963). We recall that the Mee, Eberhardt and Reeve (1991) procedure is for the computation of STI's, i.e, intervals satisfying (2.7). In this example the models (2.8) and (2.9) are applicable. The response variable y represents the speed of a missile (in miles per hour) and the explanatory variable x is the dimension of the orifice opening (measured in inches) of the valve which admits the fuel. In Section 7 of Lieberman and Miller (1963), values of the speed, namely the y_i's, are given corresponding to the values x_i's of x for $i = 1, 2, ..., 15$. The x_i values were in the range 1.310 inches to 1.400 inches with $\bar{x} = 1.3531$ and $\sum_{i=1}^{15}(x_i - \bar{x})^2 = 0.011966$; see Lieberman and Miller (1963, p. 166). Table 27.3 gives three values of x and the corresponding values of c and $k(c)$. In the table $k(c)$ denotes the value that satisfies (2.16) and $k_{\mathrm{MER}}(c)$ denotes the value obtained using the Mee, Eberhardt and Reeve (1991) procedure. As mentioned earlier, a MATLAB program was used for obtaining $k(c)$. As expected, the $k(c)$ values are substantially smaller compared to the $k_{\mathrm{MER}}(c)$ values. However, as already pointed out, our $k(c)$ values do not satisfy the stronger condition (27.2.9), whereas $k_{\mathrm{MER}}(c)$ does satisfy this condition.

TABLE 27.3 Values of $k(c)$ satisfying (27.2.18) and $k_{\mathrm{MER}}(c)$ for $n = 15$

	$\alpha = 0.95,\ \gamma = 0.99$			$\alpha = 0.75,\ \gamma = 0.95$		
x	1.310	1.353	1.400	1.310	1.353	1.400
c	0.4713	0.2582	0.5002	0.4713	0.2582	0.5002
$k(c)$	3.9218	3.6239	3.9852	2.0102	1.7795	2.0554
$k_{\mathrm{MER}}(c)$	4.4927	3.8330	4.5822	2.3933	1.9059	2.4594

27.4 CALIBRATION

Consider the models (27.2.10) and (27.2.11) and suppose x in (27.2.11) is unknown. The calibration problem consists of statistical inference concerning x. Here we shall discuss the problem of constructing confidence regions for x. More specifically, we shall discuss the construction of multiple use confidence regions. The problem arises when the y_i's in (27.2.10), referred to as the calibration data, are used repeatedly to construct a sequence of confidence regions for a sequence of unknown and possibly different x values after observing the corresponding $y(x)$, following the model (27.2.11). Multiple use confidence regions are derived subject to the following coverage and confidence level requirements. Given that the confidence regions are constructed using the same calibration data, the proportion of confidence regions that include the corresponding true x values is to be at least α. The probability that the calibration data will provide such a $100\alpha\%$ coverage is to be at least γ. For further details on this criterion, we refer to Mee and Eberhardt (1996). This article also contains other relevant references.

One approach for obtaining a multiple use confidence region for the calibration problem is to invert an STI. However, what is required is the inversion of an interval satisfying the condition (27.2.19). That is, observe

$y(x)$ in (27.2.11) and invert the interval $(1, x)\hat{\boldsymbol{\beta}} \pm k(c)s$ in order to obtain the region for x, where $k(c)$ satisfies (27.2.19). Here $\hat{\boldsymbol{\beta}} = (\hat{\beta}_0, \hat{\beta}_1)'$ and $k(c)$ is a function of x; see (27.2.16). Note that in order to do this, the functional form of $k(c)$ is required. Since the numerical results in the previous section indicate that $k(c)$ satisfying (27.2.18) [or, equivalently, (27.3.21)], also satisfies (27.2.19), at least approximately, our recommendation is to compute $k(c)$ satisfying (27.2.18) and then invert the interval $(1, x)\hat{\boldsymbol{\beta}} \pm k(c)s$ in order to obtain a confidence region for x. We shall briefly illustrate the procedure using the line width calibration problem discussed in Croarkin and Varner (1982), and also in Mee, Eberhardt and Reeve (1991) and Mee and Eberhardt (1996). Here the response variable represents line-width measurements in a certain range on integrated circuit photomasks. The explanatory variable is the line-width certified by the National Institutes of Standards and Technology (NIST). The data given in Croarkin and Varner (1982) consist of 40 values of the response variable, say y_i, corresponding to known values x_i of the explanatory variable ($i = 1, 2, ..., 40$). The assumed model is $y_i \sim N(\beta_0 + \beta_1 x_i, \sigma^2)$, $i = 1, 2, ..., 40$, where the y_i's are also assumed to be independent. Let $y(x)$ denote the line-width measurement corresponding to an unknown value of x of the explanatory variable. Then $y(x) \sim N(\beta_0 + \beta_1 x, \sigma^2)$, where $y(x)$ is assumed to be independent of the y_i's ($i = 1, 2, ..., 40$). The 40 y_i values are to be used repeatedly to obtain confidence regions for the unknown x as and when the corresponding line-width measurement becomes available.

Based on the data in Croarkin and Varner (1982), we have $\hat{\beta}_0 = 0.282$, $\hat{\beta}_1 = 0.977$ and $s = 0.0683$. For the line-width measurement $y(x) = 8$, $\alpha = 0.95$ and $\gamma = 0.99$, we can use $k(c)$ given in Appendix A (see A5 in Appendix A) and invert the interval $(1, x)\hat{\boldsymbol{\beta}} \pm k(c)s$ for $y(x)$ in order to get the confidence region for x. The resulting confidence region for x is the interval (7.709, 8.097). This is narrower than the interval (7.69, 8.12) obtained by the Mee, Eberhardt and Reeve (1991) procedure.

For a multivariate calibration problem, Mathew and Zha (1997) have derived confidence regions using a criterion similar to (27.2.18). However, they did not verify if such a confidence region will satisfy the requirement (27.2.19).

27.5   CONCLUSIONS

Our work suggests that $k(c)$ satisfying the TI condition can be used to obtain intervals satisfying the property (27.2.8). This has resulted in less conservative multiple use confidence regions for the calibration problem.

However, it appears quite difficult to theoretically prove that the TI condition (27.2.18) will imply the condition (27.2.19). The main difficulty in proving (27.2.19) is that the random variables $C(c_i; u_1, u_2, v)$, $i = 1, 2, \ldots$, are not independent.

It does take a certain amount of computational time in order to arrive at the tables like Table 27.1 and Table 27.2. However, the numerical computation of $k(c)$ satisfying (27.2.18) is quite simple and fast. All it takes is the evaluation of $k(c)$ for a few values of c. We could easily accomplish this using a MATLAB program. A polynomial of appropriate degree can then be fitted to such $k(c)$ values. In practice, c may vary in a rather narrow range and hence $k(c)$ may not vary too much. Consequently, it may not be necessary to evaluate $k(c)$ for a large number of values of c; about 100 values should be quite adequate. Note that in problems that call for the construction of a sequence of intervals, for example, multiple use confidence intervals in the calibration problem, such intervals will be constructed for $y(\mathbf{x})$ in (27.1.2) for many values of $\mathbf{x}$. However, the function $k(c)$ required for this purpose need to be evaluated only once and it can then be used repeatedly.

APPENDIX A: SOME FITTED FUNCTIONS $k(c)$

A1 $df = 5$, $\alpha = 0.90$ and $\gamma = 0.95$

$$
\begin{aligned}
k(c) = \ & 3.661834395044096 - 1.084185283581492 c^{0.5} \\
& + 17.68204764229182 c - 109.6797466632338 c^{1.5} \\
& + 336.2904577225682 c^2 - 523.4726825194977 c^{2.5} \\
& + 313.3880955450217 c^3 + 287.0237310106820 c^{3.5} \\
& - 741.7750855567490 c^4 + 694.4604897134429 c^{4.5} \\
& - 369.7252613665112 c^5 + 117.4741508668473 c^{5.5} \\
& - 20.82382277339587 c^6 + 1.590442930845107 c^{6.5}.
\end{aligned}
$$

A2 $df = 13$, $\alpha = 0.90$ and $\gamma = 0.95$

$$
\begin{aligned}
k(c) = \ & 2.546683413004918 - 0.3691102519682013 c^{0.5} \\
& + 7.779996267734723 c - 60.40992979166536 c^{1.5} \\
& + 241.9812167219166 c^2 - 563.9616940176060 c^{2.5} \\
& + 835.3111449692373 c^3 - 799.0992269824635 c^{3.5} \\
& + 483.7552929736485 c^4 - 169.4624298461488 c^{4.5}
\end{aligned}
$$

$$+22.2918850602545c^5 + 5.723184215541130c^{5.5}$$
$$-2.556303166463672c^6 + 0.2815580648973948c^{6.5}.$$

A3 *df = 38, $\alpha = 0.90$ and $\gamma = 0.95$*

$$
\begin{aligned}
k(c) = \ & 2.069620494350655 + 0.1158707174421790c^{0.5} \\
& -2.889039283908703c + 26.02846474714621c^{1.5} \\
& -108.2654502390272c^2 + 232.1392174207692c^{2.5} \\
& -226.1476529130201c^3 + 8.82155888543924c^{3.5} \\
& +216.2299262465762c^4 - 247.5022859580667c^{4.5} \\
& +141.4577577314878c^5 - 46.27202224638564c^{5.5} \\
& +8.277981082257501c^6 - 0.6310232495202666c^{6.5}.
\end{aligned}
$$

A4 *df = 100, $\alpha = 0.90$ and $\gamma = 0.95$*

$$
\begin{aligned}
k(c) = \ & 1.872765486267810 + 0.5404951712473437c^{0.5} \\
& -13.07811384864294c + 120.6875007114894c^{1.5} \\
& -572.9906008827039c^2 + 1599.896395989058c^{2.5} \\
& -2809.602838218623c^3 + 3276.24191171360c^{3.5} \\
& -2613.384240648143c^4 + 1437.157415885933c^{4.5} \\
& -537.0295241442310c^5 + 130.412963219429c^{5.5} \\
& -18.58135740146952c^6 + 1.179624481188825c^{6.5}.
\end{aligned}
$$

A5 *df = 38, $\alpha = 0.95$ and $\gamma = 0.99$*

$$
\begin{aligned}
k(c) = \ & 2.679911240571308 + 2.606656839404740c^{0.5} \\
& -53.25322021539181c + 443.7337351928734c^{1.5} \\
& -1981.29300685469c^2 + 5357.64834056361c^{2.5} \\
& -9351.052415037213c^3 + 10991.65397853717c^{3.5} \\
& + - 8891.94594913713c^4 4965.435648115365c^{4.5} \\
& -1881.40908955501c^5 + 461.9305911906758c^{5.5} \\
& -66.30338747055679c^6 + 4.223942117759620c^{6.5}.
\end{aligned}
$$

Acknowledgement This research was supported in part by National Science Foundation Grant DMS 95-30932. We are grateful to a reviewer whose suggestions resulted in the clarification of several ideas and improved presentation of the results.

REFERENCES

Croarkin, C. and Varner, R. N. (1982). Measurement assurance for dimensional measurements on integrated-circuit photomasks, *Technical Note 1164*, National Bureau of Standards.

Eberhardt, K. R., Mee, R. W. and Reeve, C. P. (1989). Computing factors for exact two-sided tolerance intervals for a normal distribution, *Communications in Statistics—Simulation and Computation*, **18**, 397–413.

Lieberman, G. J. and Miller, R. G., Jr. (1963). Simultaneous tolerance intervals in regression, *Biometrika*, **50**, 155–168.

Mathew, T. and Zha, W. (1996). Conservative confidence regions in multivariate calibration, *The Annals of Statistics*, **24**, 707–725.

Mathew, T. and Zha, W. (1997). Multiple use confidence regions in multivariate calibration, *Journal of the American Statistical Association*, **92**, 1141–1150.

Mathew, T., Sharma, M. K. and Nordström, K. (1998). Tolerance regions and multiple use confidence regions in multivariate calibration, *The Annals of Statistics*, **26**, 1989–2013.

Mee, R. W. and Eberhardt, K. R. (1996). A comparison of uncertainty criteria for calibration, *Technometrics*, **38**, 221–229.

Mee, R. W., Eberhardt, K. R. and Reeve, C. P. (1991). Calibration and simultaneous tolerance intervals for regression, *Technometrics*, **33**, 211–219.

CHAPTER 28

AN OVERVIEW OF SEQUENTIAL AND MULTISTAGE METHODS IN REGRESSION MODELS

SUJAY DATTA

Northern Michigan University, Marquette, MI

Abstract: This chapter presents a brief survey of the wide variety of sequential and multistage inference-procedures used in linear regression and other related models. The primary motivation behind using such methodologies in these models comes from, among other things, fixed precision inference and online (or adaptive) data-processing. After a general introduction to these models and methodologies, here we explore the significant developments in sequential and multistage fixed-precision inference in deterministic regression models, sequential shrinkage estimation in such models, and the Bayes sequential approach. These are followed by sequential inference in stochastic regression, inverse regression, and errors-in-variables models. Throughout, the emphasis is on first and second-order asymptotic properties of the procedures involved. An updated list of references is provided at the end.

Keywords and phrases: Linear regression, Fixed precision inference, Sequential and multistage procedures, Bayes sequential procedures, Shrinkage estimation, Stochastic regression models, Inverse regression

28.1 INTRODUCTION

This chapter is intended to provide a brief overview of the wide variety of sequential and multistage methodologies that are used for drawing inferences in linear regression and related models. A linear regression model is

427

one of the oldest in statistical data analysis. Also, sequential and multi-stage inference-procedures have come a long way since the days of Wald's Sequential Probability Ratio Test (SPRT). The 'marriage' of these two was primarily motivated by, among other things, *fixed precision inference* (for example, fixed-width interval estimation, bounded-risk or minimum-risk point estimation, fixed proportional accuracy estimation, etc.) and the increasing demand for online (or adaptive) data-processing techniques. The result has been an astonishingly rich array of procedures encompass-ing almost all aspects of regression analysis and some related models as well. Here, following a concise introduction to these models and a gen-eral discussion of sequential and multistage methodologies in Section 28.2, Section 28.3 addresses the issue of fixed-precision inference in regression models with deterministic regressors. In the classical (that is, frequentist) framework, few-stage, sequential and accelerated sequential procedures for point estimation, confidence set estimation and hypothesis-testing are de-scribed with a predetermined measure of *accuracy* in each case. Section 28.4 provides a survey of sequential shrinkage estimation methodologies in regression. Then in Section 28.5, Bayes sequential inference in a regression model is considered and the idea of an *asymptotically pointwise optimal* (APO) estimator is introduced. Section 28.6 deals with fixed precision inference in stochastic regression models where the regressors themselves have their own probability distributions. Section 28.7 explores sequential solutions to inverse regression (or calibration) problems. Finally, Section 28.8 presents some miscellaneous topics which are relevant in this context. Throughout, asymptotic optimality properties of the procedures involved are examined, with special emphasis on second-order asymptotic charac-teristics. An extensive and updated (but by no means exhaustive) list of references are provided at the end. For additional information and further details regarding some of the topics mentioned here, the interested reader should see, for example, Ghosh and Sen(1991) or Ghosh, Mukhopadhyay and Sen (1997).

28.2   THE MODELS AND THE METHODOLOGIES—A GENERAL DISCUSSION

Here we provide a brief description of the models we will be dealing with in the later sections, and also a preview of sequential and multistage inference-procedures in general. We also indicate, by means of an example, why one would use such procedures in the context of these models.

28.2.1   Linear Regression and Related Models

In the following sections, we will primarily be concerned with the model:

$$\mathbf{Y}_{n\times1} = \mathbf{X}_{n\times p}\boldsymbol{\beta}_{p\times1} + \boldsymbol{\epsilon}_{n\times1}, \qquad (28.2.1)$$

where $\mathbf{X}_{n\times p}$ is the known *design matrix* of rank p, $\boldsymbol{\beta}_{p\times1}$ is the unknown vector of *regression coefficients*, and $\boldsymbol{\epsilon}_{n\times1}$ is the vector of random errors having some joint distribution F_n with zero mean and variance-covariance matrix $\boldsymbol{\Sigma}_{n\times n}$. Under this model, typically we would be interested in inferences regarding $\boldsymbol{\beta}$, or $\mathbf{H}\boldsymbol{\beta}$ for some known $r \times p$ matrix $\mathbf{H}$ of rank r. Depending on the situation, certain special cases or variants of the above model may be appropriate. For example, $\boldsymbol{\epsilon}$ might be assumed to have a multivariate Gaussian distribution, $\boldsymbol{\Sigma}$ could have the special configuration $\sigma^2\mathbf{I}_{n\times n}$ for some unknown $\sigma^2 > 0$ or $\sigma^2\boldsymbol{\Lambda}_{n\times n}$ for some known positive definite matrix $\boldsymbol{\Lambda}$, or it could have the equicorrelation structure $\{(\sigma^2 - \delta)\mathbf{I}_{n\times n} + \delta\mathbf{J}_{n\times n}\}$ for some unknown $\sigma^2 > 0$ and $\delta \in (-\infty, \infty)$. On the other hand, a more general version of this model might assume that the i-th row of $\mathbf{X}$ is $\{\phi_1(x_{i1}), \ldots, \phi_p(x_{ip})\}$ for some known real-valued functions $\phi_1, \ldots, \phi_p$, instead of just $\{x_{i1}, \ldots, x_{ip}\}$. Another direction for generalization would be to consider the *generalized linear model*: $E(y_i) = \Psi(\beta_1 x_{i1}, \ldots, \beta_p x_{ip})$ for some known real-valued '*link-function*' Ψ. It might even be justified, depending on the nature of the underlying experiment, to assume that the rows of $\mathbf{X}$ are themselves independent and identically distributed (i.i.d.) random vectors coming from a certain p-variate joint distribution which could be independent of the error-distribution. In that case we would be dealing with a *stochastic* regression model. In a deterministic or stochastic regression model, the classical (i.e. frequentist) approach to inference would assume that $\boldsymbol{\beta}$ is a vector of (unknown) constants. However, the Bayesian approach would consider it as another random vector having an *a priori* distribution (say, p-variate Gaussian with mean-vector $\boldsymbol{\mu}$ and dispersion-matrix $\boldsymbol{\Delta}$). While an *empirical* Bayesian approach would try to estimate $\boldsymbol{\mu}$ and $\boldsymbol{\Delta}$ from the observed data, a *hierarchical* Bayesian approach would assume certain (known) probability distributions on these parameters.

While inference on $\boldsymbol{\beta}$, the vector of unknown regression coefficients, is our primary objective in all the above models, the classical univariate *inverse regression* (or *calibration*) problem is concerned with determining the value (say, x_0) taken by the regressor X in the model: $y = \beta_0 + \beta_1 x + \epsilon$, by using one or more observed values of the response-variable Y corresponding to $X = x_0$. Generally, the parameters β_0 and β_1 are assumed unknown and are estimated by fitting the regression model to a set of observations independent of those taken at $X = x_0$.

In an *errors-in-variables* (EIV) model, also known as a *measurement error* model, the basic setup is as follows: one observes pairs $\{(X_i, Y_i), i = 1, \ldots, n\}$ where $Y_i = \gamma + \beta\xi_i + \epsilon_i$; $X_i = \xi_i + \delta_i$, with γ, β being real constants, ϵ_i's i.i.d. $\sim N(0, \sigma_\epsilon^2)$, δ_i's i.i.d. $\sim N(0, \sigma_\delta^2)$ and ϵ_i's being independent of δ_i's. If the ξ_i's are fixed constants, this is the so-called *functional relationship* model, as opposed to a *structural relationship* between X and Y where ξ_i's would also be assumed to have come from a probability distribution.

28.2.2 Sequential and Multistage Methodologies

In classical parametric inference, the probability distributions of the random variables involved are completely characterized by *parameters* whose values are usually unknown to us. Typically the information available regarding the parameters is in the form of random samples, often obtained through a chosen experimental design or a controlled random mechanism. Based on these samples, we draw inferences about the unknown parameters in many different ways, such as point estimation, confidence-set estimation, hypothesis-testing, etc.. Corresponding to each of these approaches, there are many competing inference-procedures and to be able to choose one out of them, we must use some criterion for comparing their performances. Statistical decision theory provides several such criteria. One starts with a suitable *loss function*, intended to quantify the loss incurred when a 'wrong' inference is drawn, and subsequently chooses the decision-rule that minimizes the *risk* (i.e. the expected loss). But the problem is, often such *optimal* decision-rules may not exist and even if they do, their own performances may be far from satisfactory. For example, if it is desirable in a point estimation problem to bound the maximal risk from above by a given number, or in a hypothesis-testing problem to achieve a preassigned power, even the decision-theoretically 'best' procedures may fail to achieve these goals for a fixed number of sample observations. Dantzig(1940) showed the nonexistence of a fixed sample-size solution to a certain version of the problem of constructing fixed-width confidence intervals for a normal mean μ such that the coverage-probability is at least some preassigned quantity $(1 - \alpha)$, irrespective of the value of the unknown population variance σ^2. Inferences of this sort with preassigned precision-constraints are called *fixed-precision* inferences.

On the other hand, sometimes the 'optimal' decision-rules for fixed sample-size inference problems may turn out to be 'inefficient' in the sense that the same level of performance could be achieved with possibly fewer observations by following an *adaptive sampling scheme* in which, the 'necessity' of taking any further sample for the inference-drawing purpose is re-examined after every step. A classic example of this phenomenon is the SPRT of Wald(1947) which is essentially a likelihood-ratio test based on a *sequential* sampling scheme governed by a '*stopping rule*'. The *average sample number* (ASN) for this test turns out to be smaller than the number of samples needed to achieve the same power via the fixed sample-size UMP test. Almost simultaneously with the emergence of Wald's SPRT, a version of the problem that Dantzig(1940) proved to be unsolvable in the fixed sample-size scenario, was solved by Stein(1945) using a *two-stage sampling scheme*—the first stage intended to provide an estimate of the unknown nuisance-parameter σ^2 and the final inference intended to be based on both stages combined.

Apart from the two examples cited above where an adaptive multistage sampling scheme either helps save samples or solves an inference problem otherwise unsolvable via fixed sample-size methodologies, there are situations where such a sampling scheme is appealing from other viewpoints as well. For example, in industrial process control where detection of *change-points* in the production system is an important issue and one observes the products one by one (or batch by batch) as they come out of the assembly-line, a sequential sampling scheme is intrinsic and it is difficult to deny it preference over any fixed sample-size procedure. Also, in clinical and pharmaceutical research involving humans or other living beings, any plan to apply fixed sample-size methodologies almost always runs the risk of becoming unethical. This is because some subjects already treated with the drug under study might suddenly develop fatal side-effects forbidding any further experimentation with it, or conclusive evidence regarding the drug's performance may already be there even before all the subjects are treated with it. Adaptive sequential and multistage designs are often tailor-made to take into account such contingencies.

As the scope of applying sequential and multistage methodologies has gradually widened, several criteria have emerged over the years to assess their performances:

- *Consistency* or *Exact Consistency* in the sense of Stein (1945)— when the procedure *exactly* achieves the underlying inferential goal;

- *Asymptotic Consistency* in the sense of Chow and Robbins (1965)— when the procedure achieves the underlying inferential goal only *asymptotically* (i.e. in the limit);

- *Asymptotic First Order Efficiency* in the sense of Chow and Robbins (1965) or Ghosh and Mukhopadhyay (1981) respectively— if the *ratio* of the ASN of the multistage procedure to the corresponding 'optimal' fixed sample-size approaches 1 asymptotically;

- *Asymptotic Second Order Efficiency* in the sense of Ghosh and Mukhopadhyay (1981)—if the *difference* between the ASN of the multistage procedure and the corresponding 'optimal' fixed sample-size remains bounded asymptotically.

It turns out that Stein-type two-stage procedures, although exactly consistent in most cases, lack even asymptotic first order efficiency—a drawback that makes them less attractive. Mukhopadhyay (1980) came up with a modified version of it which is both exactly consistent and asymptotically first-order efficient. On the other hand, purely sequential procedures are in general only asymptotically consistent whereas their asymptotic efficiency is often of the second order. But the biggest shortcoming of a purely sequential sampling scheme is its operational inconvenience. The trade-off between asymptotic second order efficiency and operational convenience came in the form of a *triple-sampling* scheme, first considered by Mukhopadhyay and later recommended by Hall (1981), which cuts the number of sampling operations to bare bones without sacrificing second order optimality. Subsequently, Hall (1983) tried to impart ease of implementation to a purely sequential procedure by suggesting an *accelerated* version of it. The idea was to go purely sequentially part of the way, and then resort to a single batch-sampling. Mukhopadhyay (1990) provided a unified theory behind three-stage sampling while Mukhopadhyay and Solanky (1991) generalized Hall's idea of acceleration to a variety of inference problems. Recently, Mukhopadhyay (1993) modified their earlier accelerating technique in order to simplify the underlying asymptotic theory.

28.2.3 Sequential Inference in Regression: A Motivating Example

Suppose, in the context of a deterministic multiple linear regression model $\mathbf{Y} = \mathbf{X}\boldsymbol{\beta} + \boldsymbol{\epsilon}$, that the vector of random errors has a $N(\mathbf{O}, \sigma^2 \mathbf{I}_{n \times n})$ distribution for some unknown $\sigma^2 > 0$, and $\boldsymbol{\beta} \in R^p$ is unknown $(n \geq p)$. Given two numbers $d(> 0)$ and $0 < \alpha < 1$, let it be our goal to construct a fixed-volume confidence set for $\boldsymbol{\beta}$ whose volume is a given function of d and confidence coefficient is at least $(1 - \alpha)$. One might consider confidence ellipsoids of the form:

$$\Omega_n = \{\omega \in R^p : (\hat{\boldsymbol{\beta}} - \omega)'(n^{-1}\mathbf{X}'\mathbf{X})(\hat{\boldsymbol{\beta}} - \omega) \leq d^2\}, \tag{28.2.2}$$

where $\hat{\beta} = (\mathbf{X'X})^{-1}\mathbf{X'Y}$. It is easy to see that $P(\beta \in \Omega_n) = F(nd^2/\sigma^2)$ where $F(x) = P(\chi_p^2 \leq x)$ for $x > 0$. Hence the confidence coefficient associated with the above ellipsoid will be at least $(1-\alpha)$ if n is the smallest integer $> a^2\sigma^2/d^2$ $(= C$, say$)$ where $F(a^2) = 1 - \alpha$. So in order to achieve the desired confidence coefficient and, at the same time, satisfy the given volume constraint, one needs to take at least C samples. In other words, C is the 'optimal' fixed sample-size for this problem. But C involves the unknown parameter σ^2 and is, therefore, unknown— implying that there can be no fixed sample-size solution to this problem that will work for all values of σ^2.

A way of getting around this stumbling block was suggested by Mukhopadhyay and Abid (1986) who proposed the following purely sequential sampling scheme: start with $\{(y_i, \mathbf{x}_i) : i = 1,\ldots, k_0\}$ for some $k_0 > p$ and continue to take one more sample at a time until there are N of them where

$$N = \inf \{n \geq k_0 : n \geq a^2 S_n^2/d^2\}, \qquad (28.2.3)$$

where $S_n^2 = (n-p)^{-1}(\mathbf{Y} - \mathbf{X}\hat{\beta})'(\mathbf{Y} - \mathbf{X}\hat{\beta})$. At the stopped stage, construct the confidence set Ω_N as defined in (28.2.2). Using some nice properties of a multivariate Gaussian error-distribution, one can show that $P(\beta \in \Omega_N) = E\{g(N/C)\}$ where $g(x) = F(a^2x)$ for $x > 0$. The N defined in (28.2.3) is called a *stopping variable* for this sampling scheme. Now it remains to verify whether this procedure really achieves the targetted confidence level of $(1 - \alpha)$ and how efficient it is compared to its fixed sample-size counterpart (which uses C samples). Mukhopadhyay and Abid (1986), in their Theorem 4, have shown that:

Theorem 28.2.1 *For the above procedure, as $d \to 0$ $(\Longleftrightarrow C \to \infty)$,*
(i) *$N/C \to 1$ and $E(N/C) \to 1$;*
(ii) *$E(N - C) = A + o(1)$; and*
(iii) *$P(\beta \in \Omega_n) = 1 - \alpha + BC^{-1} + o(C^{-1})$,*
for some computable real constants A and B. In other words, the procedure is asymptotically consistent and second order efficient in the sense of the previous subsection. Some of the procedures described in the following sections share only the first one of these three properties and are, therefore, called first order efficient.

28.3   FIXED-PRECISION INFERENCE IN DETERMINISTIC REGRESSION MODELS

Here we set out to explore the plethora of sequential and multistage methodologies used for drawing inferences under precision-constraints. We consider point estimation, confidence-set estimation and hypothesis-testing separately, although the basic techniques involved share a great deal of similarity.

28.3.1 Confidence Set Estimation

For inference on $\mathbf{H}\beta$ ($\mathbf{H}_{r \times p}$ being a known matrix of rank r) under the 'Aitken setup' where, in the model (28.2.1), ϵ has a $N(\mathbf{O}, \sigma^2 \Lambda)$ distribution with a known positive definite (p.d.) matrix $\Lambda_{n \times n}$ and an unknown $\sigma^2 > 0$, Stein(1945) proposes a *two-stage* procedure. Actually, Stein does it for the special case $\Lambda = \mathbf{I}_{n \times n}$, and one should see Chatterjee (1959a,b) or Chatterjee (1991) for extension to a general p.d. matrix Λ. The procedure is essentially as follows: start with n_0 *pilot samples* (with $nn_0 - p > 1$) and compute the usual unbiased estimator $S_0^2 = (n_0 - p)^{-1} (\mathbf{Y} - \mathbf{X}\hat{\beta})'(\mathbf{Y} - \mathbf{X}\hat{\beta})$ of σ^2 from it ($\hat{\beta}$ being the least-squares estimator of β). Then take $(N - n_0 + 1)$ more samples where $N = \max \{n_0 + 1, \ < z_\alpha^{-1} S_0^2 > +1\}$, with z_α being the upper α-percentage point of a standard normal distribution and $<x>$ being the largest integer $< x$. Finally, use the least-squares estimator of β based on N samples in the point estimator of $\mathbf{H}\beta$, or in order to construct confidence-ellipsoids of the form (28.2.2) for it, or for testing $H_0 : \mathbf{H}\beta = \Theta$ vs. $H_1 : \mathbf{H}\beta \neq \Theta$. It turns out that Stein's procedure is not even asymptotically first order efficient. However, Bishop (1978) and Wilcox (1985) considered Stein-type two-stage procedures for constructing fixed-width confidence intervals for the parameters of a general linear model assuming unequal and equal variances respectively.

Albert (1966) and Srivastava (1967, 1971) proposed sequential procedures for constructing fixed-volume ellipsoidal confidence regions for β (the former actually did it for $\mathbf{H}\beta$ for a known matrix $\mathbf{H}$ of full row-rank). Neither of them required the error-distribution to be Gaussian, and their methods involve eigenvalues of $n(\mathbf{X}'\mathbf{X})^{-1}$. They considered confidence ellipsoids of the form (28.2.2) with only d^2 replaced by d^2/λ_n, where λ_n is the largest eigenvalue of $n(\mathbf{X}'\mathbf{X})^{-1}$. In order to obtain a confidence ellipsoid with maximum diameter $\leq 2d$ which achieves the targetted confidence level $(1 - \alpha)$ at least asymptotically, they suggested the following sampling scheme: start with $n_0(> p)$ sample-vectors and continue taking one more vector at a time until there are N of them where

$$N = \inf \{n \geq n_0 : n \geq a^2 \lambda_n (S_n^2 + n^{-1})/d^2\}, \tag{28.3.4}$$

where S_n^2 and a^2 are as in Subsection 28.2.3. At the stopped stage, construct the ellipsoid Ω_N with d^2 replaced by d^2/λ_N. Assuming that $\lim_{n \to \infty} n^{-1}(\mathbf{X}'\mathbf{X})$ is a $p \times p$ positive definite matrix, they prove the asymptotic consistency and first order efficiency of their procedures. Gleser (1965, 1966) provide similar results for this problem, also without assuming a Gaussian error-distribution. But he considered spherical (instead of ellipsoidal) confidence sets.

Mukhopadhyay and Abid (1986) put forward a two-stage, a three-stage and a sequential sampling scheme for constructing fixed-size confidence ellipsoids for β under the assumption of Gaussian errors. Their sequential procedure, which is asymptotically second order efficient, can be found in our Subsection 28.2.3. Their two-stage procedure is a modified version of Stein's, in the sense that unlike Stein(1945), they start with an initial sample-size of n_0 which is no longer a fixed positive integer but an appropriately chosen monotone *function* of C (the corresponding 'optimal' fixed sample-size). This gives their double-sampling scheme asymptotic first order efficiency. For details, see also Mukhopadhyay(1980). Their triple-sampling strategy is along the line of Hall (1981) and is second order efficient. The idea here is to start with a first-stage sample-size that is a certain monotone function of C, and then determine the second-stage sample-size in much the same way as N was determined in Stein's two-stage procedure(see above)—except that $z_\alpha^{-1} S_0^2$ will now be replaced by $\rho z_\alpha^{-1} S_0^2$ for a suitably chosen $\rho \in (0, 1)$. The third-stage sample-size is then figured out in a similar manner, this time without the ρ. See Mukhopadhyay (1990) for the general theory. As mentioned earlier, a triple-sampling procedure is a nice trade-off between second order efficiency and ease of implementation. In order for the sequential procedure of Subsection 28.2.3 (which has better second order properties than triple-sampling) to be operationally convenient, Mukhopadhyay and Solanky (1991) proposed an 'acceleration' technique for it following Hall(1983), which was later simplified by Mukhopadhyay (1996). The idea is to replace the $a^2 S_n^2 / d^2$ in Equation (28.2.3) by $\rho a^2 S_n^2 / d^2$ for a suitably chosen $\rho \in (0, 1)$ such that ρ^{-1} is an integer and when sequential sampling stops (say, with N^* samples), take $(\rho^{-1} - 1)N^*$ more samples in a single batch. This modification significantly reduces the number of sampling operations involved, without sacrificing second order efficiency. Recently, Mukhopadhyay and Datta (1995) 'fine-tuned' the procedure in Subsection 28.2.3 by modifying the stopping variable (28.2.3) so that the second order expansion of $P(\beta \in \Omega_N)$ in part(iii) of Theorem 1 no longer has the unwanted middle-term BC^{-1}. For many of the procedures described above, Sinha (1991) is a useful reference. See also Sriram and Bose (1988) in this context.

Chaturvedi (1988) also came up with a modified sequential procedure, which uses an idea somewhat similar to the acceleration technique mentioned above, for fixed-size confidence set estimation of β. He derives conditions for his procedure to be asymptotically second order efficient, but only provides an expansion of the form $P(\beta \in \Omega_N) = 1 - \alpha + O(d^2)$ of the coverage-probability. Martinsek (1989), on the other hand, suggested a sequential procedure for obtaining fixed-size confidence regions for β which uses certain regression-analogues of *trimmed means* [as formulated by Welsh (1987)], instead of the least-squares estimate. He assumes a continuous and symmetric error-distribution with $(4 + \delta)$-th moment finite for some $\delta > 0$, and proves asymptotic consistency and efficiency for his procedure. Coleman (1995) considered the slightly different problem of constructing fixed-width confidence intervals for $1/\beta$ in the simple linear regression model where it is known that $\beta \in (0, \beta^*)$ for some $\beta^* > 0$. His sequential procedure ensures that the coverage-probability $\rightarrow \gamma$ uniformly over $\beta \in (0, \beta^*)$ as the width of the interval tends to 0 (γ being pre-specified).

As a departure from the usual regression model (28.2.1), Rahbar (1995) considered constructing fixed-width confidence intervals sequentially for the parameters of a simple linear regression having only discrete regressors and randomly right-censored responses. Timofeev (1991) addressed the issue of bounded-width interval estimation (with confidence coefficient at least $1 - \alpha$ for a given $0 < \alpha < 1$) for the parameter θ^* in the model :

$$y(x_t) = f_t(x_t, \theta^*) + \xi_t \; ; \; t \in \{1, 2, \ldots\}, \qquad (28.3.5)$$

where $\{x_t\}_{t \geq 1}$ are inputs (such that $a \leq x_t \leq b$ for all t and for some real constants a, b), $\{f_t\}_{t \geq 1}$ are known functions, $\{\xi_t\}_{t \geq 1}$ are independent random variables with $E(\xi_t) = 0$ and $E(\xi_t^4) < \sigma$ for some known $\sigma > 0$.

An interesting variation to the conventional way of constructing fixed-size confidence sets in the regression context is the idea of 'accurate estimation' introduced by Finster (1985). He considers estimating β by $\hat{\beta}$ so as to ensure that $P(\hat{\beta} - \beta \in \mathbf{A}) \geq \gamma$ for all β and for all σ^2 (the common error-variance) for a pre-specified set $\mathbf{A}$ and a given $0 < \gamma < 1$. He uses a sequential version of the maximum-probability estimator of Weiss and Wolfowitz (1974). Later, Lohr (1990, 1993) introduced three-stage and two-stage procedures for *accurate estimation* of β (in the above sense). However, she used accuracy-sets $\mathbf{A}$ that are compact, orientable manifolds with boundary and star-shaped with respect to 0. Asymptotic second order properties of these procedures are provided.

28.3.2 Point Estimation

As mentioned earlier, Chatterjee (1959 a,b) uses his Stein-type two-stage sampling scheme for point estimation as well. However, such procedures are grossly inefficient compared to their fixed sample-size counterparts. Later,

Chatterjee (1962) considers a sequential procedure for bounded-risk point estimation of β under the loss $L(\hat{\beta}, \beta) = \phi((\hat{\beta} - \beta)'\Lambda(\hat{\beta} - \beta))$, with ϕ a 'nicely behaved' function and $\Lambda_{p \times p}$ a known p.d. matrix. He proves the asymptotic first order efficiency of his procedure. Mukhopadhyay (1974) proposed a sequential sampling scheme for minimum-risk point estimation of β under the loss

$$L_n(\hat{\beta} - \beta) = (\hat{\beta} - \beta)'\mathbf{A}(\hat{\beta} - \beta) + cn, \qquad (28.3.6)$$

for some known p.d. matrix $\mathbf{A}_{p \times p}$. This loss penalizes for both wrong estimates and sampling extravagance (the cost per unit sample being c). His overall approach and the stopping variable are quite similar to those in our Subsection 28.2.3. Since the risk under this loss ($= p\sigma^2 n^{-1} + cn$) is minimized for $n =< (p\sigma^2 c^{-1})^{1/2} > +1$, it should be the 'optimal' fixed sample-size for this problem. But it involves the unknown parameter σ^2 and so he formulates a sequential stopping rule whose boundary-crossing condition 'mimics' $(p\sigma^2 c^{-1})^{1/2}$. He proved this procedure to be asymptotically first order efficient and bounded-regret, but Finster (1983) later derived conditions for its second order efficiency and Nickerson (1987) improved upon those conditions. Actually, Finster (1983) addressed this problem with a more general loss-function and even allowed for the responses to be vectors. See also Sinha (1991) in this context. Chaturvedi (1987), on the other hand, puts forward a sequential procedure for point estimating β under the loss $L_n(a, \alpha, c, t) = a\{n^{-1}(\hat{\beta} - \beta)'(\mathbf{X'X})(\hat{\beta} - \beta)\}^{\alpha/2} + cn^t$ where $a > 0$, $\alpha > 0$, $c > 0$ and $t > 0$ are known constants. This generalizes Mukhopadhyay's (1974) results in some sense, and sharpens the regret-bounds obtained therein.

Once again, departing from the usual regression model (28.2.1), Konev and Pergamenshchikov (1997) consider what they call 'guaranteed mean-squared accuracy' estimation (which is nothing but bounded-risk point estimation under the squared error loss) of the parameters $\theta_1, \ldots, \theta_p$ in the model:

$$y_n = \Sigma_{j=1}^p \theta_j \phi_j(n) + \xi(n) \text{ for } n \geq 1; \; \xi(n) = \lambda\xi(n-1) + \epsilon_n, \qquad (28.3.7)$$

where $|\lambda| < 1$, ϕ_j's are known deterministic functions and ϵ_i's are i.i.d. $\sim F$ with mean 0 and variance 1. They come up with a sequential procedure and indicate its applications in control and identification of dynamic systems under random perturbations.

28.3.3   Hypotheses Testing

Chatterjee (1962) considered a sequential test for testing $H_0 : \boldsymbol{\beta} = \boldsymbol{\beta_0}$ vs. $H_1 : \boldsymbol{\beta} \neq \boldsymbol{\beta_0}$ with Gaussian errors and stochastic regressors which are independent of the errors. The test asymptotically guarantees a minimum power at any specified alternative. The corresponding test for deterministic regressors follows as a special case.

Sen (1981) proposed a *repeated significance* test (RST) based on a *general linear rank statistic* for testing $H_0 : \beta_1 = 0$ vs. $H_1 : \beta_1 > 0$ in the simple linear regression model with slope β_1 and intercept β_0 where the error-distribution is unknown and continuous almost everywhere. Kim (1994) considered another RST for the coefficients of a linear regression model in the context of sequential testing for the difference between two medical treatments whose effectiveness is influenced by prognostic factors. For more details on a RST, see Siegmund (1985).

Arghami and Billard (1981) introduced two *partial sequential tests* separately for the slope β_1 and the intercept β_0 of a simple linear regression. The basic idea is to eliminate the nuisance parameters by means of a suitable transformation to the original data and then run an SPRT using the transformed data. When testing $H_0 : \beta_1 = b_1$ vs. $H_1 : \beta_1 = b_2$, the nuisance parameters are β_0 and σ^2 (the common variance of the Gaussian errors). After transforming the data to get rid of them, the SPRT based on the new variables has an Operating Characteristic (OC) function free from β_0 and σ^2, and an ASN function free from β_0. Similar is the case when testing for β_0, except that β_1 is now a nuisance parameter.

Khan (1984) used his idea of *confidence sequences* to put forward sequential tests with power 1 for the parameters in simple linear regression.

28.4   SEQUENTIAL SHRINKAGE ESTIMATION IN REGRESSION

Ever since the famous James-Stein estimator was shown to dominate the sample-mean in terms of having a smaller risk under the squared-error loss while estimating the mean-vector $\boldsymbol{\mu}$ of a p-variate Gaussian distribution (with $p \geq 3$), a significant amount of research has gone into the area of *shrinkage estimation*, and doing it sequentially was a natural option to consider. Sclove (1968) explicitly derived a class of James-Stein type estimators dominating the least-squares estimator $\hat{\boldsymbol{\beta}}$ under the loss (28.3.6) with $\mathbf{A}$ replaced by $n^{-1}\mathbf{X'X}$. Nickerson (1987) explored the sequential versions of these estimators and obtained asymptotic (as $c \rightarrow 0$) second order risk-expansions for them. Comparing these second order risk-expansions with that of the least-squares estimator, he also provided the optimal choice for the shrinkage-factor for small c. Later Sriram and Bose (1988) provided a significant generalization of these procedures and simplified the derivation of the asymptotic risk-expansions, by considering vector-valued responses and stochastic regressors (independent of the errors). For more details in this regard, see Sinha (1991) and Mukhopadhyay (1991).

On a slightly different note, Kubokawa and Saleh (1990) considered sequential shrinkage estimation for the coefficient-matrix Γ in the following model: let $\mathbf{Y}_1, \mathbf{Y}_2, \ldots$ be independent random vectors with $\mathbf{Y}_i \sim N(\Gamma a_i, \Sigma)$ where a_i's are known $r \times 1$ vectors and $\Gamma_{p \times r}$ and $\Sigma_{p \times p}$ are unknown matrices. Under the loss-function $L(\hat{\Gamma}, \Gamma, c) = n^{-1} \operatorname{tr}\{(\hat{\Gamma} - \Gamma)\mathbf{A}_n \mathbf{A}_n' (\hat{\Gamma} - \Gamma)'\} + cn$, where c = the cost per unit sample and $\mathbf{A}_n$ is the matrix $\{a_1 \mid a_2 \mid \ldots \mid a_n\}$, they developed a sequential shrinkage estimator of Γ that exactly dominates its usual sequential estimator in each of the following two cases: (i) when Σ is of the form $diag\{\sigma_1, \ldots, \sigma_p\}$ and (ii) when Σ is totally unknown. In each case, they provide an asymptotic second order risk-expansion for their estimator.

28.5   BAYES SEQUENTIAL INFERENCE IN REGRESSION

In Bayesian inference, the unknown population-parameters (in this case β and Σ, the dispersion matrix of the errors) are considered random and are assumed to have certain *a priori* probability distributions. Depending on whether the parameters of these *a priori* distributions are again assumed to have distributions of their own or are estimated from the data, it is called a *hierarchical* Bayes or an *empirical* Bayes model. It is well-known that in Bayes sequential inference problems, once a Bayes stopping rule tells us to stop, the Bayes action taken at the stopped stage is *independent* of the stopping rule and is the same as what would have been its fixed sample-size counterpart. However, despite the fact that Bayes stopping rules exist under fairly general conditions, their exact determination is quite a daunting task. As a result, numerous ways of approximating Bayes rules are found in the literature. Berger (1985) and Ghosh (1991) are useful references in this context. One of them is the concept of an *asymptotically pointwise optimal* (APO) stopping rule introduced by Bickel and Yahav (1967, 1968, 1969a,b) which is nothing but approximating the Bayes stopping rules asymptotically as c (the cost per unit sample) tends to 0. See Ghosh and Hoekstra (1991) for an elaborate account. Bickel and Yahav developed several asymptotic optimality properties of the APO rules in the context of sequential estimation and hypothesis-testing for a fairly general class of distributions, including the one-parameter exponential family. For this special family, Woodroofe (1981) showed that the APO rules were *asymptotically nondeficient*—in the sense that the difference between the Bayes risk of a Bayes estimator under the optimal Bayes rule and that of the APO rule is $o(c)$ as $c \to 0$. Finster (1987) extended Woodroofe's results to a normal regression model.

Ghosh (1991) develops an APO sequential estimation procedure for β in the following hierarchical Bayes setup: conditional on $R = r$ and β, Y_i's are independent observations with $Y_i \sim N(z_i'\beta, r^{-1})$ where $Z_n Z_n' = (z_1 \mid \ldots \mid z_n)(z_1 \mid \ldots \mid z_n)'$ is assumed to be invertible for all n. Conditional on $M = m$ and $R = r$, $\beta \sim N(m1_{p\times p}, (\lambda r)^{-1}I_{p\times p})$. Marginally, M and R are independent with $M \sim g(m)$ for a certain known density $g(m)$, and $R \sim Gamma(a/2, b/2)$ for known constants a, b and λ. Under the loss $L(\hat{\beta}, \beta) = (\hat{\beta} - \beta)'Z_n Z_n'(\hat{\beta} - \beta)$, the APO stopping variable N_{APO} is just a variant of Mukhopadhyay's (1974) stopping variable. At the stopped stage, the Bayes estimator of β based on N_{APO} samples is used.

Finster (1987) considers an APO estimation rule in Zellner's economic regression model, and it is based on a single-stage (informative) prior. Hoekstra (1989) suggests an APO rule in a hierarchical Bayes setup starting with Finster's (1983) model and using an expansion of the above hierarchical structure mentioned in Ghosh (1991).

28.6 SEQUENTIAL INFERENCE IN STOCHASTIC REGRESSION MODELS

Let us now switch to the case with stochastic regressors which may be natural under many experimental circumstances. As mentioned earlier, Chatterjee (1959, 1962) considered stochastic regressors when developing sequential procedures for the construction of confidence sets with pre-specified contours for β or point estimation of β under the loss mentioned in Subsection 28.3.2. Finster (1983) also considered stochastic regressors.

Martinsek (1995), however, looked at the slope-estimation problem in simple linear regression with stochastic regressors from a slightly different viewpoint. He considered estimation of β (the slope) with a prescribed *proportional accuracy*. In other words, he wanted to ensure that $P(|\hat{\beta} - \beta| \le \rho \mid \beta|) = 1 - \gamma$ (or, at least, approximately equal to $1 - \gamma$) for two preassigned constants $\rho > 0$ and $\gamma \in (0, 1)$—a loss-function which is appropriate when β is either very close to 0 or very large in magnitude. Assuming that the regressors and the errors are independently distributed, with the regressor-distribution having finite $(6 + \delta)$-th moment for some $\delta > 0$ and the error-distribution having finite 4th moment, he devised a sequential procedure for this purpose and showed his procedure to be both asymptotically consistent and efficient as the degree of accuracy increases (i.e. $\rho \to 0$). He also obtains a central limit theorem for the associated stopping variable. In an attempt to generalize his results, Datta (1996) proposed sequential procedures for fixed-size confidence set estimation and fixed proportional accuracy estimation of β in a simple, as well as multiple, linear regression model with stochastic regressors (distributed independently of the errors) when (i) both the distributions are Gaussian and (ii) both the distributions are unknown satisfying certain moment-conditions. In each case, the asymptotic consistency and second order efficiency of the procedure are proved using nonlinear renewal theoretic techniques from Aras and Woodroofe (1993) and these theoretical results are supported by extensive simulation-studies. Etemadi, Sriram and Vidyashankar (1997) also considered stochastic regressors in their sequential estimation methodology for β in a simple linear regression model (without an intercept) as an application of their results on L_p-convergence of reciprocals of sample-means. They too consider fixed proportional accuracy estimation (calling it estimation under *relative squared error* loss) and establish the asymptotic second order efficiency of their procedure, in addition to providing an asymptotic second order risk-expansion, under the assumption of finite 4th moments for the regressors as well as the errors.

In estimating the parameters of a linear regression model with arbitrary noise, Goldenshluger and Polyak (1993) also considered stochastic regressors. Their regressors are zero-mean random vectors independent of the noise-variables which, however, are allowed to be nonzero-mean or correlated or even nonrandom. Their methodology is stochastic approximation as well as least-squares.

28.7 SEQUENTIAL INFERENCE IN INVERSE LINEAR REGRESSION AND ERRORS-IN-VARIABLES MODELS

Levy and Samaranayake (1988) proposed a sequential sampling scheme as a solution to the multiple-response extension of the univariate calibration problem. In doing so, they generalized the existing univariate calibration results of Perng and Tong (1974). The technique used is primarily based on the Chow and Robbins (1965) theory. Tahir (1989), on the other hand, derived an asymptotic second order expansion for the coverage-probability of a fixed-width confidence interval for the x-variable in an inverse linear regression model via a two-stage sampling scheme similar to Perng and Tong (1974).

Datta (1996) considered fixed-width interval estimation of the parameter β in the errors-in-variables model (with a *structural relationship*) described in Subsection 28.2.1. He used approximate $100(1 - \alpha)$ % confidence intervals for β with a prescribed semiwidth d, along the line of Gleser (1987). For his sequential procedure, once again he establishes asymptotic second order efficiency and provides a second order expasion for the coverage-probability using tools from Aras and Woodroofe (1993).

28.8 SOME MISCELLANEOUS TOPICS

In addition to the various categories of research mentioned in the earlier sections, there are a number of other interesting pieces of research relevant to the theme of this review-article. Due to space-constraints, we must resist the temptation of elaborating on their contents, and so we provide a short list of them here. The exact citations are available in the reference-list.

Ellingsen and Leathrum (1975) applied on-line *ridge regression* (biased estimation) to sequential estimation in order to come up with an improved estimation technique when the $(\mathbf{X'X})$ matrix is *ill-conditioned*. They treat the problem of a singular $(\mathbf{X'X})$ matrix by both biased estimation and the generalized inverse method. An algorithm for nonlinear parameter estimation in ill-conditioned systems is obtained by a parallel application of sequential biased and sequential least-squares estimation. Perl (1977) derived weak convergence results for a sequential regression algorithm that arises in the identification of non-linear memoryless systems and the adaptive design of moving average filters. He shows the algorithm to be weakly consistent if the system-input is a stationary sequence (in the wide sense) of order 4 that satisfies certain covariance and fourth cumulant conditions. Gyorfi (1980) uses a Robbins-Monro type stochastic approximation procedure $x_{n+1} = x_n - (n + 1)^{-1}(A_{n+1}x_n - y_{n+1})$ to solve the linear equation $Ax = y$ in a Hilbert space, where y_n and A_n are estimators such that their arithmetic means converge to y and A respectively. Eichhorn and Zacks (1981) suggest a Bayes sequential search procedure for an optimal dosage under the following model: assuming that $Y(x)$, the log-toxicity at dosage x, has a Gaussian distribution with mean $\alpha + \beta x$ and variance $\sigma^2(x)$, an *optimal dosage* ξ_γ is defined as the largest value of x for which, $P(Y(x) \leq \eta \mid x) \geq \gamma$ with η being a specified threshold. For each of the two cases (i) $\sigma^2(x) = \sigma^2$ for all x and (ii) $\sigma^2(x) = \sigma^2 x^2$ for all x, a Bayes sequential search procedure is devised by using a bivariate Gaussian prior distribution for the unknown parameters-vector (α, β), and it is proved to be ϵ-consistent for any given (α, β, σ). Hsu and Huang (1982) consider a sequential selection procedure to select a subset of random size which includes all 'good' regressor-variables in a regression model. Diaz, O'Reilly and Rincon-Gallardo (1983). introduce a set of sequential residuals for the multivariate linear regression model and show that they are with *known* distributions not involving the parameters of the regression model. Irle (1999) proposed estimating the conditional expectation $E(X_{n+1} \mid X_n = x)$ for a discrete-time Markov process by means of a sequential nearest-neighbor regression estimator.

REFERENCES

Albert, A. (1966). Fixed-size confidence ellipsoids for linear regression parameters, *Annals of Mathematical Statistics*, **37**, 1602–1630.

Aras, G. and Woodroofe, M. (1993). Asymptotic expansions for the moments of a randomly stopped average, *Annals of Statistics*, **21**, 503–519.

Arghami, N. R. and Billard, L. (1981). A sequential test for a normal mean with unknown variance, *Bulletin of the International Statistical Institute*, **43**, 113–115.

Berger, J. O. (1985). *Statistical Decision Theory and Bayesian Analysis*, Springer-Verlag, New York.

Bickel, P. and Yahav, J. (1967). Asymptotically pointwise optimal procedures in sequential analysis, In *Proceedings of the 5th Berkeley Symposium*, **V1**, 401–413.

Bickel, P. and Yahav, J. (1968). Asymptotically optimal Bayes and minimax procedure in sequential estimation, *Annals of Mathematical Statistics*, **39**,442-456.

Bickel, P. and Yahav, J. (1969a). An A.P.O. rule in sequential estimation with quadratic loss, *Annals of Mathematical Statistics*, **40**, 417–426.

Bickel, P. and Yahav, J. (1969b). Some contributions to the asymptotic theory of Bayes solutions, *Z. Wahr. Ver. und Gebiete*, **11**, 257–276.

Bishop, T. A. (1978). A Stein two-sample procedure for the general linear model with unequal error-variances, *Communications in Statistics—Theory and Methods*, **7**, 495–507.

Chatterjee, S. K. (1959a). On an extension of Stein's two-sample procedure to the multinormal problem, *Calcutta Statistical Association Bulletin*, **8**, 121–148.

Chatterjee, S. K. (1959b). Some further results on the multinormal extension of Stein's two-sample procedure, *Calcutta Statistical Association Bulletin*, **9**, 20–28.

Chatterjee, S. K. (1962). Sequential inference procedures of Stein's type for a class of multivariate regression problems, *Annals of Mathematical Statistics*, **33**, 1039-1064.

Chatterjee, S. K. (1991). Two-stage and multistage procedures, In *Handbook of Sequential Analysis*, Marcel Dekker, New York.

Chaturvedi, A. (1987). Sequential point estimation of regression parameters in a linear model, *Annals of the Institute of Statistical Mathematics*, **39(A)**, 55–67.

Chaturvedi, A. (1988). On a modified sequential procedure to construct confidence regions for regression parameters, *Journal of the Indian Society for Agricultural Statistics*, **XL**, **2**, 111–115.

Chow, Y. S. and Robbins, H. (1965). On the asymptotic theory of fixed-width sequential confidence intervals for the mean, *Annals of Mathematical Statistics*, **36**, 457–462.

Chukwu, W. I. E. (1985). The asymptotic theory of fixed-width sequential confidence bounds for linear regression models, *Statistica*, **XLV**, **2**, 265–270.

Coleman, D. A. (1995). A fixed-width interval for $1/\beta$ in simple linear regression, *Journal of Statistical Planning and Inference*, **44**, 291–312.

Datta, S. (1996). Sequential fixed-precision estimation in stochastic linear regression and errors-in-variables models, *Technical Report*, Department of Statistics, University of Michigan, Ann Arbor.

Dantzig, R. B. (1940). On the non-existence of tests of Student's hypothesis having power functions independent of σ, *Annals of Mathematical Statistics*, **11**, 186–192.

Diaz, J., O'Reilly, F. J. and Rincon-Gallardo, S.(1983). A set of independent sequential residuals for the multivariate regression model, *Journal of Statistical Planning and Inference*, **8**, 21–25.

Eichhorn, B. H. and Zacks, S. (1981). Bayes sequential search of an optimal dosage: linear regression with both parameters unknown, *Communications in Statistics—Theory and Methods*, **10**, 931–953.

Ellingsen, W. R. and Leathrum, J. F. (1975). On-line ridge regression: sequential biased estimation for nonorthogonal problems, *Journal of Statistical Computing and Simulation*, **3**, 249–264.

Etemadi, N., Sriram, T. N. and Vidyashankar, A. (1997). L_p-convergence of reciprocals of sample-means with applications to sequential estimation in linear regression, *Journal of Statistical Planning and Inference*, **46**.

Finster, M. (1983). A frequentistic approach to sequential estimation in the general linear model, *Journal of the American Statistical association*, **78**, 403–407.

Finster, M. (1985). Estimation in the general linear model when the accuracy is specified before data collection, *Annals of Statistics*, **13**, 663–675.

Finster, M. (1987). A frequentist and Bayesian analysis of Zellner's economic regression model under an informative prior, *Sequential Analysis*, **6**, 139–153.

Ghosh, M. (1991). Hierarchical and empirical Bayes sequential estimation, In *Handbook of Sequential Analysis*, pp. 441–458, Marcel Dekker, New York.

Ghosh, B. K. and Sen, P. K. (Eds.) (1991). *Handbook of Sequential Analysis*, Marcel Dekker, New York.

Ghosh, M. and Hoekstra, R. M. (1991). Asymptotically pointwise optimal stopping rules, In *Handbook of Sequential Analysis*, Marcel Dekker, New York.

Ghosh, M. and Mukhopadhyay, N. (1981) Consistency and asymptotic efficiency of two-stage and sequential estimation procedures, *Sankhyā, Series A*, **43**, 220–227.

Ghosh, M., Mukhopadhyay, N. and Sen, P. K. (1997). *Sequential Estimation*, John Wiley & Sons, New York.

Gleser, L. J. (1965). On the asymptotic theory of fixed-size confidence bounds for linear regression parameters, *Annals of Mathematical Statistics*, **36**, 463–467.

Gleser, L. J (1966). Correction to the above, *Annals of Mathematical statistics*, **37**, 1053–1055.

Gleser, L. J. (1987). Confidence intervals for the slope in a linear errors-in-variables regression model, In *Advances in Multivariate Statistical Analysis* (Ed., K. Gupta), pp. 85–109, D. Reidel Publishing Company, Dordrecht, The Netherlands.

Goldenshluger, A. V. and Polyak, B. T. (1993). Estimation of regression parameters with arbitrary noise, *Mathematical Methods of Statistics*, **2(1)**, 18–29.

Gyorfi, L. (1980). Stochastic approximation from ergodic sample for linear regression, *Zeitschrift fur Wahrscheinlichkeitstheorie und verwandte Gebiete*, **54**, 47–55.

Hall, P. (1981). Asymptotic theory of triple sampling for estimation of a mean, *Annals of Statistics*, **9**, 1229–1238,

Hall, P. (1983). Sequential estimation saving sampling operations, *Journal of the Royal Statistical Society, Series B*, **45**, 219–223.

Hoekstra, M. (1989). A.P.O. stopping rules in multiparameter estimation, *Ph.D. Dissertation*, University of Florida, Gainesville, FL.

Hsu, T.-A. and Huang, D.-Y. (1982). Some sequential selection procedures for good regression models, *Communications in Statistics—Theory and Methods*, **11**, 411–421.

Irle, A. (1999). On sequential nearest neighbor regression estimation, *Sequential Analysis*, **18**, 33–42.

Khan, R. A. (1984). On confidence sequences for the parameters of a linear regression, *Sequential Analysis*, **3**, 23–37.

Kim, H.-J. (1994). A repeated significance test in a linear model, *Sequential Analysis*, **13**, 113–126.

Konev, V. V. and Pergamenshchikov, S. M. (1997). Guaranteed estimation of linear regression parameters under dependent disturbances, *Automation and Remote Control*, **58**, 213–223.

Kubokawa, T. and Saleh, A. K. Md. E. (1990). Sequential shrinkage estimation for a coefficient matrix in a multivariate regression model, *Journal of the Japan Statistical Society*, **20**, 33–42.

Lai, T. L. (1986). Asymptotically efficient adaptive control in stochastic regression models, *Advances in Applied Mathematics*, **7**, 23–45.

Levy, M. and Samaranayake, V. A. (1988). Fixed width interval estimation for the multiple response calibration problem, *Sequential Analysis*, **7**, 283–306.

Lohr, S. L. (1990). Accurate multivariate estimation using triple sampling, *Annals of Statistics*, **18**, 1615–1633.

Lohr, S. L. (1993a). Three-stage accurate estimation in the general linear model, *Journal of Statistical Planning and Inference*, **34**, 317–331.

Lohr, S. L. (1993b). Two-stage accurate estimation in the general linear model, *Sequential Analysis*, **12**, 1–23.

Martinsek, A. T. (1989). Sequential estimation in regression models using analogues of trimmed means, *Annals of the Institute of Statistical Mathematics*, **41**, 521–540.

Martinsek, A. T. (1995). Estimating a slope-parameter in regression with prescribed proportional accuracy, *Statistics and Decisions*, **13**, 363–377.

Mukhopadhyay, N. (1974). Sequential estimation of regression parameters in Gauss-Markoff setup, *Journal of the Indian Statistical Association*, **12**, 39–43.

Mukhopadhyay, N. (1980). A consistent and asymptotically efficient two-stage procedure to construct fixed-width confidence intervals for the mean, *Metrika*, **27**, 281–284.

Mukhopadhyay, N. (1990). Some properties of a three-stage procedure with applications in sequential analysis, *Sankhyā, Series A*, **52**, 218–231.

Mukhopadhyay, N. (1991). Parametric sequential point estimation, In *Handbook of Sequential Analysis* (Eds., B. K. Shah and P. K. Sen), pp. 245–267, Marcel Dekker, New York.

Mukhopadhyay, N. (1996). An alternative formulation of accelerated procedures with applications, *Sequential Analysis*, **15**, 253–269.

Mukhopadhyay, N. and Abid, A. D. (1986). On fixed-size confidence regions for the regression parameters, *Metron*, **31**, 297–306.

Mukhopadhyay, N. and Datta, S. (1995). On fine-tuning a purely sequential procedure and associated second-order properties, *Sankhyā, Series A*, **57**, 100–117.

Mukhopadhyay, N. and Solanky, T. K. S. (1991). Second order properties of accelerated sequential stopping times with applications in sequential estimation, *Sequential Analysis*, **10**, 99–123.

Murali, T. and Rao, B. V. (1987). A sequential regression based algorithm for estimating sinusoids in white noise, *Computer and Electrical Engineering*, **13**, 1–6.

Nickerson, D. (1987). Sequential shrinkage estimation of linear regression parameters, *Sequential Analysis*, **6**, 93–117.

Perl, J. (1977). Weak convergence results for sequential regression in memoryless systems, *International Journal of Systems Science*, **8**, 1243–1247.

Perng, S. K. and Tong, Y. L. (1974). A sequential solution to the inverse regression problem, *Annals of Statistics*, **2**, 535–539.

Rahbar, M. H. (1995). Sequential fixed-width confidence intervals for regression parameters from censored data with a discrete covariate, *Sequential Analysis*, **14**, 143–156.

Sclove, S. L. (1968). Improved estimators for coefficients of linear regression, *Journal of the American Statistical Association*, **63**, 596–606.

Sen, P. K. (1981). *Sequential Nonparametrics: Invariance Principles and Statistical Inference*, John Wiley & Sons, New York.

Siegmund, D. (1985). *Sequential Analysis: Tests and Confidence Intervals*, Springer-Verlag, New York.

Sinha, B. K. (1991). Multivariate problems, In *Handbook of Sequential Analysis* (Eds., B. K. Ghosh and P. K. Sen), Marcel Dekker, New York.

Sriram, T. N. and Bose, A. (1988). Sequential shrinkage estimation in the general linear model, *Sequential Analysis*, **7**, 149–163.

Srivastava, M. S. (1967). On fixed-width confidence bounds for regression and the mean vector, *Journal of the Royal Statistical Society, Series B*, **29**, 132–140.

Srivastava, M. S. (1971). On fixed-width confidence bounds for regression parameters, *Annals of Mathematical Statistics*, **42**, 1403–1411.

Stein, C. (1945). A two-sample test for a linear hypothesis whose power is independent of the variance, *Annals of Mathematical Statistics*, **16**, 243–258.

Timofeev, A. V. (1991). Non-asymptotic solution of confidence-estimation parameter task of a non-linear regression by means of sequential analysis, *Problems of Control and Information Theory*, **20(5)**, 341–351.

Tahir, M. (1989). Asymptotic expansions for sequential confidence levels in inverse linear regression, *Communications in Statistics—Theory and Methods*, **18**, 4501–4509.

Wald, A. (1947). *Sequential Analysis*, John Wiley & Sons, New York.

Weiss, L. and Wolfowitz, J. (1974). Maximum probability estimators and related topics, *Lecture Notes in Mathematics*, **424**, Springer-Verlag, New York.

Welsh, A. H. (1987). The trimmed mean in the linear model, *Annals of Statistics*, **15**, 20–45.

Wilcox, R. R. (1985). On a Stein-type two-stage procedure for the general linear model, *British Journal of Mathematical and Statistical Psychology*, **38**, 222–226.

Woodroofe, M. (1981). A.P.O. rules are asymptotically non-deficient for estimation with squared error loss, *Z. Wahr. Ver. und Gebiete*, **58**, 331–341.

CHAPTER 29

BAYESIAN INFERENCE FOR A CHANGE-POINT IN NONLINEAR MODELING

VENKATA K. JANDHYALA

Washington State University, Pullman, WA

JAMAL A. ALSALEH

Kuwait University, Kuwait

Abstract: Under the setting of a nonlinear model with normal additive errors, the nonlinear change-point problem is formulated and inferential procedures are developed adapting the Bayesian approach. Assuming an exponential type regressor, numerical methods are implemented for approximating the marginal posterior densities. Calculations are carried out for approximating these posterior densities by means of the Gibbs sampler. The outcome of this Bayesian analysis are the calculation and graphical display of posterior densities of the parameters of interest. Results of simulation studies are presented.

Keywords and phrases: Nonlinear regression, change-point, Bayesian analysis, Gibbs sampler, simulation

29.1 INTRODUCTION

Nonlinear models arise in a variety of applications. With the rapid growth of software capacity and desk-top computing power, nonlinear methods have become more accessible to practitioners as well as researchers for the modeling and analysis of statistical data. As applications of these nonlinear

451

procedures become more common, the need for their careful use becomes that much more important. Several of the phenomena in areas such as engineering, biology, medicine, business and economics are known to be inherently nonlinear. For example, the effects of pollutants on crop yield, and the growths of trees and animals are best explained by nonlinear models.

One encounters quite commonly time-ordered data for the purposes of modeling and analysis. The models explaining such data are invariably dynamic in nature. Such dynamic effects may or may not influence the parameters of the underlying model. When dynamic effects influence the parameters also, then, model fitting and diagnostics become much more complicated even when the parameters are linear in the model. Change-point modeling is one of the ways by which dynamic effects on the parameters of a model may be incorporated. In a model with a single change-point, the parameters remain same for all the observations up to an unknown change-point. Subsequent to the change-point, the observations are explained by a different set of parameters. Inferential problems associated with a change-point model are both testing for the occurrence of an unknown change-point as well as estimation of the unknown change-point and the parameters. While both of these inferential problems are inherently complicated, there has been substantial progress in the literature for the case of linear change-point models. Some recent references in the area include Smith and Cook (1980) and Carlin, Gelfand and Smith (1992) on the Bayesian approach and Jandhyala and MacNeill (1989, 1991, 1997) and Jandhyala and Minogue (1993) on the Bayes-type approach. For advances on likelihood based methods, one may refer to Worsley (1986) and Kim and Siegmund (1989). It should be noted that the Bayes-type approach was first introduced by Chernoff and Zacks (1964) for deriving change-detection statistics.

To our knowledge, the change-point problem has not been considered in the literature for the case of nonlinear models. Our goal in this paper is to formulate the nonlinear change-point problem and then develop inferential procedures adapting the Bayesian approach. This work, thus, is a first attempt at developing inferential methods for nonlinear change-point models. In the Bayesian approach, one is concerned with the derivation of the marginal posterior densities for the parameters of interest. Often, there are no closed form expressions for these posterior densities and many times these are not easily evaluated by numerical integration. In the case of nonlinear change-point models, the marginal posterior densities are quite involved and are beyond direct numerical computation.

Recently, Markov Chain Monte Carlo methods such as the Gibbs sampler have been found to be extremely useful for evaluating Bayesian posterior densities. In this paper, we approximate the marginal posterior densities of the model parameters including that of the unknown change-point, by means of the Gibbs sampler iterative procedure. The outcome of the analysis is the computation and graphical display of marginal posterior densities of the parameters of interest.

The Gibbs sampler procedure requires that all of the full conditional distributions be available. The method, then, evolves by simulating observations from these full conditional distributions based on a Markovian updating scheme. The method is easily implemented when the full conditional distributions are from well known families. However, when the full conditional densities are not from well known families, implementation of the Gibbs sampler procedure becomes much more complicated. In the nonlinear change-point analysis, one unavoidably encounters such difficulties. We, however, implement methods in this paper for simulating observations from general discrete as well as continuous random variables with no specific families attached to their respective probability functions and densities. This enables us to implement the Gibbs sampling procedure for the nonlinear change-point analysis.

The nonlinear model that we consider involves exponential type regressor functions. Simulations suggest that the inferential procedure along with the Gibbs sampler perform extremely well for the nonlinear change-point models.

29.2 GIBBS SAMPLER

The Gibbs sampler is an iterative Monte Carlo integration method which proceeds by a Markovian updating scheme. It was introduced by Geman and Geman (1984) in the context of image processing. It is essentially a modification of the Metropolis algorithm, see Metropolis et al. (1953). The Metropolis algorithm has been used in many application areas including pattern analysis, image restoration, and in the implementation of simulated annealing, see also Hastings (1970).

In the statistical frame work, a closely related technique is the 'data augmentation' algorithm of Tanner and Wong (1987), who adopted a substitution sampling approach. The data augmentation problem arises naturally in missing data problems. Data augmentation as such refers to a scheme closely related to the Gibbs sampler where augmentation of the given data facilitates data analysis. Subsequently, the Gibbs sampler was proposed as a general method for Bayesian calculations by Gelfand and Smith (1990). They gave an empirical illustration of the Gibbs sampler and other sampling based schemes as alternatives for the calculation of marginal probability distributions. In particular, the relevance of these approaches is in the calculation of Bayesian posterior densities for a variety of structured models. The method has already been successfully applied for the analysis of linear variance components models by Gelfand *et al.* (1990).

Carlin and Polson (1991) used the Gibbs sampler for evaluating the influence diagnostics in a parametric setting. Their conclusion is that the influence diagnostics obtained, performed well in flagging an aberrant subset of the data, exemplified in the cases of a two-stage linear model, a hierarchical model, and a nonlinear Michaelis-Menten model. The application of their influence measure to nonlinear models is based on a non-informative prior for the nuisance parameters. Zeger and Karim (1991) considered the Gibbs sampler for estimating parameters in a generalized linear model with Gaussian random effects. They focused on the logistic Gaussian case because it poses some numerical difficulties.

Gelfand *et al.* (1992) introduced methods of using Gibbs sampler in constrained parameter and truncated data problems. Carlin *et al.* (1992) provided solutions to problems of multivariate state-space modeling using the Gibbs sampler. In their modeling, they allowed for the possibility of nonnormal errors as well as nonlinear functionals in the state equations. The methodology, thus provides a general strategy for computing marginal densities of unknown parameters in the state-space modeling. Gelman and Rubin (1992) illustrated the use of Gibbs sampler for the analysis of random effect mixture models and then applied it to analyze measurements of reaction times of normal and schizophrenic patients.

Carlin, Gelfand and Smith (1992) first introduced the Gibbs sampler as an alternative for the calculation of marginal posterior densities in change-point problems. Their study mainly considers hierarchical models and in particular the changing Poisson and linear models. Our study extends their work to the case of nonlinear change-point models.

More recently, Ingrassia (1994) derived bounds on the spectral gap for the transition matrices associated with the Gibbs sampler and the Metropolis algorithm. Furthermore, it was shown that the random updating dynamics of sites based on the Gibbs sampler and the Metropolis algorithm have the same rate of convergence. Raghunathan and Grizzle (1995) developed a survey design and analysis for data with imputations using the Gibbs sampler. Rosenthal (1995) considered the question of the number of iterations required for convergence when applying the Gibbs sampler.

We shall now give a brief introduction to the Gibbs sampler. Let $V_1, \cdots, V_p$ be random variables such that their full conditional distributions can be determined. It may or may not be that these full conditional distributions belong to well known families. We shall assume for the present that it is possible to generate random samples from these full conditional distributions. Denote for each random variable V_k, its full conditional density by

$$h_k(v_k|v_1, \cdots, v_{k-1}, v_{k+1}, \cdots, v_p) . \qquad (29.2.1)$$

Now, given an arbitrary set of starting constants $v_1^{(0)}, \cdots, v_p^{(0)}$ respectively for $V_1, \cdots, V_p$, we implement the following steps:

$$(1): \quad \text{draw } v_1^{(1)} \text{ from } h_1(v_1|v_2^{(0)}, \cdots, v_p^{(0)})$$

$$(2): \quad \text{draw } v_2^{(1)} \text{ from } h_2(v_2|v_1^{(1)}, v_3^{(0)}, \cdots, v_p^{(0)})$$

$$(29.2.2)$$

$$\vdots \quad \vdots$$

$$(p): \quad \text{draw } v_p^{(1)} \text{ from } h_p(v_p|v_1^{(1)}, v_2^{(1)}, \cdots, v_{p-1}^{(1)}) .$$

This completes one cycle. Next, we repeat the above cycle with $v_1^{(1)}, \cdots, v_p^{(1)}$ as the starting set of constants. After t such cycles, we obtain $(v_1^{(t)}, \cdots, v_p^{(t)})$. Geman and Geman (1984) showed that under mild conditions, as $t \to \infty$, the p-tuple $(v_1^{(t)}, \cdots, v_p^{(t)})$ converges in distribution to a random observation from $g(v_1, \cdots, v_p)$. Thus, if we repeat the process m times, we obtain m independent observations from $(V_1, \cdots, V_p)$:

$$(v_{1j}^{(t)}, \cdots, v_{pj}^{(t)}), \quad j = 1, \cdots, m . \qquad (29.2.3)$$

We may then use the above pseudo-random data to estimate the marginal densities. Gelfand and Smith (1990) recommend a density estimate of the form:

$$g_k(v_k) = \frac{1}{m} \sum_{j=1}^{m} h_k(v_k|v_{ij}^{(t)}, \cdots, v_{k-1j}^{(t)}, v_{k+1j}^{(t)}, \cdots, v_{pj}^{(t)}). \qquad (29.2.4)$$

We consider the above density estimate to be a suitable choice for our study. For more discussion on the density estimation, see Gelman and Rubin (1992) and Carlin, Gelfand and Smith (1992).

Finally, note that complete implementation of the Gibbs sampler requires determination of suitable values for t and m. Both Gelman and Rubin (1992) and Raftery and Lewis (1992) suggested approaches for choosing t and m. For example Carlin, Gelfand and Smith (1992) chose $t = 50$ and

$m = 100$. For our nonlinear models, we found their choices to be inadequate. After much experimentation, we were satisfied with the choice of $t = 100$ and $m = 150$. In the sequel, these will be our choices whenever we implement the Gibbs sampler.

29.3 BAYESIAN PRELIMINARIES AND THE NONLINEAR CHANGE-POINT MODEL

Bayesian analysis for inference regarding change-point problems started with the work of Chernoff and Zacks (1964). Smith (1975) presents a Bayesian formulation of the change-point problem for sequences of independent random variables. In the linear models frame work, the problem was considered among others by Chin Choy and Broemeling (1980), Smith and Cook (1980) and Moen et al. (1985). One may also refer to Booth and Smith (1982) for related Bayesian approaches to the change-point problem.

In the above references, the models are all linear in the parameters. Our interest, however, is in the Bayesian analysis of nonlinear change-point models. In the Bayesian analysis, once subjective opinion or further information about a parameter can be expressed as a suitable prior, then the posterior distribution of the parameter given the data may be obtained by applying the Bayes Theorem. As a natural estimate of the unknown parameter, one usually considers the posterior mode, it being the most probable value of the parameter given the data. Furthermore, a prior is said to be a conjugate prior when the prior and the posterior have the same functional form for their densities. When conjugate priors are available, then marginal posterior densities may be found analytically rather than by numerical integration. However, in complicated models such as nonlinear models, conjugate priors are not available and numerical methods must be used for finding the marginal posterior densities.

We introduce our model and the analysis in the frame work of an exponential type nonlinear regressor function with additive errors. This is among the more commonly encountered nonlinear models in many application areas.

Let $Y_1, \cdots, Y_n$ be a sequence of time-ordered observations observed at time-points $t_1, \cdots, t_n$ respectively. One may assume without loss of generality the entire sampling interval to be [0,1]. Then, we have $0 \leq t_1 < t_2 < \cdots < t_n \leq 1$. Let $Y_1, \cdots, Y_n$ satisfy the nonlinear change-point model given by:

$$
\begin{aligned}
Y_i &= \theta_0 \, e^{\theta_1 t_i} + \epsilon_i, & i = 1, \cdots, k \\
&= \theta_0 \, e^{\theta_2 t_i} + \epsilon_i, & i = k+1, \cdots, n
\end{aligned}
\qquad (29.3.5)
$$

where the parameters $\theta_0, \theta_1, \theta_2$ are unknown and $k\epsilon\{1, \cdots, n\}$ is unknown. In the literature, k is called the change-point. When $k = n$, the model (29.3.5) is said to be time homogeneous in its parameters. As for the errors, we assume $\epsilon_1, \cdots, \epsilon_n$ to be independent and distributed normally each with mean 0 and variance σ^2.

In the sequel, we let the parameter space for the nonlinear parameter to be [0,1]. As long as the original parameter space is finite, one may easily achieve this by a suitable reparameterization. Thus, we assume that $\theta_1\epsilon[0, 1]$ as well as $\theta_2\epsilon[0, 1]$. For convenience, we also let $\theta_0\epsilon[0, 1]$.

Given the observations, the primary objective is to draw inferences regarding the unknown change-point k and the parameters θ_0, θ_1 and θ_2. It should be noted that at the change-point k in the model (29.3.5), we could have allowed a change in θ_0 as well. However, since θ_0 is a parameter which is linear in the model, for simplicity, we let θ_0 remain time homogeneous. In all, the parameters associated with the model 29.3.5) are $k, \theta_0, \theta_1, \theta_2$ and σ^2. We present our inference procedures in the form of the following two different cases.

(i): We let $k, \theta_0, \theta_1, \theta_2$ to be unknown and assume σ^2 to be known.

(ii): We let $k, \theta_1, \theta_2, \sigma^2$ to be unknown and assume θ_0 to be known.

Initially, prior to case (i), we assumed θ_0 also to be known and carried out inferences upon simulated data based on selected choices of k, θ_1 and θ_2. We subsequently let θ_0 to be unknown and repeated the same inferences under case (i). A comparison of the two inferences revealed the assumption of 'θ_0 known' to have negligible effect on the accuracy of both the change-point estimate as well as the estimates of the nonlinear parameters. Thus, we are satisfied in assuming θ_0 to be known under case (ii). It was of paramount importance for us to be able to carry out inferences by letting σ^2 to be unknown, and this has been achieved through case (ii).

29.4 BAYESIAN INFERENTIAL METHODS

Here, we first begin with case (i) and present our inferential method in sufficient detail. A less detailed version of the method for case (ii) will be presented subsequently. The specific implementations and the corresponding outputs for both the cases will be presented in the next section.

Case (i): Here, we let $k, \theta_0, \theta_1, \theta_2$ to be unknown and assume σ^2 to be known. Under this case, as given by model (29.3.5), we have:

$$Y_i \sim N(\theta_0 e^{\theta_1 t_i}, \sigma^2) , \quad i = 1, \cdots, k$$

and

$$Y_i \sim N(\theta_0 e^{\theta_2 t_i}, \sigma^2) , \qquad i = k+1, \cdots, n \qquad (29.4.6)$$

where $k, \theta_0, \theta_1, \theta_2$ are unknown and σ^2 is known.

Let $\rho(k), \xi_0(\theta_0), \xi_1(\theta_1)$ and $\xi_2(\theta_2)$ represent independent priors for the parameters k, θ_0, θ_1 and θ_2 respectively and let $\boldsymbol{y} = (y_1, \cdots, y_n)$ represent given data for $Y_1, \cdots, Y_n$. Furthermore let $p(k|\boldsymbol{y})$ be the marginal posterior probability function of the change-point k and let $u_0(\theta_0|\boldsymbol{y}), u_1(\theta_1|\boldsymbol{y})$ and $u_2(\theta_2|\boldsymbol{y})$ represent the respective marginal posterior density functions for θ_0, θ_1 and θ_2. Then, from (29.4.6) and the Bayes' theorem, we have:

$$p(k|\boldsymbol{y})$$
$$= \frac{\int \int \int \exp\left[-\frac{1}{2}\sigma^2\left\{\sum_{i=1}^k (y_i - \theta_0 e^{\theta_1 t_i})^2 + \sum_{i=k+1}^n (y_i - \theta_0 e^{\theta_2 t_i})^2\right\}\right] \times \rho(k)\xi_0(\theta_0)\xi_1(\theta_1)\xi_2(\theta_2)d\theta_0 d\theta_1 d\theta_2}{\sum_{k=1}^n \int \int \int \exp\left[-\frac{1}{2\sigma^2}\left\{\sum_{i=1}^k (y_i - \theta_0 e^{\theta_1 t_i})^2 + \sum_{i=k+1}^n (y_i - \theta_0 e^{\theta_2 t_i})^2\right\}\right] \times \rho(k)\xi_0(\theta_0)\xi_1(\theta_1)\xi_2(\theta_2)d\theta_0 d\theta_1 d\theta_2} .$$

$$(29.4.7)$$

The marginal posterior densities $u_0(\theta_0|\boldsymbol{y}), u_1(\theta_1|\boldsymbol{y})$ and $u_2(\theta_2|\boldsymbol{y})$ may also be derived similarly. The form of $p(k|\boldsymbol{y})$ above suggests clearly that all of the marginal posteriors are quite complicated for computation. There are no conjugate choices for priors and neither are there any easier choices for $\rho(k), \xi_0(\theta_0), \xi_1(\theta_1)$ and $\xi_2(\theta_2)$ that easily enable us to compute these marginal posteriors. We are thus motivated to implement the Gibbs sampler for the calculation of the required marginal posteriors.

The Gibbs sampler, however, requires the full conditional probability distributions of all the parameters. Accordingly, let $q(k|\theta_0, \theta_1, \theta_2, \boldsymbol{y})$ be the full conditional probability function of the change-point k and, let $v_0(\theta_0|k, \theta_1, \theta_2, \boldsymbol{y}), v_1(\theta_1|k, \theta_0, \theta_2, \boldsymbol{y})$ and $v_2(\theta_2|k, \theta_0, \theta_1, \boldsymbol{y})$ be the full conditional density functions of θ_0, θ_1 and θ_2 respectively.

Again, implementing the Bayes' rule, the full conditionals are found to be as given below:

$$q(k|\theta_0, \theta_1, \theta_2, \boldsymbol{y})$$
$$= \frac{\exp\left[-\frac{1}{2\sigma^2}\left\{\sum_{i=1}^k (y_i - \theta_0 e^{\theta_1 t_i})^2 + \sum_{i=k+1}^n (y_i - \theta_0 e^{\theta_2 t_i})^2\right\}\right]\rho(k)}{\sum_{k=1}^n \left[\exp\left[-\frac{1}{2\sigma^2}\left\{\sum_{i=1}^k (y_i - \theta_0 e^{\theta_1 t_i})^2 + \sum_{i=k+1}^n (y_i - \theta_0 e^{\theta_2 t_i})^2\right\}\right]\rho(k)\right]}$$

$$(29.4.8)$$

$$v_0(\theta_0|k, \theta_1, \theta_2, \boldsymbol{y})$$

$$= \frac{\exp\left[-\frac{1}{2\sigma^2}\left\{\sum_{i=1}^{k}(y_i - \theta_0 e^{\theta_1 t_i})^2 + \sum_{i=k+1}^{n}(y_i - \theta_0 e^{\theta_2 t_i})^2\right\}\right]\xi_0(\theta_0)}{\int \exp\left[-\frac{1}{2\sigma^2}\left\{\sum_{i=1}^{k}(y_i - \theta_0 e^{\theta_1 t_i})^2 + \sum_{i=k+1}^{n}(y_i - \theta_0 e^{\theta_2 t_i})^2\right\}\right]\xi_0(\theta_0)d\theta_0}$$

$$(29.4.9)$$

$$v_1(\theta_1|k, \theta_0, \theta_2, \boldsymbol{y}) = \frac{\exp\left[-\frac{1}{2\sigma^2}\left\{\sum_{i=1}^{k}(y_i - \theta_0 e^{\theta_1 t_i})^2\right\}\right]\xi_1(\theta_1)}{\int \exp\left[-\frac{1}{2\sigma^2}\left\{\sum_{i=1}^{k}(y_i - \theta_0 e^{\theta_1 t_i})^2\right\}\right]\xi_1(\theta_1)d\theta_1}$$

$$(29.4.10)$$

$$v_2(\theta_2|k, \theta_0, \theta_1, \boldsymbol{y}) = \frac{\exp\left[-\frac{1}{2\sigma^2}\left\{\sum_{i=k+1}^{n}(y_i - \theta_0 e^{\theta_2 t_i})^2\right\}\right]\xi_2(\theta_2)}{\int \exp\left[-\frac{1}{2\sigma^2}\left\{\sum_{i=k+1}^{n}(y_i - \theta_0 e^{\theta_2 t_i})^2\right\}\right]\xi_2(\theta_2)d\theta_2}.$$

$$(29.4.11)$$

With these full conditionals, one may implement the Gibbs sampler procedure as described in Section 29.2. The implementation, however, is not immediately straightforward. It requires drawing random samples from the full conditionals given by (29.4.8)–(29.4.11) respectively. These conditionals, however, do not belong to any of the well known parametric families of distributions. Consequently, generation of pseudo random observations from these distributions is not so easy. The methods by which we accomplish this task are briefly elaborated in the next section. We shall now move on to the next case.

Case (ii): Here we let $k, \theta_1, \theta_2, \sigma^2$ to be unknown and assume θ_0 to be known.

As in the previous case, here also, we assume independent priors for the unknown parameters k, θ_1, θ_2 and σ^2. Specifically, let $\rho(k), \xi_1(\theta_1), \xi_2(\theta_2)$ and $\psi(\sigma^2)$ be the independent priors. Let the full conditionals for the parameters k, θ_1, θ_2 and σ^2 be denoted by $r(k|\theta_1, \theta_2, \sigma^2, \boldsymbol{y}), w_1(\theta_1|k, \theta_2, \sigma^2, \boldsymbol{y})$, $w_2(\theta_2|k, \theta_1, \sigma^2, \boldsymbol{y})$ and $t(\sigma^2|k, \theta_1, \theta_2, \boldsymbol{y})$ respectively. The derivation of these full conditionals is quite analogous to the previous case. Hence, we present only the form of the full conditional for the variance σ^2:

$t(\sigma^2 | k, \theta_1, \theta_2, \boldsymbol{y})$

$$= \frac{\exp\left[-\frac{1}{2\sigma^2}\left\{\sum\limits_{i=1}^{k}(y_i - \theta_0 e^{\theta_1 t_i})^2 + \sum\limits_{i=k+1}^{n}(y_i - \theta_0 e^{\theta_2 t_i})^2\right\}\right]\psi(\sigma^2)}{\int \exp\left[-\frac{1}{2\sigma^2}\left\{\sum\limits_{i=1}^{k}(y_i - \theta_0 e^{\theta_1 t_i})^2 + \sum\limits_{i=k+1}^{n}(y_i - \theta_0 e^{\theta_2 t_i})^2\right\}\right]\psi(\sigma^2)d\sigma^2}.$$

$$(29.4.12)$$

The full conditionals all being available, we are again ready to implement the Gibbs sampler. The specifics of the implementation are presented in the next section.

29.5 IMPLEMENTATION AND THE RESULTS

Implementation of the Gibbs sampler for both the cases requires the drawing of random samples from their respective full conditional distributions. However, these full conditionals do not belong to any of the well known parametric families of distributions. Hence, drawing random samples from these full conditionals requires careful attention to the choice of methods and their implementation. Thus, we first briefly refer to the methods by which we accomplish this task.

First note that the full conditionals $q(k|\theta_1, \theta_1, \theta_2, \boldsymbol{y})$ and $r(k|\theta_1, \theta_2, \sigma^2, \boldsymbol{y})$ are both discrete with their support set being $\{1, \cdots, n\}$. The remaining conditionals are all continuous. Thus, we require methods of drawing pseudo random samples from a general discrete probability distribution as well as a general continuous probability distribution. We describe below our approach for both the cases, beginning first with the discrete case.

In a general framework, let $p(i) = P(k = i), i = 1, \cdots, n$ be a probability distribution for k. The method that we implement is called the Alias method and is based on the work of Brately, Fox and Schrage (1983). The method requires one uniform variate, one comparison, at most two memory references per sample and two tables of length n. According to the method, first transform a uniform variable U into an integer I uniformly distributed over $\{1, \cdots, n\}$. Now, I is a tentative value for k. Then, with a certain probability $R(I)$, replace the value I by its 'alias' $A(I)$. If we choose the aliases and the aliasing probabilities properly, then, k has the desired distribution. The Alias method is a stream lined composition method. We decompose the original distribution into a uniform mixture of two-point distributions. One of the two points is an alias. Having found a particular two-point distribution from which to generate, the rest is implemented easily. The actual generator to be followed and the proper selection methods of $R(I)$ and $A(I)$ are presented in the Appendix.

We implement the Akima algorithm in order to generate pseudo random samples from the continuous full conditional densities. The method uses the inverse CDF technique by interpolation of points of the distribution function given in a table. The interpolation we use is a technique due to Guerra, Tapia and Thompson (1976). They give a description of the Akima algorithm and an accuracy comparison between this technique and linear interpolation. The relative error associated with the Akima interpolation is generally considered very good. A small description of how we implement the Akima algorithm is presented in the Appendix.

With the above two procedures in place, the Gibbs sampler as described in Section 29.2 is ready for implementation.

Case (i): We let $k, \theta_0, \theta_1, \theta_2$ unknown and assume σ^2 to be known. From (29.4.6), we have

$$Y_i \sim N(\theta_0 e^{\theta_1 t_i}, \sigma^2), \quad i = 1, \cdots, k$$

and

$$Y_i \sim N(\theta_0 e^{\theta_2 t_i}, \sigma^2), \quad i = k+1, \cdots, n .$$

First, we generate 100 data points ($n = 100$) from the above model by setting the parameters to be:

(a_1) $k = 50, \ \theta_0 = 0.95, \ \theta_1 = 0.2, \ \theta_2 = 0.7 \ $ and $\ \sigma^2 = 0.25 .$

The values of $t_1, \cdots, t_n$ are set to be $t_i = \frac{i}{n}, i = 1, \cdots, n$. This amounts to sampling at equi-spaced intervals. We then implement our Bayesian inference by assuming the following priors on the unknown parameters:

$$k \ \sim \ \text{Uniform} \ \{1, \cdots, n\} \ ,$$
$$\theta_0 \ \sim \ \text{Uniform} \ [0, 1]$$
$$\theta_1 \ \sim \ \text{Beta} \ (p_1, q_1)$$
$$\theta_2 \ \sim \ \text{Beta} \ (p_2, q_2)$$

with all of k, θ_0, θ_1 and θ_2 being independent. The choices we consider for (p_1, q_1) and (p_2, q_2) are $p_1 = 0.2, q_1 = 2.0$ and $p_2 = 8.0, q_2 = 2.0$. Next, we implement the Gibbs sampler on the simulated data as per our description. We present graphically the corresponding posteriors $p(k|\boldsymbol{y}), u_0(\theta_0|\boldsymbol{y})$, $u_1(\theta_1|\boldsymbol{y})$ and $u_2(\theta_2|\boldsymbol{y})$ in Figures 29.1a–29.1d.

In order to assess the robustness of the inferential procedure, we consider two more cases as below:

(a_2)　　　$k = 65,$　$\theta_0 = 0.95,$　$\theta_1 = 0.3,$　$\theta_2 = 0.7,$　$\sigma^2 = 0.49$

(a_3)　　　$k = 65,$　$\theta_0 = 0.95,$　$\theta_1 = 0.3,$　$\theta_2 = 0.7,$　$\sigma^2 = 0.25$

In both (a_2) and (a_3) above, we let $p_1 = 0.5,$ $q_1 = 2.0;$　$p_2 = 8.0,$ $q_2 = 2.0$. Since estimation of the unknown change-point is of primary interest, we present the graphs of the posterior $p(k|\boldsymbol{y})$ only. The graphs appear in Figures 29.2a and 29.2b.

Case (ii): We let $k, \theta_1, \theta_2, \sigma^2$ unknown and assume θ_0 to be known. Here, we perform our analysis on the two data sets generated in (a_2) and (a_3) above treating $\theta_0 = 0.95$ to be known. As for the priors, we assume the same priors on k, θ_1 and θ_2 and on σ^2, we assume:

$$\sigma^2 \sim \text{Inverted Gamma } (c, \lambda).$$

For both data sets, the choices for (p_1, q_1) and (p_2, q_2) remain the same as before and for (c, λ), we choose $c = 5, \lambda = 1$. The graphs of the posterior are presented in Figures 29.3a and 29.3b.

As indicated by Figures 29.1a–29.1d, the marginal posterior modes estimate the true parameter values extremely well. Fig. 29.1a indicates $\hat{k} = 48$ where the true parameter is $k = 50$. Further, we note that the three highest probabilities are at $k = 48, 49$ and 50. As for θ_0, θ_1 and θ_2, their true values are also captured quite well by their respective marginal posterior modes.

Figures 29.2a and 29.2b correspond to the cases where $\sigma^2 = 0.49$ and $\sigma^2 = 0.25$ respectively and for both cases true $k = 65$. First note that the change-point is again estimated extremely well despite the true value being shifted away from the middle. There is higher amount of spread in the case of Figure 29.2a as compared with Figure 29.2b. This is quite what we would expect because the true variance is lower ($\sigma^2 = 25$) for the case of Figure 29.2b as compared to its value of ($\sigma^2 = 0.49$) for Figure 29.2a.

Figures 29.3a and 29.correspond to the case where σ^2 is assumed unknown. Figure 29.3a corresponds to the case where the true $\sigma^2 = .49$ and Figure 29.3b with case where $\sigma^2 = 0.25$. Despite σ^2 being assumed unknown, the true change-point again is estimated extremely well. A closer comparison between Figures 29.2a, 29.2b and 29.3a, 29.3b indicates higher posterior modes (heights) in Figures 29.2a and 29.2b than in Figures 29.3a and 29.3b. This again is to be expected because Figures 29.2a and 29.2b are the marginal posteriors with σ^2 known while Figures 29.3a and v3b are the posteriors based on σ^2 being unknown.

APPENDIX

1. **Generating pseudo random numbers from a general discrete distribution:**

   We implement the Alias algorithm for this purpose. The algorithm is as below:

   (1) Generate U uniform on the continuous interval $(0, n)$.

   (2) Set $I \longrightarrow [U]$ (Thus, I is uniform on the integers $\{1, \cdots, n\}$).

   (3) Set $W \longrightarrow I - U$.

   (4) If $W \leq R(I)$, then

          (a) out put $k = I$; otherwise,

          (b) out put $k = A(I)$, where $A(I)$ and $R(I)$ are tabulated values.

   The remaining part is the proper selection of $R(t)$ and A (t). For this, we let $P_n(W \leq R(I), I = i) = R(i)/n$ and $P_n(W > R(I), I = j) = [1 - R(j)]/n$. Summing the mutually exclusive probabilities to get $k = i$, the generator gives

   $$P_n(k = i) = R(i)/n + \sum_{j:A(j)=i} [1 - R(j)]/n.$$

   To make this equal $p(i)$, select $R(i)$ and $A(i)$ using the following set-up algorithms:

   0. Set $H \longrightarrow \Phi, L \longrightarrow \Phi$, [$\Phi$ denotes the empty set].

   1. For $i = 1, n$:

          (a) set $R(i) \longrightarrow np(i)$;

          (b) if $R(i) > 1$, then add i to H.

          (c) if $R(i) < 1$, then add i to L.

   2. (a) if $H = \Phi$, stop;

      (b) [Claim: $H \neq \Phi \implies L \neq \Phi$]: otherwise select an index j from L and an index l

from H.

3. (a) set $A(j) \longrightarrow 1$;

   (b) $R(l) \longrightarrow R(l) + R(j) - 1$;

   (c) if $R(l) \leq 1$, remove l from H;

   (d) if $R(l) < 1$, add l to L;

   (e) remove j from L;

4. Go to step 2.

To generate a random deviate from the full conditional general discrete distribution, the routine [RNGDS] in the IMSL (1988) library has been used to set-up a table which made it possible for us to apply the method discussed above. The alias algorithm for set-up and random generation can be found in Kronmal and Peterson (1979); see also, Brately, Fox and Schrage (1983) and Ahrens and Dieter (1982).

2. Generating pseudo random numbers from a general continuous distribution:

Here we adopt the Akima algorithm. The steps implemented are as follows:

(1) Set up a table $\{(x_i, F(x_i)), i = 1, \cdots, n; x_1 < \cdots < x_n\}$

(2) Compute and store the coefficient $x = F^{-1}(y)$

(3) Obtain a pseudo random-number y from $U(0, 1)$.

(4)

      (a) If $y \leq F(x_1)$, the random number is x_1.

      (b) If $y \geq F(x_n)$, the random number is x_n.

      (c) If $F(x_1) < y < F(x_n)$, the random number is $F^{-1}(y)$.

The routine [GCDF] in the IMSL (1988), is used to set up a table to evaluate a general continuous distribution function, given ordinates of the probability density function. This approach requires that the range of the distribution be specified. Also, end-points must be chosen in such a way that most of the probability mass is included within the end-points. The routine then uses a C^1 cubic spline interpolation while computing the distribution function.

REFERENCES

Ahrens, J. H. and Dieter, U. (1982). Generating gamma variate by modified rejection technique, *Communications of the ACM*, **25**, 47–54.

Booth, N. B. and Smith, A. F. M. (1982). A Bayesian approach to retrospective identification of change-points, *Journal of Econometrics*, **19**, 7–22.

Bratly, P., Fox, B. and Schrage, L. (1983). *A Guide to Statistical Simulation*, Springer-Verlag, New York.

Carlin, B. P. and Polson, N. G. (1991). Inference for nonconjugate Bayesian models using the Gibbs sampler, *Canadian Journal of Statistics*, **19**, 399–405.

Carlin, B. P., Gelfand, A. E. and Smith, A. F. M. (1992). Heirarchical Bayesian analysis of change-point problems, *Applied Statistics*, **41**, 389–405.

Carlin, B. P., Polson, N. G. and Stoffer, D. S. (1992). A Monte Carlo approach to nonnormal and nonlinear state-space modeling, *Journal of the American Statistical Association*, **87**, 493–500.

Chernoff, H. and Zacks, S. (1964). Estimating the current mean of a normal distribution which is subject to change in time, *Annals of Mathematical Statistics*, **34**, 999–1018.

Chin Choy, J. H. and Broemeling, L. D. (1980). Some Bayesian inferences for a changing linear model, *Technometrics*, **22**, 71–78.

Gelfand, A. E. and Smith, A. F. M. (1990). Sampling based approaches to calculating marginal posterior densities, *Journal of the American Statistical Association*, **85**, 398–409.

Gelfand, A. E., Hills, S. E., Racien-Poon, A. and Smith, A. F. M. (1990). Illustration of Bayesian inferences in normal data models using Gibbs sampling, *Journal of the American Statistical Association*, **85**, 972–985.

Gelfand, A. E., Smith, A. F. M. and Lee, T. M. (1992). Bayesian analysis of constrained parameter and truncated data problems, *Journal of the American Statistical Association*, **87**, 523–532.

Gelman, A. and Rubin, D. B. (1992). A single series from the Gibbs sampler provides a false sense of security, In *Bayesian Statistics 4* (Eds., J. M. Bernardo, J. O. Berger, A. P. Dawid and A. F. M. Smith),pp. 625–632, Oxford University Press, London, England.

Geman, D. and Geman, S. (1984). Stochastic relaxation, Gibbs distribution and the Bayesian restoration of images, *IEEE Transactions in Pattern Analysis and Machine Intelligence*, **6**, 721–741.

Guerra, V. R., Tapia, A. and Thompson, J. R. (1976). A random number generator for continuous random variables based on an interpolation procedure by Akima, In *Proceedings of the Ninth Interface Symposium in Computer Science and Statistics*, pp. 228–230, Prindle, Weber and Schmidt, Boston.

Hastings, W. K. (1970). Monte Carlo sampling methods using Markov chains and their applications, *Biometrika*, **57**, 97–109.

Jandhyala, V. K. and MacNeill, I. B. (1989). Residual partial sum limit processes for regression models with applications to detecting parameter changes at unknown times, *Stochastic Processes and Their Applications*, **33**, 309–323.

Jandhyala, V. K. and MacNeill, I. B. (1991). Tests for parameter changes at unknown times in linear regression models, *Journal of Statistical Planning and Inference*, **27**, 291–316.

Jandhyala, V. K. and MacNeill, I. B. (1997). Iterated partial sum sequences of regression residuals and tests for change-points with continuity constraints, *Journal of the Royal Statistical Society, Series B*, **59**, 147–156.

Jandhyala, V. K. and Minogue, C. D. (1993). Distributions of Bayes-type statistics under polynomial regression, *Journal of Statistical Planning and Inference*, **37**, 271–290.

Kim, H. J. and Siegmund, D. (1989). Likelihood ratio test for a change-point in simple linear regression, *Biometrika*, **76**, 409–423.

Metropolis, N., Rosenbluth, A. W., Rosenbluth, M. N., Teller, A. H. and Teller, E. (1953), Equations of state calculations by fast computing machines, *Journal of Chemical Physics*, **21**, 1087–1091.

Moen, D. H. and Broemeling, L. D. (1985). The uncertainty of forecasting: Models with structural change versus those without changing parameters, *Communications in Statistics—Theory and Methods*, **14**, 2029–2040.

Raftery, A. E. and Lewis, S. (1992). How many iterations in the Gibbs sampler?, In *Bayesian Statistics 4* (Eds., J. M. Bernardo, J. O. Berger, A. P. Dawid and A. F. M. Smith), pp. 763–773, Oxford University Press, London, England.

Raghunathan, T. E. and Grizzle J. E. (1995). A split questionnaire survey design, *Journal of the American Statistical Association*, **90**, 54–63.

Rosenthal, J. S. (1995). Minorization conditions and convergence rates for Markov chain Monte Carlo, *Journal of the American Statistical Association*, **90**, 558–566.

Smith, A. F. M. and Cook, R. D. (1980). Straight lines with a change-point: A Bayesian analysis of some renal transplant data, *Applied Statistics*, **29**, 180–189.

Tanner, M. and Wong, W. H. (1987). The calculation of posterior distributions by data augmentation (with discussion), *Journal of the American Statistical Association*, **82**, 528–550.

Worsley, K. J. (1986). Confidence regions and tests for a change-point in a sequence of exponential family random variables, *Biometrika*, **73**, 91–104.

Zeger, S. L. and Karim, M. R. (1991). Generalized linear models with random effects: A Gibbs sampling approach, *Journal of the American Statistical Association*, **86**, 79–86.

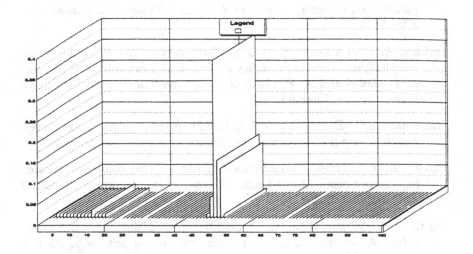

FIGURE 29.1a Graph of $p(k|y)$ for case (a_1) when $\sigma^2 = 0.25$ is known

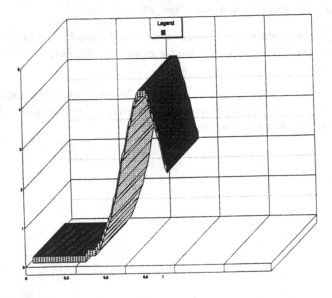

FIGURE 29.1b Graph of $u_0(\theta_0|y)$ for case (a_1) when $\sigma^2 = 0.25$ is known

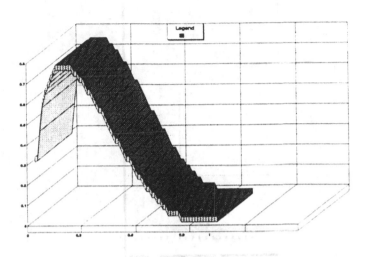

FIGURE 29.1c Graph of $u_1(\theta_1|y)$ for case (a_1) when $\sigma^2 = 0.25$ is known

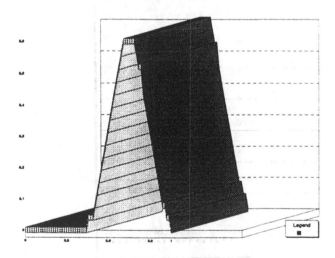

FIGURE 29.1d Graph of $u_2(\theta_2|y)$ for case (a_1) when $\sigma^2 = 0.25$ is known

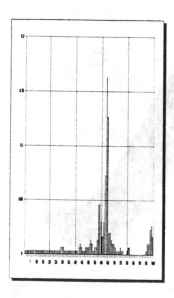

FIGURE 29.2a Graph of $p(k|y)$ for case (a_2) when $\sigma^2 = 0.49$ is known

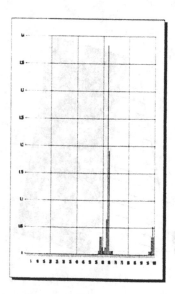

FIGURE 29.2b Graph of $p(k|y)$ for case (a_3) when $\sigma^2 = 0.25$ is known

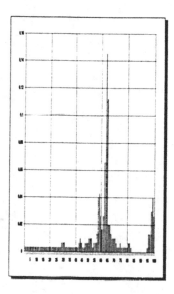

FIGURE 29.3a Graph of $p(k|y)$ for case (a_2) when $\sigma^2 = 0.49$ is known

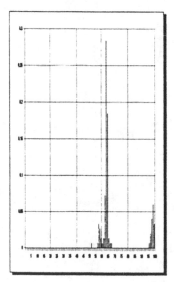

FIGURE 29.3b Graph of $p(k|y)$ for case (a_3) when $\sigma^2 = 0.25$ is known

FIGURE 29.2a.

FIGURE 29.2b.

CHAPTER 30

CONVERGENCE TO TWEEDIE MODELS AND RELATED TOPICS[1]

BENT JØRGENSEN

University of British Columbia, Vancouver, Canada

VLADIMIR VINOGRADOV

Ohio University, Athens, OH

Abstract: The class of Tweedie exponential dispersion models generalizes positive and extreme stable distributions. They were used in the statistical analysis of car insurance claims by Jørgensen and Souza (1994), and a new type of weak convergence theorem to the Tweedie models was proved by Jørgensen, Martínez and Tsao (1994). We discuss some analogies between the development of this theory and certain phase-transition phenomena in the theory of large deviations, branching processes, mathematical physics and actuarial mathematics. Thus, we characterize a new part of the domains of attraction to the Tweedie families, and suggest certain probability models for analyzing some insurance claims data. Specifically, distributions with exponential-power tails belong to the domains of attraction to certain Tweedie models, and the same class of distributions reveals some critical-point properties.

Keywords and phrases: Critical points, weak convergence, domains of attraction, Tweedie models

---

[1]An expanded version of this paper including all proofs will be submitted elsewhere [see Jørgensen, Martínez and Vinogradov (1999)].

30.1   INTRODUCTION

In this work, we primarily study some properties of the following univariate nonnegative natural exponential family:

$$p(y; \theta) = \begin{cases} \frac{1}{b_\delta(\theta)} \cdot c(y) \cdot e^{\theta y} \cdot y^{\delta - 1} & \text{if } y > 0, \\ 0 & \text{otherwise,} \end{cases} \qquad (30.1.1)$$

where the canonical parameter $\theta \leq 0$, parameter $\delta < 0$ is fixed,

$$c(y) \to 1$$

as $y \to \infty$, and this function $c(y)$ is such that

$$b_\delta(\theta) := \int_0^\infty c(y) \cdot e^{\theta y} \cdot y^{\delta - 1} \cdot dy < \infty.$$

Our main result, Theorem 30.1.1 of this section, deals with weak convergence to the so-called *Tweedie models* for some probability densities from class (30.1.1). Section 30.2 is devoted to a detailed consideration of some special cases of the probability densities satisfying condition (30.1.1). Also, in Section 30.3 we review some results and methods relevant to the use of the densities satisfying condition (30.1.1) in various problems of the theory of large deviations, branching processes, mathematical physics and actuarial mathematics. In addition, in the concluding Section 30.4 we suggest an approach to analyzing conditional distributions of the insurance risk process under the assumption that the common probability density of the amounts of claims satisfies condition (30.1.1). Note that in that section, we also determine the *critical value* for the ratio of the initial capital and the time period during which the risk process is being considered.

Now, we briefly review some basic concepts of the theory of dispersion models which will be relevant for the further consideration. Note that dispersion models were introduced in statistics in an attempt to weaken the requirement on normality of data. In particular, Nelder and Wedderburn (1972) introduced the so-called method of the *analysis of deviance* that generalizes the classical analysis of variance approach to a wide class of not necessarily normal data. We refer to Chapter 1 of Jørgensen (1997) for a comprehensive review of the main concepts and basic properties of the dispersion models.

In general, a *natural exponential family* is defined by densities for a random variable Y having the following form:

$$\frac{1}{b(\theta)} \cdot h(y) \cdot e^{\theta y}, \qquad (30.1.2)$$

where $h(y)$ is a given function and $b(\theta)$ is a normalizing constant. Simple calculations show that the mean

$$\mu := b'(\theta)/b(\theta) \tag{30.1.3}$$

is a one-to-one function of θ, which in turn implies that given μ, the unique solution $\theta(\mu)$ to (30.1.3) is such that for this particular value of parameter μ, and $\theta = \theta(\mu)$, the variance of Y may be expressed as a function of parameter μ only, namely,

$$\mathrm{Var}(Y) = V(\mu).$$

Here, V is the *variance function*. It should be noted that the variance function V together with its open domain Ω characterize family (30.1.2) within the class of all natural exponential families [see Section 2.3 of Jørgensen (1997) for more detail]. Also, the given natural exponential family (30.1.2) generates, uniquely, an exponential dispersion model, hereinafter denoted by $\mathrm{ED}(\mu, \sigma^2)$, where σ^2 is the dispersion parameter and μ continues to be the mean. This model is characterized as the class of natural exponential families with variance $\mathrm{Var}(Y)$ being *proportional* to the given variance function $V(\cdot)$. In particular, $\mathbf{E}Y = \mu$ and

$$\mathrm{Var}(Y) = \sigma^2 \cdot V(\mu).$$

Here, the domain for σ^2 may be $\mathbb{R}_+^1$ or some subset thereof [see Chapter 3 of Jørgensen (1997) for more detail]. Dispersion models are *reproductive*, that is, the sample mean of an i.i.d. sample from $\mathrm{ED}(\mu, \sigma^2)$ has distribution $\mathrm{ED}(\mu, \sigma^2/n)$, where n is the sample size.

It is interesting to note that exponential dispersion models were introduced by Tweedie (1947), but remained unnoticed for about 25 years. An important special case that can be viewed as a generalization of infinitely divisible and in particular stable distributions, is the so-called class of Tweedie models which were introduced independently by Tweedie (1984), Hougaard (1986), and Bar-Lev and Enis (1986). This class is characterized by a particularly simple form of the variance function, namely,

$$\mathrm{Var}(Y) = \sigma^2 \cdot \mu^p,$$

where the parameter $p \in \mathbb{R}^1 \backslash (0, 1)$. Hereinafter, we denote a specific Tweedie model characterized by parameters p, μ and σ^2 by $\mathrm{Tw}_p(\mu, \sigma^2)$.

Recall that Tweedie models generalize the classical stable laws. In particular, the following scaling property holds: for each fixed $b > 0$,

$$b \cdot \mathrm{Tw}_p(\mu, \sigma^2) \stackrel{\mathrm{d}}{=} \mathrm{Tw}_p(b \cdot \mu, b^{2-p} \cdot \sigma^2), \tag{30.1.4}$$

where '$\overset{d}{=}$' means that *distributions* of random variables coincide. In addition, each Tweedie distribution possesses its own domain of attraction. In fact, our main result, Theorem 30.1.1, that generalizes Theorem 4.2 and Example 5.7 of Jørgensen, Martínez and Tsao (1994), deals with weak convergence of our exponential family (30.1.1) to the Tweedie models. See also the end of Section 4.4 of Jørgensen (1997), where Theorem 4.2 of Jørgensen, Martínez and Tsao (1994) is formulated using a notation more consistent with ours. Also, note that condition (30.1.4) characterizes only a particular class of Tweedie models, namely, those which do not depend on a location parameter. See Jørgensen (1997, Section 4.5) for a more complete description.

We now proceed with the formulation of

Theorem 30.1.1 *Consider an exponential family* $Y \sim \text{ED}(\mu, \sigma^2)$ *satisfying condition (30.1.1) with* $\delta < -1$. *Then*

(a) *If* $-2 < \delta < -1$, *we have*

   i)

$$V(\mu) \sim c_1(\mu_0 - \mu)^{p(\delta)}$$

   as $\mu \uparrow \mu_0$, *where* c_1 *is a certain positive constant,* μ_0 *is defined by (30.1.7), and*

$$p(\delta) = (\delta + 2)/(\delta + 1).$$

   ii)

$$c^{-1} \cdot \left\{ \mu_0 - \text{ED}(\mu_0 - c\mu, c^{2-p}\sigma^2) \right\} \overset{d}{\to} \text{Tw}_p(\mu, c_1\sigma^2) \quad (30.1.5)$$

   as $c \to 0$.

(b) *If* $\delta \leq -2$, *we have*

   i)

$$V(\mu) \to c_2$$

   as $\mu \uparrow \mu_0$, *where* c_2 *is a certain positive constant.*

   ii)

$$c^{-1} \cdot \left\{ \mu_0 - \text{ED}(\mu_0 - c\mu, c^2\sigma^2) \right\} \overset{d}{\to} \text{N}(\mu, c_2\sigma^2) \quad (30.1.6)$$

   as $c \to 0$, *where* $\text{N}(\mu, c_2\sigma^2)$ *is a normal random variable having the specified mean and variance.*

Remark 30.1.1 It is easily seen that under fulfillment of (30.1.1),

$$\mu_0 := \lim_{\theta\uparrow 0} \frac{1}{b_\delta(\theta)} \cdot \int_0^\infty c(y) \cdot e^{\theta y} \cdot y^\delta \cdot dy = \lim_{\theta\uparrow 0} \frac{b_{\delta+1}(\theta)}{b_\delta(\theta)} = \frac{b_{\delta+1}(0)}{b_\delta(0)} < \infty.$$

$$(30.1.7)$$

Remark 30.1.2 Point (a.ii) of Theorem 30.1.1 can be viewed as a generalization of the well-known fact on weak convergence to a certain stable distribution with index $\alpha = -\delta \in (1, 2)$, whereas point (b.ii) generalizes the central limit theorem. The interested reader is referred to Vinogradov (1999) for more detail.

Remark 30.1.3 The natural exponential family $Y \sim ED(\mu, \sigma^2)$ that satisfies condition (30.1.1) with $\delta < -1$ is *not steep* in the sense of Definition 2.2 of Jørgensen (1997). Indeed, it follows from Remark 30.1.1 that the *open* domain Ω of the means is an open interval $(0, \mu_0)$ with $\mu_0 < \infty$, whereas the interior of the support of the family Y is the entire semi-axis $(0, \infty)$, and therefore is different from Ω.

Remark 30.1.4 The well-known natural exponential family of the inverse Gaussian distributions that will be described in the next Section 30.2, satisfies condition (30.1.1) with $\delta = -1/2$ and hence our Theorem 30.1.1 is not applicable to that family. However, the above Theorem 30.1.1 covers an important subclass of the class of *generalized inverse Gaussian distributions*. The main purpose of this paper is to consider the nonsteep case, whereas one will have steepness in the case of fulfillment of condition (30.1.1) with $-1 \leq \delta < 0$. Hence, the consideration of the latter case is beyond the scope of this paper and will be carried out separately [see Jørgensen, Martínez and Vinogradov (1999)]. However, a trivial convergence result for the inverse Gaussian family is obtained by combining the fact that the inverse Gaussian distributions are in fact Tw_3-models and the scaling relationship (30.1.4). In turn, this suggests that under fulfillment of condition (30.1.1) with $\delta = -1/2$, the corresponding natural exponential family would converge to a certain inverse Gaussian distribution, since $p(\delta) = 3$ (see also the second-to-last paragraph of this section). However, $\mu_0 = \infty$ in that case.

Since an expanded version of this paper including all proofs will appear as Jørgensen, Martínez and Vinogradov (1999), we confine ourselves to making just a few remarks regarding the proof of Theorem 30.1.1. In particular, the derivation of the asymptotics for $V(\mu)$ stipulated in points (a.i) and (b.i) is similar to the proof of Theorem 4.2 of Jørgensen, Martínez and Tsao (1994), and essentially repeats the calculations relevant to Example 5.7 of that paper (see pp. 236–237 therein). Finally, the results on weak convergence follow from those on the asymptotic behaviour of the variance function $V(\mu)$ by an application of Mora's theorem [see, e.g., Theorem 2.12 of Jørgensen (1997)].

Now, in order to better explain the result of Theorem 30.1.1, we proceed with a more detailed description of the results on weak convergence to Tweedie models as well as the characterization of some important special cases.

First, it should be pointed out that in the Tweedie class, the case of $p = 0$ corresponds to the normal distribution, $p = 1$ to Poisson, $p = 2$ to gamma, and $p = 3$ to inverse Gaussian distribution (see Table 4.1 of Jørgensen (1997) for more detail). In particular, random variable $N(\mu, c_2\sigma^2)$ which emerged on the right-hand side of (30.1.6) can thus be rewritten as $Tw_0(\mu, c_2 \cdot \sigma^2)$. However, neither Poisson, nor gamma, nor inverse Gaussian distributions emerge as the limit on the right-hand side of (30.1.5). This is because we have that

$$p(\delta) = (\delta + 2)/(\delta + 1) < 0$$

for $-2 < \delta < -1$, and the corresponding Tweedie model $Tw_p(\mu, c_1 \cdot \sigma^2)$ is obtained by exponential tilting of a certain extreme stable distribution with index $1 < \alpha = -\delta < 2$ [see Jørgensen (1997, p. 136) and also the above Remark 30.1.2].

Recall that Tw_3-models, i.e., the inverse Gaussian distributions, as well as their natural generalizations, namely, the class of generalized inverse Gaussian distributions, are typical representatives of the class of distributions satisfying condition (30.1.1). Since the latter ones will play an important role in the further consideration, we now proceed with a brief description of the main properties of the most studied class—the inverse Gaussian and generalized inverse Gaussian distributions. Our opinion is that the history of discovery of these distributions well illustrates some interesting connections between probability and statistics.

30.2 SPECIAL CASES: INVERSE GAUSSIAN AND GENERALIZED INVERSE GAUSSIAN DISTRIBUTIONS

First, note that the inverse Gaussian distribution with particular values of parameters μ and σ^2, hereinafter denoted by $IG(\mu, \sigma^2)$ can be defined in terms of its density $p_{IG}(y; \mu, \sigma^2)$ as follows:

$$p_{IG}(y; \mu, \sigma^2) = \begin{cases} \frac{1}{(2\pi\sigma^2)^{1/2}} \cdot y^{-3/2} \cdot \exp\left\{-\frac{1}{2\sigma^2} \cdot \frac{(y-\mu)^2}{y\mu^2}\right\} & \text{if } y > 0, \\ 0 & \text{otherwise.} \end{cases}$$
(30.2.8)

Also, it is often more convenient to represent density (30.2.8) in the following equivalent form:

$$p_{IG}(y; \theta, \kappa)$$
$$= \begin{cases} (\kappa/2\pi y^3)^{1/2} \cdot \exp\left\{-\frac{\kappa}{2y} + \kappa \cdot \left(\theta y + (-2\theta)^{1/2}\right)\right\} & \text{if } y > 0, \\ 0 & \text{otherwise,} \end{cases}$$
(30.2.9)

where the parameters from formulas (30.2.8) and (30.2.9) are related as follows:

$$\mu = (-2\theta)^{-1/2}$$

and

$$\sigma^2 = 1/\kappa.$$

This distribution was originally derived as the distribution of the first hitting time of a univariate Brownian motion with drift $1/\mu$ and unit variance to level $1/\sigma$ independently and simultaneously by Schrödinger (1915) and by Smoluchowsky (1915). Subsequently, the importance of $IG(\mu, \sigma^2)$ in statistics was realized [see, e.g., Tweedie (1957a, 1957b), and also Seshadri (1993) for a comprehensive review of recent developments]. Later on, the *generalized* inverse Gaussian distributions, hereinafter denoted by $GIG(\delta, \xi, \psi)$, were discovered and also used in statistics [see, e.g., Halphen (published in 1941 through Dugué), Sichel (1974, 1975), Barndorff-Nielsen (1977, 1978), and also Jørgensen (1982)]. The class of GIG distributions is obtained from the class of IG distributions simply by replacing the power $-3/2$ of y by the power $\delta - 1$ (compare formulas (30.2.9) and (30.2.10)). It can be characterized in terms of the following density:

$$p_{GIG}(y; \delta, \chi, \psi)$$
$$= \begin{cases} \frac{(\psi/\chi)^{\delta/2}}{2K_\delta((\chi\psi)^{1/2})} \cdot y^{\delta-1} \cdot \exp\left\{-\frac{1}{2} \cdot (\chi y^{-1} + \psi y)\right\} & \text{if } y > 0, \\ 0 & \text{otherwise,} \end{cases}$$
(30.2.10)

where K_δ denotes the modified Bessel function of the third kind with index $\delta \in \mathbb{R}^1$. Namely, for each real $\omega > 0$,

$$K_\delta(\omega) := \frac{1}{2} \cdot \int_0^\infty x^{\delta-1} \cdot \exp\left\{-\frac{1}{2} \cdot \omega \cdot (x + x^{-1})\right\} \cdot dx$$

[see, e.g., p. 170 of Jørgensen (1982)].

It is interesting to note that upon the discovery of this class of GIG distributions, an attempt was made to associate them with some probability models. Thus, Theorem 2.1 of Barndorff-Nielsen, Blæsild and Halgreen (1978) stipulates that for each GIG distribution with a nonpositive value of parameter δ, there exists a certain time-homogeneous diffusion process such that its first hitting time follows the specified GIG distribution.

Remark 30.2.1 It is clear that for $\delta < 0$, the class of GIG distributions belongs to the class of natural exponential families satisfying condition (30.1.1).

We will now turn from the consideration of the special cases, i.e., IG and GIG distributions, to the consideration of other interesting properties of our class of natural exponential families characterized by condition (30.1.1).

30.3   CRITICAL POINTS IN THE FORMATION OF LARGE DEVIATIONS

In this section, we fix a particular density function $p(y; \theta_0)$ from our exponential family (30.1.1) with the value of the canonical parameter $\theta = \theta_0$ *strictly less than zero*. Let $\{X_n, n \geq 1\}$ be independent random variables with common density function $p(y; \theta_0)$ introduced above. This class of probability densities as well as its extensions were considered in a number of works. It should be noted that various *phase-transition-type* phenomena were discovered for these classes of distributions in connection to some probability models.

We will now discuss some of them in more detail.

In particular, Chover, Ney and Wainger (1973) considered the following class $\mathcal{C}(d)$ of distribution functions: $G(\cdot) \in \mathcal{C}(d)$ if

$$\frac{1 - G^{*2}(x)}{1 - G(x)} \to 2d \qquad (30.3.11)$$

as $x \to \infty$, where real $d > 1$ is fixed, and $G^{*2}(\cdot)$ denotes the two-fold convolution of the distribution function $G(\cdot)$. One can check that each member of our exponential family (30.1.1) with $\theta = \theta_0 < 0$ satisfies condition (30.3.11) with

$$1 < d = \frac{1}{b_\delta(\theta_0)} \cdot \int_0^\infty c(y) \cdot y^{\delta-1} \cdot dy = b_\delta(0)/b_\delta(\theta_0) < \infty.$$

It was then established that if the particle lifetime distribution function $G(t)$ of a certain *subcritical* branching process satisfies condition (30.3.11) then the value

$$m_0 = 1/d < 1$$

will be a *critical point* for the average number m of children born at each birth epoch. Namely, Chover, Ney and Wainger (1973) established that the mechanisms of survival of this process conditioned by nonextinction will be of different nature depending on whether $m < m_0$ or not. One should note that conditioning a *subcritical* branching process by non-extinction constitutes the consideration of a *rare event* (large deviation).

A result of a similar character in the context of non-life insurance was obtained by Klüppelberg (1989). In particular, she revealed a non-classical long-time behaviour of the ruin probability for the simplest risk process under fulfillment of the net-profit condition (see formula (30.4.13) below), and under the assumption that the common distribution function of the amounts of claims satisfies condition (30.3.11).

Phase-transition-type phenomena which we believe are of a similar nature were also established in mathematical physics. Thus, Dobrushin and Shlosman (1994, Theorem 1.5.2) established the existence of a certain critical value and determined the range of a non-classical behaviour of the probabilities of large deviations for the Ising model. The main argument behind their result on the presence of two distinct, classical and non-classical, regimes for the behaviour of large deviations is based on the fact that the rate function $H(\cdot)$ that determines the rate of exponential decay of the probabilities of large deviations, degenerates after a certain *critical* point. In other words, it is *strictly convex* in one region and *nonstrictly convex* in the other region.

It should also be noted that the rate function $H(\mu)$ that corresponds to the considered density function $p(y; \theta_0)$ possesses a similar degeneracy property [see, e.g., Vinogradov (1993, 1994, 1996)]. Now, we introduce some auxiliary notation to make this rigorous. Thus, one can show that under fulfillment of (30.1.1) with $\theta = \theta_0 < 0$ and $\delta < 0$, the moment-generating function

$$\mathbf{E}\exp\{\theta \cdot X_i\} = \frac{b_\delta(\theta + \theta_0)}{b_\delta(\theta_0)}$$

is monotonically increasing and finite within the semi-infinite interval $\theta \in$

$(-\infty, -\theta_0]$. Also, its maximum is

$$\mathbf{E}\exp\{-\theta_0 \cdot X_i\} \quad \left(= \frac{b_\delta(0)}{b_\delta(\theta_0)}\right) \quad = d.$$

Subsequently, the rate function $H(\mu)$, which is the Legendre transform of

$$\log\left(b_\delta(\theta + \theta_0)/b_\delta(\theta_0)\right),$$

can be represented as follows:

$$
\begin{aligned}
H(\mu): \quad &= \sup_\theta\left(\mu \cdot \theta - \log\frac{b_\delta(\theta + \theta_0)}{b_\delta(\theta_0)}\right) \\
&= \begin{cases} \mu \cdot \theta(\mu) - \log\left(b_\delta(\theta(\mu) + \theta_0)/b_\delta(\theta_0)\right) & \text{if } \mu \leq \mu_0, \\ H(\mu_0) - \theta_0 \cdot (\mu - \mu_0) & \text{otherwise.} \end{cases}
\end{aligned}
$$

Here, the *critical value* μ_0 defined in Remark 30.1.1 is finite (see also Remark 30.1.3), and $\theta(\mu)$ is the unique solution to equation (30.1.3) with

$$b(\theta) = \frac{b_\delta(\theta + \theta_0)}{b_\delta(\theta_0)}.$$

It is known that solution $\theta(\mu)$ to equation (30.1.3) exists for each $\mu \leq \mu_0$ and fails to exist for $\mu > \mu_0$ [see, e.g., Chapter 5 of Vinogradov (1994)]. Recall (see the above Remark 30.1.3) that the same phenomenon is referred to in statistics as *nonsteepness*.

At this stage, let us point out that under fulfillment of the conditions specified in the first paragraph of this section, the exact asymptotics of the probabilities of large deviations in the range greater than μ_0 times the number of terms *is not of Cramér's type* [see Vinogradov (1993, 1994, 1996) for more detail]. In particular, we have that if $\delta < -2$ then for each fixed $\mu > \mu_0$,

$$\mathbf{P}\{X_1 + \cdots + X_n > \mu \cdot n\} \sim \text{Const}(\mu, \mu_0) \cdot (b_\delta(0)/b_\delta(\theta_0))^{n-1} \cdot n^\delta \quad (30.3.12)$$

as $n \to \infty$. In other words, the asymptotic behaviour of the latter probability is again of the exponential-power type. Similar results also hold for $-2 \leq \delta < -1$ [see, e.g., Theorem 5.1.1 of Vinogradov (1994)].

A probabilistic interpretation of the just described results of the phase-transition type is of a sufficient interest. Thus, Section 5.3 of Vinogradov (1994) and Theorems 1.1 and 2.1 of Vinogradov (1996) deal with the probabilistic interpretation of representation (30.3.12) using the techniques of conditioning. In addition, in the next, concluding section of this paper, we describe a relevant phenomenon for the classical risk process.

30.4 DIFFERENT MECHANISMS OF RUIN IN NON-LIFE INSURANCE

Recall that Klüppelberg (1989) revealed that under fulfillment of (30.3.11) and some additional assumptions, the asymptotics of the ruin probability of the classical risk process is not given by the classical Cramér–Lundberg formula. Here, we consider a similar model, but in contrast to Klüppelberg (1989), we impose a more restrictive assumption (30.1.1)—recall that (30.1.1) implies (30.3.11) but not vice versa. The fact that our conditions are more restrictive along with the consideration of *finite* rather than *infinite* time intervals as in Klüppelberg (1989) and the use of an approach based on conditioning enable us to reach rather interesting (in our opinion) conclusions.

To be rigorous, we now introduce the classical insurance risk process as follows:

$$R(t) := R_0 + C \cdot t - \sum_{i=1}^{N(t)} X_i$$

for $t \geq 0$, where $R_0 > 0$ is the *initial capital*, $C > 0$ is the *constant premium rate*, and $\{N(t), t \geq 0\}$ is a time-homogeneous Poisson process with intensity $\lambda > 0$ that counts the number of claims up to time t. Here, the random variables $\{X_n, n \geq 1\}$ are the same as in Section 30.3, i.e., they are independent with common density function $p(y; \theta_0)$ satisfying condition (30.1.1) with $\theta = \theta_0 < 0$ and $\delta < -2$. These variables $\{X_n, n \geq 1\}$ are interpreted as the amounts of the subsequent positive claims which are independent of each other and also of the Poisson process $N(t)$ that describes the flow of arrivals of the claims. Let

$$M := \mathbf{E}X_i > 0.$$

Then the following condition

$$C > \lambda \cdot M \tag{30.4.13}$$

is commonly known as the *net-profit* condition and stipulates the prevalence of the insurance premiums over the accumulated claims. Subsequently, it can be shown that under fulfillment of condition (30.4.13), the ruin probability

$$\phi(R_0) := \mathbf{P}\{R(t) < 0 \text{ for some } t > 0\}$$

approaches zero as $R_0 \to \infty$ which justifies referring to the ruin as a *rare event* (large deviation). Also, note that the quantity

$$\rho := \frac{C}{\lambda M} - 1$$

is usually referred to as the *relative safety loading* coefficient.

It is well known that under our assumptions, the interarrival times $\{E_n, n \geq 1\}$ between subsequent claims are independent and identically distributed exponential random variables with mean $1/\lambda$, and that the ruin, i.e., the occurrence of a negative value of $R(t)$, can only be observed at claim times. This motivates the consideration of the following random walk $\{S_n, n \geq 0\}$ generated by the sequence of partial sums of independent random variables

$$Y_i := X_i - C \cdot E_i.$$

Namely,

$$S_0 := 0,$$

and

$$S_n := \sum_{i=1}^{n} Y_i = \sum_{i=1}^{n}(X_i - C \cdot E_i).$$

Here, random variable $C \cdot E_i$ is interpreted as the amount of premiums collected between the subsequent claims.

Using the just introduced notation, one can represent the ruin probability, that characterizes the measure of the long-term stability of the risk process, as follows:

$$\phi(R_0) = \mathbf{P}\{S_n > R_0 \text{ for some } n \in \mathbb{N}\} = \mathbf{P}\left\{\max_{n \geq 1} S_n > R_0\right\}.$$

Note that the study of the asymptotic decay of the ruin probability $\phi(R_0)$ as $R_0 \to \infty$ is an important problem of the risk theory [see, e.g., Grandell (1991)].

Here, we study a similar, but perhaps more realistic problem of the behaviour of

$$
\begin{aligned}
\phi(R_0, T(R_0)) : \quad &= \quad \mathbf{P}\{R(t) < 0 \text{ for some } 0 < t \leq T(R_0) \} \\
&= \quad \mathbf{P}\{S(n) > R_0 \text{ for some } 1 \leq n \leq N(T(R_0))\},
\end{aligned}
$$

$$(30.4.14)$$

where $N(T(R_0))$ stands for the total number of claims arrived over the *finite* time period $[0, T(R_0)]$. Recall that this number has Poisson distribution with parameter $\lambda \cdot T(R_0)$. Apparently, the probability on the right-hand side of (30.4.14) can be represented as

$$\mathbf{P}\left\{\max_{1 \leq k \leq N(T(R_0))} S_k > R_0\right\}.$$

Let us point out that the latter probability is in fact the probability of a *large deviation*, since the random walk $\{S_n, n \geq 0\}$ has a negative drift under fulfillment of the net-profit condition (30.4.13). In particular,

$$\mathbf{E}Y_i = M - C/\lambda < 0.$$

In addition, one gets that the probability density of random variable

$$Y_i = X_i - C \cdot E_i$$

is also of the exponential-power type with the same values of parameters θ_0 and δ. Indeed,

$$
\begin{aligned}
p_{Y_i}(y) &= \int_0^\infty p_{X_i}(x+y) \cdot p_{cE_i}(x) \cdot dx \\
&= \int_0^\infty \frac{c(x+y)}{b_\delta(\theta_0)} \cdot e^{\theta_0(x+y)} \cdot (x+y)^{\delta-1} \cdot \frac{\lambda}{C} \cdot e^{-\lambda x/C} \cdot dx \\
&= \frac{\lambda}{C} \cdot \frac{\hat{c}(y)}{b_\delta(\theta)} \cdot \frac{1}{-\theta_0 + \lambda/C} \cdot e^{\theta_0 y} \cdot y^{\delta-1},
\end{aligned}
\tag{30.4.15}
$$

where

$$\hat{c}(y) \to 1$$

as $y \to \infty$ [compare to Remark 5.1.6 of Vinogradov (1994)].

At this stage, note that the validity of representation (1.4.15) makes possible an application of the results and methods from Vinogradov (1993, 1994, 1996). Recall that some of them were briefly described in the previous section [see, e.g., formula (30.3.12)]. In particular, the ratio $R_0/T(R_0)$ will play an essential role. It turns out that for this ratio, there is a *critical value* which enables one to distinguish between two opposite types of the mechanism of ruin. However, this critical value, hereinafter denoted by μ_0^*, slightly differs from μ_0 (which is defined by (30.1.7)). Namely,

$$\mu_0^* := \lambda \cdot \lim_{\theta \uparrow 0} \frac{\int_0^\infty \hat{c}(y) \cdot e^{\theta y} \cdot y^\delta \cdot dy}{\int_0^\infty \hat{c}(y) \cdot e^{\theta y} \cdot y^{\delta-1} \cdot dy} < \infty.$$

In particular, if $R_0 \to \infty$, $T(R_0) \to \infty$, and

$$R_0/T(R_0) \to \mu > \mu_0^*,$$

then the order of decay of $\phi(R_0, T(R_0))$ will be as follows:

$$\text{Const}(\mu, \mu_0, \lambda, \theta_0, \delta) \cdot \left(\frac{b_\delta(0)}{b_\delta(\theta_0)} \right)^{\lambda T(R_0)} \cdot T(R_0)^\delta$$

[compare to (30.3.12)].

One is also capable to describe the mechanism of ruin within the time period $[0, T(R_0)]$ provided that the ruin has occurred. Thus, if

$$R_0/T(R_0) \to \mu \leq \mu_0^*$$

then the ruin is likely to occur on the reason that the difference between the amount X_i of ith claim and the amount of the insurance premiums $C \cdot E_i$ collected over the time period between the occurrences of $(i-1)$-st and ith claims would be approximately equal to μ/λ, for each $1 \leq i \leq N(T(R_0))$. One can speculate that observing substantial deviations from this mechanism might necessitate auditing some of the claims.

On the other hand, if

$$R_0/T(R_0) \to \mu > \mu_0^*$$

then the ruin is likely to occur on the reason that the difference between the amount of ith claim and the amount of the insurance premiums $C \cdot E_i$ collected over the time period between $(i-1)$-st and ith claims would be approximately equal to μ_0^*/λ, for each $1 \leq i \leq N(T(R_0))$ except for the largest of the differences $\{X_i - C \cdot E_i\}$, $1 \leq i \leq N(T(R_0))$. In turn, the largest difference should be of magnitude

$$\lambda \cdot T(R_0) \cdot (\mu/\lambda - \mu_0^*/\lambda).$$

Similarly to the previous case, one can speculate that observing substantial deviations from this mechanism, e.g., two or more sufficiently large values of Y_i's, might necessitate auditing some of the claims (compare to Section 5.3 of Vinogradov (1994) and Vinogradov (1996)).

An interesting open problem is the study of the asymptotics of $\phi(R_0, T(R_0))$ under various regimes of growth of R_0 and $T(R_0)$.

Acknowledgement Research of the first author was partially supported by an NSERC grant.

REFERENCES

Bar-Lev, S. K. and Enis, P. (1986). Reproducibility and natural exponential families with power variance functions, *Annals of Statistics*, **14**, 1507–1522.

Barndorff-Nielsen, O. E. (1977). Exponentially decreasing log-size distributions, *Proceedings of the Royal Society, Series A*, **353**, 401–419.

Barndorff-Nielsen, O. E. (1978). Hyperbolic distributions and distributions on hyperbolae, *Scandinavian Journal of Statistics*, **5**, 151–157.

Barndorff-Nielsen, O. E., Blæsild, P. and Halgreen, C. (1978). First hitting time models for the generalized inverse Gaussian distribution, *Stochastic Processes and Their Applications*, **7**, 49–54.

Chover, J., Ney, P. and Wainger, S. (1973). Degeneracy properties of subcritical branching processes, *Annals of Probability*, **1**, 663–673.

Dobrushin, R. L. and Shlosman, S. B. (1994). Large and moderate deviations in the Ising model, In *Probability Contributions to Statistical Mechanics* (Ed., R. L. Dobrushin), *Advances in Soviet Mathematics*, **20**, 91–219.

Dugué D. (1941). Sur un nouveau type de courbe de fréquence, *Comptes Rendus de l'Academie des Sciences, Paris*, Tome 213, 634–635.

Grandell, J. (1991). *Aspects of Risk Theory*, Springer-Verlag, New York.

Hougaard, P. (1986). Survival models for heterogeneous populations derived from stable distributions, *Biometrika*, **73**, 387–396.

Jørgensen, B. (1982). *Statistical Properties of the Generalized Inverse Gaussian Distribution*, Lecture Notes in Statistics, – 9, Springer-Verlag, New York.

Jørgensen, B. (1997). *The Theory of Dispersion Models*, Chapman and Hall, London, England.

Jørgensen, B., Martínez, J. R. and Tsao, M. (1994). Asymptotic behaviour of the variance function, *Scandinavian Journal of Statistics*, **21**, 223–243.

Jørgensen, B., Martínez, J. R. and Vinogradov, V. (1999). On weak convergence to Tweedie laws and regular variation of natural exponential families, *Under preparation*.

Jørgensen, B. and de Souza, M. C. P. (1994). Fitting Tweedie's compound Poisson model to insurance claim data, *Scandinavian Actuarial Journal*, 69–93.

Klüppelberg, C. (1989). Estimation of ruin probabilities by means of hazard rates, *Insurance: Mathematics and Economics*, **8**, 279–285.

Nelder, J. A. and Wedderburn, R. W. M. (1972). Generalized linear models, *Journal of the Royal Statistical Society, Series A*, **135**, 370–384.

Schrödinger, E. (1915). Zur theorie der fall-und steigversuche an teilchen mit Brownscher bewegung, *Physikalische Zeitschrift*, **16**, 289–295.

Seshadri, V. (1993). *The Inverse Gaussian Distribution: A Case Study in Exponential Families*. Clarendon Press, Oxford.

Sichel, H. S. (1974). On a distribution law for word frequencies, *Journal of the Royal Statistical Society, Series A*, **137**, 25–34.

Sichel, H. S. (1975). On a distribution law for word sequences, *Journal of American Statistical Association*, **70**, 542–547.

Smoluchowsky, M. V. (1915). Notiz über die Berechnung der Brownschen Molekularbewegung bei der Ehrenhaft-Millikanchen Versuchsanordnung, *Physikalische Zeitschrift*, **16**, 318–321.

Tweedie, M. C. K. (1947). Functions of a statistical variate with given means, with special reference to Laplacian distributions, *Proceedings of the Cambridge Philosophical Society*, **49**, 41–49.

Tweedie, M. C. K. (1957a). Statistical properties of inverse Gaussian distributions, I, *Annals of Mathematical Statistics* **28**, 362–377.

Tweedie, M. C. K. (1957b). Statistical properties of inverse Gaussian distributions, II, *Annals of Mathematical Statistics* **28**, 696–705.

Tweedie, M. C. K. (1984). An index which distinguishes between some important exponential families, In *Statistics: Applications and New Directions. Proceedings of the Indian Statistical Institute Golden Jubilee International Conference* (Eds., J. K. Ghosh and J. Roy), pp. 579–604, Indian Statistical Institute, Calcutta.

Vinogradov, V. (1993). Large deviations for i.i.d. random sums when Cramér's condition is fulfilled only on a finite interval, *Comptes Rendus Mathematical Reports of the Academy of Science of Canada*, **15**, 229–234.

Vinogradov, V. (1994). *Refined Large Deviation Limit Theorems*, Pitman Research Notes in Mathematics, **315**, Longman, Burnt Mill.

Vinogradov, V. (1996). On weak convergence for two families of conditioned random functions, In *Stochastic Analysis: Random Fields and*

Measure-Valued Processes (Eds., J.-P. Fouque *et al.*), Israel Mathematical Conference Proceedings, Vol. 10, pp. 203–214, American Mathematical Society, Providence, RI.

Vinogradov, V. (1999). On a conjecture of B. Jørgensen and A. D. Wentzell: from extreme stable laws to Tweedie exponential dispersion models, In *Proceedings Volume for the International Conference on Stochastic Models in Honour of D. A. Dawson* (Eds., L. Gorostiza and G. Ivanoff), Canadian Mathematical Society Conference Proceedings Series, pp. 435–444, American Mathematical Society, Providence, RI.

Part VII
Methods in Health Research

CHAPTER 31

ESTIMATION OF STAGE OCCUPATION PROBABILITIES IN MULTISTAGE MODELS

SOMNATH DATTA

University of Georgia, Athens, GA

GLEN A. SATTEN

Centers for Disease Control and Prevention, Atlanta, GA

SUSMITA DATTA

Georgia State University, Atlanta, GA

Abstract: In this paper, we review recent developments in nonparametric estimation of stage occupation probabilities for the three stage irreversible illness-death model. Fractional risk set estimators of the stage occupation probabilities under independent right censoring recently proposed by the authors are discussed. Closed form expressions of the asymptotic standard errors of the proposed estimators are presented which were previously unavailable. Extensions of these estimators to more general multistage model are also considered.

Keywords and phrases: Competing risks, fractional risk set, illness-death model, nonparametric estimation, Kaplan-Meier estimator, right-censoring, survival analysis

31.1   INTRODUCTION

The three stage illness death model is a model for simultaneous investigation of the occurrence of an intermediate event (illness) and a subsequent failure event of interest (death). In this model, all individuals are assumed to be alive and free of disease (i.e., in the 'well' state) at time zero; their subsequent evolution is described schematically in Figure 31.1. Specifically, individuals may die before ever developing disease, or may first develop disease before death. The model is irreversible in that an individual in the 'disease' state will eventually die without ever returning to the well state. Typical questions which can be addressed using the 3-stage illness-death model are: What proportion of individuals are alive but ill at time t; what proportion of individuals have died by time t having had illness; and what proportion of individuals have died by time t without having become ill. These questions amount to estimating the stage occupation probabilities for the model.

In real applications, data are usually subject to right censoring. Because the onsets of illness and death can each be considered as failure events, it would appear that estimating the stage occupation probabilities for the three stage illness death model is equivalent to estimating the joint distribution function of times to illness and death. However, this is not the case, and two nonparametric approaches have been recently proposed which make use of the special restricted structure of this model [Hoover et al. (1996a, 1996b)] [Hoover's (1996b) estimators make very strong assumptions on censoring and are not considered further here]. Note that it is also possible to estimate stage occupation probabilities under additional structural assumptions such as a semi-Markov model [Lagakos, Sommer and Zelen (1978)] using a time inhomogeneous Markov model [Aalen and Johansen (1978)].

In a recent paper, Datta et al. (1999) proposed a simple new non-parametric estimators for the stage occupation probabilities of the three stage illness death model which are valid without additional structural assumption. They introduce the concept of a fractional or estimated risk set to obtain a Kaplan-Meier estimator of the conditional distribution of the death times given that an individual dies following illness. The novelty of this approach is that they make use of the illness time information in assessing the death time distribution of such individuals that is ignored by Hoover's estimators. These estimators are reviewed in the next section.

The rest of the paper is organized as follows. In Section 31.2, we introduce the fractional risk set estimator of Datta et al. (1999). A small sample correction is proposed which can be used to ensure that all stage occupation probability estimates are proper (i.e., they lie in $[0, 1]$). Closed form estimators of the standard error of these estimators are presented in Section 31.3 that were previously unpublished. An extension of the fractional risk set approach to more general multistage models is indicated in Section 31.4.

31.2   THE FRACTIONAL RISK SET ESTIMATORS

We first rewrite the three stage illness-death model, shown in Figure 31.1, as a four stage model as shown in Figure 31.2. Let $P_j(t)$ denote the probability of being in stage j at time t, for $j = 0, \cdots, 3$. (The stage 3 occupation probability for the original three-stage model can be obtained by combining the stage 1 and 3 occupation probabilities in the four stage model.) For $1 \leq j \leq 3$, let T_{ij}^* be the random variable denoting the time the ith subject enters stage j ($= \infty$ if stage j is never entered) and let $T_{i0}^* = \min(T_{i1}^*, T_{i2}^*)$ denote the time stage 0 is left. Let C_i denote the censoring time of the ith subject, and let $T_{ij} = \min(T_{ij}^*, C_i)$ if $T_{ij}^* < \infty$, and equals ∞ if stage j is never entered. For $1 \leq j \leq 3$, let $\delta_{ij} = 1$ if the ith subject is seen to have entered stage j and 0 otherwise, and let $\delta_{i0} = 1$ if the ith subject is seen to have left stage 0 and 0 otherwise. Define the indicator variables $X_{ij} = 1$ if the ith subject ever enters stage j in the uncensored experiment and 0 otherwise ($X_{i0} = 1$ for all i). Because of censoring, not all of the X_{ij}'s or event times will be observed for each individual, and the values of δ_{ij} contain all the information about the X_{ij}'s that is observed. Finally, let $t_1 < \cdots t_k \cdots < t_K$ denote the set of all the distinct observed event times (i.e. all observed illness and death times).

If all values of X_{ij} were known, stage occupation probabilities could be easily calculated in the following way. For $0 \leq j \leq 3$, let $n_j = \sum_{i=1}^{n} I[X_{ij} = 1]$, and define counting processes

$$N_j(t) = \sum_{i=1}^{n} I[T_{ij} \leq t, X_{ij} = 1, \delta_{ij} = 1] \qquad (31.2.1)$$

and risk set indicators

$$Y_j(t) = \sum_{i=1}^{n} I[T_{ij} \geq t, X_{ij} = 1]. \qquad (31.2.2)$$

If we define Kaplan-Meier estimators $\widehat{S}_j(t)$ of the conditional survival functions $S_j(t) = Pr[T_{ij} > t | X_{ij} = 1]$ by

$$\widehat{S}_j(t) \equiv 1 - \widehat{F}_j(t) = \prod_{t_k \leq t} \left(1 - \frac{\Delta N_j(t_k)}{Y_j(t_k)}\right) \qquad (31.2.3)$$

where $\Delta N_j(t_k) = N_j(t_k) - N_j(t_k-)$, then we could estimate the stage occupation probabilities using

$$\widehat{P}_0(t) = \widehat{S}_0(t), \qquad (31.2.4)$$

$$\widehat{P}_j(t) = \frac{n_j}{n}\widehat{F}_j(t), \ j = 1, 3, \qquad (31.2.5)$$

and

$$\widehat{P}_2(t) = \frac{n_2}{n}\widehat{F}_2(t) - \frac{n_3}{n}\widehat{F}_3(t). \qquad (31.2.6)$$

Note that $Y_0(t)$ and the $N_j(t)$'s are calculable using the observed data (because $\delta_{ij} = 1$ implies $X_{ij} = 1$). However, the values of $Y_j(t)$ for $j \geq 1$ are not calculable using the censored data because some values of X_{ij} may be missing. Hence, only $\widehat{P}_0(t)$ can be calculated using censored data.

Next we define new estimators of $P_j(t)$, denoted $\tilde{P}_j(t)$ which are similar in form to $\widehat{P}_j(t)$ but can be calculated from the observed data as they use estimates of the size of the population at risk $Y_j(t)$. These estimates of $Y_j(t)$, denoted by $\tilde{Y}_j(t)$ are based on a self-consistency property of the competing risks estimators noted by Satten and Datta (1999)

As mentioned earlier, because both $Y_0(t)$ and $N_0(t)$ are determined from the observed data, the estimator of stage 0 does not require any modification, that is, we take $\tilde{P}_0(t) = \widehat{P}_0(t)$. Following Hoover et al. (1996a), note that we may estimate $P_1(t)$ and $P_2(t) + P_3(t)$ using standard competing risks methodology [Aalen (1976) and Kalbfleisch and Prentice (1980)]. Hence, it only remains to estimate either $P_2(t)$ or $P_3(t)$ and obtain the other by subtraction.

Estimation of $P_1(t)$ and $P_2(t) + P_3(t)$ is accomplished by pooling stages two and three (in the four-stage version shown in Figure 31.2). For $j = 1, 2$, let $P_{0j}(s,t)$ denote the probability that an individual who is in state 0 at time s will move to state j at or before time t. Note that in this notation, $P_{01}(0,t) = P_1(t)$ and $P_{02}(0,t) = P_2(t) + P_3(t)$. A competing risk estimate of $P_{0j}(s,t)$ [see, e.g., Andersen et al. (1993)] is given by

$$\widehat{P}_{0j}(s,t) = \sum_{s < t_k \leq t} \left\{ \prod_{s < t_{k'} < t_k} \left(1 - \frac{\Delta N_0(t_{k'})}{Y_0(t_{k'})} \right) \right\} \frac{\Delta N_j(t_k)}{Y_0(t_k)}, \ j = 1, 2. \qquad (31.2.7)$$

Hence we take

$$\tilde{P}_1(t) = \widehat{P}_{01}(0,t) \qquad (31.2.8)$$

and

$$\tilde{P}_2(t) + \tilde{P}_3(t) = \widehat{P}_{02}(0,t). \qquad (31.2.9)$$

However, Satten and Datta (1999) show that, if we define for $j = 1, 2$,

$$\widehat{\phi}_{ij} = (1 - \delta_{i0})\widehat{P}_{0j}(C_i, \infty) + \delta_{i0}\delta_{ij}, \tag{31.2.10}$$

$$\widehat{n}_j = \sum_{i=1}^{n} \widehat{\phi}_{ij}, \tag{31.2.11}$$

$$\tilde{Y}_j(t) = \sum_{i=1}^{n} \widehat{\phi}_{ij} I[T_{ij} \geq t], \tag{31.2.12}$$

and

$$\tilde{S}_j(t) \equiv 1 - \tilde{F}_j(t) = \prod_{t_k \leq t} \left(1 - \frac{\Delta N_j(t_k)}{\tilde{Y}_j(t_k)}\right) \tag{31.2.13}$$

then

$$\widehat{P}_{0j}(0, t) = \frac{\widehat{n}_j}{n} \tilde{F}_j(t) \quad j = 1, 2 . \tag{31.2.14}$$

The interpretation of (31.2.10)–(31.2.14) is that $\widehat{\phi}_{ij}$ is an estimate of $\phi_{ij} := \Pr[X_{ij} = 1 | T_{ij}, \delta_{ij}, \delta_{i0}]$ i.e. $\widehat{\phi}_{i1}$ is an estimate of the probability that the ith subject is in the group which experiences death before illness, while $\widehat{\phi}_{i2}$ is an estimate of the probability that the ith subject is in the group which experiences illness before death. Hence $\widehat{n}_1$ is an estimate of the number of subjects experiencing death before illness and $\widehat{n}_2$ is an estimate of the number of subjects experiencing illness before death. Equation (31.2.14) states that we may consider the $\widehat{n}_j$ individuals to be in separate groups, and that calculation of separate Kaplan-Meier survival function estimates $\tilde{S}_j(t)$ in each group is exactly equivalent to the competing risks estimators $\widehat{P}_{0j}(0, t)$ calculated using equation (31.2.7).

Equations (31.2.10)–(31.2.14) and their interpretation now suggest the following fractional risk set estimator of $P_3(t)$. Among the $\widehat{n}_2$ (fractional) persons estimated to develop illness before death, calculate $\tilde{P}_3(t)$ as the Kaplan-Meier estimator of times of death. Specifically, because $X_{i2} = 1$ is equivalent to $X_{i3} = 1$, we take $\widehat{\phi}_{i3} = \widehat{\phi}_{i2}$ for each i, $\widehat{n}_3 = \widehat{n}_2$ and write

$$\tilde{Y}_3(t) = \sum_{i=1}^{n} \widehat{\phi}_{i3} I[T_{i3} \geq t] \tag{31.2.15}$$

$$\tilde{S}_3(t) \equiv 1 - \tilde{F}_3(t) = \prod_{t_k \leq t} \left(1 - \frac{\Delta N_3(t_k)}{\tilde{Y}_3(t_k)}\right) \tag{31.2.16}$$

and

$$\tilde{P}_3(t) = \frac{\widehat{n}_3}{n} \tilde{F}_3(t) \tag{31.2.17}$$

and hence

$$\tilde{P}_2(t) = \frac{\hat{n}_2}{n}\,\tilde{F}_2(t) - \frac{\hat{n}_3}{n}\tilde{F}_3(t). \tag{31.2.18}$$

Note that the resulting estimates (31.2.14), (31.2.17) and (31.2.18) are analogous to (31.2.5) and (31.2.6) except that the unknown at-risk functions $Y_j(t)$ have been replaced by the estimable fractional risk set functions $\tilde{Y}_j(t)$ which denote the estimated fraction of mass at risk for death before illness ($j = 1$), illness before death ($j = 2$) or death after illness ($j = 3$) at time t, and the sample fractions n_j/n have been replaced by their fractional risk set estimates.

Because $\tilde{P}_j(t)$ can be expressed as well-studied quantities (Kaplan-Meier estimators or Aalen-Johansen competing risks estimators) it is known that they are valid under any independent censoring mechanism in which the instantaneous hazard of moving between stages in the censored experiment is the same as in the corresponding uncensored experiment. From an operational viewpoint, if we were simulating complete data T_{ij}^* and censoring times C_i assume that $(C_i, T_{i1}^*, T_{i2}^*, T_{i3}^*)$ are independent and identically distributed, and C_i is independent of $(T_{i1}^*, T_{i2}^*, T_{i3}^*)$.

Because the representation of (31.2.14) is exact, consistency and asymptotic normality of $\tilde{P}_1(t)$ and $\tilde{P}_2(t) + \tilde{P}_3(t)$ follows from standard results on the Aalen-Johansen model [see e.g. Anderson et al. (1993)]. To establish consistency of $\tilde{P}_3(t)$ and hence $\tilde{P}_2(t)$, note that because $\tilde{Y}_3(t)$ is asymptotically equivalent to a sum of independent, identically distributed quantities (obtained by replacing $\hat{\phi}_{i3}$ by ϕ_{i3}), the law of large numbers implies that

$$n^{-1}\tilde{Y}_3(t) \xrightarrow{P} E(\phi_{i3}\,I[T_{i3} \geq t]). \tag{31.2.19}$$

Because $\phi_{i3} = E[X_{i3} = 1 \,|\, T_{i3}, \delta_{i0}, \delta_{i2}]$, $1 \leq i \leq n$, the expected value in (31.2.19) equals

$$E\Big\{ I[T_{i3} \geq t] E(I[X_{i3}$$
$$= 1 | T_{i3}, \delta_{i0}, \delta_{i2})\Big\} = E(I[T_{i3} \geq t]I[X_{i3} = 1]$$
$$= P[T_{i3} \geq t, X_{i3} = 1]$$

which is the in-probability limit of $n^{-1}Y_3(t)$ by the law of large numbers. Therefore we obtain

$$\tilde{Y}_3(t)/Y_3(t) \xrightarrow{P} 1, \quad \text{as } n \to \infty. \tag{31.2.20}$$

By the Duhamel equation [cf. (4.3.4) in Andersen *et al.* (1993)],

$$\frac{(\tilde{S}_3(t) - S_3(t))}{S_3(t)} = -\int_0^t \frac{r(s-)}{Y_3(s)} dM_3(s) + \int_0^t \frac{r(s-)}{Y_3(s)}\left\{1 - \frac{Y_3(s)}{\tilde{Y}_3(s)}\right\} dN_3(s)$$

(31.2.21)

where $r(s-) = \tilde{S}_3(s-)/S_3(s)$, $S_3(t) = \exp[-A_3(t)]$ and where $M_3(t) = N_3(t) - \int_0^t Y_3(s)dA_3(s)$ is a (zero mean) Martingale. Using standard Martingale methods, the first term on the right hand side of (31.2.21) converges to zero in probability, while the second term is $o_p(1)$ by (31.2.20), establishing that $\tilde{S}_3(t)$ is a consistent estimator of $S_3(t)$. Consistency of $\hat{n}_3/n$ follows from (31.2.14) because $\hat{n}_3 = \hat{n}_2$. Hence, $\tilde{P}_3(t)$ is a consistent estimator of $P_3(t)$.

Remark 31.2.1 Because $\tilde{P}_2(t)$ is estimated by subtraction, it is necessary to consider whether it is always non-negative. Although $\tilde{P}_2(t)$ is the difference between two Kaplan-Meiers in which time-ordered data from the same individuals is used, this unfortunately does not guarantee positivity of $\tilde{P}_2(t)$ as the following example illustrates. Consider hypothetical data from a population in which death cannot occur before illness. For three individuals, the following illness times τ_i and death times T_i are observed: $(\tau_i, T_i) = (1,5), (2,3), (6,7)$ while a fourth individual was censored before illness at time 4. Standard calculation shows that the Kaplan-Meier estimator of the proportion remaining free of illness is greater than the Kaplan-Meier estimator of the proportion alive in the interval (5, 6), even though $\tau_i \leq T_i$ for each individual. A similar phenomenon was also observed in the one-sample case by Oakes (1993).

To develop estimators which are always lie between 0 and 1 and are also normalized, we replace $\tilde{F}_3(t)$ by min $\tilde{F}_2(t), \tilde{F}_3(t))$ in the formulas (31.2.17) and (31.2.18); it then follows immediately that $\tilde{P}_j(t) \geq 0$, $\sum_{j=0}^3 \tilde{P}_j(t) = 1$ for all t. Furthermore, it is a small sample correction that does not affect their large sample properties such as consistency and asymptotic normality.

An alternative approach to achieve non-negativity that is more complex but intuitively appealing is as follows. It follows from standard results on Kaplan Meier estimation theory that $\tilde{S}_3(t) = 1 - \tilde{F}_3(t)$ can be represented as the function which maximizes the likelihood

$$L = \prod_{k=1}^m [S_3(t_k-) - S_3(t_k)]^{\Delta N_3(t_k)} S_3(t_k)^{\Delta \tilde{Y}_3(t_k) - \Delta N_3(t_k)}$$

(31.2.22)

with respect to $S_3(t)$, where $\Delta \tilde{Y}_3(t) = \tilde{Y}_3(t) - \tilde{Y}_3(t+)$. Choose $\tilde{S}_3(t)$ to maximize (31.2.22) subject to the condition $\tilde{S}_3(t) \geq \tilde{S}_2(t) \equiv 1 - \tilde{F}_2(t)$.

Remark 31.2.2 As argued before the fractional risk set estimators are valid (e.g., consistent) provided the censoring mechanism is independent of the transitions in the chain. Most of the existing survival analysis literature is based on such assumptions of independent censoring. However, in practice, this assumption may be hard to justify. It is quite conceivable that the censoring pattern changes from the well stage to the illness stage. Datta *et al.* (1999) considered a modification of their estimator to adjust for dependent censoring by reweighting both the counting process $N_3(t)$ and the at fractional risk process $\tilde{Y}_3(t)$ by inverse probability of censoring. They showed that the resulting estimator would be valid even under dependent censoring caused by a change in censoring hazard from the well to the illness stage. See Datta *et al.* (1999) for the details.

31.3   VARIANCE ESTIMATION

In this section, we obtain estimates for the asymptotic variances of the estimators $\tilde{P}_j(t)$, $j = 0, \cdots, 3$.

Because $\tilde{P}_0(t)$ is a regular Kaplan-Meier and $\tilde{P}_1(t)$ is obtained using standard competing risks methodology, their asymptotic variances estimates are known [see e.g. Andersen *et al.* (1993)]; we report them here for the sake of completeness. In our notation, we have

$$\widehat{\mathrm{Var}}\left(\tilde{P}_0(t)\right) = \left(\tilde{P}_0(t)\right)^2 \sum_{t_i \leq t} \frac{\Delta N_0(t_i)}{Y_0(t_i)(Y_0(t_i) - \Delta N_0(t_i))}, \qquad (31.3.23)$$

and

$$\widehat{\mathrm{Var}}\left(\tilde{P}_1(t)\right)$$
$$= \sum_{t_i \leq t} \frac{\left\{(\tilde{P}_1(t) - \tilde{P}_1(t_i))^2 \Delta N_0(t_i) + \tilde{P}_0(t_i)(\tilde{P}_0(t_i) - 2\tilde{P}_1(t) + 2\tilde{P}_1(t_i))\Delta N_1(t_i)\right\}}{Y_0(t_i)(Y_0(t_i) - \Delta N_0(t_i))}.$$

$$(31.3.24)$$

The asymptotic variance of $\widehat{P}_3(t) = (\widehat{n}_3/n)\tilde{F}_3(t)$ is obtained from the asymptotic i.i.d. representations of $(\widehat{n}_3/n) = \widehat{P}_{02}(0, \infty)$ and $\tilde{F}_3(t)$. It can be shown that

$$n^{1/2}\left(\tilde{F}_3(t) - F_3(t)\right) \approx S_3(t)\, n^{-1/2} \sum_{i=1}^{n} D_i(t),$$

where $D_i(t)$ are zero mean i.i.d. random variables. Replacing the various population quantities in the definition of $D_i(t)$ by their natural estimators one can obtain

$$\widehat{D}_i(t) = -\frac{nI(T_{i3} \leq t, \Delta N_3(T_{i3}) > 0)}{\widetilde{Y}_3(T_{i3}) - \Delta N_3(T_{i3})}$$

$$+ \widehat{\phi}_{i3} \sum_{t_j \leq T_{i3} \wedge t} \frac{n \Delta N_3(t_j)}{\widetilde{Y}_3(t_j)(\widetilde{Y}_3(t_j) - \Delta N_3(t_j))}$$

$$+ \frac{n\widehat{V}(t, T_{i0})\{-\widehat{P}_{02}(T_{i0}, \infty)I(\Delta N_0(T_{i0}) > 0) + I(\Delta N_2(T_{i0}) > 0)\}}{Y_0(T_{i0})}$$

$$- \sum_{t_j \leq T_{i0}} \frac{n\widehat{V}(t, t_j)(-\widehat{P}_{02}(t_j, \infty)\Delta N_0(t_j) + \Delta N_2(t_j))}{Y_0^2(t_j)},$$

where

$$\widehat{V}(t, s) = \frac{\widetilde{P}_0(s-)}{n} \sum_{t_j < s} \frac{\widehat{H}_3(t \wedge t_j)\Delta N_c(t_j)}{\widetilde{P}_0(t_j)},$$

and

$$\widehat{H}_3(s) = n \sum_{t_j \leq s} \frac{\Delta N_3(t_j)}{\widetilde{Y}_3(t_j)(\widetilde{Y}_3(t_j) - \Delta N_3(t_j))}.$$

In above, N_c is the counting process corresponding to the censoring event. Similarly,

$$n^{1/2}(\widehat{P}_{02}(0, \infty) - P_{02}(0, \infty)) \approx n^{-1/2} \sum_{i=1}^{n} B_i(\infty),$$

with

$$\widehat{B}_i(t) = \frac{n[T_{i0} \leq t]\widetilde{P}_0(T_{i0}-)\{-\widehat{P}_{02}(T_{i0}, t)I(\Delta N_0(T_{i0}) > 0) + I(\Delta N_2(T_{i0}) > 0)\}}{Y_0(T_{i0})}$$

$$- \sum_{t_j \leq T_{i0} \wedge t} \frac{n\widetilde{P}_0(t_j-)(-\widehat{P}_{02}(t_j, t)\Delta N_0(t_j) + \Delta N_2(t_j))}{Y_0^2(t_j)}.$$

Therefore, by delta method, we can estimate $\text{Var}(\widetilde{P}_3(t))$ as

$$\widehat{\text{Var}}\left(\widetilde{P}_3(t)\right) = n^{-2} \sum_{i=1}^{n} \{(\widehat{n}_3/n)\widetilde{S}_3(t)\widehat{D}_i(t) + \widetilde{F}_3(t)\widehat{B}_i(\infty)\}^2. \quad (31.3.25)$$

In the same way, one obtains

$$\widehat{\text{Var}}\left(\widetilde{P}_2(t)\right) = n^{-2} \sum_{i=1}^{n} \{\widehat{B}_i(t) - \widetilde{F}_3(t)\widehat{B}_i(\infty) - (\widehat{n}_3/n)\widetilde{S}_3(t)\widehat{D}_i(t)\}^2,$$

$$(31.3.26)$$

using the representation $\widetilde{P}_2(t) = \widehat{P}_{02}(0, t) - (\widehat{n}_3/n)\widetilde{F}_3(t)$.

31.4   EXTENSION TO MULTISTAGE MODELS

The fractional risk set methodology can be extended to calculate stage occupation probabilities of general multistage models without any structural assumption. Here we briefly indicate how such an extension can be achieved. The details of the results will be available in a future paper.

We assume the network of stages has a tree structure. This assumption is possible because any acyclic network is equivalent to a tree (e.g., Figure 31.2 shows the tree version of the model in Figure 31.1). Cyclic networks are equivalent to infinite trees; we assume the laws governing transition between stages are such that any finite amount of data set from a cyclic network can be described by a finite tree. Hence, we consider a network of $J + 1$ stages with a tree topology in which all individuals begin in stage 0 at time 0. Let t_{ij}^* be the (possibly unobserved) times the ith person enters stage j ($= \infty$ if the ith person never enters stage j), $1 \leq i \leq n$, $1 \leq j \leq J$. We take $t_{i0}^* \equiv 0$, for all i. Let u_{ij}^* denote the (possibly unobserved) time the ith person leaves stage j ($= \infty$ if the ith person never enters stage j or if stage j is a terminal node of the tree). Let C_i be the censoring time for the ith person. Let s_i^* denote the last stage a person was seen at. Define the censoring indicators δ_{ij} and η_{ij} for the entry and departure times to and from stage j, respectively, as

$$\delta_{ij} = \begin{cases} 2 & \text{if } P(t_{ij}^* < \infty | s_i^*) = 0 \\ 1 & \text{if } t_{ij}^* \leq C_i \\ 0 & \text{otherwise}, \end{cases} \qquad (31.4.27)$$

and

$$\eta_{ij} = I(u_{ij}^* \leq C_i). \qquad (31.4.28)$$

Note that the different values of δ_{ij} correspond to $\delta_{ij} = 2$ if person i is known not to have entered stage j, $\delta_{ij} = 1$ if person i has been seen to have entered stage j and $\delta_{ij} = 0$ if person i has been censored at a stage from which it was possible to move to stage j in a number of steps, respectively. Finally, let $t_{ij} = t_{ij}^* \wedge C_i$ if $\delta_{ij} \neq 2$ and $t_{ij} = \infty$ otherwise, and $u_{ij} = u_{ij}^* \wedge C_i$ if $\delta_{ij} \neq 2$ and $u_{ij} = \infty$ otherwise be the censored entry and departure times, respectively of stage j for the ith person. The basic underlying assumption is that the data vectors $Y_i = (\delta_{ij}, \eta_{ij}, t_{ij}, u_{ij}; 0 \leq j \leq J)$ for $1 \leq i \leq n$ are independent and identically distributed.

The goal of this section is to estimate the stage occupation probabilities $P_j(t) = P(s(t) = j)$, $1 \leq j \leq J$, where $s(t)$ denotes the stage an individual occupies at time t, based on $\{Y_i, 1 \leq i \leq n\}$.

Let $N_{\cdot j}(t) = \sum_{i=1}^{n} I(t_{ij} \leq t, \delta_{ij} = 1)$ and $N_{j\cdot}(t) = \sum_{i=1}^{n} I(u_{ij} \leq t,$

$\delta_{ij} = 1$) be the counting processes that count the number of times we observe stage j be entered, and left, respectively, in the time interval $[0, t]$.

Let $X_{ij} = I[t^*_{ij} < \infty]$, $1 \le j \le J$ be random variables that track the stages visited by individual i. Note that due to censoring the X_i's are only partially observed.

The probability of being in stage j at time t is estimated through the relationship

$$P_j(t) = \frac{n_j}{n}\{F_j(t) - G_j(t)\} \qquad (31.4.29)$$

where $n_j = nP[X_{1j} = 1]$ is the expected number of persons out of n who ever enter stage j, $F_j(t) = Pr[t^*_{1j} \le t | X_{1j} = 1]$ and $G_j(t) = Pr[u^*_{1j} \le t | X_{1j} = 1]$ are the distribution functions of the entry and the departure times, respectively, among all individuals in the population who would ever enter stage j. We take $G_j(t) \equiv 0$ if j is a terminal node of the tree, while $F_j(t) \equiv 1$ if $j = 0$ for all $0 \le t < \infty$. The functions F_j and G_j are estimated through the Kaplan-Meier formulas but with corresponding fractional risk sets

$$\tilde{Y}_j(t) = \sum_{i=1}^{n} \hat{\phi}_{ij} I(t_{ij} \ge t), \qquad (31.4.30)$$

$$\tilde{Z}_j(t) = \sum_{i=1}^{n} \hat{\phi}_{ij} I(u_{ij} \ge t), \qquad (31.4.31)$$

$$1 - \tilde{F}_j(t) = \prod_{(0,t]}(1 - \frac{dN._j(s)}{\tilde{Y}_j(s)}), \qquad (31.4.32)$$

$$1 - \tilde{G}_j(t) = \prod_{(0,t]}(1 - \frac{dN_j.(s)}{\tilde{Z}_j(s)}), \qquad (31.4.33)$$

where $\hat{\phi}_{ij}$ is an estimate of the probability that individual would eventually pass through stage j given its censoring information. It is possible to calculate the $\hat{\phi}_{ij}$ recursively from the censored data. Finally, estimate n_j by $\tilde{Y}_j(0)$ to obtain a fractional risk estimator of $P_j(t)$ via (31.4.29).

REFERENCES

Aalen, O. O. (1976). Nonparametric inference in connection with multiple decrement models, *Scandinavian Journal of Statistics*, **3**, 15–27.

Aalen, O. O. and Johansen, S. (1978). An empirical transition matrix for nonhomogeneous Markov chains based on censored observations, *Scandinavian Journal of Statistics*, **5**, 141–150.

Andersen, P. K., Borgan, Ø, Gill, R. D. and Keiding, N. (1993). *Statistical Models Based on Counting Processes*, Springer-Verlag, New York.

Datta, S., Satten, G. A. and Datta, S. (1999). Nonparametric estimation for the three stage irreversible illness-death model, *Biometrics* (to appear).

Dykstra, R. L. (1982). Maximum likelihood estimation of the survival functions of stochastically ordered random variables, *Journal of the American Statistical Society*, **77**, 621–628.

Hoover, D. R., Peng, Y., Saah, A. J., Detels, R. R., Day, R. S. and Phair, J. P. (1996a). Using multiple decrement models to estimate risk and morbidity from specific AIDS illness, *Statistics in Medicine*, **15**, 2307–2321.

Hoover, D. R., Peng, Y., Saah, A. J., Detels, R. R., Rinaldo, C. R. Jr. and Phair, J. P. (1996b). Projecting disease when death is likely, *American Journal of Epidemiology*, **143**, 943–952.

Kalbfleisch, J. D. and Prentice, R. L. (1980). *The Statistical Analysis of Failure Time Data*, John Wiley & Sons, New York.

Lagakos, S. W., Sommer, C. J. and Zelen, M. (1978). Semi-Markov models for partially censored data, *Biometrika*, **65**, 311–318.

Oakes, D. (1993). A note on the Kaplan-Meier estimator, *The American Statistician*, **47**, 39–40.

Satten, G. A. and Datta, S. (1999). A Kaplan-Meier representation of the competing risks estimates, *Statistics & Probability Letters*, **42**, 299–304.

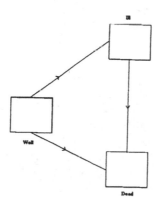

FIGURE 31.1   The irreversible illness-death model

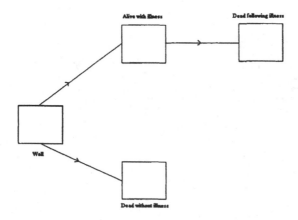

FIGURE 31.2   A tree representation for the illness-death model

CHAPTER 32

STATISTICAL METHODS IN THE VALIDATION PROCESS OF A HEALTH RELATED QUALITY OF LIFE QUESTIONNAIRE: CLASSICAL AND MODERN THEORY

MOUNIR MESBAH   AGNÉS HAMON

Université de Bretagne Sud, Vannes, France

Abstract: In this chapter, we review various statistical methods that have been developed in the literature for the validation process of a health related quality of life questionnaire. The classical methods are based on mixed linear models while modern methods are based on mixed generalized linear models.

Keywords and phrases: Quality of life (QoL), linear models, mixed linear models, Rasch model, QoL questionnaire

32.1 INTRODUCTION

Statistical methods used during the validation process of health related Quality of Life (QoL) questionnaires are, in fact, also commonly used in a wide variety of other fields. In educational science, tests can be used to indicate possible promotion of students from one course to the next or evaluating the impact of new teaching methodologies. In the social and

507

behavioral sciences, interest may focus on such diverse subjects such as intelligence, sociability or political attitudes. In the medical field, psychiatric questionnaires are commonly used to measure, for example, depressive status. Recently health related QoL scales have been developed and they are increasingly used in different contexts: comparing drugs or interventions in clinical trial or measuring health of specific population in epidemiologic studies.

Measuring a variable needs an instrument. QoL in clinical trial or epidemiological studies is generally assessed by self-rated questionnaires consisting of a number of items (questions). Responses to these items are generally dichotomous (Yes, No; Agree, Disagree) or ordinal (Not at All, A little, Quite a bit, Very much). This QoL instrument must produce scalar scores (numbers) for each item (item measure). Each item of the instrument contributes to produce a sub-dimensional score (sub-scale) in common with a sub-group of items measuring the same underlying trait. Item scores and sub-scales scores constitute the measurement of the instrument. The purpose of this paper is to investigate how to evaluate the quality of such instruments, i.e.,which statistical methods and which measurement models to use for that purpose. A crucial question is what is a good item and what is a bad item in a questionnaire.

Many standard instruments are currently available. Among others, two of the most well known are the Sickness Impact Profile (SIP) which comprises 136 items distributed into 12 dimensions [Bergner et al. (1976)] and the Nottingham Health Profile (NHP) [Hunt et al. (1981)] with 38 dichotomous items for 6 dimensions: specifically, Physical Mobility (8 items), Social Isolation (5 items), Pain (8 items), Emotional Reaction (9 items), Energy (3 items) and Sleeping (5 items). These two questionnaires are known as generic instruments as opposed to the EORTC QLQ-C30 questionnaire [Aaronson (1993)] which is a specific QoL measure of cancer patients. This contains 38 items forming 15 dimensions. Annex 1 presents the items of the Communication dimension of the SIP.

The contribution of statisticians in developing and validating QoL instruments is small but increasing. The validation process of a health related QoL questionnaire involve multidisciplinary collaboration of physicians, nurses, psychologists, linguists and so on. Such multidisciplinary participation is essential in the face or contents validation of the instrument [Nunnally and Bernstein (1994)]. Our main interest in this paper concerns statistical validation of an instrument and in particular, internal statistical validation.

The common statistical methods used to evaluate the measurement properties of an instrument depend on what data is available. If, in addition to a sample from the questionnaire, we have, jointly, data from an external criterion: expert, gold standard, etc. , we can externally validate the new QoL instrument by correlating it to that external criterion. Classical methods of regression and discriminant analysis can be used for that purpose. The statistical method to evaluate the quality of the instrument is simply an evaluation of the association between the expert or gold standard and the measurements given by the questionnaire. In the QoL field, this is unusual. Only responses of the questionnaire are observed and the gold standard or expert responses could be considered as unobserved latent variable.

Concurrent validity involves indirect evaluation of the association between measurements given by the questionnaire and some surrogate variables expected close to the latent trait measured.

Test-retest correlation is often used to evaluate the reproducibility of the questionnaire, when the experimental design includes for each subject two assessments at two time points so close together to consider that the QoL is unchanged yet so far apart as to avoid memory bias.

From now, we focus on statistical internal validation methods. Factorial analysis and related methods are probably the most commonly used, essentially for dividing the items in separate dimensions. Principal component analysis with varimax rotation is the factorial analysis method which is often used in QoL fields. These methods give best results when the item responses are normally distributed, or truly quantitative.

From now, we consider only unidimensional scales for which we present classical and modern methods. Classical methods are based on mixed linear models, while modern methods use mixed generalized linear models.

32.2   CLASSICAL PSYCHOMETRIC THEORY

32.2.1   The Strictly Parallel Model

We note X_{ij} the response of person i $(i = 1, .., n)$ to item j $(j = 1, .., k)$. The first model is just a mixed one way model with the measure or item as random factor

$$X_{ij} = \mu + a_i + e_{ij}$$

with

1. μ is a constant fixed effect.

2. a_i is a random effect with zero mean and standard error σ_a corresponding to the person variability. It produces the variance of the true measure.

3. e_{ij} is a random effect with zero mean and standard error σ_e corresponding to the additional measurement error.

4. The true measure and the error are uncorrelated, cov $(a_i, e_{ij}) = 0$.

5. $\forall j = 1..k$ (a_i, e_{ij}) and $(a_{i'}, e_{i'j})$ are independent for $i \neq i'$

These assumptions are classical in experimental design. This model defines relationships between different kinds of variables: the observed score X_{ij}, the true score a_i and the error e_{ij}. It is interesting to make some remarks about assumptions underlying this model:

i) e_{ij} are measurement errors of item j

ii) the "true" measure of person i is $\tau_i = \mu + a_i$ and it is assumed to be independent of item j

iii) the measurement errors e_{ij} are assumed uncorrelated with the true measure τ_i.

iv) X_{ij} is observed whilst τ_i is not

v) it is easy to show that E $(X_{ij}) = \mu$; Var $(X_{ij}) = \sigma_a^2 + \sigma_e^2$; cov$(X_{ij}, X_{ij'}) = \sigma_a^2$

Clearly, this model is good for biological repeated measures, when the same measure is repeated at different steps, assuming no systematic change ($\mu = constant$). In the case of QoL, we can assume that all items measure the same thing, but they do it probably at a different level or difficulty. Thus, a more realistic model is the parallel model, with the true score being:

$$\tau_{ij} = \mu_j + a_i$$

We can, then, review the previous remarks to adjust them. This model allows the same covariance structure for the data, but with a slight different mean structure.

32.2.2 Reliability of an Instrument

Definition of reliability coefficient

A measurement instrument gives us values that we call observed measure. The reliability ρ of the instrument is defined as the ratio

$$\frac{\text{Var (true measure)}}{\text{Var (observed measure)}}$$

If the parallel model (or the strictly parallel one) is assumed, we can show that

$$\rho = \frac{\sigma_a^2}{\sigma_a^2 + \sigma_e^2}$$

which is also the constant correlation between any two items. This coefficient is also known as the intra-class coefficient.

Interpretation

The reliability coefficient ρ can be easily interpreted as a correlation coefficient between the true and the observed measure. The k straight regression lines (X_{ij}, τ_{ij}), corresponding to the items $(j = 1, .., k)$ are parallel and

$$\text{Corr}(X_{ij}, \tau_{ij}) = \rho^{1/2}$$

When the parallel model is assumed, the reliability $\tilde{\rho}$ of the sum of k items equals

$$\tilde{\rho} = \frac{k\rho}{(k-1)\rho + 1}$$

This formula is known as the Spearman-Brown formula, it shows that, under the parallel model, when the number of items increase the reliability tends to 1. Its maximum likelihood estimator, under the assumption of a normal distribution of the error and parallel model, is known as the Cronbach Alpha Coefficient (CAC) [Kristof (1963)]:

$$\alpha = \frac{k}{k-1}\left[1 - \frac{\sum_{j=1}^k S_j^2}{S_{tot}^2}\right]$$

with

$$S_j^2 = \frac{1}{n-1}\sum_{i=1}^n (x_{ij} - \bar{x}_j)^2$$

$$S_{tot}^2 = \frac{1}{nk-1}\sum_{i=1}^n\sum_{j=1}^k (x_{ij} - \bar{x})^2$$

CAC and principal components analysis

It is easy to show a direct connection between CAC and the percentage of variance of the first component in PCA, which is, with factor analysis,

often used in the validation process of a questionnaire [Moret *et al.* (1993)].
The PCA is based on the correlation matrix of the items R. Here

$$R = \begin{bmatrix} 1 & & \rho \\ & 1 & \\ \rho & & 1 \end{bmatrix}$$

This matrix has only two different latent roots, the greater root is

$$\lambda_1 = (k-1)\rho + 1$$

and the other multiple roots are

$$\lambda_2 = 1 - \rho = \frac{(k - \lambda_1)}{(k-1)}$$

This clearly indicate a monotone relationship between $\tilde{\rho}$, which is estimated
by α (CAC) and the first latent root λ_1, which is in practice estimated by
the corresponding value of the observed correlation matrix, and thus, the
percentage of variance of the first principal component in a PCA. Thus,
CAC is often also considered as a measure of unidimensionality.

Step by step procedure to select items with CAC

CAC can be computed to find the most reliable subset of items [Curt *et al.*
(1997) and Moret *et al.* (1993)]. At a first, all items are used to compute the
CAC. Then at every step, one item is removed from the scale. The removed
item is the one which gives for the scale without the item the maximum
CAC. This procedure is repeated until only two items remained. If the
parallel model is true, it can be shown, using Spearman-Brown formula
that increasing the number of items increase the reliability of the total score
which is estimated by Cronbach alpha. Thus, a decrease of such curve when
adding an item could strongly bring us to suspect that the given item is a
bad one (in term of goodness of fit of the model). If an instrument is already
validated, the curve is monotonously increasing (Figure 32.1, example with
the Communication dimension of the SIP). Figure 32.2 (Social Interaction
dimension of the SIP) shows a non-increasing curve. We can choose to
increase the reliability of the instrument by deleting 4 items: numbers 2, 13,
18 and 10 (see annex for content of these items). In fact, this instrument
is already validated in previous studies with different populations and is
considered as a standard. Moreover, we can see that the decreasing of
the curve is very slight. These arguments bring us to choose to keep the
instrument without removing any items.

A supplementary and popular way to assess the influence of the item on the goodness of fit of the parallel model is by examining the empirical correlations item to total (or to total minus the given item). Assuming the parallel model, these correlations must be equal. A low correlation indicates a bad item. All popular software (SAS, SPSS, SYSTAT, etc.) include computation of these statistics and additional goodness of fit assessment. Unfortunately none include the stepwise built curve of Cronbach Alpha or even any other parsimonious criteria.

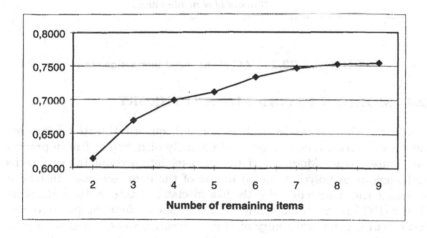

FIGURE 32.1 CAC of the communication scale

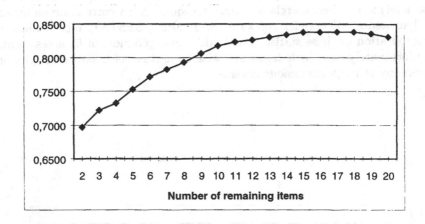

FIGURE 32.2   CAC of the social interaction scale

32.3   MODERN PSYCHOMETRIC THEORY

Clearly, the classical model deals better with quantitative item responses, and much better if it can be assumed normally distributed. But in practice this is rarely true. Modern test theory takes into account the qualitative (dichotomous or polychotomous) nature of the item response. Instead of modelling the item score directly like in classical theory, Item Response Theory (IRT) relies on modelling the item response function (or characteristic curve) i.e., the probability of "taking positive value" conditionally on the actual value of the latent trait. More precisely classical methods use classical linear models, in fact mixed linear models and modern methods use mixed generalized linear models.

Let us consider a set of n people having answered a test of k dichotomous items. Denote X_{ij} like previously, be the answer of individual i to item j and θ_i the unobserved latent variable. The three assumptions, common to all unidimensional IRT models, are :

1. the latent variable θ_i is unidimensional (scalar)

2. variables X_{ij} are independent conditionally on θ_i, this hypothesis is also called "local independence".

3. $P(X_{ij} = 1/\theta_i) = f(\theta_i, \gamma_j)$ where $\gamma_j \in \mathbb{R}^p$ is a vector of unknown item parameters and f is a monotone increasing function in θ (in most cases the logistic or probit link function).

The coded answer 1 is called "correct" answer; this term comes from the educational science vocabulary. When θ_i is very large (towards $+\infty$), and thus when one consider an individual with a very good aptitude, his probability of answering 1 is very close to 1.

The local independence property, weaker than pairwise independence, is useful for deriving the likelihood. It define such Rasch model as a graphical model [Edwards (1995)] with unobserved variable θ.

The marginal distribution of θ is often assumed Gaussian. The developer of a QoL instrument generally wishes to produce normally distributed scores, thus Conditional Gaussian model [Lauritzen and Wermuth (1988)] is rarely used. Cox, in a discussion of the papers by Edwards and Wermuth and Lauritzen (1990) have already remarked : "The conditional Gaussian distributions provide an elegant basis for dealing with mixed and continuous responses, but it should be emphasized that there are other possibilities, especially simple linear logistic models conditional on a marginally multivariate normal response or a fully probit model, have simple properties for some purposes."

32.3.1   The Rasch Model

This model was first developed by the Danish mathematician Rasch (1960) and it is also sometimes called the one parameter logistic model because of its formulation

$$\text{logit}(P(X_{ij} = 1/\theta_i)) = \theta_i - \beta_j \quad \Leftrightarrow \quad P(X_{ij} = 1/\theta_i) = \frac{e^{\theta_i - \beta_j}}{1 + e^{\theta_i - \beta_j}}$$

$$\text{where } \beta_j \in \mathbb{R}$$

The probability in this formula, viewed as a function of θ, is called the item characteristic curve. It is $1/2$ when $\theta_i = \beta_j$ and at fixed θ_i, it is decreasing. So the larger β_j, the more the probability of a good answer approaches 0. The parameter β_j is thus called the item difficulty parameter. We present in Figure 32.3 the estimated characteristic curves of the Rasch model for six items of the Communication scale of the SIP (β_j estimates are given in section 32.3.1). First, we remark that the central parts of the curves are straight lines. Another interesting property is that the different curves corresponding to different items are "parallel" in the sense that they don't cross.

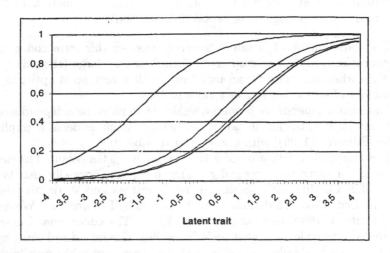

FIGURE 32.3 Estimated characteristic curves for the communication scale

Sufficiency of the total subject score

The Rasch model distinguishes itself from the other IRT models by the following main property : the individual's total score is a sufficient statistic for the ability parameter. Hence when we use the total score as the measure produced by the instrument, we assume that the Rasch model is true. If the Rasch model is not true, we can fit and use a more complicated model like a two parameter model [Birnbaum (1968)], but then we must estimate the latent trait by another method. We cannot use the total person score as an estimate of the latent trait. If we want to produce a simple score, we have to estimate weights, using data from a representative sample, to construct weighted sum for items as a simple score.

Specific objectivity

One interesting property of the Rasch model is the specific objectivity. This is a measurement property which means that comparison of any two persons v and w does not involves instrument parameter. A natural comparator is the probability that person v passes the item j and that a second person w does not pass it conditionally on the event that one of the two succeeds, that is

$$C(v, w) = P\left(X_{vj} = 1 \quad X_{wj} = 0 / X_{vj} + X_{wj} = 1\right)$$
$$= \frac{P\left(X_{vj} = 1 \quad X_{wj} = 0\right)}{P\left(X_{vj} + X_{wj} = 1\right)}$$

This can be written

$$\frac{e^{\theta v - \beta_j}}{1 + e^{\theta v - \beta_j}} \times \frac{1}{1 + e^{\theta w - \beta_j}} \times \left[\frac{e^{\theta v - \beta_j} + e^{\theta w - \beta_j}}{(1 + e^{\theta v - \beta_j})(1 + e^{\theta w - \beta_j})}\right]^{-1}$$

$$= \frac{e^{\theta v}}{e^{\theta v} + e^{\theta w}}$$

We can see that this probability does not depend on the item parameters (β_j). This is true only for the Rasch model. This is not true for another link function. [Fischer and Molenaar (1995)]

Parameter estimation

Three methods of estimation are commonly used : Joint (or unconditional) Maximum Likelihood (JML), Conditional Maximum Likelihood (CML) and Marginal Maximum Likelihood (MML).

The Joint Maximum Likelihood is just the classical maximum likelihood method with θ_i and β_j considered as unknown fixed parameters. Using the local independence assumption and independence between individuals, we can write the likelihood

$$L\left(x^1, ..., x^n; \theta_1, ..., \theta_n, \beta_1, ..., \beta_k\right)$$
$$= \prod_{i=1}^{n} \prod_{j=1}^{k} \frac{\exp\left[x_{ij}\left(\theta_i - \beta_j\right)\right]}{1 + \exp\left(\theta_i - \beta_j\right)}$$

where $x^i = (x_{i1}, .., x_{ik})$ is the response vector of person i. Noting $S_i = \sum_{j=1}^{k} x_{ij}$ and $T_j = \sum_{i=1}^{n} x_{ij}$, the log-likelihood can be written

$$\log L = \sum_{i=1}^{n} S_i \theta_i - \sum_{j=1}^{k} T_j \beta_j - \sum_{i,j} \log\left(1 + e^{\theta_i - \beta_j}\right)$$

Taking derivatives with respect to θ_i and β_j, we find the following equations

$$\begin{cases} S_i = \sum_{j=1}^{k} \frac{e^{\theta_i - \beta_j}}{1 + e^{\theta_i - \beta_j}} \\ T_j = \sum_{i=1}^{n} \frac{e^{\theta_i - \beta_j}}{1 + e^{\theta_i - \beta_j}} \end{cases}$$

Unfortunately the estimates of θ_i and β_j thus obtained are asymptotically biased [Andersen (1973a)] and inconsistent [Haberman (1977)] when $n \to \infty$ and k remains moderate.

A second method consists of maximizing the conditional likelihood given the sufficient statistic of the ability parameters. Under mild conditions, this method gives consistent and asymptotically normally distributed estimators of the item parameters [Andersen (1973a) and Pfanzagl (1993)]. Fischer (1981) derived a set of necessary conditions for existence and uniqueness of the CML estimates.

The last method, called marginal likelihood estimation, is directly related to the interpretation of the Rasch model as a mixed model with θ as a random effect with a distribution g with unknown parameters ϕ. The likelihood is then

$$L\left(x^1, ..., x^n; \phi, \beta_1, ..., \beta_k\right)$$
$$= \int \prod_{j=1}^{k} P\left(X_{ij} = x_{ij}/\theta_i, \beta_j\right) \times g\left(\theta_i, \phi\right) \, d\theta_i$$

The problem of this method is that we have no any explicit form of this integral and maximum likelihood solutions are not straightforward. A first approach implements an EM algorithm [Thissen (1982)]. Considering θ as a missing variable, the complete likelihood is

$$l_c\left(x, \theta, \beta, \phi\right) = \prod_{i=1}^{n} \left\{ \prod_{j=1}^{k} \frac{\exp\left[x_{ij}\left(\theta_i - \beta_j\right)\right]}{1 + \exp\left(\theta_i - \beta_j\right)} \times g\left(\theta_i, \phi\right) \right\}$$

and the EM can be decomposed into two steps : having previous estimations $\left(\beta^{(p)}, \phi^{(p)}\right)$

E-Step Compute the mean of the complete log-likelihood with respect to the distribution of the unobserved θ conditionally to the observations.

$$Q\left(\beta, \phi/\beta^{(p)}, \phi^{(p)}\right)$$

$$= \mathrm{E}\left(\log l_c\left(x, \theta, \beta, \phi\right)/x, \beta^{(p)}, \phi^{(p)}\right)$$

$$= \int \log l_c\left(x, \theta, \beta, \phi\right) f\left(\theta/x, \beta^{(p)}, \phi^{(p)}\right) d\theta$$

$$= \sum_{i=1}^{n} \int \log\left(g\left(\theta_i, \phi\right)\right) f\left(\theta_i/x_i, \beta^{(p)}, \phi^{(p)}\right) d\theta_i$$

$$+ \sum_{i=1}^{n} \sum_{j=1}^{k} \int \log\left(p\left(x_{ij}/\theta_i, \beta_j\right)\right) f\left(\theta_i/x_i, \beta^{(p)}, \phi^{(p)}\right) d\theta_i$$

M-Step Update estimations by computing the point $\left(\beta^{(p+1)}, \phi^{(p+1)}\right)$ that maximize $Q\left(\beta, \phi/\beta^{(p)}, \phi^{(p)}\right)$

The difficulty of this algorithm is the E-step because there is no explicit form of the integral. A possible solution is to use the Gauss-Hermite quadrature, to obtain a numerical approximation. Another solution is to implement a stochastic EM algorithm [Hamon (1998)]. In this case, the E-step of the EM algorithm is replaced by a simulation step to impute missing values of θ and then one perform an EM-step which consist on maximizing the complete log-likelihood.

Goodness of fit tests

The use of goodness-of-fit tests checks the nature of the departure of the data from the IRT assumptions [Dupuy (1999)]. In the following, difficulties of the items are estimated using the conditional maximum likelihood method. We present three test statistics, which will be used to assess the properties of the quality of life questionnaire. The first one is an overall measure of how all the items of the questionnaire fit the Rasch model. It is based on the fact that under the Rasch model, we should expect the overall CML estimates of the difficulties to be approximately equal to the CML estimates computed in each score group [Andersen (1973b)]. Let

- $\hat{\beta}_r$ denote the vector of items difficulties estimated using the subsample of respondents scoring r on the questionnaire

- $l_r\left(\hat{\beta}_r\right)$ denote the likelihood function computed using the subsample of respondents scoring r on the questionnaire.

Under the null hypothesis that the assumptions of the Rasch model hold, the test statistic:

$$Z = -2\ln \frac{l\left(\hat{\beta}\right)}{\prod\limits_{r=1}^{k-1} l_r\left(\hat{\beta_r}\right)}$$

is asymptotically distributed as χ^2 with $(k-1)(k-2)$ degrees of freedom [Andersen (1973b)].

The R_{1c} statistic tests the null hypothesis against the alternative that some of the item response functions are not monotone increasing functions. Suppose that this assumption is not satisfied for an item. The observed response function for this item should diverge from its expected response function. For each score, one calculates how close these two response functions are, making the differences between observed and expected proportions of individuals who succeed on the item. Let n_r be the number of respondents scoring r, n_{ri} be the observed number of respondents scoring r and who answer *yes* to item i, and $\hat{n}_{ri}$ be its expected value with the item parameter CML estimates inserted. The test is based on the difference $d_{ri} = (n_{ri} - \hat{n}_{ri})/\sqrt{n_r}$. Let $\mathbf{d_r}$ denote the vector of elements d_{ri} and V_r denote the matrix of covariance of $\mathbf{d_r}$. Glas (1988) has shown that

$$R_{1c} = \sum_{r=1}^{k-1} \mathbf{d_r'} \cdot \hat{\mathbf{V}}_{\mathbf{r}}^{-1} \cdot \mathbf{d_r}$$

has an asymptotic χ^2 distribution with $(k-1)(k-2)$ degrees of freedom.

The R_{2c} statistic tests the null hypothesis against the alternative hypothesis that local independence does not hold. If two items are not locally independent, say a difficult item is easier for individuals succeeding on an easier one, the observed number of individuals succeeding on both items will be larger than the expected number of such individuals. For each score, computing the difference between observed and expected numbers of individuals succeeding on a couple of items should reveal such a pattern. Computing differences for every couples of items result in an overall measure of local independence. Let n_{rij} denote the observed number of respondents scoring r and who answer *yes* simultaneously to items i and j and $\hat{n}_{rij}$ be its CML expected value. Let f_{rij} denote the difference $(n_{rij} - \hat{n}_{rij})/\sqrt{n}$ and $\mathbf{f}$ denote the vector of elements f_{rij}. Let $\mathbf{U}$ be the covariance matrix of $\mathbf{f}$. Glas (1988) has shown that

$$R_{2c} = \mathbf{d_1'} \cdot \hat{\mathbf{V}}_1^{-1} \cdot \mathbf{d_1} + \mathbf{f'} \cdot \hat{\mathbf{U}}^{-1} \cdot \mathbf{f}$$

has an asymptotic χ^2 distribution with $\frac{k(k-1)}{2}$ degrees of freedom.

In fact this statistic tests the independence of the observed items conditional to the total score observed, not conditional to the true unobserved latent score. Clearly this hypothesis depends on the form of its distribution, which is unidentifiable. Nevertheless, the sufficiency property of the total score allows us to reasonably expect that the null hypothesis tested by this statistic is probably close to the true one expected. Hatzinger (1989) shows how to derive equivalent tests using log-linear models for contingency tables and GLIM software.

Assessment of reliability

In section 32.2, we showed that reliability is an important feature of a psychometric questionnaire. Although the Rasch model is more and more widely used to study QoL questionnaire, reliability is, in practice, almost always estimated with alpha coefficient. In fact, in the mixed Rasch model, it is possible to define a true score like in classical theory and in the same manner, to define reliability coefficient as the ratio of the true score variance and the observed score variance. More precisely if we define the true score $T(\theta_i)$ as the expected score for individual with latent trait θ_i

$$T(\theta_i) = \sum_{j=1}^{k} P(X_{ij} = 1/\theta_i)$$

then we have

$$\text{Var}(T(\theta_i)) = \int T(\theta_i)^2 g(\theta_i, \sigma^2)\, d\theta_i$$

With the general formula

$$\text{Var}(X) = \text{Var}(\text{E}(X/Y) + \text{E}(\text{Var}(X/Y))$$

and if we suppose that the random variable θ is normally distributed, we have

$$\text{Var}\left(\sum X_{ij}\right)$$
$$= \text{Var}(T(\theta_i)) + \text{E}(\text{Var}(T(\theta_i)/\theta_i))$$
$$= \int T(\theta_i)^2 g(\theta_i, \sigma^2)\, d\theta_i$$
$$+ \sum_{j=1}^{k} \int P(X_{ij} = 1/\theta_i)[1 - P(X_{ij} = 1/\theta_i)] g(\theta_i, \sigma^2)\, d\theta_i$$

where $g\left(\theta_i, \sigma^2\right)$ is the normal distribution with variance σ^2. To calculate this coefficient, one has to evaluate complex integrals. An approximation with a Taylor expansion when $\sigma^2 \to 0$ (σ^2 variance of θ) gives an expression of reliability coefficient :

$$\rho = \sigma^2 \sum_{j=1}^{k} P_j \left(1 - P_j\right) + o\left(\sigma^2\right)$$

where

$$P_j = \frac{\exp\left(-\beta_j\right)}{1 + \exp\left(-\beta_j\right)}$$

Unfortunately, in applications σ^2 is too large and the approximation is no more valid. Another way to compute this coefficient, when parameters estimations are known, is to approximate the integrals with numerical procedures like Monte Carlo algorithm.

In the IRT literature, no such reliability coefficient is clearly defined with link to classical theory. The most common approach is to study the Fisher information for θ [Lord (1980)]. This function provides an accuracy measure of ability estimations at any point along the latent trait. So it is not a global measure and practical applications are not evident. A step by step procedure is not possible to implement with the test information function, because removing an item implies necessary a decrease of information. And then no operational parsimonious criteria is available to select items.

Example, the communication scale of the SIP

TABLE 32.1   Estimation of the difficulty parameters for the communication scale

Item number	$\hat{\beta}_j$	SE$\left(\hat{\beta}_j\right)$
2	1.06	0.14
3	0.94	0.13
4	-0.41	0.11
6	-0.33	0.11
8	0.48	0.12
9	-1.75	0.12

We present here estimates of the difficulty parameters of 6 items from the Communication dimension of the SIP. Estimations are obtained with

RSP software [Glas and Ellis (1993)] and we use the CML method. Items 1, 5 and 7 are removed because they contributed to the lack of fit of the model. The remaining items give a p-value for the R_{1c} statistic of 0.16 which indicates that the hypothesis of parallel curves is not rejected.

32.4 CONCLUSION

As we move towards the end of the millennium QoL assessments are becoming of increased importance and interest to a variety of users. "Increasing and sometimes indiscriminate use of QoL measures has provoked concern about these methods in the (health) context, especially when important consequences, such as treatment decisions or resource allocation, depend on them" [Cox *et al.* (1992)].

Two kinds of scientists usually deals with QoL data. Developers of questionnaires are mainly social scientists and, more often than not, psychologists. Analysis of data from clinical trials or other medical data is generally conducted by statisticians and in particular biostatisticians. The problem of non sampling error and particularly the problem of measurement error became an important problem for statisticians. Analyzing data without taking into account the model underlying the production of the measure could bring us to make the wrong decisions. In this paper we focus on statistical methods based on simple measurement models for unidimensional scales. We present the Rasch model for dichotomous items. Generalization to polychotomous items is straightforward, and equivalently to polychotomous logistic models. However care must be taken in the choice of parameters needed for success at different levels of the item [van der Linden and Hambleton (1996)]. Composite or multidimensional instruments are commonly used in QoL. Multi-Trait Multi-Item analysis methods [Ware *et al.* (1997)] are commonly used to validate multidimensional questionnaires when the grouping of items is already established. These methods are related to classical models. No equivalent methods, related to Rasch, exist or are commonly used.

Acknowledgement This research was partly supported by the Conseil Régional de Bretagne.

REFERENCES

Aaronson, N. K., *et al.* (1993). The European Organization for Research and Treatment of Cancer QLQ-C30 : A Quality of life instrument for use in international clinical trials in oncology, *Journal of the National Cancer Institute*, **85**.

Andersen, E. B. (1973a). Conditional inference and multiple-choice questionnaires, *The British Journal of Mathematical and Statistical Psychology*, **26**, 31–44.

Andersen, E. B. (1973b). A goodness of fit test for the Rasch model, *Psychometrika*, **38**, 123–140.

Birnbaum, A. (1968). Some latent trait models and their use in inferring an examinee's ability, In *Statistical Theories of Mental Test Scores* (Eds., F. M. Lord and M. R. Novick), pp. 395–479, Addison-Wesley, Reading, MA.

Bergner, M. , Bobbit, R. A., Pollard, W. E., Martin, D. P. and Gilson, B. S. (1976). The Sickness Impact Profile : validation of a health status measure, *Medical Care*, **14**, 57–67.

Cochran, W. G. and Cox, G. M. (1957). *Experimental Design.* John Wiley & Sons, New York.

Cox, D. (1990). Discussion of the papers by Edwards, and Wermuth and Lauritzen, *Journal of the Royal Statistical Society, Series B*, **52**, 56.

Cox, D., Fitzpatrick, R., Fletcher, A. E., Gore, S. M., Spiegelhalter, D. J. and Jones, D. R. (1992). Quality-of-life assesment: Can we keep it simple?, *Journal of the Royal Statistical Society, Series A*, **155**, 353–393.

Curt, F., Mesbah, M., Lellouch, J. and Dellatolas, G. (1997). Handedness scale: How many and which items?, *Laterality*, **2**, 137–154.

Dupuy, J. F. (1999). Validating a questionnaire of quality of life using the Rasch model, *Technical Report No. 9903*, Laboratoire SABRES, Université de Bretagne Sud.

Edwards, D. (1995). *Introduction to Graphical Modelling*, Springer-Verlag, New York.

Fischer, G. H. (1981). On the existence and uniqueness of maximum-likelihood estimates of the Rasch model, *Psychometrika*, **46**, 59–77.

Fischer, G. H. and Molenaar, I. W. (1995). *Rasch Models, Foundations, Recent Developments, and Applications*, Springer-Verlag, New York.

Ghosh, M. (1995). Inconsistent maximum likelihood for the Rasch model, *Statistics & Probability Letters*, *23*, 165–170.

Glas, C. A. W. (1988). The derivation of some tests for the Rasch model from the multinomial distribution, *Psychometrika*, **53**, 525–546.

Glas, C. A. and Ellis, J. (1993). *Rasch Scaling Program: User's Manual Guide*, Iec-Progamma, Groningen.

Haberman, S. J. (1977) Maximum likelihood estimates in exponential response models, *Annals of Statistics*, **5**, 815–841.

Hamon, A. (1998). A stochastic EM algorithm for estimation in the Rasch model, *Technical Report No. 9804*, Laboratoire SABRES, Université de Bretagne Sud.

Hatzinger, R. (1989). The Rasch model, some extensions and their relation to the class of generalized linear models, *Lecture Notes in Statistics*, – **57**, pp. 172–179, Springer-Verlag, Berlin.

Hunt, S. M., McEwen, J. and Mc Kenna. S. P. (1981). *The Nottingham Health Profile User's Manual*.

Kristof, W. (1963). The statistical theory of stepped-up reliability coefficients when a test has been divided into several equivalent parts, *Psychometrika*, **28**, 221–238.

Lauritzen, S. L. and Wermuth, N. (1989). Graphical models for associations between variables, some of which are qualitative and some quantitative, *Annals of Statistics*, **17**, 31–54.

Lord, F. M. (1980). *Applications of Item Response Theory to Practical Testing Problems*, Erlbaum, Hillsdale, NJ.

Moret, L., Mesbah, M., Chawlow, J. and Lellouch, J. (1993). Validation interne d'une échelle de mesure : relation entre analyse en composantes principales, coefficient α de Cronbach et coefficient de corrélation intra-classe, *Rev. Epidém. et Santé Publ.*, **41**, 179–186.

Nunnally, J. C. and Bernstein, I. H. (1994). *Psychometric Theory*, Third Edition, McGraw-Hill, New York.

Pfanzagl, J. (1993). On the consistency of conditional maximum likelihood estimators, *Annals of the Institute of Statistical Mathematics*, **45**, 703–719.

Rasch, G. (1960, 1993). *Probabilistic Models for Some Intelligence and Attainment Tests*, MESA Press, Chicago. Original Edition, The Danish Institute for Educational Research.

Thissen, D. (1982). Marginal maximum likelihood estimation for the one-parameter logistic model, *Psychometrika,* **47**, 175–186.

van der Linden, W. and Hambleton, R. K. (1996). *Handbook of Modern Item Response Theory*, Springer-Verlag, New-York.

Ware, J. E., Harris, W. J., Gandek, B., Rogers, B. W. and Reese, P. R. (1997). *MAP-R for Windows: Multitrait/Multi-Item Ananlysis Program-Revised User's Guide*, Health Assesment Lab, Boston, MA.

ANNEX 1: COMMUNICATION DIMENSION OF THE SIP (9 ITEMS)

1. I am having trouble writing or typing

2. I communicate mostly by gestures,for examples, moving head, pointing, sign language

3. I am having trouble writing or typing

4. My speech is understood only by e few people who know me well

5. I often lose control of my voice when i talk, for example, my voice gets louder or softer, trembles, changes unexpectedly

6. I don't write except to sign my name

7. I carry on a conversation only when very close to the other person or looking at him

8. I have difficulty speaking, for example, get stuck, stutter, stammer, slur my words

9. I am understood with difficulty

10. I do not speak clearly when I am under stress

ANNEX 2: SOCIAL INTERACTION DIMENSION OF THE SIP (20 ITEMS)

1. I am going out less to visit people

2. I am not going out to visit people at all

3. I show less interest in other people's problems, for example, don't listen when they tell me about their problems, don't offer to help

4. I often act irritable toward those around me, for example, snap people, give sharp answers, criticize easily

5. I show less affection

6. I am doing fewer social activities with groups of people

7. I am cutting down the length of visits with friends

8. I am avoiding social visits from others

9. My sexual activity is decreased

10. I often express concern over what might be happening to my health

11. I talk less with those around me

12. I make many demands, for example, insist that people do things for me, tell them how to do things

13. I stay alone much of the time

14. I act disagreeable to family members, for example, I act spiteful, I am stubborn

15. I have frequent outbursts of anger at family members, for example, strike at them, scream, throw things at them

16. I isolate myself as much as I can from the rest of the family

17. I am playing less attention to the children

18. I refuse contact with family members, for example, turn away from them

19. I am not doing the things I usually do to take care of my children or family

20. I am not joking with family members as I usually do

Printed in the United States
by Baker & Taylor Publisher Services